Electro-Optical Instrumentation

Sensing and Measuring with Lasers

Silvano Donati
University of Pavia

PRENTICE HALL PTR
UPPER SADDLE RIVER, NJ 07458
WWW.PHPTR.COM

Library of Congress Cataloging-in-Publication Data

Donati, Silvano.
    Electro-optical instrumentation: sensing and measuring with lasers / Silvano Donati.
       p. cm.
    Includes bibliographical references and index.
    ISBN 0-13-061610-9
      1. Lasers. 2. Measuring instruments. 3. Remote sensing--Equipment and supplies. I. Title.
III. Series.

TA1677.D63 2004
681'.2--dc22

                                                                         2004043865

Editorial/production supervision: *Nicholas Radhuber*
Publisher: *Bernard Goodwin*
Cover design director: *Jerry Votta*
Cover design: *Talar Boorujy*
Manufacturing manager: *Maura Zaldivar*
Editorial assistant: *Michelle Vincenti*
Marketing manager: *Dan DePasquale*

© 2004 Pearson Education, Inc.
Published by Prentice Hall Professional Technical Reference
Pearson Education, Inc.
Upper Saddle River, New Jersey 07458

Prentice Hall books are widely used by corporations and government agencies for training, marketing, and resale.

Prentice Hall offers excellent discounts on this book when ordered in quantity for bulk purchases or special sales. For more information, please contact:
U.S. Corporate and Government Sales
1-800-382-3419
corpsales@pearsontechgroup.com

For sales outside of the U.S., please contact:
International Sales
1-317-581-3793
international@pearsontechgroup.com

Other product or company names mentioned herein are the trademarks or registered trademarks of their respective owners.

Printed in the United States of America

1st Printing

ISBN 0-13-061610-9

Pearson Education LTD.
Pearson Education Australia PTY, Limited
Pearson Education Singapore, Pte. Ltd.
Pearson Education North Asia Ltd.
Pearson Education Canada, Ltd.
Pearson Educación de Mexico, S.A. de C.V.
Pearson Education — Japan
Pearson Education Malaysia, Pte. Ltd.

Contents

Preface vii

Chapter 1 Introduction 1

    1.1 Looking Back to Milestones 2
    References 10

Chapter 2 Alignment, Pointing, and Sizing Instruments 11

    2.1 Alignment 12
    2.2 Pointing and Tracking 16
        2.2.1 The Quadrant Photodiode 16
        2.2.2 The Position Sensing Detector 19
        2.2.3 Position Sensing with Reticles 22
    2.3 Laser Level 24
    2.4 Wire Diameter Sensor 27
    2.5 Particle Sizing 30
    References 38

Chapter 3 Laser Telemeters 39

    3.1 Triangulation 40
    3.2 Time-of-Flight Telemeters 43
        3.2.1 Power Budget 44
        3.2.2 System Equation 46
        3.2.3 Accuracy of the Pulsed Telemeter 51
        3.2.4 Accuracy of the Sine-Wave Telemeter 59
        3.2.5 The Ambiguity Problem 61
        3.2.6 Intrinsic Precision and Calibration 63
        3.2.7 Transmitter and Receiver Optics 64
    3.3 Instrumental Developments of Telemeters 66
        3.3.1 Pulsed Telemeter 66
        3.3.2 Sine-Wave Telemeter 74
    3.4 Imaging Telemeters 80
    3.5 The LIDAR 81
    References 87

Chapter 4 Laser Interferometry   89

    4.1 Overview of Interferometry Applications   91
    4.2 The Basic Laser Interferometers   93
        4.2.1 The Two-Beam Laser Interferometer   94
        4.2.2 The Two-Frequency Laser Interferometer   101
        4.2.3 Measuring with the Laser Interferometer   107
    4.3 Performance Parameters   112
    4.4 Ultimate Limits of Performance   113
        4.4.1 Quantum Noise Limit   114
        4.4.2 Temporal Coherence   116
        4.4.3 Spatial Coherence and Polarization State   118
        4.4.4 Dispersion of the Medium   118
        4.4.5 Thermodynamic Phase Noise   119
        4.4.6 Brownian Motion, Speckle-Related Errors   120
    4.5 Read-Out Configurations of Interferometry   122
        4.5.1 Internal Configuration   124
        4.5.2 Injection (or Self-Mixing) Configuration   126
    4.6 Laser Vibrometry   146
        4.6.1 Short-Distance Vibrometry   146
        4.6.2 Medium-Distance Vibrometry   151
        4.6.3 Long-Distance Vibrometry   157
    4.7 Other Applications of Injection Interferometry   163
        4.7.1 Absolute Distance Measurements   163
        4.7.2 Angle Measurements   166
        4.7.3 Detection of Weak Echos   168
    4.8 White Light Interferometry   172
        4.8.1 Application to Profilometry   175
        References   178

Chapter 5 Speckle-Pattern Instruments   181

    5.1 Speckle Properties   182
        5.1.1 Basic Description   182
        5.1.2 Statistical Analysis   186
        5.1.3 Speckle Size from Acceptance   191
        5.1.4 Joint Distributions of Speckle Statistics   192
        5.1.5 Speckle Phase Errors   197
        5.1.6 Speckle Errors due to Target Movement   198
        5.1.7 Speckle Errors due to Beam Movement   199
        5.1.8 Speckle Errors with a Focussing Lens   199
        5.1.9 Phase and Speckle Errors due to Detector Size   200
    5.2 Speckle in Single-Point Interferometers   201
        5.2.1 Speckle Regime in Vibration Measurements   201
        5.2.2 Speckle Regime in Displacement Measurements   203
        5.2.3 The Problem of Speckle Phase Error Correction   210
    5.3 Electronic Speckle Pattern Interferometry   214

| Contents | v |

References	221

Chapter 6 Laser Doppler Velocimetry — 223

6.1 Principle of Operation	224
6.1.1 The Velocimeter as an Interferometer	227
6.2 Performance Parameters	229
6.2.1 Scale Factor Relative Error	229
6.2.2 Accuracy of the Doppler Frequency	230
6.2.3 Size of the Sensing Region	231
6.2.4 Alignment and Positioning Errors	233
6.2.5 Placement of the Photodetector	235
6.2.6 Direction Discrimination	237
6.2.7 Particle Seeding	238
6.3 Electronic Processing of the Doppler Signal	239
6.3.1 Time-Domain Processing	240
6.3.2 Frequency-Domain Processing	242
6.4 Optical Configurations	244
References	246

Chapter 7 Gyroscopes — 247

7.1 Overview	248
7.2 The Sagnac Effect	252
7.2.1 The Sagnac Effect and Relativity	253
7.2.2 Sagnac Phase Signal and Phase Noise	255
7.3 Basic Gyro Configurations	259
7.4 Development of the RLG	265
7.4.1 The Dithered Laser Ring Gyro	268
7.4.2 The Ring Zeeman Laser Gyro	270
7.4.3 Performances of RLGs	274
7.5 Development of the Fiber Optics Gyro	277
7.5.1 The Open-Loop Fiber Optic Gyro	277
7.5.2 Requirements on FOG Components	282
7.5.3 Technology to Implement the FOG	289
7.5.4 The Closed-Loop FOG	291
7.6 The Resonant FOG and Other Configurations	295
7.7 The 3x3 FOG for the Automotive	297
7.8 The MEMS Gyro and Other Approaches	301
7.8.1 MEMS	303
7.8.2 Piezoelectric Gyro	307
References	308

Chapter 8 Optical Fiber Sensors — 311

8.1 Introduction	312
8.1.1 OFS Classification	313
8.1.2 Outline of OFS	314
8.2 The Optical Strain Gage: A Case Study	316

8.3 Readout Configuration	318
8.3.1 Intensity Readout	318
8.3.2 Polarimetric Readout	328
8.3.3 Interferometric Readout	345
8.4 Multiplexed and Distributed OFS	352
8.4.1 Multiplexing	353
8.4.2 Distributed Sensors	357
References	361

Appendix A0 Nomenclature   363

Appendix A1 Lasers for Instrumentation   365

A1.1 Laser Basics	367
A1.1.1 Conditions of Oscillation	370
A1.1.2 Coherence	371
A1.1.3 Types of He-Ne Lasers	372
A1.2 Frequency Stabilization of the He-Ne Laser	375
A1.2.1 Frequency Reference and Error Signal	376
A1.2.2 Actuation of the Cavity Length	381
A1.2.3 Ultimate Frequency Stability Limits	382
A1.2.4 A Final Remark on He-Ne Lasers	382
A1.3 Semiconductor Narrow-Line and Frequency Stabilized Lasers	383
A1.3.1 Types and Parameters of Semiconductor Lasers	383
A1.3.2 Frequency-Stabilized Semiconductor Lasers	388
A1.4 Diode-Pumped Solid-State Lasers	390
A1.5 Laser Safety Issues	393
References	395

Appendix A2 Basic Optical Interferometers   397

A2.1 Configurations and Performances	397
A2.2 Choice of Optical Components	403
References	406

Appendix A3 Propagation through the Atmosphere   407

A3.1 Turbidity	407
A3.2 Turbulence	415
References	418

Appendix A4 Optimum Filter for Timing   419

Appendix A5 Propagation and Diffraction   421

A5.1 Propagation	421
A5.2 The Fresnel Approximation	423
A5.3 Examples	424
References	428

Appendix A6 Source of Information on Electro-Optical Instrumentation   429

Index   431

Preface

This book is an outgrowth of the lecture notes for a semester course that the author has given at the University of Pavia for several years. The course was tailored for electronic engineer graduates in their fifth and final year of the curriculum (the 18th grade). The course is designed for students of the optoelectronic engineering section and is also offered as an elective to students of the instrumentation and microelectronics engineering sections.

During the years, I have gone through several versions of the text, trying to improve and expand the material. I have also added new topics taken from the literature or from my own research on interferometers, laser telemeters, speckle-pattern and optical fiber sensors.

About terminology, *electro-optics* is a very interdisciplinary science that receives contributions from researchers whose cultural roots originate in electronics, as well in optics, laser physics, and electromagnetism.

Other researchers may refer to the content of this book as *measurements by lasers*, *coherent techniques*, *optoelectronic measurements*, *optical metrology*, and perhaps some more names equally acceptable.

I think that *electro-optical instrumentation* is preferable to give this field a name of its own and to underline the connection between optics and electronics, a very fruitful synergy we have actually observed in this field through the years.

In distributing the material between text and appendixes, the rationale used here is that core arguments of the book are text, whereas arguments either complementary, common to other disciplines, or recalls of basics are reviewed in the appendixes.

As a general scope, in this book I have tried to (i) illustrate the basic principles behind the application; (ii) outline the guidelines for the design; and (iii) discuss the basic performance achievable and the ultimate limit set by noise.

I have attempted to make the chapters and the accompanying appendixes self-containing, so that a selection of them can be used for a shorter course as well.

As a guideline, in 35 hours of classroom lessons I cover most of the material presented in the text, starting from basic ideas and going on to clarify performance limits and give development hints. To lessons, I add some 10 hours of lab and exercises. In the classes, I usually skip mathematical details and some advanced topics that are most difficult. I have anyway included them in the text because they illustrate the state of the art of electro-optical instrumentation and are useful for advanced study.

Based on my experience, students with a limited background in optoelectronics can follow the course profitably if they are given a primer of an additional 6 to 8 hours on the fundamentals of lasers and fibers that are collected in the appendixes.

About derivations, I have tried to keep the mathematical details to a minimum and report just the very straight derivations. Purposely, to report just simple expressions, I have typed the equations all on one line to save precious typographical space. This kind of typing may look odd at first but will be easier as the reader becomes acquainted with it.

Throughout the text, special attention has been paid to complement the understanding of the basic principle with the treatment of development issues. Thus, as compared with other books on the subject, the reader will find a lot of electronic schematics about signal handling, discussions on the impact of practical components, comments about the ultimate limits of performance, etc. Of course, this is just the attitude we try to develop in an electronic engineer, but it is also interesting for the generic reader to grasp the engineering problems of instrumentation.

Presently, the book does not contain problems. Actually, I prefer to publish the book first and follow up with problems afterwards. Problems will be made available to the reader a few months after publication at the author's website, ele.unipv.it/~donati. At the same location, instructors adopting the book for their course will be able to find a selection of viewgraphs in PowerPoint.

Many individuals have helped through the years in collecting the lecture notes from which the book has started. I wish to especially thank Tiziana Tambosso for the help in shaping the arguments of the book and for suggesting many good hints.

It is also a pleasure to thank Risto Myllyla, Gordon Day, Peter deGroot, Jesse S. Greever, and Thierry Bosch for reading the manuscript and providing useful feedback.

I hope this book will be useful to the young student as a guide and motivation to work in the exciting field of electro-optical instrumentation and to the designer as a reference to the state of the art in this field. If the book will stimulate new ideas or development of products, my effort in writing it will be amply rewarded.

Silvano Donati
Pavia, December 2003
ele.unipv.it/~donati

CHAPTER 1

Introduction

The coming of age of electro-optical instrumentation dates back to a few years after the invention of the laser, that is, during the mid 1960s. The enormous potential of lasers in measurements was soon recognized by the scientific community as one capable of providing new approaches with unparalleled performances. Plenty of successful examples of several instruments and sensors started to appear in the 1970s, including gyroscopes, laser interferometers, pulsed telemeters, and laser Doppler velocimeters. These instruments were at first just bright scientific ideas, but, after several years of research and development, they grew up to industrial products of great success.

Initially, military applications provided a big push to the research and development effort, and, in the system balance, the laser played the role of the enabling component. After a few decades, parallel with the growth of a civil application market, electro-optical instrumentation became to stand on its own feet.

The focus of attention then moved from the source to the philosophy of measurement, thus boosting the synergy of optics and electronics and merging the new technology in the frame of measurement science.

1.1 LOOKING BACK TO MILESTONES

The album of successes of electro-optical instrumentation soon became very rich and astonishing.

About telemeters, we may recall the famous LUnar Ranging Experiment (LURE) carried out in 1969 (Fig.1-1) when the astronauts of Apollo 11 brought on the moon a 1-m by 1-m array of corner-cubes [1].

Several telescopes of astronomical observatories on earth aimed their own ruby laser, Q-switched pulsed beam on the array. The task was difficult, because the target was invisible and only its lunar coordinates were available. Three out of five telescopes succeeded in hitting the target and getting the very small (about 10 photons) return. The pulse delay, or time-of-flight (about 2 seconds), was measured with nanosecond resolution, thus sampling the 384,000-kilometer distance to about 30-centimeter.

This was not just a mere big-science exhibition, as this early laser ranging experiment was the forerunner of the modern earth-to-satellite telemeter network. Now long-distance measurement by laser telemeters is an established, powerful tool for geodesy survey.

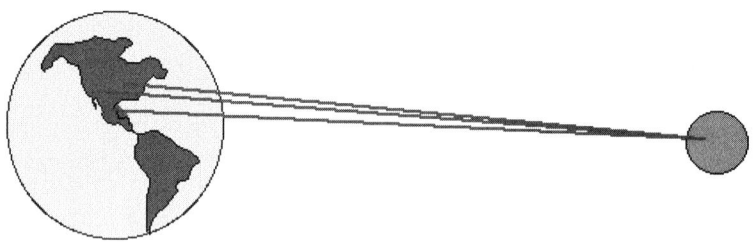

Fig.1-1 The LURE experiment in 1969 set the record of long-distance pulse telemetry. The baseline was the 384,000-km earth-to-moon distance.

The experiment also opened the way to airborne telemeters and altimeters (Fig.1-2), based mainly on the Nd:YAG Q-switched laser and occasionally on CO_2 lasers for better transmission through haze and fog. The most recent versions employ compact semiconductor-laser telemeters and fit in the headlight compartment of a car (Fig.1-3) [2].

This telemeter is now sold as a standard product of automotive anticollision systems. It fosters next years' automatic–guide systems, which is not a dream of science fiction anymore, but a technical reality being actively pursued.

The sine-wave-modulated diode-laser telemeter introduced in the early 1970s is now an instrument widespread in construction works and marks the retirement of the old theodolite and associated set of rulers (Fig.1-4) [3].

As a device directly evolved from the early concept, the topographic telemeter provides 1-cm accuracy over several hundred- to thousand-meter distance and accounts for a healthy $200 million per year market segment.

1.1 Looking Back to Milestones

Fig.1-2 Pulsed telemeters (based on Q-switched solid-state lasers) have been routinely mounted aboard military aircraft since the 1970s . . .

Fig.1-3 . . . and their compact variants (based on semiconductor diode lasers) entered the automotive market in recent years, as the sensor of anti-collision systems [2]. They are likely to become the key of future automatic-drive systems (photo courtesy of CRF-Fiat, Turin).

Fig.1-4 Meanwhile, in the early 1970s [3], sine-wave modulated telemeters started replacing the old theodolite in civil engineering applications. Distance measurement is now carried out instantly in the field on ranges up to several hundred meters and cm resolution.

The LURE experiment sensation was renewed and amplified recently by the Mars Orbiter Laser Altimeter (MOLA) flown aboard the Mars Global Surveyor in 1997, which orbited the red planet for over a year to collect height data on the Mars surface.

The MOLA altimeter sent the amazing map in Fig.1-5 back to earth. In this map, pixels are 100-m×100-m wide, and their relative height has been mapped with a 10-m accuracy [4] from the average 400-km distance from the planet surface [5]. The telemeter aboard the Mars Global Surveyor uses a diode-laser pumped Cr,Nd:YAG laser that operates in Q-switching regime as the source and supplies 8-nanosecond (ns) pulses at 10-Hertz (Hz) repetition rate. The receiver is a 50-cm diameter telescope ending on a Si-avalanche photodiode detector and includes a 2-nanometer (nm) band pass filter to reject solar background.

Similar to planetary telemeters, a number of Laser Ranger Observatories (LRO) have been built around the world to target geodetic satellites in earth orbit. One of these, the Matera LRO [6], reaches 25-ps of timing accuracy, or approximately 3 millimeter of range accuracy.

Another version of the basic telemetry scheme is provided by the Laser Identification, Detection, and Ranging (LIDAR, see Sect.3.5). This instrument is again a pulsed telemeter, but has a λ-tunable laser as the source. By analyzing the echoes collected from the atmosphere, according to different conceptual schemes (DIAL, CARS, etc.), pollutant concentrations can be measured down to part-per-billion (ppb) concentration with distance-resolved plots of the pollutant concentration (Fig.1-6).

1.1 Looking Back to Milestones

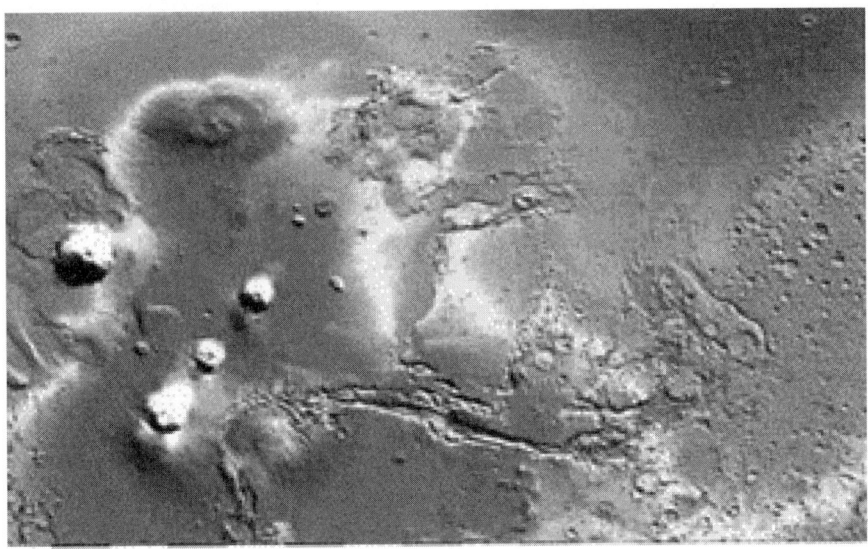

Fig.1-5 Map of Mars as provided by the MOLA orbiting telemeter (1998). Pixels are 100 m in size, and their height is measured to a 10-m accuracy by the orbiting telemeter (courtesy of MOLA Science Team).

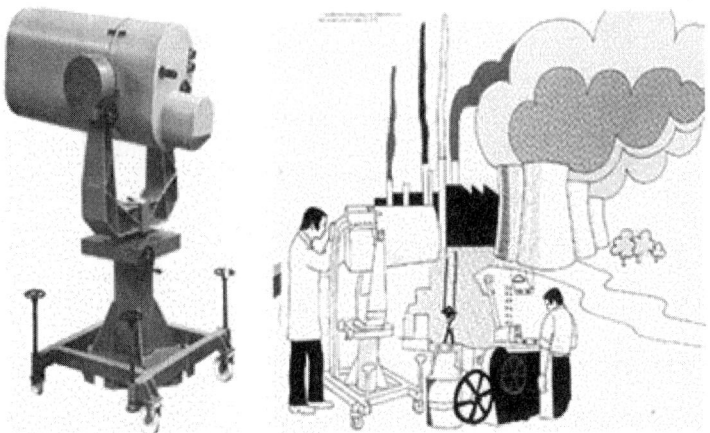

Fig.1-6 The LIDAR is a pulsed telemeter, but with a λ-tunable laser as the source. Echoes from the atmosphere at different wavelengths allow measuring pollutants down to ppb concentration and get distance-resolved plots (from [7], courtesy of Ferranti UK).

The gyroscope is another big success, both scientific and industrial, of electro-optical instrumentation. The laser gyroscope concept was demonstrated as early as 1962, just one year after the discovery of the He-Ne laser. The seminal experiment was carried out in the laboratories of Sperry Corp.[8], with a 4-arms He-Ne, 1-m by side, square ring configuration capable of detecting the earth rotation rate (15 degree/hour). See Fig.1-7. From this encouraging start, the way to a commercial gyroscope suitable for field deployment was then long and full of obstacles, like the locking effect, which was very hard to overcome.

Fig.1-7　Early demonstration of the Sagnac effect in a square cavity He-Ne laser by Macek of Sperry Corp. in 1962 (from [9], by courtesy of the Institute of Electrical and Electronic Engineers [IEEE]).

Fig.1-8　A modern 4-inch side ring laser gyro (RLG) is the heart of the linertial Navigation Units (INUs) of today's airliners (from [9], by courtesy of IEEE).

1.1 Looking Back to Milestones

Only after a decade of worldwide research efforts by the scientific community, clever conceptual solutions and improved technology finally evolved, which led to the modern top-class RLG with the amazing accuracy of $0.001°/h$. This device (Fig.1-8) has been used in all newly fabricated airliners since the 1980s. It is the heart of modern Inertial Navigation Units (INUs), electronics box about 10 inches on a side that costs \$1.5 million. The INU is capable of telling the actual position of the aircraft with 1-mile accuracy after a few hours of flight.

The gyro story did not come to an end, however. In 1978, by taking advantage of the newly developed single-mode fibers, the Fiber Optics Gyroscope (FOG) was proposed, and the quest restarted with another technology. The aim was to produce a medium-class light, cheap, and more reliable device especially for space applications.

Again, about a decade of efforts has been necessary to cure a series of small idiosyncrasies of the new approach. Units were finally ready in the 1990s (Fig.1-9) for use as a reference aboard telecommunication satellites as an attitude system as well as for military applications. Because of its low cost, the FOG was also initially used in the automotive industry to demonstrate the car navigation concept (Fig.1-10). Today, it has become clear that, despite the attempts to squeeze the production cost, the FOG is too complex to meet the desired target price.

Because the automotive market is very strong, a last generation effort of the electro-optical gyroscope is being pursued using the Micro-Optical-Electro-Mechanical System (MOEMS) technology [10]. Once more, it is likely that we will record a new breakthrough in gyro in the next years.

Fig.1-9   A 3-inch diameter fiber optics gyroscope (right) and its printed circuit board for signal processing (left) (model FOG-1B introduced by SEL Alcatel in 1990).

Fig.1-10 The gyroscope has opened the way to car navigators and modern adaptive cruise control (by courtesy of Magneti Marelli Sistemi Elettronici, Torino).

Interferometry is another big chapter in electro-optical instrumentation. As soon as the first He-Ne lasers were frequency-stabilized in the mid-1960s, many scientific as well as industrial applications were created, taking advantage of the unprecedented coherence length made available by this source. Well-known examples, and a huge commercial success, are the so-called laser interferometer and Doppler velocimeter. These soon penetrated and captured the markets of mechanical metrology and machine-tool calibration and the fluidics and anemometry engineering segments, respectively. These instruments sell at levels of 10,000 units per year and by themselves would justify a place for electro-optical instrumentation.

In addition, the synergy of optics and electronics has pushed the field of interferometry well beyond the fringe performance of the classical eighteenth century optics. Pico-meters or atto-meters are well resolved in displacement, as well as 10^{-9} strains in relative $\Delta l/l$ variation. After the early detection of earth-crust tides (in the 1960s, at a 10^{-9} strain), a big scientific enterprise is being completed in these years, namely the interferometer detection of gravitational waves coming from remote galaxies and massive collapsing stars.

A number of teams are developing gravitational antennas around the world (in the United States, Europe and Japan) under the names of Laser Interferometer Gravitational Observatory (LIGO) [11], Laser Interferometer Gravitational Antenna (LIGA), etc. (see Fig.1-11). Recently, a proposal to build a 5-million-km arm space interferometer is under study [12]. These interferometers push the resolution to unprecedented (10^{-20}m in LIGO), yet theoretically achievable, limits (as set by the quantum noise). The LIGO instrument may soon become a new window for probing the cosmos that will be comparable in importance and observational results, perhaps, with the advent of the radio telescope.

1.1 Looking Back to Milestones

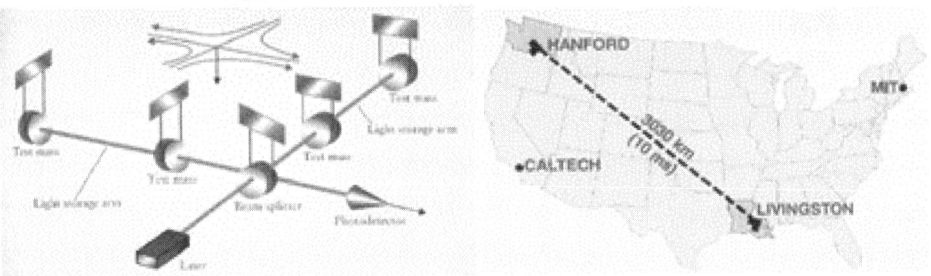

Fig.1-11 The LIGO interferometer has 2-km arms (top: the facility; bottom left: the optical layout). The instrument should be able to detect the gravitational collapse of a star into a black hole a Megaparsec away. To rule out spurious pulses due to earth perturbations, two interferometers located 3000-km apart will operate in coincidence (bottom right). Photos reprinted by courtesy of Caltech/LIGO.

Another ambitious project involving interferometry is the Terrestial Planet Finder (TPF), which was recommended by the Astrophysics Survey Committee. The instrument will comprise four 3.5-m telescope mirrors placed in space a few km apart and should be able to detect earth-size planets around stars up to 50 light-years away.

The initial effort to validate the concept is the Space Interferometer Mission (SIM) scheduled for launch in 2009. It consists of an optical stellar interferometer with 10-m baseline, and will be capable of measuring angular coordinates of stars with approximately 50 pico-radian accuracy.

The test facility of the instrument is a 10-pm resolution interferometer [13] that operates on an approximately 10-m arm length.

In summary, just from the few examples previously reported, we may conclude that the field of electro-optical instrumentation has been very exciting since its inception.

Though only a few decades old, electro-optical instrumentation has already achieved remarkable scientific and technical successes and has proven the advantages of a synergy and cross-fertilization of optics and electronics.

A healthy field not licked by the bubble-technology syndrome, electro-optical instrumentation will continue to grow in ideas and achievements. In the years to come, this field will certainly continue to offer significant potential to researchers and engineers from the point of view of advancement in science, as well as of engineering development of new products.

REFERENCES

[1] C.O. Alley et al., *"The LURE Experiment: Preliminary Results"*, Science, vol.167 (1970), p.368.
[2] L. Ampane, E. Balocco, E. Borrello, and G. Innocenti, *"Laser Telemetry for Automotive Applications"*, in Proceedings of LEOS Conference ODIMAP II, edited by S. Donati, Pavia, 20-22 May 1999, pp.179-189.
[3] AGA Geodimeter model 710, Publication 571.30006 2k8.71 (1971).
[4] M.T. Zuber and D.E. Smith, *"The Mars Orbrserver Laser Altimeter Investigation"*, J. Geophys. Res., vol.97 (1992), pp.7781-7797.
[5] D.E. Smith and M.T. Zuber, *"The MOLA Investigation of the Shape and Topography of Mars"*, Proceedings of LEOS Conference ODIMAP III, edited by S. Donati, Pavia, 20-22 September 2001, pp.1-4.
[6] G. Bianco and M.D. Selden, *"The Matera Ranging Observatory"*, in Proceedings of LEOS Conference ODIMAP II, edited by S. Donati, Pavia, 20-22 May 1999, pp.253-260.
[7] Ferranti 700 Series Lidar System, Publication DDF/524/675 (June 1975).
[8] W.M. Macek and D.I.M. Davis, *"Rotation Rate Sensing with a Travelling wave Laser"*, Appl. Phys. Lett., vol.2 (1963), p.67-68.
[9] G.J. Martin, *"Gyroscopes May Cease Spinning"*, IEEE Spectrum, vol.23 (October 1986), pp.48-53.
[10] E.A. Brez, *"Technology 2000: Transportation"*, IEEE Spectrum, vol.37 (January 2000), pp.91-96.
[11] B.C. Barish and R. Weiss, *"LIGO and the Detection of Gravitational Waves"*, Physics Today (Oct.1999), pp.44-50.
[12] J. Riordon, *"An Early Look at LISA"*, Sky and Telescope (October 2000), pp.46-47.
[13] P.G. Halverson et al., *"Characterization of Picometer Repeatability Displacement Metrology Gauges"*, in Proceedings of LEOS Conference ODIMAP III, edited by S. Donati, Pavia, 20-22 September 2001, pp.63-68; see also huey.jpl.nasa.gov.

CHAPTER **2**

Alignment, Pointing, and Sizing Instruments

One of the specific properties of laser sources is the ability to supply a well-collimated light beam. In free propagation, the beam of a common laser like the He-Ne laser (see Appendix A1) can be visualized at a distance much larger than an ordinary flashlight can attain. This is a consequence of the low angular divergence of the laser beam and has the well-known application to laser pointers.

The divergence of the laser beam is usually the small value dictated by the diffraction limit. The diffraction limit is reached because the source generates light in the single-mode spatial regime, which is a condition easy to meet, especially with He-Ne lasers. The property of the laser emitting a particularly collimated, directional beam gives instruments used for alignment an advantage, as an extension of the plumb line on a generic direction, which is not necessarily vertical.

Alignment is used for positioning objects along a desired direction, as indicated by the propagation direction of the laser beam. Applications range from construction work and laying pipelines to checking terrain planarity in agriculture. In this last application, the concept is extended by making a laser level that uses a rotating fan beam and defines the reference plane for alignment. Another class of instruments exploiting the good spatial quality of the laser beam, as obtained by single transverse-mode operation, is dimensional measurement by diffraction. The object under measurement is positioned in the beam waist of the laser beam so that the wave front of illumination is plane. Light scattered by the object has a far-field diffracted profile precisely related to the dimensions of the object, so the measurement of it can be developed by analyzing the signal collected by a detector scanning the far-field distribution. Examples of noncontact measurement discussed later in this chapter are wire diameter and fine-particle sizing.

2.1 ALIGNMENT

In an alignment instrument, we want to project the beam at a distance of interest and be able to keep its size the smallest possible along the longest possible path.

Because a single-transversal TEM_{00} mode is the distribution with the least diffraction in propagation, the He-Ne laser has traditionally been the preferred source. The He-Ne laser readily operates in single-transversal mode regime and has a wavelength $\lambda=633$ nm in the visible.

Semiconductor lasers may be used as well if the elliptical near-field spot is properly corrected by an anamorphic objective lens circularizing the output beam.

Let us consider the propagation of the Gaussian beam emitted by the laser (see Appendix A1.1), as in Fig.2-1.
The Gaussian beam keeps its distribution unaltered in free propagation and in imaging through lenses, and the spot-size w is the parameter that describes it.
In free propagation (Fig.2-1, top) w evolves with the distance z from the beam waist w_0 as:

$$w^2(z) = w_0^2 + (\lambda z/\pi w_0)^2 \qquad (2.1)$$

In optical conjugation through a lens, the spot size is multiplied by the magnification factor $m=r_1/r_2$, with r_1 and r_2 being the radii of curvature of the wave fronts at the lens surfaces. The radii obey a formula similar to Newton's lens formula $1/p+1/q=1/f$, that is:

$$1/r_1 + 1/r_2 = 1/f \qquad (2.2)$$

On its turn, the radius of curvature of the wave front, as a function of the distance z from the beam waist, is given by:

$$r = z\,[1+ (\pi w_0^2/\lambda z)^2] \qquad (2.3)$$

From Eq.2.3, we have r=∞ for z=0 (at the beam waist), and r= z for $z \gg \pi w_0^2/\lambda$ (in the far field, where the Newton formula holds).

2.1 Alignment

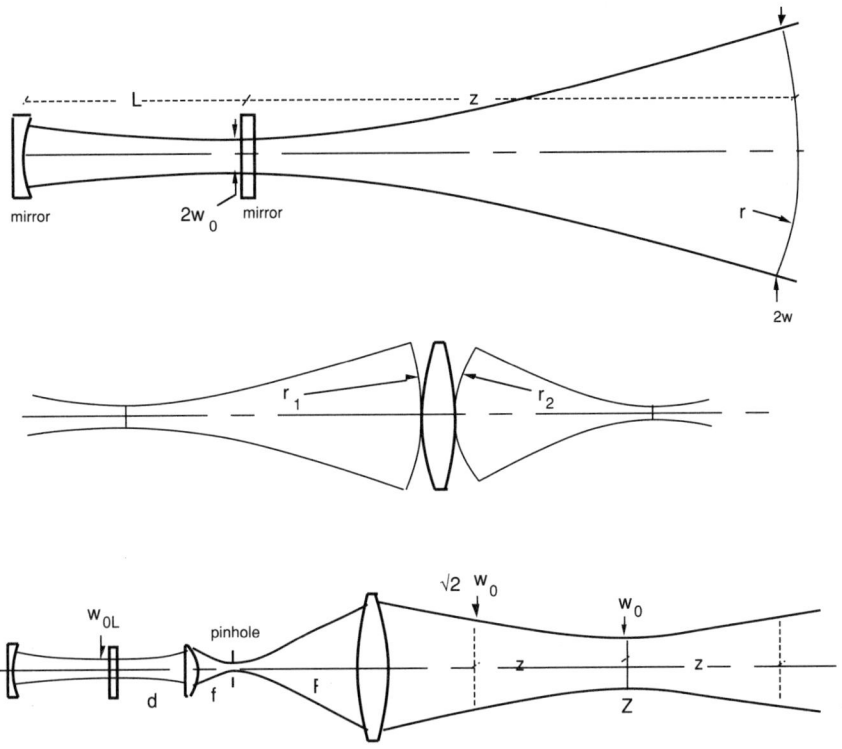

Fig. 2-1 Alignment using the Gaussian beam of a laser. Top: free propagation out of the laser; middle: conjugation through a lens; bottom: projection of the beam through a collimating telescope.

Now, we may look for the optimum waist w_0 that minimizes the spot size w on a given span of distance, let's say between $-z$ and $+z$. To do that, we find the minimum of w^2 in Eq.2.1 by differentiating respect to w_0^2 and equating to zero. We thus obtain:

$$\partial w^2/\partial w_0^2 = 1 - (\lambda z/\pi w_0^2)^2 = 0,$$

whence the optimum value of beam waist to be projected at a distance Z from the laser is:

$$w_0 = \sqrt{(\lambda z/\pi)} \qquad (2.4)$$

By comparing Eq.2.4 and Eq.A1.5, it turns out that the beam waist minimizing the size of the beam on the desired distance $-z \,..+z$ is the same of a laser oscillating on mirrors placed at a distance $2z$ apart.

In the span $-z..+z$, the beam size is minimum at the midpoint, $w=w_0$, and, from Eq.2.1, it becomes $w=\sqrt{2}w_0$ at the ends of the span. To be able to project a beam waist w_0 at distance Z from the laser, a collimating telescope is used (Fig.2.1 bottom). The telescope works better than a single lens and can easily adjust the distance Z by small focusing of its ocular. The magnification $m=w_0/w_{0L}$ is easily calculated as $m=(f/d)(Z/F)=Z/Md$ where M is the magnification of the telescope.

Example. For a laser with a plano-concave cavity and L=20 cm, K=5, we get from Eq.A1.5 w_{0L} =0.282 mm as the laser beam waist. The location of the waist is just at the output mirror.

If we want to cover a span of $z=\pm20$m, then Eq.2.4 gives w_0 =2 mm as the necessary beam waist. Then we need a magnification of $m=w_0/w_{0L}=7.1$. Taking Z=z=20m and d=10cm, the collimating telescope shall have a magnification M=F/f=Z/md=20/7.1×0.1=28.

In Fig.2-1 bottom, the pinhole in the telescope filters out spots extraneous to the laser mode, for example, originated by the reflection at the second surface of the laser output mirror. This mirror flat is slightly wedged to prevent the second surface from contributing to reflection into the cavity.

An alignment instrument usually includes a 0.5-2-mW He-Ne laser and a 50-mm diameter telescope providing a magnification of 20-30. The collimating telescope carries a viewfinder for target checking. The assembly is mounted on a tripod (see Fig.2-2) and is used for a number of applications in construction work with checks by sight.

Fig. 2-2 A typical alignment instrument, based on a low-power He-Ne laser and collimating telescope, and the application to construction of pipelines. This unit was introduced in 1978 by LaserLicht AG, Munchen.

2.1 Alignment

Nearly the same instrument (see Fig.2-3) has been used in harbors to establish a *marine channel light* that helps boats in their approach to maneuver to the harbor and to avoid dangerous sand bars.

In this application, the beam is switched periodically by a small angle in a horizontal plane through a prism inserted in the beam path. This arrangement provides an error signal that helps the boat to keep the correct homing direction.

Using a He-Ne laser emitting approximately 10 mW in the red (633 nm) and a 10-cm diameter telescope, the beam can be seen by eye at a distance of several miles. The actual range is limited primarily by weather conditions, which affect atmospheric attenuation and turbulence (see App.A3). Of course, in presence of haze and fog, the optical power is attenuated and the useful range drops to about a few times the visibility L_{vis} (see Eq.A3-4).

Another problem with the marine channel light is laser safety (see Appendix A1.5). To ensure safe view of the laser beam, power density is designed to be less than the Maximum Permissible Exposure (MPE) from a minimum distance onward. If personnel aboard the ship stare at the beam through binoculars, the power density increases by one or two orders of magnitude and ocular hazard may occur.

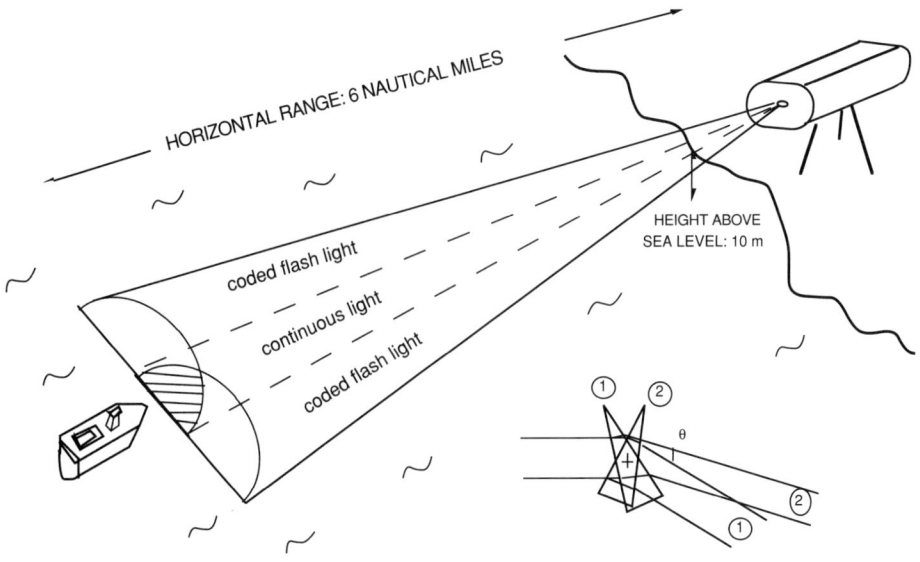

Fig. 2-3 Alignment instrument used as a marine channel light to help ships maneuver by homing to the harbor. The beam is switched by an angle θ by a prism rotating at approximately 3Hz and is transmitted by a telescope above the sea surface to the ship. In the region of superposition, light appears steady, whereas outside it flashes. Coding semiperiod duration allows discriminating between left and right beams.

2.2 POINTING AND TRACKING

Centering of the laser spot can be performed by sight, as in construction applications where we just need a resolution of a fraction of the beam spot size, or approximately 1 mm.

For more exacting applications, we may use a photodetector to generate an error signal proportional to the alignment error. The photodetector may either be a special type designed for position sensing or pointing purposes or may be a normal one acting in combination to a reticle or mask, which establishes a spatial reference for the photodetector.

2.2.1 The Quadrant Photodiode

This device has the usual structure of a normal photodiode (see Ref.[1], Sect.5.2), but one of the access electrodes, for example, the anode in a pn or pin junction, is sectioned in four sectors (Fig.2-4, left). Light input is from the chip backside when sections are defined by metallization, and from the top when a p^+ layers with side-ring contact is used (Fig.2-4, right). Size of the quadrant photodiode may go from 0.2 to 2 mm in diameter. The gap between sectors may be as small as 5-10 µm.

Light impinging on a certain sector originates a photocurrent collected by the correspondent electrode. Thus, we have four signals, S_1, S_2, S_3, and S_4, out of the four-quadrant photodiode.

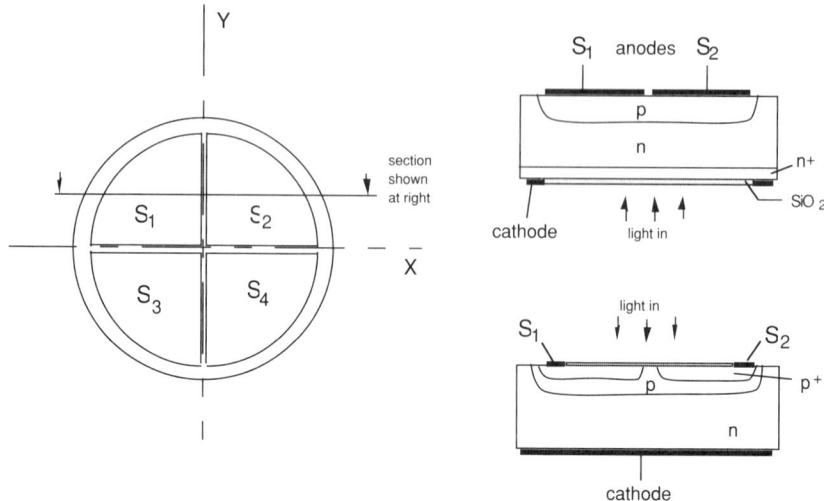

Fig.2-4 A quadrant photodiode for pointing applications. Left: the segmented electrodes structure; right: section of the pn photodiode structure with metal electrodes for backside light input (top) and section of a p^+pn structure for front light input (bottom).

2.2 Pointing and Tracking

We can compute two coordinate signals S_X and S_Y as:

$$S_X = (S_2+S_4)-(S_1+S_3), \text{ and } S_Y = (S_1+S_2)-(S_3+S_4), \qquad (2.5)$$

A practical circuit that interfaces the four-quadrant photodiode and computes the S_X and S_Y signals is shown in Fig.2-5. Based on operational amplifiers, the circuit terminates the four anodes of the photodiode on the virtual ground of transresistance preamplifiers ([1], Sect.5.3), that is, on a low-impedance value useful for enhancing the high-frequency performance. The common cathode is fed by the positive voltage ($+V_{bb}$) to reverse-bias the photodiodes. Each transresistance stage converts the photodetected current I_{ph} in a voltage signal $V_u = -RI_{ph}$, where R is the feedback resistance of the op-amp. With the four voltages available, it is easy to carry out the sum and difference operation (Eq.2.5) with the op-amps arranged as shown in the second stage.

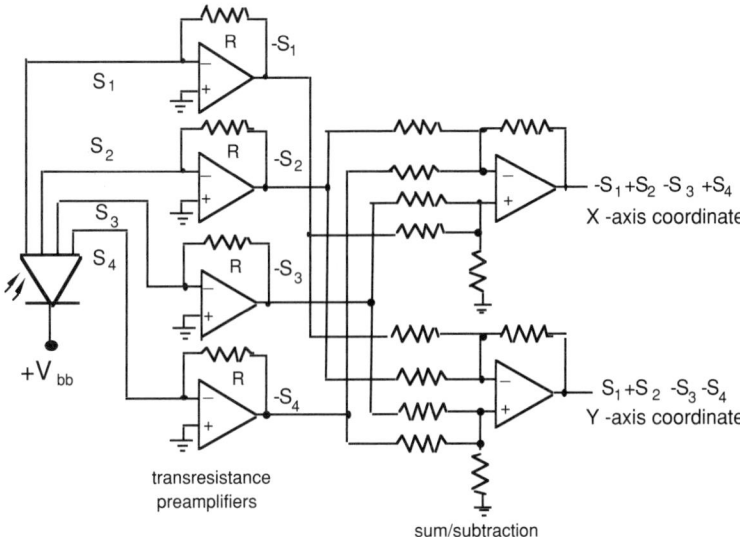

Fig.2-5 Op-amp circuit for computing the coordinate signals S_X and S_Y from the outputs of the four-quadrant photodiode.

To analyze the dependence of signals S_X (or S_Y) from the coordinate X (or Y), let us suppose that a probe spot scans the photodiode along a line parallel to the X axis. The spot will be smaller than the photodiode usually, but finite in size. Let X represent the coordinate of the spot center, and let R_{ph}, R_s be the photodiode and spot radii, while w_{dz} is the dead-zone width.

In Fig.2-6 the coordinate signal S_X is plotted versus X. When X is outside the photodiode by more than the spot radius, the signal is zero. When X is well inside the photodiode, the signal attains its maximum value. When X approaches zero, the signal decreases and reverses its

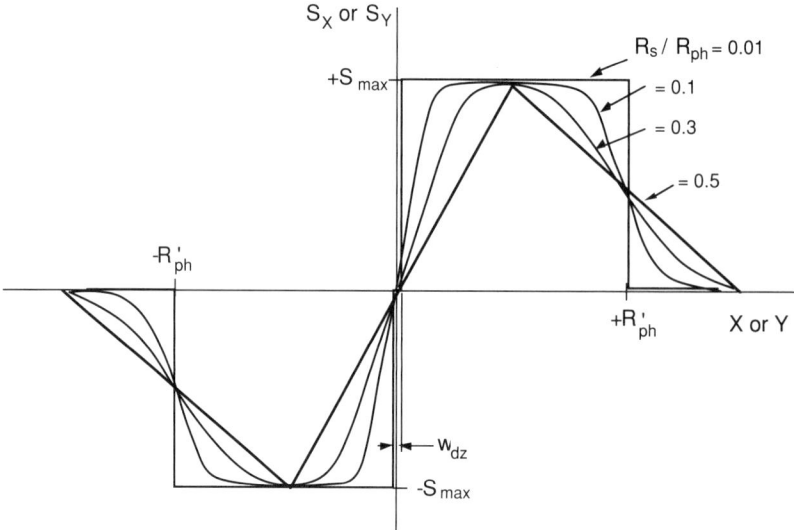

Fig.2-6 Dependence of the coordinate signal S_X (or S_Y) from the true coordinate X (or Y) of a light spot falling on the device. When the spot size R_s is very small (R_s/R_{ph}=0.01 or less), the response is a square wave with a small dead band near X=0 (or Y=0). When R_s is not so small, the response is smoothed, and we find an almost linear response close to X=0 (or Y=0). The width of the linear region is $\pm R_s$. The dead-zone width w_{dz} only affects the response for a very small R_s/R_{ph} ratio.

sign, as indicated by Eq.2.5.

Thus, the response around the X=0 is nearly linear (as shown in Fig.2-6), and the width of linear regime, before S_X reaches saturation, is about $\pm R_s$.

In *tracking* applications, the outputs from the position sensor are used as the error signals to control the system motion. In this case, the dependence shown in Fig.2-6 is just the desired one to get a good control action, both in acquisition and regulation regions:

- in the *regulation region* (near X=0), we need a corrective action proportional to the error so that the equilibrium point near X=0 is kept smoothly (without ripple).
- in the *acquisition region* (large X), the signal shall saturate so that the tracker is driven toward X=0 at the maximum speed.

The quadrant photodiode and other position-sensitive devices can be used as a *position sensor* that measure the X and Y coordinates of the impinging light spot. In this case, we will mount the quadrant photodiode on a positioning stage with reference marks to set the X=0 and Y=0 position (Fig.2-7). The accuracy (σ_X or σ_y) of the coordinate measurement depends on several parameters, such as optical power in the spot, spot shape and its fluctuations, and photodiode dimensions like R_{ph} and w_{dz}.

2.2 Pointing and Tracking

In a well-designed device, typical values are in the range $\sigma_X = 0.03$ to $0.1\ R_{ph}$.

The quadrant photodiode becomes an *angle sensor* when we place it in the focal plane of an objective lens (Fig.2-7). It measures the angular coordinates θ_x and θ_y of the spot presented in its field of view. Because of the optical conjugation, the coordinates θ_x and θ_y are tied to the spatial coordinate by $X = \theta_x F$ and $Y = \theta_y F$, where F is the focal length of the objective lens. The accuracy of the angle measurement follows as $\sigma_\theta = \sigma_X / F$.

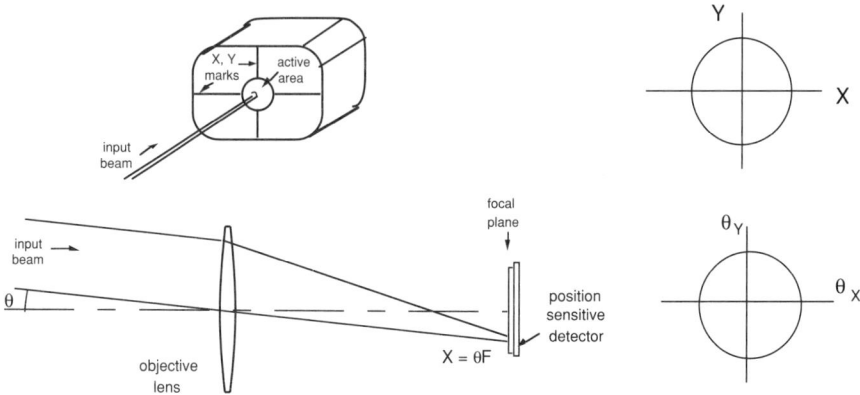

Fig.2-7 A position-sensing photodiode (either four-quadrant or reticle) can be used as an X-Y coordinate detector (top) or, when it is placed in the focal plane of an objective lens, as an angle-coordinate detector (bottom). In this case, the field of view is $\theta = R_{ph}/F$.

2.2.2 The Position Sensing Detector

The Position Sensing Detector, or PSD, is a multi-electrode photodiode similar to the four-quadrant photodiode. Developed in recent years as an evolution of the four-quadrant concept, the PSD offers a much better linearity over the entire active area and a response curve independent from the spot size and shape.

Actually, there are mono- and bidimensional PSDs for sensing one or two coordinates. In the following, we describe the bidimensional or array PSD only, because the monodimensional or linear PSD follows as a simpler case of the bidimensional.

The basic structure of a PSD (see Fig.2-8) is that of a normal pin junction with thin p and n regions. The electrodes for the X and Y coordinates are metal stripes deposited on opposite faces of the chip. Electrically, the stripes are like the normal anode and cathode of a photodiode, yet they are placed at the edges of the p and n regions and sectioned. The doping level of p and n regions is kept much lower than in a normal photodiode to enhance the series resistance of the undepleted p and n regions facing the junction, contrary to what is common.

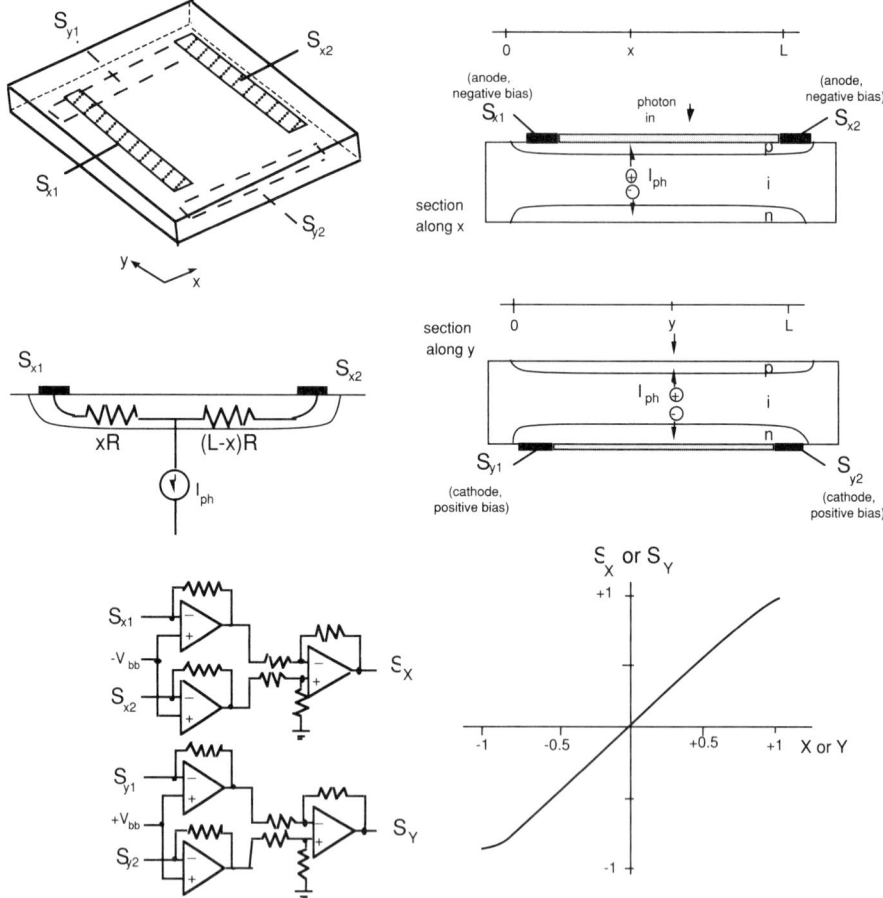

Fig.2-8 Structure of the PSD photodiode is a normal pin junction, with thin p and n regions, lightly doped or implanted. Light impinging at the point x,y generates a photocurrent that is shared between the X and Y electrodes. Coordinate signals are obtained as the difference between the currents exiting from the electrodes, $S_{x2}-S_{x1}$ and $S_{y2}-S_{y1}$. To ensure good linearity and to have a negligible dependence from R, the currents are sunk by the virtual ground of a transresistance stage. The non inverting inputs of the op-amp stage are used to set the bias, $+V_{bb}$ and $-V_{bb}$, of the X and Y stripes.

The typical resistance R between the stripe contacts (S_{x2} and S_{x1}) is about 10-100 kΩ.

Following the absorption of photons at point x,y (Fig.2-8), photogenerated electrons and holes are swept away quickly by the electric field at the junction. Thus, the current I_{ph} to be collected reaches the undepleted p and n regions without appreciable error in transversal (x,y) position.

2.2 Pointing and Tracking

The hole current arrives at point x in the p region (Fig.2-8, top) and shall cross series resistances xR and (L-x)R before reaching the electrodes S_{x1} and S_{x2}. The electron current has a similar response with respect to the electrodes S_{y1} and S_{y2}, but has formally the opposite sign ($-I_{ph}$) as it conventionally enters the electrodes.

Because of the series resistance, the current I_{ph} is shared by the electrodes proportional to the divider ratio of the respective conductances $(xR)^{-1}$ and $(L-x)^{-1}R^{-1}$. Therefore, the currents from S_{x1} and S_{x2} are:

$$I_{x1} = (xR)^{-1}/[(xR)^{-1}+(L-x)^{-1}R^{-1}] \, I_{ph} = (1-x/L) \, I_{ph}$$

$$I_{x2} = (L-x)^{-1}R^{-1}/[(xR)^{-1}+(L-x)^{-1}R^{-1}] \, I_{ph} = x/L \, I_{ph} \qquad (2.6)$$

Similarly, the current $-I_{ph}$ arriving in the n region is shared by the S_{y1} and S_{y2} electrodes, and Eq.2.6 holds with y in place of x and $-I_{ph}$ in place of I_{ph}:

$$I_{y1} = (1-y/L)(-I_{ph}), \quad I_{y2} = y/L(-I_{ph}) \qquad (2.6')$$

To recover the currents without errors induced by a finite load resistance, the best approach is to terminate the photodiode outputs (S_{y1}, etc.) on the virtual ground of op-amps, as shown in Fig.2-8. With this scheme, we get a very low-resistance termination, and the current I_x is deviated in the feedback resistance where it develops an output voltage $-R'I_x$.

In addition, the use of the op-amp allows an easy biasing of the PSD electrodes. Because of the virtual ground, the inverting input voltage is dynamically locked to that of the non-inverting input, and here we will apply the appropriate bias, positive ($+V_{bb}$) at the cathode and negative ($-V_{bb}$) at the anode. The best for V_{bb} is half the value of the supply voltage ($+V_{BB}$ and $-V_{BB}$) feeding to the op-amps.

With op-amps, it is easy to compute the coordinate signals S_X and S_Y as the differences of the transresistance-stage outputs (Fig.2-8, bottom left):

$$S_X = -R(I_{x2}-I_{x1})=-(2x/L-1)RI_{ph}, \; S_Y = R(I_{y2}-I_{y1})= -(2y/L-1)RI_{ph} \qquad (2.7)$$

Even better, if we add a voltage-controlled gain stage (not shown in Fig.2.8), we can divide outputs S_X and S_Y by I_{ph} and obtain coordinate signals independent from the impinging optical power $P= I_{ph}/\sigma$ (or PSD spectral sensitivity σ).

Often, instead of a fine spot we may have an extended source as the optical input to the PSD. In this case, the coordinates supplied by the PSD are the center of mass of the extended source. This can be easily seen by considering that the power p(x) between x and x+dx, which gives a contribution xp(x) to current $\int xp(x)dx = \langle x \rangle$, equal to the mean value of p(x).

Commercially available PSDs have a square active area with 0.5-5mm side and are mostly in silicon for a spectral response from 400 to 1100 nm.

Linearity error is less than 0.5% within the 80% of the active area, and cross-talk of coordinate signal is usually less than 10^{-4}. Response time is usually limited by the large junction capacitance, $C_b \approx 1$-50 pF, and by the op-amp frequency response. With 100-MHz op-amps, the response time of coordinate signals may go down to 3-10 µs.

22 Alignment, Pointing, and Sizing Instruments Chapter 2

2.2.3 Position Sensing with Reticles

Sometimes, we may not be able to use four-quadrant or PSD photodiodes and may merely have available a normal single-point detector. Nevertheless, we can realize a position-sensing or angular-sensing sensor by placing the detector behind a rotating reticle. Reticles have been developed and used since the early times of thermal-infrared technique [2] to track hot spots with a single-point detector.

In its simplest form, the reticle is a half-transparent and half-opaque disk, as illustrated in Fig.2-9. When put into rotation, it actually works as a chopper of the radiation reaching the photodiode located in the focal plane.

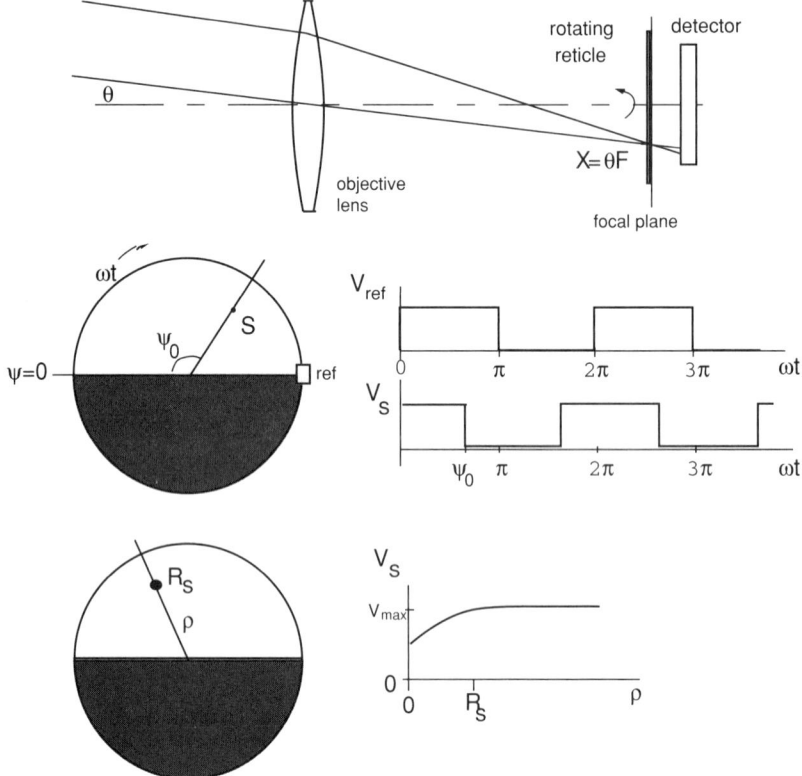

Fig.2-9 Position sensing by a rotating reticle. Light from a bright spot at the angle θ is imaged by the objective lens on the focal plane, where it is chopped by the reticle placed in front of the photodetector. By comparing the phase shift of the square waveform from the photodetector and of a reference, the position ψ_0 of the source is determined. The amplitude of the signal carries information on the polar coordinate ρ similar to Fig.2-6.

2.2 Pointing and Tracking

Thus, if we have a single bright spot in the field of view, the photodiode current will be modulated into a square wave, as shown in Fig.2-9. The information about the angular coordinate ψ_0 of the source is contained in the phase shift of the signal waveform with respect to a reference square wave at $\psi_0=0$.

It is easy to get this reference square wave, for example, by a microswitch or with a Light Emitting Diode (LED)-and-photodiode combination placed on the edge of the reticle, as shown in Fig.2-9. The radial coordinate ρ can also be determined by looking at the amplitude V_s of the square wave. With the same argument leading to the dependence plotted in Fig.2-6, we have the diagram of V_s versus ρ reported at the bottom of Fig.2-9.
The plain black/white reticle works well with point-like sources, but is disturbed by the presence of extraneous extended sources in the field of view.

Rejection to extended sources is improved by the *rising sun reticle* (Fig.2-10, top). In this reticle, half of the circle has 50% transmission, and half has 0 or 100%-transmission sectors, averaging 50% on the entire reticle. As a result, a point source gives a response with large amplitude V_s, whereas details larger than the sector width have $V_s \approx 0$ and only add a continuous component to the output signal.

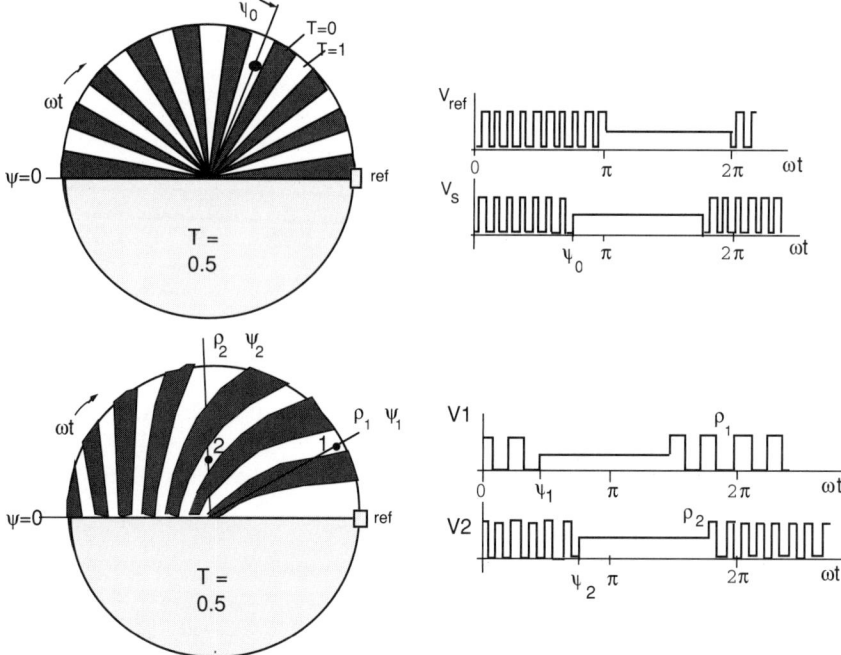

Fig.2-10 The rising-sun reticle provides an improved suppression to extended sources (top). The digital readout reticle (bottom) supplies both ρ and ψ coordinates in digital format.

The readout of the angle coordinate ψ can be made digital by counting the periods elapsed between the signal down-going edge (electrical angle ψ_0) and the reference down-going edge (electrical angle π). See the waveforms in Fig.2-10.

Another interesting reticle allowing an all-digital readout of coordinates is *the digitizing –sector reticle* shown in Fig.2-10, bottom. Here, the sectors are shaped according to a special radial profile. We choose the profile so that a point source at a radius ρ intersects a number of sectors proportional to ρ.

Thus, by counting the number of periods contained in a complete 2π-rotation of the reticle, we get the digits of the radial coordinate. The angular coordinate ψ is determined, as before, by the delay between the downgoing edges of signal and reference and is digitized when the delay is sorted out by period counting. As an example, in Fig.2-10, the signal for point 1 has 7 periods in a half turn while the signal for point 2 has 4 because its radius is smaller. The angular coordinate $\psi_2 > \psi_1$ is seen to correspond to a different number of periods between ψ and π.

For sake of clarity, reticles in Fig.2-10 have been shown with a small number of sectors. In a real reticle, we want to increase the number of sectors as much as possible to improve resolution and dynamic range of the readout.

The limits are set by the smallest detail we can write into the optical layer of the reticle, usually a metal film on a glass substrate. Using photolithographic techniques, we can easily attain a few-μm detail on a few cm-side reticle, or a 10^{-3} to 10^{-4} relative resolution or potential number of sectors. This argument explains how a 10-bit resolution in the digital readout of ρ,ψ coordinates is achieved with rotating reticles.

2.3 LASER LEVEL

The laser level is an extension of the concept of point-like alignment described in Section 2.1. When we do not need a direction to follow, or two coordinates (spatial x,y or angular ρ,ψ) as exemplified by the pipeline alignment, but rather a single coordinate (the height z or the horizontal ϕ) with the others free, the laser level is the instrument we need.

The most common application of the laser level is the distribution of the horizontal plane in construction works (Fig.2-11) and in leveling (or slope removing) of cultivation terrain subject to intensive watering, like rice. In this case, leveling is very important to save huge quantities of water.

As shown in Fig.2-11, a laser level can be implemented by a tripod-mounted laser. The laser source carries a collimating telescope to minimize the spot size on the range to be covered, as explained in Section 2.1. In addition, we transform the beam in the desired fan shape. This may be accomplished by reflecting the beam with a rotating mirror, oriented at 45 degrees to the vertical (Fig.2-12).

To generate a horizontal plane, the mirror is carefully oriented at 45 degree, and the beam from the laser is oriented vertically. Errors from the nominal angles result in a tilt error of the fan-beam plane.

2.3 Laser Level

Fig. 2-11 Alignment with a laser level. Left: use of the level for height relief in construction works. A fan beam is distributed down a 20-50 m radius area, and its position is checked at sight with the graduated stick. Right: a typical tripod-mounted, laser-level instrument developed by Spectra Physics.

The mirror angle-error can be eliminated by using a pentaprism in place of the mirror. The pentaprism has a dihedral angle of 45 degrees and therefore steers the input beam of 90 degrees, irrespective of the angle of incidence (Section A2.2).

Now, we have to adjust the laser beam to the vertical. A spirit level may be used for low cost. For maximum simplicity, the spirit level can be incorporated in one tripod leg, but in doing so, the resolution is modest ($\approx$10mrad or 30arc-min).

Otherwise, we may provide the instrument with a mechanical reference plane perpendicular to the laser beam, and the user will adjust the legs to verticality using a spirit level sliding on this reference plane.

For a good resolution, however, we may need a relatively bulky spirit level (50-cm side for $\approx$1mrad). Alternatively, we may use an optical method to check the verticality of the laser beam. One method to do so is comparing the beam wave vector to the perpendicular of a liquid surface.

As illustrated in Fig.2-12, the liquid (water) is kept in a hermetic box, and we use a small reflection ($\approx$3%) of the main laser beam passing through it toward the output. The other glass surfaces of the bow are antireflection coated.

Another small ($\approx$3%) portion of the laser beam is picked at the beamsplitter (Fig.2-12 right), which is also antireflection coated at the second surface.

The two beams are recombined by the beamsplitter and come to an objective lens acting as the input of an angle-sensing detector (Fig.2-7). An angular error of verticality produces a tilt in the wave vector reflected by the liquid surface, and then the two beams come to

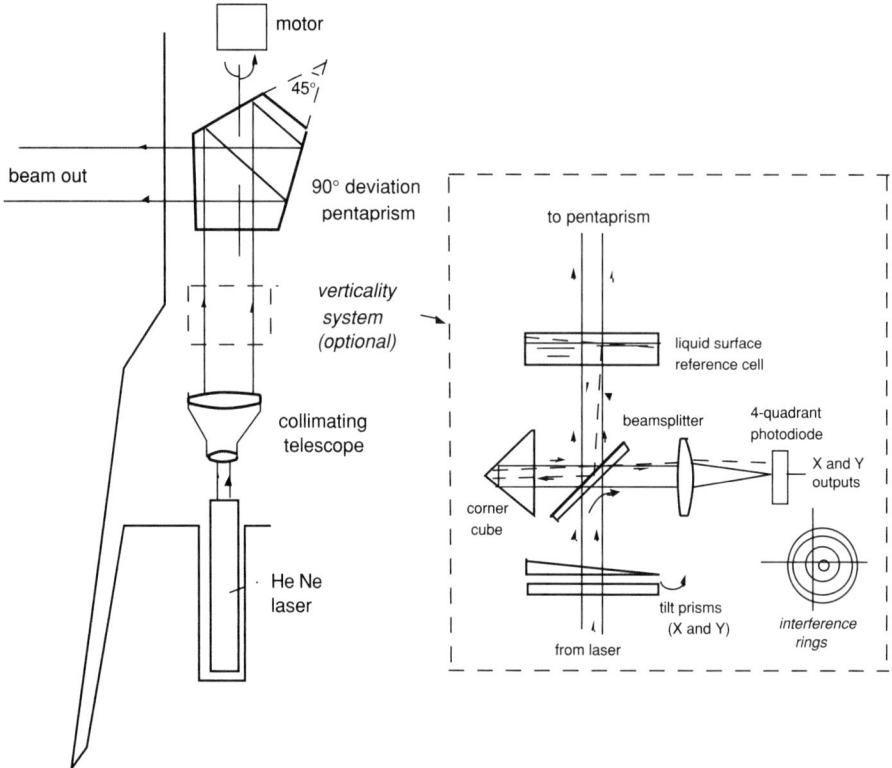

Fig. 2-12 A laser level uses a pentaprism (left, top) to deflect the incoming beam exactly by 90 degrees, independent from the angle of incidence. To check verticality, a partial reflection of the laser beam from the surface of a liquid in a cell is superimposed to part of the beam itself and is reflected backward. Looking at the interference rings generated by the superposition with a four-quadrant detector brings out the error (right). Deflecting prisms may be used for a servo providing the automatic compensation of the error.

the focal plane of the objective with different directions. Thus, they will produce interference rings in an offset position with respect to the optical axis of the angle-sensing detector (Fig.2-12, bottom right). The detector supplies the coordinate signals θ_X and θ_Y of the center of the interference figure. These signals can be used directly, displaying them to the user, for a manual adjustment of the legs of the tripod, or, with added sophistication, they may be fed to actuators moving a pair of deflecting prisms to adjust the beam and dynamically lock its direction to the vertical.

2.4 Wire Diameter Sensor

In industrial applications and manufacturing, instruments for measuring dimension are common for testing and online control purposes. In particular, two laser-based instruments for noncontact sensing are successfully employed: the wire diameter sensor and the particle size analyzer. Both instruments deal with relatively small-dimension D, in a range such that the ratio λ/D is favorable for exploiting diffraction effects. Then, the laser is the ideal source because, in the beam waist, it supplies a wave front with good spatial quality, as well as a high irradiance.

Diameter instruments are used mostly as the sensor of a process control in wire manufacturing. The wire may be a metal or a plastic wire extruded from a strainer or drawplate, or an optical fiber as well. The online measurement allows a feedback action on process parameters, and the result is a better, tight control of tolerance of the fabricated wire.

The wire to be measured is allowed to pass inside an aperture of the instrument, where it is illuminated by the beam waist of an expanded laser beam. As usual, we use a He-Ne laser and a collimating telescope (Fig. 2-13).

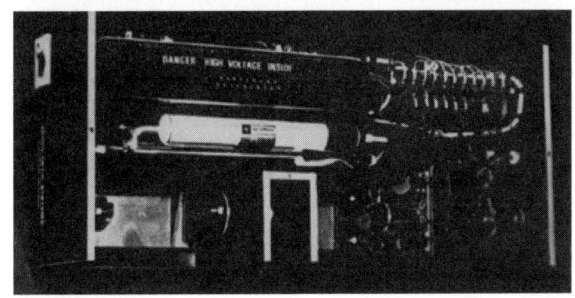

Fig. 2-13 A noncontact sensor for diameter measurement. Top: the instrument envelope, with the U-slit for wire passage; bottom: internal details include the He-Ne laser (center left), the collimating telescope (bottom left), the receiver (bottom right), and the electronics box (top left).

For the illuminating field, the wire is an obstruction or diffraction stop, and, in view of the Babinet principle, it is equivalent to a diffraction aperture. The diffracted field is collected by a lens (Fig.2-14) and its distribution is converted into an electrical signal by a photodetector placed in the focal plane of the lens. From this signal, the unknown diameter D can be calculated.

As well known from diffraction theory and recalled in App.A5, the distribution of diffraction from the wire is the Fourier transform of the wire aperture. Assuming a constant illuminating field E_0, the field and the radiant power density diffracted at the angle θ are given by (see Sect.A5.3):

$$E(\theta) = E_0 \; \text{sinc} \; \theta D/\lambda \qquad (2.8)$$

$$P(\theta) = E_0^2 \; \text{sinc}^2 \; \theta D/\lambda, \qquad (2.8')$$

In these expressions, sinc ξ =(sin $\pi\xi$)/$\pi\xi$. The first zero of the sinc distribution is at $\xi=\pm 1$, which corresponds to a diffraction angle $\theta_{zero} =\pm \lambda/D$.

To work out D, we first manage to make available the $P(\theta)$ distribution and then look for the measurement strategy that is well suited to extract D from the distribution.

We use a lens (Fig.2-14) to convert the angular distribution $P(\theta)$ in a focal plane distribution $P(X)$ to be scanned by the photodetector. The scale factor is $X=F\theta$, with F being the focal length of the lens.

Usually, we require a long focal length F for a sizeable X, yet we want to keep the overall length short (much less than F).

As an example, if D=200μm and λ=0.5μm, we need F=400 mm to have a zero at X=±1mm. A good choice is a telephoto lens [3] with two lenses, a front positive and a rear negative (or

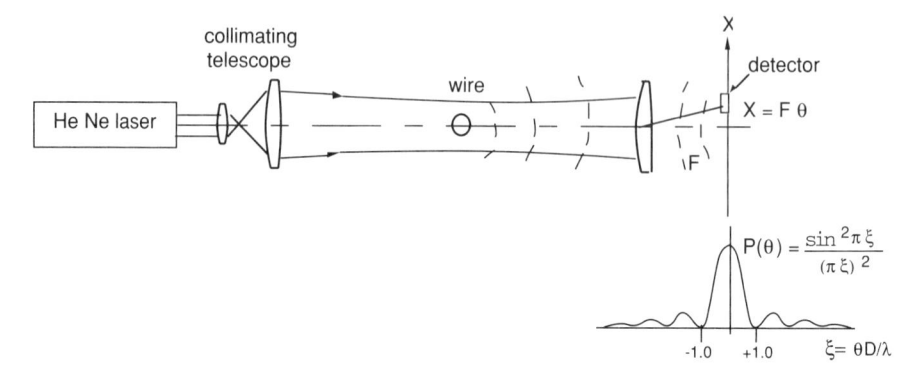

Fig. 2-14 In a wire-diameter sensor, the wire is illuminated with a laser beam, properly expanded and collimated. The far-field distribution of diffracted light is converted by a lens to a spatial distribution in the focal plane. Scanning the focal plane by a photodetector reveals the zeros of the distribution, located at X/F=θ=±λ/D, hence the measurement of D.

2.4 Wire Diameter Sensor

Barlow's focal multiplier). By this, an F=400 mm lens may be only 50-mm long.

On the source side (Fig.2-14), a collimating telescope is used to expand and collimate the laser beam, which makes a nearly plane wave front available in the measurement region. The size of the expanded beam D_{beam} is usually kept several mm to a few cm, according to user's specifications.

With a larger size, a larger transversal movement is tolerated, and the errors associated to beam waist nonidealities are reduced. On the other hand, a larger size means less power on the wire and therefore a weaker signal and S/N ratio at the detector.

To measure D from the experimental distribution $P(X) = P_0 \text{sinc}^2 XD/F\lambda$, we may in principle look at several different features in the waveform P(X), for example, the main pulse width at half maximum, the secondary-maximum positions, the flex points, etc.

However, we shall take into account two nonidealities in the P(X) detection: (i) the finite size of the detector W_{det}, by which the P(X) distribution is averaged on W_{det} and (ii) the relatively strong peak coming from the non diffracted beam and superposed to the useful signal near the origin X=0 (this contribution is not shown in Fig.2-14).

If D_{beam} is the beam diameter at the waist illuminating the wire, the non diffracted peak has a width of the order of $F\lambda/D_{beam}$ and an amplitude D_{beam}^2/D^2 times higher than the wire diffraction amplitude.

Then, a better feature to look at in the P(X) distribution is the location of the first zeroes near the peak. These are positioned at $\theta = \pm\lambda/D$, or $X = \pm F\lambda/D$. We may detect them conveniently with a linear-array Charge Coupled Device (CCD) [4] placed in the focal plane.

The CCD is a self-scanned detector, that provides a signal $i(t) = i_0 \text{sinc}^2 vtD/F\lambda$, where v is the scanning speed. By time differentiating the signal i(t), we get an i'(t) signal, which passes through the zero level in correspondence to the minima of the sinc^2 function.

Further, we can arrange a digital processing of the measurement as follows: a threshold discrimination of the i'(t) signal picks up the zeroes at time $t = \pm F\lambda/vD$ and gates a counter to count clock transitions between two start and stop triggers. If the clock frequency is f_c, the number of counted pulses is $N_{count} = 2f_c F\lambda/vD$, and we have a digital readout of the diameter D. However, we still require calculating a reciprocal.

In commercially available instruments, the range of measurable diameters extends from 10 µm to over 2 mm, and the corresponding accuracy is 1% and 5%, respectively, at the border of the measurement range.

Reasons for limits in the measurement are as follows. At large diameters, the diffraction angle becomes small, and the $P(\theta)$ is difficult to sort out, whereas errors due to illuminating wave front planarity and detector finite size become increasingly important. At the lower end of the diameter range, where diffraction angle is conveniently large, limits are due to the approximations in $P(\theta)$ and to vignetting-effect in the collecting lenses. It is worth noting that user's specifications set a requirement on small diameter at ≈50 µm because very small wires are uncommon in industrial manufacturing.

In commercially available instruments, the allowed transversal movement of the wire is typically 2 to 5 mm with no degradation of other specifications. A typical overall size of the diameter-sensing instrument (Fig.2-14) is 50 by 20 by 10 cm.

2.5 PARTICLE SIZING

Another instrument based on diffraction is the particle-size analyzer. With this instrument, we can measure the distribution of particle diameters in a fine-conglomerate mixture. A lot of mixtures are found interesting for particle sizing, ranging from manufacturing and industrial processes (ceramic, cement, and alloys powders) to particulate monitoring in air and water and sorting of living material for medical and biological purposes.

The first particle-size analyzers date back to the 1970s [5]. They were developed originally to keep the size distribution of ceramic and alloy powders under control in industrial manufacturing. Since then, the field has steadily matured, and the new products accepted in a variety of applications. The measurement range now extends from the submicron particulate to relatively large particles, up to about 1 mm, that cover a variety of surrounding media, while accuracy and resolution are improved and the sample size is reduced.

A basic scheme of a particle-size analyzer is shown in Fig.2-15. This scheme is similar to the wire sensor, and with it we look at the small-angle diffraction or elastic scattering of light from the particles. Because of this, the scheme is classified as Low-Angle Elastic Light Scattering (LAELS) [6].

As a source, we may use a low power (typically 5mW) semiconductor laser or He-Ne laser. The collimating telescope expands the beam and projects a nearly plane wave front on the measurement cell that contains the particles to be measured. In the cell, the particles are kept in suspension by means of a mechanical stirrer or an ultrasonic shaker (the PZT ceramic indicated in Fig.2-15). In this way, we avoid flocculation, or, the sticking together of particles.

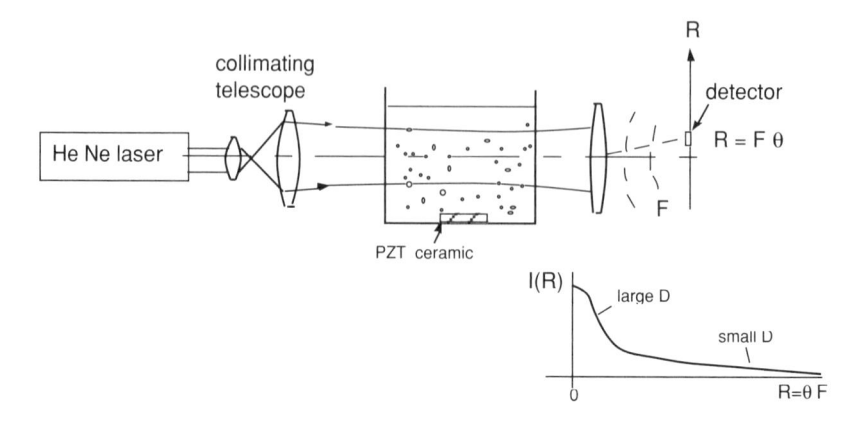

Fig. 2-15 A particle-size analyzer is based on the diffraction of light projected on a cell containing the particles in a liquid suspension. At the output of the cell, a lens collects light diffracted at angle θ and converts it in a focal plane distribution I(R), with R=θF. Then, I(R) is measured by the photodetector, and the desired diameter distribution p(D) is computed.

2.5 Particle Sizing

If the cell is relatively thin, the wave front can be assumed plane throughout the cell, and particles are illuminated by the same wave vector. Each particle contributes to the far field distribution collected at the angle θ according to Fourier transform of an aperture of diameter D, as given by Eqs.A5-16 and A5.16'. The contribution is A_0 somb $[(D/\lambda)\sin\theta]$ in field amplitude and I_0 somb$^2[(D/\lambda)\sin\theta]$ in power density, where somb $x = 2J_1(\pi x)/\pi x$ (see App.A5.3). The total field amplitude and power density are the integrals of somb and somb^2 terms, weighted by p(D), the distribution of diameters in the particle ensemble.

In summing up the contributions, we shall recall that particles are randomly distributed in the cell, and their positions fluctuate well in excess of a wavelength. Because of this, the interference terms originated by the superposition of field amplitudes have zero average.

Thus, the square of the field sum is equal to the sum of the squares. In other words, in the far field of the cell, we shall add the intensities or the squares of the fields contributed by individual particles.

Using Eq.A5.16' and integrating on the diameter distribution, the far field intensity I(θ) at the angle θ is written as:

$$I(\theta) = I_0 \int_{0-\infty} \text{somb}^2[(D/\lambda)\sin\theta]\, p(D)\, dD \qquad (2.9)$$

Now, we want to calculate from Eq.2.9 the unknown distribution p(D) of particle size, after I(θ) is determined through a measurement of the far field intensity. The kernel of the Fredholm's integral, somb$^2[(D/\lambda)\sin\theta]$, is a given term because we can compute it for any pair of given θ and D values.

However, the problem is known as an ill-conditioned one because p(D) is multiplied by the kernel, and the result is integrated on D so that details in p(D) are smoothed out in the result I(θ). For example, very large particles affect only a small interval in I(θ) near θ≈0, whereas very small ones give only a little increase of I(θ) at large θ.

Another source of error is the constant scattering coefficient tacitly assumed in writing Eq.2.9. This holds for large-particles, D>>λ, whereas for intermediate and small particles Q_{ext} varies appreciably (see Fig.A3-6 and Eq.A3.2).

From the previous considerations, the measurement range best suited for the LAELS will normally go from a few μm to perhaps several mm in diameter. At small D, the limit comes from the Rayleigh approximation, whereas at large diameter it depends on the size of the photodiode, due to measure at θ≈0 without disturbance from the undiffracted beam.

Because we are unable to solve Eq.2.9 exactly in all cases, we will try to maximize the information content picked out from the experiment through the measurements of I(θ). To do so, we will repeat the measurement at as many angles θ as the outcomes I(θ) are found appreciably different. With information provided by the I(θ) set, the kernel integral will limit resolution to a certain value, beyond which the solution becomes affected by a large oscillating error. We can then optimize the solution by increasing resolution (or number of diameters) until the computed p(D) starts being affected by oscillation errors.

Let us now briefly discuss the mathematical approaches that have been considered and applied to solve the inversion problem given by Eq.2.9.

Analytical Inversion. Although Eq.2.9 looks like intractable, Chin et al.[7] have been able to solve it for p(D) using hypergeometric functions, and the result is:

$$p(D) = -[(4\pi/D)^2/\lambda] \int_{\theta=0-\infty} K(\pi D \sin\theta/\lambda) \, d[\theta^3 I(\pi D \sin\theta/\lambda)] / I_0 \quad (2.10)$$

In this equation we have let $K(x) = xJ_1(x)Y_1(x)$, J and Y are the usual Bessel functions, and $d[.] = (\partial[.]/\partial\theta) \, d\theta$ are the differential of the product $[.] = \theta^3 I(.)$, in the variable θ.

Despite being a remarkable mathematical result, Eq.2.10 is difficult to use because it requires integrating $I(\theta)$ on θ from 0 to ∞. Unfortunately, the Fraunhofer approximation and the assumption of Rayleigh regime are valid only for $\theta \approx 0$. To avoid an abrupt truncation of $I(\theta)$ and the resulting oscillation error in p(D), the data $I(\theta)$ is apodized to zero. A typical smoothing profile is $P(\theta) = 1 - (\theta/\theta_{max})^2$, and the maximum θ is chosen as $\theta_{max} \approx (2-5)\lambda/D$ to cover the range of interest in diffraction.

In the range of diameter 10 to 50 µm and with $\lambda = 0.633$ µm, the analytical inversion works nicely if measured data is very clean and accurate, that is, affected by an error of 1% or less.

Least Square Method. Using a discrete approximation, we may bring Eq.2.9 to a set of linear equations and then apply the Least-Square Method (LSM) to improve the stability of solution, i.e., reduce spurious oscillations.

To do so, we write the integral as a summation on the index k of the diameter variable D_k and repeat the equations for as many variables n of the angle of measurement θ_n. The matrix of known coefficients connecting θ_n and D_k is the discrete kernel S_{nk}:

$$S_{nk} = \text{somb}^2[(D_k/\lambda)\sin\theta_n] = 4J_1^2[(\pi D_k/\lambda)\sin\theta_n] / [(\pi D_k/\lambda)\sin\theta_n]^2 \quad (2.11)$$

The range of variables n and k is n=1..N and k=1..K, and the associated variables are $I_n = I(\theta_n)$ and $p_k = p(D_k)$. With this, Eq.2.9 becomes:

$$I_n = \sum_{k=1..K} S_{nk} \, p_k \quad (n=1..N) \quad (2.12)$$

In this set of equations, the number of equations N may be different from the number of unknown K. Usually, N is the number of angular measurements performed by the photodetector. If we use an array or a CCD to obtain a number of separate channels in angle θ, N is fixed and may be usually 50 to 100. The number of unknown diameters K is variable because we want to determine the calculated distribution p(D) on the largest number of separate diameters that are compatible with the absence of oscillating errors. Then usually we may have K=6-20, depending on the average D and the shape of p(D).

To solve for the unknown p_k, we need to bring the number of unknowns to coincide with the number of equations, whatever N and K are, and we do so by applying the LSM. With the LSM, we require that the solution p_k is such that the error ε^2, or mean square deviation of the computed values $\Sigma S_{nk} p_k$ and measured ones I_n, is a minimum:

$$\varepsilon^2 = \sum_{n=1..N} [I_n - \sum_{k=1..K} S_{nk} p_k]^2 = \min \quad (2.13)$$

2.5 Particle Sizing

The condition of minimum is obtained by setting to zero the partial derivative of the error ε^2, with respect to each value p_k of the unknown:

$$0 = \partial(\varepsilon^2)/\partial p_k = \Sigma_{n=1..N} \, [\, I_n - \Sigma_{k=1..K} \, S_{nk} \, p_k]^2$$
$$= \Sigma_{n=1..N} \, 2 \, [\, I_n - \Sigma_{k=1..K} \, S_{nk} \, p_k] \, (-S_{nk}) \quad (2.13')$$

By rearranging the terms in this expression, we get:

$$J_h = \Sigma_{k=1..K} \, Z_{hk} \, p_k \qquad (h=1..K) \quad (2.14)$$

where we have let:

$$J_h = \Sigma_{n=1..N} \, I_n S_{nh} \quad \text{and} \quad Z_{nk} = \Sigma_{n=1..N} \, S_{nk}^2 \quad (2.14')$$

Now, the number of equations in 2.14 is equal to the number of unknowns and we can solve for p_k with standard algebra.

Using the LSM, the accuracy of solution improves, and the range of diameter distribution is larger with respect to the analytical inversion method. However, if data are affected by errors or the number of diameters is excessive, the error of the solution increases. Also, the error may be large if the waveform p(D) is bimodal or contains secondary peaks.

Thus, one has to choose a reasonable number of diameters to get the best result. A strategy is to increase the number K of diameters and repeat the calculation until in the new result oscillations or wrong features start to show up in the new result.

Usually, the range of diameter of interest may be large (for example, two decades from 2 to 200μm), but the number of affordable diameter is modest (e.g., K=6). Then, we get more information from the cumulative distribution P(D) than from the density function p(D). The relationship between the two is:

$$P(D) = \int_{x=0-D} p(x) \, dx \quad \text{and} \quad P_k = \Sigma_{i=1..k} \, p_i \quad (2.15)$$

Of course, using Eq.2.15 with Eq.2.14 allows us to solve for P(D) with the LSM.

Errors in measured data I_n can be reduced by working on integral quantities in place of the differential ones considered so far. In fact, the photodetector has a finite area, and we usually strive to shrink its active aperture, but with a loss of signal-to-noise ratio. Then, it is better to start the problem of solving Eq.2.9 by assuming the signal is collected in a finite angular aperture $\theta_1 .. \theta_2$.

The integrated signal is written as $Y(\theta_1, \theta_2) = \int_{\psi=\theta1..\theta2} I(\psi) d\psi$ if the photodetector is made of elements with the same area at all θ, or as $Y(\theta_1, \theta_2) = \int_{\psi=\theta1..\theta2} 2\pi\psi \, I(\psi) d\psi$ if the photodetector is made of annulus elements for which the area increases with radius. In any case, by inserting Eq.2.9 in these expressions, we can integrate the kernel of the Fredholm's integral, $\text{somb}^2[(D/\lambda)\sin\theta]$, with respect to θ, either numerically or analytically. The result, $X(\theta_1,\theta_2)$, becomes X_{nk} when we use the discrete approximation, and this value can be used to solve for p_k or P_k with the LSM expressed by Eq.2.14.

With this approach, monodispersed powders can be measured, in the range of diameter 20 to 50μm, typically with a <2% error, in K≈5 intervals of the cumulative distribution. At the expense of a reduced accuracy (≈5% error), the range can be extended to 5 to 200 μm and 7 to 10 intervals. A typical result of particle-size measurement is shown in Fig.2-16.

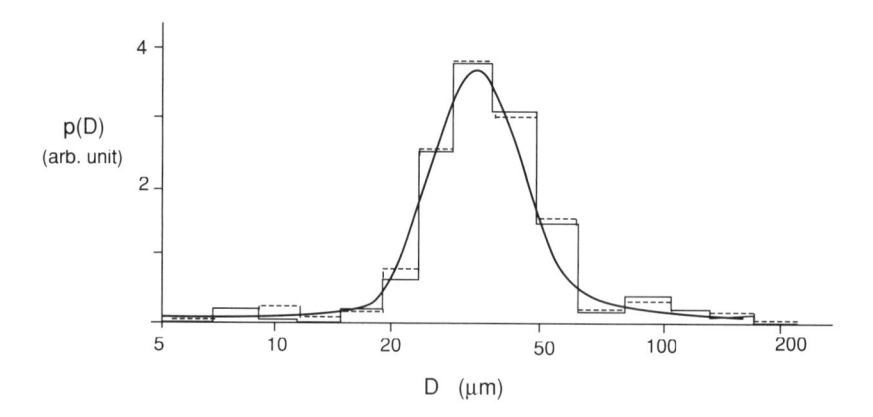

Fig. 2-16 Typical result of particle-size measurement as obtained by LAELS and LSM. Thick line: the true distribution, step-wise curve: the reconstructed distribution, dotted line: typical fluctuation for a 1% noise on data.

Other Linear Methods. The basic LSM we have outlined previously is the starting point for several extensions [8,9]. In all of them, we try to convey a physical constraint or information in a set of linear equations, just like the LSM concept leading to Eq.2.14.

A straight example is weighting of terms contributing to the total error ε^2 in Eq.2.13. To do that, we may add, under the summation term of Eq.2.13, a multiplying factor w_n that is taken =1 where the measured I_n is clean (or, has a good S/N ratio), and <1 where I_n is noisy. Of course, the choice of the exact weight is somehow arbitrary or left to our knowledge of the powder nature. Accordingly, the quality (or error) of the results will depend on the w_n values we have chosen.

Other examples of information that may be added through linear equations are (i) the nonnegative character of the distribution p(D), which leads to an inversion method known as Non-Negative LSM (NNLSM); (ii) the known shape of the p(D) waveform, for example, monodispersed or bimodal; (iii) the minimum-ripple or smoothness constraint.

With the addition of the linear constraint, both accuracy and dynamic range of the reconstruction are significantly improved. However, linear methods have the general disadvantage of a success dependent on our ability to tune the added information and to adjust free parameters describing it [7-9].

Iterative Methods. The prototype of these methods is Chahine's method [6,10] which is used to invert Fredholm's equation in a variety of applications. We start letting K=N, that is,

2.5 Particle Sizing

using a number of unknown variables p_k equal to the number of measured intensities I_n (K=N is also provided by the application of the LSM, first).

The iterative method is based on the following reasoning. If the set of diameter distribution p_k is correct, it should give the measured distribution $I_{n.calc} = \Sigma_k C_{nk} p_k$. If the set of calculated values $I_{n.calc}$ differs from the experimental values, we may expect to approach the solution by multiplying p_k by $I_{k.meas}/I_{k.calc}$, where $I_{k.meas}$ is the measured value, or explicitly:

$$p_{k+1} = I_{k.meas}/I_{k.calc}\, p_k \qquad (2.16)$$

By repeating the procedure an adequate number of times, p_k should converge to the correct solution. As a trial value of p_k, we can use either an estimated distribution or the result of a LSM calculation. Starting with an all-positive trial distribution, the results in all the iterations are necessarily positive, and we get the inherent benefit of suppressing oscillations with negative swings. The results obtained by the Chahine's method are definitely better than the normal LSM. Errors are greatly reduced (by a factor 3 to 5), and the range of diameters is significantly increased. On the other hand, the tendency exists to generate spurious spikes at small D, unfortunately, in place of the oscillations suppressed by the method.

Another problem with Chahine's and other iteration methods is that no clear sign exists when the optimal result is reached. Thus, we usually define a quality function, for example, the error ε^2 of calculated results to measured ones, and stop the iteration when a minimum of ε^2 is reached.

A refinement of Chahine's method has been recently introduced [6]. It consists in weighting the iteration by the normalized kernel, $S_{nk}/\Sigma_{n=1..N} S_{nk}$. With this position, Eq.2.16 reads:

$$p_{k+1} = (S_{nk}/\Sigma_{n=1..N} S_{nk})(I_{k.meas}/I_{k.calc})\, p_k \qquad (2.16')$$

At the generic iteration, each term p_k is weighted by S_{nk}, that is, proportionally to the efficiency of transfer of p_k to the output I_n. In consequence, spurious peaks found in Chahine's method are suppressed, and resolution and dynamic range are improved further [6].

With regard to the optical layout of the particle-size analyzer, several sources of error can influence the results. The finite size of the photodetector has already been considered, and found that it can be accounted for by reformulating Eq.2.8 in terms of integrated intensity.
Another error comes from the undiffracted beam at $\theta=0$, which requires a stop on the focal plane to be blocked out (the stop is omitted in Fig.2-15 to avoid confusion). Even using the stop, some light will inevitably scatter or leak aside on the detector, which affects the $\theta \approx 0$ measurements.

To get rid of it, we can use the filtering arrangement reported in Fig.2-17, known as reverse Fourier-transform illumination. Light projected through the cell by lens L1 converges at a focal plane position. The contribution diffracted at the angle θ reaches the focal plane of L1 at position $x=\theta F$ with rays parallel to the optical axis.

Instead of reading the intensity in the focal plane, we place a lens L2 there and perform a Fourier transform of the incoming angular distribution. All the θ-diffracted rays arriving

36   **Alignment, Pointing, and Sizing Instruments   Chapter 2**

parallel to the optical axis are focused on the axis of L2 (Fig.2-17), whereas undiffracted light arriving at an angle will reach the focal plane out of the axis. Placing a pinhole at the focal plane of L2, we can allow the diffracted contribution through and block out the undiffracted one. With another lens L3, we can then retransform the filtered distribution to one with rays parallel to the optical axis.

Because of the scattering suppression, the arrangement of Fig.2-17 is much better of a stop in the focal plane, and we can measure down to small angle θ (typ. <0.1mrad). Accordingly, the range of large diameters that can be reconstructed is expanded up to several hundred μm or even to the mm.

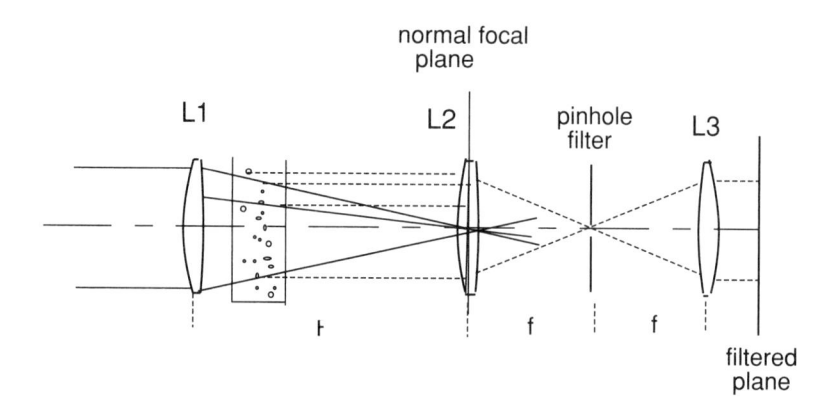

Fig. 2-17 Using a convergent beam to illuminate the cell, we are able to filter out the undiffracted beam more efficiently than with stops. Diffracted rays indicated by dotted lines are focused on axis and can pass through the pinhole, whereas undiffracted rays arrive out-of-axis and are blocked.

At the other end of the measurement range, that of large diffracted angle or small diameter, the errors are not only due to collection efficiency, vignetting, and the like, but also to the Fraunhofer approximation $D >> \lambda$. Errors start becoming significant at, say, $D \approx 2\text{-}5\lambda$, so the practical range of measurement for particle sizing based only on the LAELS is typically 2...500 μm.

We can extend the range of measurement on small particles by taking advantage of the Mie ($D \approx \lambda$) and Rayleigh ($D << \lambda$) scattering regimes. These are described by the appropriate scattering function $f(\theta)$ and extinction factor Q_{ext} (Appendix A3.1), which are known and can be calculated from Mie's theory [11].

In the Rayleigh regime, the scattering is nearly isotropic in angle, and the extinction factor Q_{ext} varies as $(D/\lambda)^4$. When D increases up to about $D \approx \lambda$, the scattering function peaks forward, and the extinction factor increases up to ≈2-4 (see Fig.A3-6).

Now, we can supplement the LAELS measurement with a measurement of light scattered from the cell at a fixed angle, i.e., large enough to be out of the range of LAELS data.

2.5 Particle Sizing

For example, a choice is to look at the cell from $\theta \approx 45°$ so that the detector is safely away the undiffracted beam leak. To collect information on the particle size distribution, we now scan the wavelength instead of the diffraction angle. By varying the ratio D/λ, the extinction factor $Q_{ext}(D,\lambda,n)$ varies and so does the scattered power too.

For a density distribution $p(D)$ of diameter, and given an extinction factor $Q_{ext}(D,\lambda,n)$ at the measurement wavelength λ, the power density collected within a solid angle $\Delta\Omega$ around $\theta=45°$ is given by:

$$I(\lambda, 45°) = f(45°) \, (\Delta\Omega/4\pi) \, I_0 \int_{0-\infty} Q_{ext}(D,\lambda,n) \, p(D) \, dD \qquad (2.17)$$

where $f(\theta)$ is the scattering function and Q_{ext} is the extinction factor defined in App. A3.1. Eq.2.17 is clearly the counterpart of Eq.2.9 for the case at hand of extinction-related measurement. All the previously discussed methods of inversion of the Fredholm's integral (Eq.2.17) can now be repeated [6] on the variables D_k and λ_n.

This method of particle sizing is called *Spectral Extinction Aerosol Sizing* (SEAS). With the computation of the distribution $p(D)$ from Eq.2.17, it goes down to 0.02–0.1μm as the minimum measurable size, whereas the maximum range overlaps the minimum of LAELS ($\approx$ 2-5 μm).

About the source, we need one with a high radiance and a wide spectral emission because the minimum detectable signal depends on radiance, and the minimum measurable size depends on the shortest wavelength available. To cover the visible and UV range, a cheap choice is a halogen lamp with 3000 to 3300-K color temperature. The lamp output is collimated by a lens and filtered by a monochromator. The output beam can either be combined with the laser beam of the LAELS by means of a wavelength-selective beamsplitter or used as stand-alone source. The detector to be used is usually a photomultiplier (see Ref.[1], Chapter 4) which is well suited for the spectral range UV-visible, and provides the best sensitivity, even in the case of small particulate concentration.
An example of a modern particle-sizing instrument is shown in Fig.2-18. This instrument is the technical evolution of a first product released in 1971.

A last method worth mentioning is the *Dynamical Scattering Size Analyzer* (DSSA), which is useful for measurements of very small (1..100-nm) particles [12]. Different from previous techniques, when particles are very small, the scattered light is frequency shifted. The shift is due to the Doppler effect $(\underline{k}_o-\underline{k}_i) \cdot \underline{v}$, or equivalently by the interferometric phase $(\underline{k}_o-\underline{k}_i) \cdot \underline{s}$, of the particle displacement $\underline{s}$ observed from the direction $\underline{k}_o$ and illuminated from $\underline{k}_I$ (all underlined quantities being intended vectors). The shift can be measured by a frequency-spectrum analysis or by a time-domain autocorrelation measurement of the detected signal. For example, the autocorrelation function $C(\tau)=(1/T)\int_{0-T} i(t)i(t+\tau)dt$ is found to depend on the diffusion constant δ of the particles, as $C(\tau)=C_0 \exp{-\delta \, (k_o-k_i)^2 \tau}$. From the measurement of the exponential decay of $C(\tau)$, the diffusion constant is determined.
Then, we use the relation $\delta = kT/2\pi\eta D$ to determine the diameter D after the viscosity η of the particle in the surrounding medium is known.

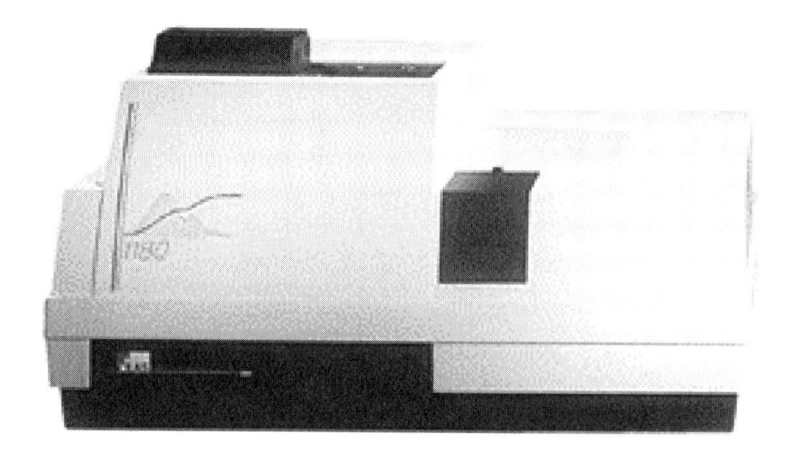

Fig. 2-18 A modern particle-size analyzer based on diffraction and extinction performs diameter measurements from 0.05 to 2500 μm (courtesy of CILAS, France)

REFERENCES

[1] S. Donati, *"Photodetectors"*, Prentice Hall: Upper Saddle River, 2000.
[2] R.D. Hudson, jr., *"Infrared System Engineering"*, Wiley Interscience: New York, 1969.
[3] D. Malacara and Z. Malacara, *"Handbook of Lens Design"*, Marcel Dekker Inc.: New York, 1994.
[4] see Ref.[1]. Section 9.4.
[5] J. Cornaillault, *"Particle Size Analyzer"*, Applied Optics vol.11 (1972), pp.265-269.
[6] F. Ferri, G. Righini, and E. Paganini, *"Inversion of Low-Angle Elastic Light Scattering Data With a New Method Modifying Chanine Algorithm"*, Applied Optics vol.36 (1997), pp.7539-50; see also: Applied Optics vol.34 (1995), pp.5829-5839.
[7] J.H. Chin, C.M. Sliepcevich, and M. Tribus, *"An Improved Least Square Method Routine to Compute Particle Disatribution"*, Journ. Phys. Chem. vol.56 (1955), pp.841-848.
[8] S. Twomey, *"Introduction to Mathematics of Inversion in Remote Sensing and Indirect Measurements"*, Elsevier: Amsterdam 1977, Chapter 7.
[9] N. Wolfson et al., *"Comparative Study of Inversion Techniques, part I and part II'"*, Applied Meteorology vol.18 (1979), pp.543-561.
[10] M.T. Chahine, *"Determination of the Temperature Profile in an Atmosphere from its Outgoing Radiance"*, Journal of Optical Soc. of Am. vol.58 (1968), pp.3074-3082.
[11] H.C. Van de Hulst, *"Light Scattering by Small Particles"*, J. Wiley: New York, 1957.
[12] B.J. Berne and R. Pecora, *"Dynamic Light Scattering"*, J. Wiley: New York, 1976.

CHAPTER **3**

Laser Telemeters

A telemeter is an instrument for measuring the distance to a remote target. Basically, the three main techniques to perform an optical measurement of distance are the following:
- *Triangulation*. The target is aimed at from two points separated by a known base D, placed perpendicular to the line of sight. By measuring the angle α formed by the two line of sights, the distance is found as $L=D/\alpha$.
- *Time of flight*. A light beam from a high-radiance source is propagated to the target and back, and the time delay $T=2L/c$ is measured (c is the speed of light). The distance follows as $L=cT/2$.
- *Interferometry*. A coherent beam is used in the propagation to the target. The returned field is detected coherently by beating with a reference field on the photodetector, and a signal of the form cos 2ks (where $k=2\pi/\lambda$) is obtained. From cos 2ks, we can count the distance increments in units of $\lambda/2$.

In view of the obtained performances, the three approaches are complementary.

Triangulation is the simplest technique to implement and may operate in daylight even without any source. Until a few decades ago, it survived in construction applications with the theodolite (to measure α) and the rulers (to set D). However, it has a poor accuracy on long distances because, when L is much larger than the base D, the angle α to be measured becomes very small and is affected by errors.

The time of flight technique requires a pulsed or a sine-wave modulated laser. In both cases, we measure the time of flight T=2L/c of light to the target at distance L. This telemeter works with a constant accuracy ΔT (or, in distance, ΔL=cΔT/2), in principle. Thus, performance is excellent on medium and long distances. Because of this, it has superseded the theodolite on medium distances (100 m to 1 km) and has proven to be a new powerful technique in a number of long-distance (greater than 1 km) applications.

Last, the interferometric technique is by far the most sensitive, but has the drawback of requiring the development of the λ/2-counts by a movement of the target. Thus, this technique provides basically an *incremental* measurement of distance, not an *absolute* one as the other two techniques. For this reason, interferometric techniques are treated separately in Chapter 4, whereas in this chapter we will concentrate mainly on time of flight telemeters.

From the point of view of the remote target features, we may have either (i) a cooperative target made of a retroreflector surface to maximize the returning signal (ii) a noncooperative target, simply diffusing back the incoming radiation with a δ<1 diffusion coefficient.

According to the measurement technique employed, time of flight telemeters are classified as one of the following:
- *Pulsed telemeters* when the measurement is performed directly as the delay T between transmitted and received pulses
- *Sine-wave* modulated *telemeters*, when the source is modulated in power by a sine wave at a frequency f, and the delay T is measured from the phase Φ=2πfT.

The first case relates to long-range telemeters and is useful for geodesy research and military applications, with operational ranges up to 100 km or more, whereas the second identifies the topographic telemeters used on medium ranges (<1km) for civil engineering and construction work applications.

The laser sources most suitable in the two cases are the following:
- Solid state lasers (like Nd, YAG) operated in the Q-switching regime and semiconductor-diode (like GaAlAs) arrays, pulsed at 1 to 10-ns durations
- Quasi continuous-wave semiconductor lasers like GaAlAs and GaInP to supply modulation up to hundreds of MHz.

3.1 TRIANGULATION

Let us consider the basic scheme for triangulation illustrated in Fig.3-1. Here, a distant object O is aimed from two observation points, A and B, along a base of width D.

The object is assumed self-luminous for the moment, which is a case referred to as passive triangulation. Active triangulation using a laser source to illuminate the target is considered later on.

The beamsplitter and rotatable mirror combination allows superimposing the images seen at A and B. The mirror M is rotated by an angle α from the initial position α=0 parallel to the beamsplitter until the object images are brought to coincide. The distance L then follows as:

3.1 Triangulation

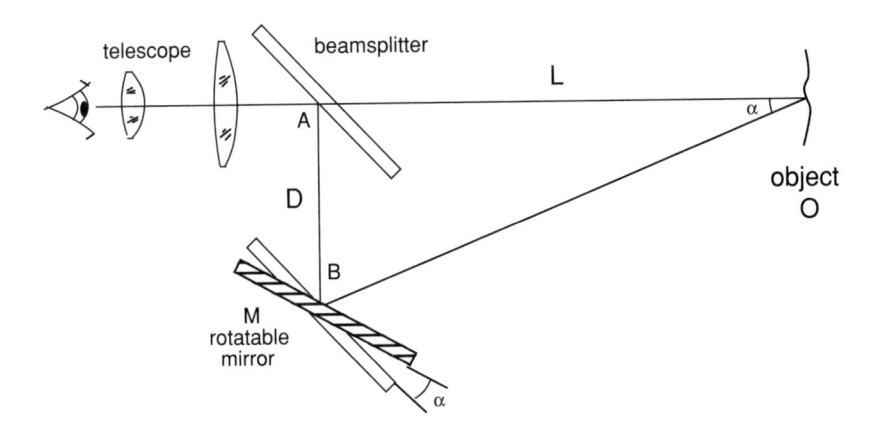

Fig. 3-1 Basic scheme of a triangulation measurement

$$L = D / \tan \alpha \approx D / \alpha \qquad (3.1)$$

Of course, we need a good measurement of small α to determine a long distance L with a reasonably short base D. The absolute and relative errors, ΔL and $\Delta L/L$, due to an angle error $\Delta \alpha$, are found from Eq.3.1 as:

$$\Delta L = - (D/\alpha^2) \Delta \alpha = - (L^2/D) \Delta \alpha \qquad (3.2)$$

$$\Delta L / L = - (L/D) \Delta \alpha \qquad (3.2')$$

and they both increase with distance. For design purposes, the errors are plotted in Fig.3-2 as a function of L and with $\Delta \alpha$ as a parameter.

With regards to the error of the angle readout, a good micrometer screw with gear reduction and backlash recovery is representative of medium-accuracy performance and may resolve $\Delta \alpha$=10 arc-min ($\approx$3 mrad). On the other hand, an angle encoder may provide a high-accuracy readout, going down to the limit of the (small) viewing telescope, typically $\Delta \alpha$=0.3 arc-min ($\approx$0.1 mrad).

In these two representative cases, it is interesting to evaluate the performance of the optical telemeter based on triangulation. Let us exemplify the results for two hypothetical telemeters, one intended for short distance (L$\approx$1m) and the other for medium distance (say 100 m).

At a distance of L=1 m, we may choose a D=10-cm base as a reasonable value for a compact instrument. From Fig.3-2, we find the intrinsic accuracy as $\Delta L/L$=3% and 0.1%, respectively for $\Delta \alpha$=3 and 0.1 mrad. Turning to the 100-m telemeter, we may expand the base within the limits allowed by the application and perhaps go to D=1m. Then, from Fig.3-2, we find $\Delta L/L$=30% and 1% for $\Delta \alpha$=3 and 0.1 mrad.

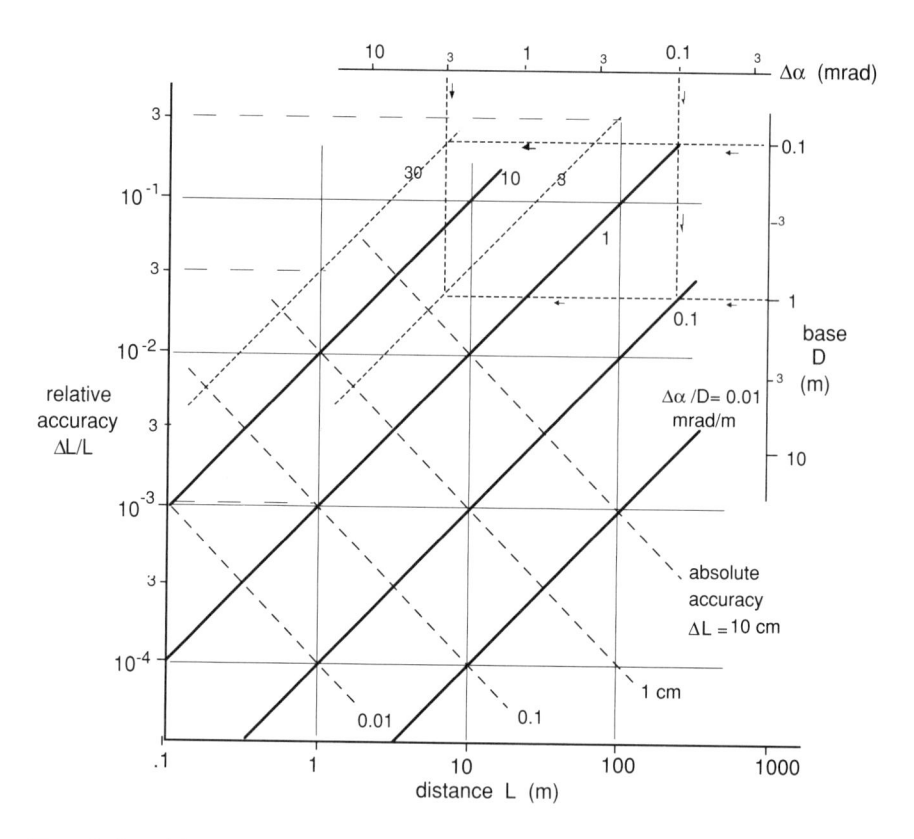

Fig. 3-2 Distance accuracy (relative ΔL/L and absolute ΔL) in triangulation measurements. Entering with Δα and D values in the upper right corner diagram (see small-dot line for 0.1 mrad and 1m) gives the parameter Δα/D and hence the relative accuracy ΔL/L versus distance. Large-dot lines supply the absolute distance accuracy ΔL. Because the approximation tanα≈α has been used, the diagram is valid for D<<L.

As it is clear from these examples, triangulation can achieve respectable performance, provided the application allows using a not-too-small base-to-distance ratio D/L.

A triangulation telemeter can be developed straight from the basic concept outlined in Fig.3-1. Such an instrument is classified as a passive optical telemeter because it does not require a source of illumination or a detector.

However, if we add a source to aim the target and a position-sensitive detector to sense the return, we can improve performance, eliminate the moving parts, and get a faster response. The best scheme for the active triangulation scheme can take very different configurations, depending on the requirements of the specific application (e.g., dynamic range, accuracy, size, and cost) [1].

To substantiate a design example, we report in Fig.3-3 the layout of an active triangulation telemeter intended for short distances (1 to 10 m) that uses a semiconductor laser and a CCD.

3.2 Time-of-Flight Telemeters

The laser wavelength is chosen in the visible for ease of target aiming, and the power is usually a few mW emitted from an elliptical near-field spot of 1×3 µm size (typically).
An anamorphic objective lens circularizes the beam to a radius w_1 (typ. 5µm) and projects an image of it on the target.
On the target, the spot size radius is then $w_1 L/F_{ill}$ (=5µm×1000/125=40µm for L=1m and F_{ill} =125mm). As a viewing objective, we use a telephoto lens with focal length F_{rec} (typically 250 mm) and get an image of the target on the CCD.
The CCD (see [2], Sect.9.2) is a silicon device composed of a linear array of N individual photosensitive elements of width w_{CCD} (typically, we may have N=1024 and w_{CCD}= 10µm). The size of the target-spot that is imaged on the CCD by the objective is easily computed as $w_1 L/F_{ill} (F_{rec}/L) = w_1 F_{rec}/F_{ill}$ =5µm×250/125=10µm, which is equal to the pixel size w_{CCD}.
We may assume that the accuracy of localization in the focal plane is limited by the pixel size (see Ref.[3] for a refinement that takes into account speckle errors). Then, the angular resolution is $\Delta\alpha = w_{CCD}/F_{rec}$ =10µm/250 mm =0.04 mrad. By taking an axis separation D=50 mm, and from the data in Fig.3-2 we have $\Delta L/L$=0.1% at L=1 m, and $\Delta L/L$=1% at L=10 m.
Converted in absolute errors, our telemeter would resolve 1mm at 1m and 10cm at 10m, and perhaps the results may be still improved.

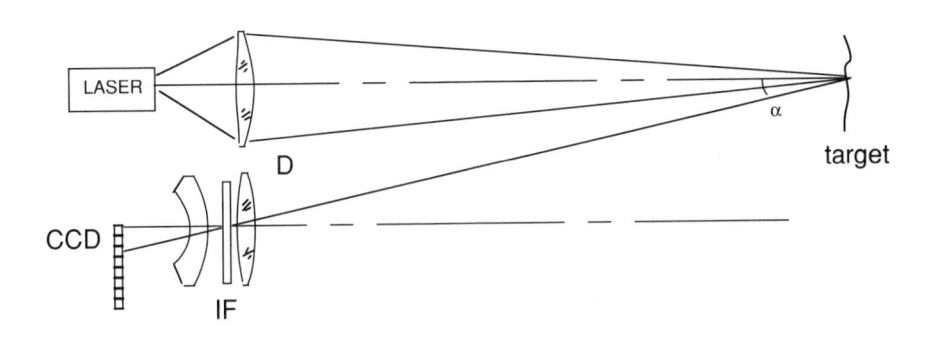

Fig. 3-3 Scheme of a triangulation telemeter with active illumination and a static angle readout. The optical axes of illuminating and viewing beams are parallel. The target-spot image is formed off-axis at a distance αF_{rec}, where F_{rec} is the focal length of the viewing objective. An Interference Filter (IF) is used for ambient light rejection.

3.2 TIME-OF-FLIGHT TELEMETERS

These telemeters are based on the measurement of the time of flight T=2L/c of light to the target at distance L and back. The uncertainty ΔT of measurement reflects itself in a distance uncertainty $\Delta L = c\Delta T/2$. Accordingly, if we are using a pulsed light source, we require

that the pulse duration τ be short enough for the desired resolution, or τ<ΔT.
Similarly, if we use a sine-wave modulated source, the frequency of modulation ω needs to be high enough, or ω>1/ΔT.

In the following sections, we first study the power budget of a generic time of flight telemeter, then evaluate the ultimate distance performance of pulsed and sine-wave modulated approaches, and conclude with the illustration of several schemes of implementation [4].

3.2.1 Power Budget

The power budget of a time of flight telemeter can be analyzed with reference to Fig.3-4. The optical source emits a power P_s, and the objective lens projects it on the target with an angular divergence $\theta_s = D_s/F_s$, where D_s is the diameter of the source and F_s is the focal length of the objective lens. A detector is aimed to the target through a receiving objective lens with diameter D_r.

We will consider the target either as *cooperative* (a corner cube for self-alignment) *or noncooperative* (to account for a normal diffusing surface, with diffusivity δ<1).

When the target is cooperative, the corner cube acts as a mirror, so the receiver sees the source as if it is at a distance 2L. Accordingly, the power fraction collected by the receiver is the ratio between the area of the collecting lens and the area of the transmitted spot at a distance 2L:

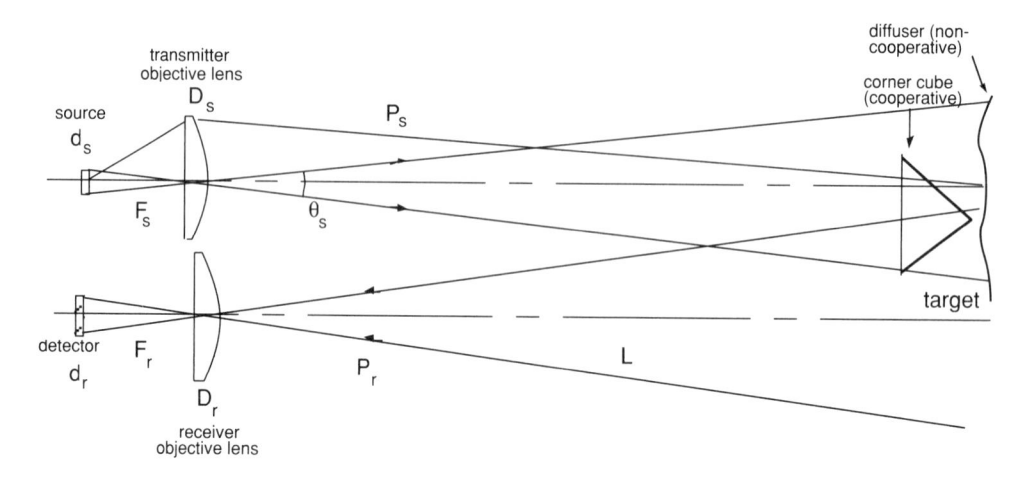

Fig. 3-4 General scheme for evaluating the received power P_r in a telemeter as a function of transmitted power P_t, distance L, and transmitter/receiver optics. The target may be either a corner cube or a diffuser. Because the distance L is much larger than other dimensions in the drawing, the field of view is actually superposed on the field of illumination.

3.2 Time-of-Flight Telemeters

$$P_r/P_s = D_r^2 /(\theta_s^2 \, 4L^2) \tag{3.3}$$

Eq.3.3 holds if the corner cube diameter D_{cc} is large enough and does not limit collection of radiation at the receiver. This requires that $D_{cc} \geq D_r/2$. If the reverse is true, we shall use D_{cc} in place of $D_r/2$ in Eq.3.3.

When the target is noncooperative, that is, a diffusing surface of area A_t, the power arriving on the target is P_s, and a fraction δ of it is rediffused back to the receiver. Assuming a Lambertian diffuser, the radiance R of the target is $1/\pi$ times the power density $\delta P_s/A_t$, or $R = \delta P_s/\pi A_t$. Accordingly, the power collected by the receiver is $RA_t\Omega$, where Ω is the solid angle of the receiver seen from the target, given by $\Omega = D_r^2/4L^2$. Thus, we have:

$$P_r/P_s = \delta \, D_r^2/4L^2 \tag{3.4}$$

Also in this case, if the target has a diameter D_{tar} smaller than $\theta_s L$, the P_r/P_s ratio is reduced by a factor $(D_{tar}/\theta_s L)^2$, and the dependence on distance would then become of the type L^{-4}, as for a microwave radar. In the following, we restrict ourselves to the case $D_{tar} > \theta_s L$.

Eqs.3.3 and 3.4 describe the attenuation due to geometrical effects only. We may also find an additional contribution from the transmittance T_{opt} of the transmitter/receiver lenses, and from the propagation through the atmosphere with an attenuation $T_{atm} = \exp{-2\alpha L}$, where α is the attenuation coefficient of the air (see Appendix A3.1). We can account for these terms by multiplying the second members of Eqs.3.3 and 3.4 by $T_{opt}T_{atm}$ so that we have in general:

$$P_r/P_s = T_{opt}T_{atm} \, D_r^2 /(\theta_s^2 \, 4L^2) \tag{3.3'}$$

$$P_r/P_s = \delta \, T_{opt}T_{atm} \, D_r^2/4L^2 \tag{3.4'}$$

As a rule of thumb, we may take $\alpha = 0.1 \text{km}^{-1}$ for an exceptionally clear atmosphere, $\alpha = 0.33 \text{ km}^{-1}$ for a limpid atmosphere, and $\alpha = 0.5 \text{km}^{-1}$ for an incipient haze. The corresponding values of T_{atm} are reported in Fig.3-5 for medium/long ranges.

The previous values are an average in the range of visible wavelengths. However, because the telemeter will operate at a well-defined wavelength, it is important to have a closer look to the spectral attenuation $\alpha = \alpha(\lambda)$.

Information on $\alpha = \alpha(\lambda)$ is provided by Fig.A3-2 of Appendix A3. In addition, comparative data can be gathered from the spectrum of the solar irradiance, see Fig.A3-3. From here, we can see that there are some wavelengths to be avoided (e.g. 0.70, 0.76, 0.80, 0.855, 0.93 and 1.13 µm) because they coincide with absorption peaks of the atmosphere. On the other hand, wavelengths like 0.633 (He-Ne), 0.82..0.88 (GaAlAs), and 1.06 µm (Nd) are acceptable for long-range telemeters.

Now, we can generalize the power-budget equations by introducing an equivalent distance $L_{eq} = L/T_{atm}$ and a gain factor G, so that:

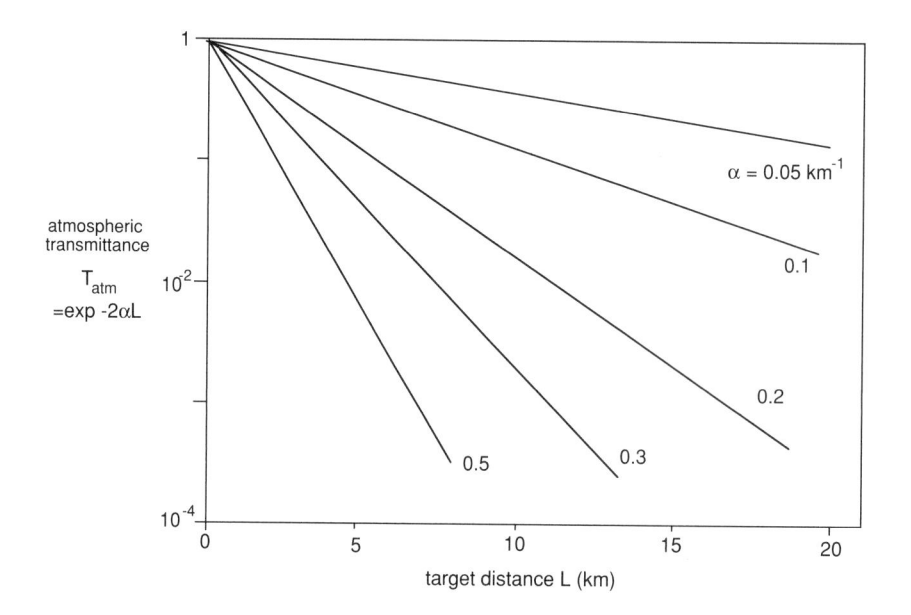

Fig. 3-5 Atmospheric transmittance as a function of target distance, with the attenuation coefficient as a parameter.

$$P_r/P_s = G\, D_r^2 / 4 L_{eq}^2 \qquad (3.5)$$

By comparing with Eqs.3.3 and 3.4, G is given by:

$$G = G_{nc} = T_{opt}\, \delta \qquad \text{(noncooperative target)} \qquad (3.6a)$$

$$G = G_c = T_{opt}/\theta_s^2 \qquad \text{(cooperative target)} \qquad (3.6b)$$

With this position, Eq.3.5 is the power-attenuation equation and tells us that the P_r/P_s ratio depends on the inverse square of target distance or, better, from the ratio $(D_r/L)^2$.

On the other hand, Eqs.3.6a and b are about the gain of the target, either cooperative or noncooperative.

The cooperative gain G_c is the counterpart of the antenna gain known in microwave, and is given by the inverse squared of the divergence angle, $1/\theta_s^2$. This factor may be very large indeed (for example, it is $G_c = 10^6$ for $\theta_s = 1$ mrad).

3.2.2 System Equation

Let us consider the system performance of the telemeter. If P_r is the received power and the receiver circuit has a noise P_n, we shall require that the ratio of the two, i.e.:

3.2 Time-of-Flight Telemeters

$$S/N = P_r/P_n, \quad (3.7)$$

is at least, say 10, to get a good measurement. By combining Eqs.3.7 with Eq.3.5 we get:

$$GP_s = P_r \, 4L_{eq}^2/D_r^2$$

$$= (S/N) \, P_n \, 4L_{eq}^2/D_r^2 \quad (3.8)$$

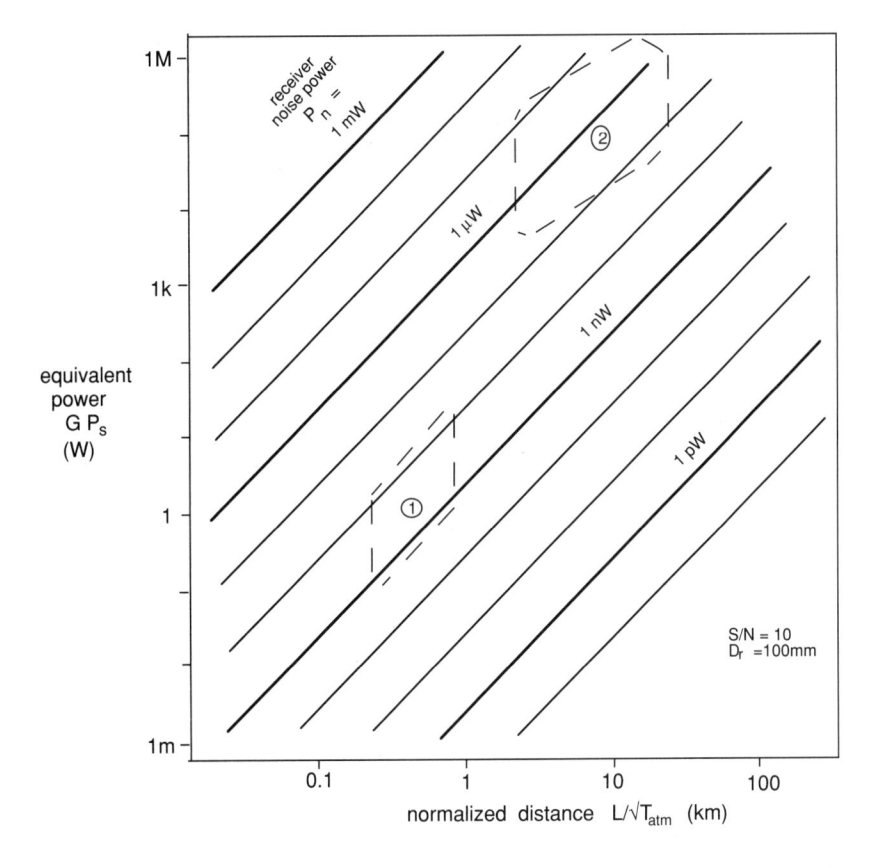

Eq.3.8 is the system equation of the telemeter because it relates the required S/N ratio to receiver noise P_n and to attenuation $4L_{eq}^2/D_r^2$. The diagram of the equivalent power GP_n versus distance L_{eq} and with the receiver noise P_n as a parameter is plotted in Fig.3-6.

Fig. 3-6 Graph of the transmitted equivalent power versus normalized distance, with the receiver noise power as a parameter. It is assumed a S/N=10 and a receiver objective with D_r=100 mm. Zone 1 is representative of a sine-wave modulated topographic meter and zone 2 of a pulsed telemeter.

Two central-design regions are indicated in Fig.3-6. One representative of a sine-wave modulated telemeter for topography, which may use a ≈1 mW transmitted power and has $GP_s \approx$ 1W in virtue of the cooperative target (corner cube) gain. Because the measurement time T can be long (e.g., 10ms to 1s), the bandwidth B=1/2T is small, and the receiver noise can go down to the nW's level.

The second case is representative of a pulsed telemeter intended for noncooperative targets. The source is likely to be a Q-switched laser with a high peak power (≈0.1-1MW, resulting from ≈mJ energy per pulse and τ≈10ns pulse duration). Using a pin/APD photodiode (see [2], Ch.4) as the detector, and taking into account that the bandwidth B≈1/τ is now much larger, the receiver noise is in the range of the µW's.

We have three contributions to add about receiver noise, namely:
- noise of the received signal P_s
- noise of the background light P_{bg} collected by the receiver
- noise of detector and front-end amplifier

To describe noise, we use the standard deviation p_n of the fluctuations referred to the detector input, a quantity called the Noise-Equivalent-Power (NEP) (see Ch.3 of Ref.[2]).

As a reasonable assumption, we assume that the photons of both signal and background light obey the Poisson statistics. Upon detection, each photon is converted into an electron with a success probability η, which is called a Bernoulli process. The photon flux F_{ph} is thus transformed in an electron flux $F_{el} = \eta F_{ph}$, where η is the quantum efficiency of the detector. It can be shown that the cascade of a Poisson and a Bernoulli process is a Poisson process. Thus, the detected electrons follow the Poisson statistics, and the associated current has fluctuations described as shot noise with a white spectral density. Now, the average detected currents are written as $I_s = \sigma P_s$ and $I_{bg} = \sigma P_{bg}$, where σ=ηe/hν is the spectral sensitivity of the detector (Ref.[2], Ch.4). Superposed to the average currents, we find shot noise fluctuations i_{ns} and i_{nbg} whose variances are given by $i_{ns}^2 = 2eI_s B$ and $i_{nbg}^2 = 2eI_{bg} B$, with B being the bandwidth of observation.

The detector and front-end noise are summarized by an equivalent characteristic current I_{ph0} (see [2], Ch.3), defined as the current value for which the shot-noise is equal to the total noise of detector and front-end, that is $i_{nph0}^2 = 2eI_{ph0}B$.

Summing up the contributions (because they are mutually uncorrelated), the total variance of the detector current is $i_{rec}^2 = i_{nph0}^2 + i_{ns}^2 + i_{nbg}^2$, or:

$$i_{rec}^2 = 2eB (I_s + I_{bg} + I_{ph0}) \qquad (3.9)$$

We can divide i_{rec}^2 by σ^2, the spectral sensitivity (coincident, in this case, with the responsivity), to get the (total) noise power variance at the input:

$$p_n^2 = (2h\nu/\eta) B (P_s + P_{bg} + P_{ph0}) \qquad (3.10)$$

In this equation, we have let $P_{ph0} = I_{ph0}/\sigma$ for the power corresponding to I_{ph0}. An extra 1/η factor is left over in Eq.3.10 after the σ's ratios of powers to currents have been cleared and 2hνσ has been transformed in 2e.

3.2 Time-of-Flight Telemeters

This is the correct result, however, because a real detector with $\eta<1$ worsens the S/N ratio of detected current by η respect to the S/N of incoming power.

The noise performances obtained by several combinations of detector and front-end preamplifiers are discussed in detail in Ref.[2], Ch.4. As a guideline for the reader, we summarize in Fig.3-7 the performance we may reasonably expect from a well-designed receiver for instrumentation applications.

Data are given in terms of the current noise spectral density $i_n/\sqrt{B}$, as a function of the maximum frequency of operation of the receiver f_2 (that is, of the cutoff frequency). To obtain p_n, the value read in Fig.3-7 shall be multiplied by the square root of the measurement bandwidth and divided by $\sigma\sqrt{\eta}$ (Eq.3.10).

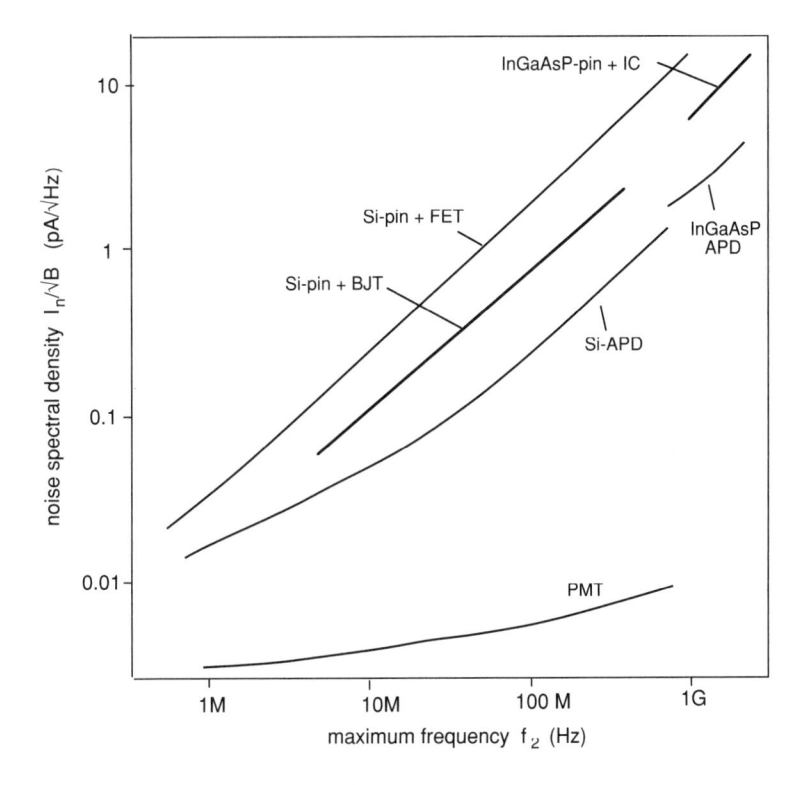

Fig. 3-7 Typical noise performance of receivers, as a function of the maximum frequency of operation f_2. Silicon-photodiodes with FET- and BJT- transistor input-stage (with eventual equalization correction) are compared to avalanche photodiodes (APD) and to photomultipliers (PMT). Si-photodiodes cover the range λ=400-1000nm, whereas InGaAsP extend up to ≈1.6μm and standard PMTs are centered in the range 300 to 900 nm. Data are for a detector with area A<0.5mm^2 and capacitance C<0.5pF, and front-ends with FET and BJT having a transition frequency $f_T \geq 2GHz$.

To evaluate the term P_{bg} in Eq.3.10, we shall consider the spectral irradiance E_s (W/m²μm) of the scene on which the telemeter is aimed.

In daytime with direct sunlight illumination at different elevation angles, E_s is given by the diagram of Fig.A3-3. In other conditions (clouds, haze, etc.), the same diagram can be used as a first approximation, by properly rescaling the curve amplitude.

From the law of photography (App.A2, Ref.[2]), we can write the power collected at the receiver objective lens as:

$$P_{bg} = \delta \, E_s \, \Delta\lambda \, NA^2 \, (\pi \, d_r^2/4) \tag{3.11}$$

where:

δ = scene diffusivity
E_s = scene spectral irradiance (W/m²μm)
$\Delta\lambda$ = spectral width of the interference filter placed in the receiver lens
$NA = \arcsin D_r/2F$ = numerical aperture of the receiver lens
d_r = diameter of the detector

To compare the relative importance of the noise terms, we can use the diagram of Fig.3-8, which is a plot of the quantity $p_n = \sqrt{[(2h\nu/\eta)PB]}$ for $\lambda = 1\mu m$ and $\eta = 0.7$.

Let us substantiate the previous considerations with the aid of numerical examples for a pulsed telemeter and a sine-wave telemeter.

For a Nd-laser pulsed telemeter operating at $\lambda = 1060$ nm, we may have (Fig.A3-3) $E_s \approx$ 500 W/m²μm in direct sunlight at AM1.5 (sun elevation 42°). Other common design values we may assume are $\Delta\lambda = 10$nm (for an 80 to 90% transmission of the filter), NA=0.5, and $d_r = 0.2$ mm. Taking $\delta = 1$ as the worst case for background and inserting in Eq.3.11, we get $P_{bg} = 40$ nW.

The attenuation is $P_r/P_s = \delta(D_r/2L)^2$. Working at L=1 km with $D_r = 100$ mm and $\delta = 0.3$ for the signal, we have $P_r/P_s = 0.075 \cdot 10^{-8}$, that is, for a typical transmitted (peak) power of $P_s = 0.3$ MW, we get a received power $P_r = 0.23$ mW. In addition, the I_{ph0} noise for an APD receiver at 100 MHz is evaluated from Fig.3-7 as 2 pA/√Hz. Multiplying by $\sqrt{B}/\eta\sqrt{\sigma}$, we get $P_{ph0} \approx 4$ nW. Thus, we obtain the points labeled 'pulsed' in Fig.3-8, and we can see that it is the received signal shot-noise to prevail.

For a sine-wave modulated telemeter operating at $\lambda = 820$ nm, we may have $E_s \approx 900$ W/m²μm and P_{bg} approximately doubles at a value of 80nW.

The attenuation is now $P_r/P_s = D_r^2/4\theta_s^2 L^2$. Working at L=100 m with $D_r = 100$mm and $\theta_s = 1$ mrad, we have $P_r/P_s = 0.25$. That is, for a typical transmitted power of $P_s = 0.1$ mW, we get a received power $P_r = 25$ μW.

Again, using an APD receiver at 10 MHz, the noise spectral density is 0.2 pA/√Hz, and multiplying by $\sqrt{B}/\eta\sqrt{\sigma}$ with a bandwidth of 100 Hz now, we get $P_{ph0} \approx 5$ pW.

In both cases, the largest term is the shot-noise of the received photons. This corresponds to a good-design result. On the contrary, if we had omitted the filter for the background or used too large a detector or noisy electronics, the I_{bg} or I_{ph0} values could easily be increased by order of magnitude and become the limiting factor of telemeter performance.

3.2 Time-of-Flight Telemeters

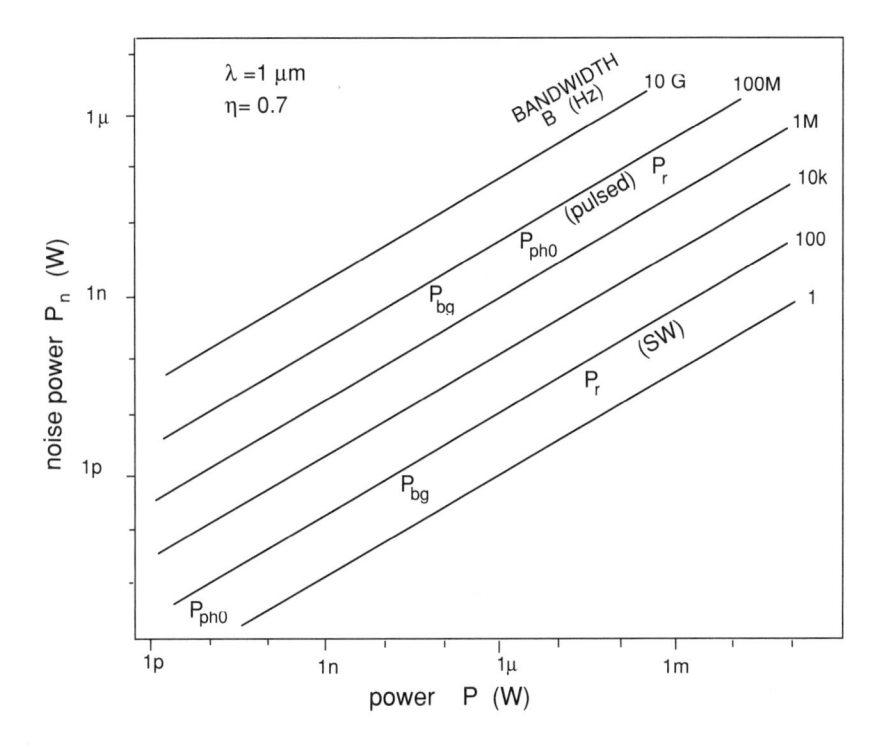

Fig. 3-8 Graph for evaluating the noise contributions associated with signal, background and detector noise as a function of power. Two examples are reported for a hypothetical sine-wave telemeter on L=100 m (bottom, SW powers) and a pulsed telemeter for L=1 km (top, pulsed powers).

3.2.3 Accuracy of the Pulsed Telemeter

Let us now examine the accuracy of a pulsed telemeter. In a pulsed telemeter, we measure the time of flight T of an optical pulse going to the target and back. Distance is obtained as L=cT/2, and a timing error ΔT reflects itself in a distance error ΔL=cΔT/2, or $\sigma_L = (c/2)\sigma_T$ in terms of the rms error. We shall therefore try to make ΔT as small as possible.

Generally, timing is performed on the electrical pulse supplied by the photodetector as a replica of the optical pulse. We may use a pair of timing circuits, one for the start and one for the stop pulse, or have a single circuit for both. The start is usually picked out as a fraction of the optical pulse leaving the transmitter and is then combined with the stop pulse arriving at the receiver from the distant target.

In any case, the time of flight is measured as T= t_{st} - t_{sp}, and the associated rms error is $\sigma_T^2 = \sigma_{st}^2 + \sigma_{sp}^2$. However, because the start pulse can be large in amplitude, σ_{st} is usually negligible compared to σ_{sp}, and we may restrict ourselves to consider the stop timing error.

Several approaches and circuit solutions are available to implement time measurements on pulses. Most of them are based on threshold crossing, as illustrated in Fig.3-9. An amplitude discriminator with a threshold S_0 is used. Well-known circuit implementations for it are the Schmitt trigger, the tunnel-diode trigger, and other fast circuits.

When the signal crosses the threshold, the circuit switches and gives a step-wise waveform at the output. The circuit adds a negligible jitter in switching time (<10ps in a well-designed circuit). Thus, the timing fluctuation σ_t around the average time t_0 is only due to pulse noise, specifically to the fluctuations σ_S on the mean amplitude waveform S(t).

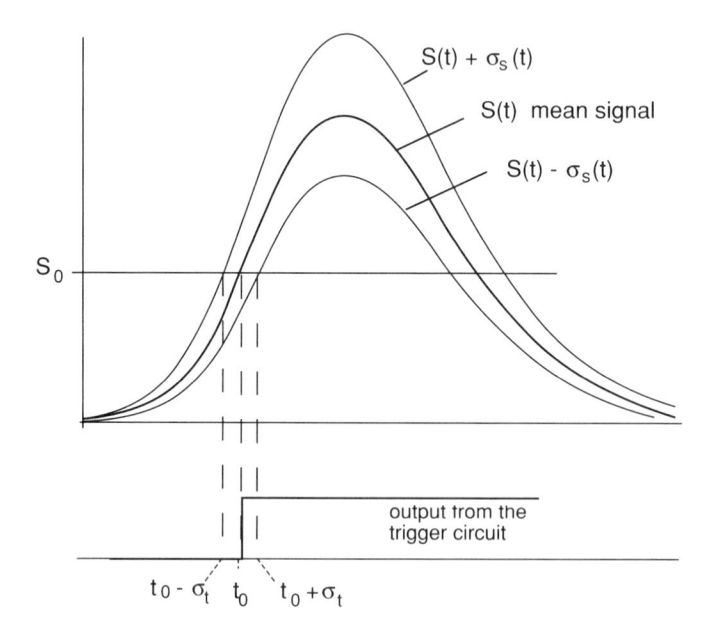

Fig. 3-9 Timing of the received pulse by means of an amplitude discriminator. When the signal crosses the threshold S_0 at time t_0, a timing signal is generated. Amplitude fluctuations $\pm\sigma_S$ produce a timing error $\pm\sigma_t$.

The error in time σ_t can be related to the error in amplitude σ_S through a linear regression, by writing:

$$\sigma_t^2 = \sigma_S^2 / |dS/dt|^2 \qquad (3.12)$$

The linear regression is accurate when the amplitude fluctuation σ_S is small and the signal slope dS/dt is nearly constant in the crossing region, which is a reasonable hypothesis in most cases.

Now, let us now introduce normalized variables to get a better insight into the problem. The received pulse is written as a power:

$$S(t) = E_r (1/\tau) s(t/\tau) = P_r(t) \qquad (3.13)$$

3.2 Time-of-Flight Telemeters

where
- τ is the time parameter (or scale) of the pulse, approximately equal to its duration;
- $s(t/\tau)$ is the dimensionless pulse waveform, with a peak value $s_{peak} \approx 1$ and an area normalized to unity, i.e., $\int_{0-\infty} (1/\tau)s(t/\tau)dt=1$. The frequency spectrum of $s(t/\tau)$ extends up to $B \approx \kappa/\tau$, with κ being a numerical factor usually not far away from $1/2\pi=0.16$;
- E_r is the energy contained in the pulse, as implied by the normalization of $s(t/\tau)$.

With these positions, the signal slope is $dP_s(t)/dt = E_r(1/\tau^2)s'(t/\tau)$. Using Eq.3.10 for the amplitude variance in Eq.3.12 and with Eq.3.13, we have:

$$\sigma_t^2 = \{(2h\nu/\eta)(\kappa/\tau)[E_r(1/\tau)s(t/\tau)+P_{bg}+P_{ph0}]\}/[E_r(1/\tau^2)s'(t/\tau)]^2$$
$$= \tau^2 \{(2h\nu/\eta)\kappa[s(t/\tau)/E_r + (P_{bg}+P_{ph0})\tau/E_r^2]/s'^2(t/\tau) \quad (3.14)$$

This expression tells us that the accuracy σ_t is primarily proportional to the pulse duration τ. In addition, by writing it in the form:

$$\sigma_t = \tau(A\, h\nu/E_r + B/E_r^2)^{1/2} \quad (3.15)$$

where A and B are appropriate terms, we can see that the first term contributes to accuracy as $\sqrt{(h\nu/E_r)}=1/\sqrt{N_r}$, which is the inverse square root of the number N_r of received photons.

In a well-designed receiver, the shot noise $h\nu/E_r$ term should dominate over the other noises, and the second term in parentheses should be negligible. In this case, we have:

$$\sigma_t \approx \tau/\sqrt{N_r} \quad (3.16)$$

By introducing N_r in Eq.3.14, we obtain the general expression:

$$\sigma_t = \tau/\sqrt{N_r}\,[(2\kappa/\eta)\,s(t/\tau)/s'^2(t/\tau)]^{1/2}\,[1+\tau(P_{bg}+P_{ph0})/s(t/\tau)E_r]^{1/2} \quad (3.16')$$

From this expression, we can see that, besides the main dependence on $\tau/\sqrt{N_r}$, the effects of quantum efficiency and pulse waveshape are summarized by the factor $\sqrt{2\kappa/\eta}$.

Also, the accuracy depends on the relative square slope $s'^2(t/\tau)/s(t/\tau)$, a dimensionless quantity less than, but not so far from unity, whose actual value is determined by the selected threshold $S_0=s(t/\tau)$. Of course, we will choose S_0 so that the threshold-dependent factor $s'^2(t/\tau)/s(t/\tau)$ is maximum.

Finally, if background and electronics noises are not negligible, their effect worsens the ideal accuracy by the factor in square parentheses in Eq.3.16'.
This factor can also be written as $[1+N_{bg+ph0}/S_0N_r]^{1/2}$, where N_r is the total number of photons in the pulse, S_0N_r is the number of photons collected at the threshold crossing time, and $N_{bg+ph0}=\tau(P_{bg}+P_{ph0})/h\nu$ is the number of photons equivalent to background front-end powers P_{bg} and P_{ph0} collected in a pulse-duration time τ.
Explicitly, we may rewrite Eq.3.16' as:

$$\sigma_t = \tau/\sqrt{N_r}\,[(2\kappa/\eta)\,S_0/s'^2(t/\tau)]^{1/2}\,[1+N_{bg+ph0}/S_0N_r]^{1/2} \quad (3.16'')$$

As a numerical example, let us assume a Gaussian waveform for the telemeter pulse, $s(t/\tau)=[2\pi]^{-1/2}\exp{-(t/\tau)^2/2}$, and a duration t=5 ns (so that the full-width-half-maximum of the pulse is 2.36×5ns=12ns]. The Gaussian has a bandwidth factor equal to κ=0.13.

The optimum level for the threshold S_0 is found from the condition (see Eq.3.16') $s(t/\tau)/s'^2(t/\tau)$=minimum, or by inserting the Gaussian function as $(t/\tau)^2[2\pi]^{-1/2}\exp{-(t/\tau)^2/2}$ = maximum. Differentiating this expression with respect to t/τ and equating to zero, we find $(t/\tau)^2$=2 as the result. Accordingly, the optimum (fixed) threshold is S_0=s($\sqrt{2}$) =1/[e$(2\pi)^{1/2}$] = 0.147 and the signal slope is s'^2=2S_0^2=0.043.

The threshold-dependent factor is therefore $s(t/\tau)/s'^2(t/\tau)=S_0/s'^2$=0.147/0.043=3.42. Assuming η=0.7 and being κ=0.13, the normalized time variance is calculated as $\sigma_t/(\tau/\sqrt{N_r})$= $[2\kappa/\eta\times3.42]^{1/2}$= $[2\times0.13/0.7\times3.42]^{1/2}$ =1.12.

The obtained timing accuracy is σ_t = 5.7ns/$\sqrt{N_r}$, which is a value very close to that (5ns/$\sqrt{N_r}$) given by the first factor in Eq.3.16.

From the data of Fig.3-8, we may expect $P_r\approx$1mW as the typical received power of the pulsed telemeter, which corresponds to a number of photons $N_r\approx P_r\tau/h\nu=10^{-3}\times5.10^{-9}/2.10^{-19}$ $\approx2.5\,10^7$. The accuracy of a single-shot measurement is then found as σ_t^2=5.7ns/$\sqrt{(2.5\,10^7)}$ = 1.1 ps, a very good theoretical limit of performance, even beyond the capabilities of electronic circuits (usually in the range 10 to 50ps).

As a more realistic pulse waveform, we may use an asymmetric Gaussian, with the leading edge τ_{le} faster that the trailing edge τ_{te}. Repeating the calculation with τ_{te}/τ_{le}=1.8 and τ_{le} =5ns, we find that the results change only by about 10%.

Regarding the weight of the last factor in square brackets of Eq.3.16'', we need the number of photons at threshold to be larger than the corresponding number of background and circuit photons if the factor has to be about unity. But, even when it is $N_{bg+pho}>S_0/N_r$, there is still a 1/$\sqrt{N_r}$ dependence, which improves resolution at increasing number of collected photons.

From statistics, we know that if the timing measurement is repeated N times on N successive pulses, the accuracy becomes $\sigma_t/\sqrt{N}$.

As the total number of photons collected is NN_r, we can see that Eq.3.16 applies in this case of a repeated measurement. If practicable, (that is, if the target is static), it is better to make several measurements rather than one with the total available NN_r.

Indeed, the single-shot accuracy may be less than the resolution of the electronic circuit. Then, it is better to accumulate several samples and get the advantage of the 1/$\sqrt{N}$ improvement of the average value obtained by summing up the single measurement results.

Last, let us comment about the nonidealities of the electronic circuits processing the pulse. One nonideality is the discriminator noise. This can be described as an amplitude jitter σ_{S0} superposed to the threshold S_0.

Let us take σ_{S0} as homogeneous to S_0, that is, a dimensionless quantity.

We may take account of this jitter by adding $(E_r/\tau)^2\sigma_{S0}^2$ to σ_s^2 in Eq.3.12, with the multiplying term being the scale factor. Propagating this quantity onward, we find that, at the right-hand side of Eq.3.16', there is an extra term, $\sigma_{S0}/s'(t/\tau)|_{S0}$. Simply stated, the amplitude jitter translates in timing jitter when divided by the signal slope.

3.2 Time-of-Flight Telemeters

A second nonideality is the finite resolution of the time-sorting circuit, usually 10 to 50 ps in a single measurement with state-of-the-art circuits. As already noted, and well known in measurement theory, by summing the outcomes of N time measurements $t_1, t_2, ..., t_N$, we get an average value $\langle t \rangle = (t_1 + t_2 + ... + t_N)/N$ whose accuracy $\sigma_{\langle t \rangle} = \sigma_{1-m}/\sqrt{N}$ is improved by $\sqrt{N}$ with respect to the single measurement. Also the number of significant digits is increased by averaging because the discretization error $\Delta\tau_{ds}$ is reduced to $\Delta\tau_{ds}/\sqrt{N}$ (provided $\Delta\tau_{ds} << \sigma_{1-m}$).

3.2.3.1 Optimum Filter for Signal Timing

The threshold-crossing timing described in previous section is a viable and widely used technique, but it is not necessarily the optimum one. Some timing information is lost because only the signal around the mean crossing time is actually used to produce the trigger, whereas all other signal portions (which could generate switching as well) are not. We may wonder if there is a better timing strategy, perhaps a variable-threshold or a pulse center of mass, or eventually something else.

To approach the problem, we look for the optimum filter [5,6] which, inserted between the detector and a threshold-crossing circuit, collects all the time-localization information contained in the pulse and makes it available at a threshold-crossing time (Fig.3-10).

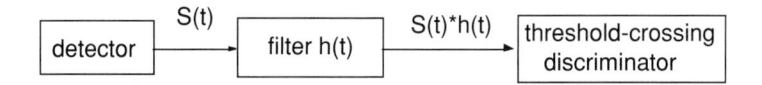

Fig. 3-10 Optimum timing of the pulse waveform. A filter with an impulse response h(t) to be determined is inserted between the detector and the threshold-crossing circuit.

If h(t) is the impulse (or Dirac-δ) response of a filter cascaded to the detector, the mean signal at the filter output is $S_{out}(t) = S(t)*h(t)$, where * stands for the convolution operation [explicitly, $a*b = \int_{0-\infty} a(\tau)b(t-\tau)d\tau$]. The slope signal is $S'(t)*h(t)$ and the amplitude variance is given by $\sigma_S^2(t)*h^2(t)$. Thus, we can rewrite the time variance as:

$$\sigma_t^2 = \sigma_S^2(t)*h^2(t) / [S'(t)*h(t)]^2 \qquad (3.17)$$

The calculation of the minimum of σ_t^2 with respect to h(t) is carried out in Appendix A4, and the result is:

$$h(t) = S'(T_m - t) / S(T_m - t) \qquad (3.18)$$

As we can see from Eq.3.18, the optimum filter response is the time-reverse of the relative slope S'/S. The time of reversal T_m is also the measurement time, at which the mean signal is found to be zero (Fig.3-11).

Indeed, by writing the mean output $S_{out}(T_m)=S(t)*h(t)|_{T_m}$ at T_m, we get:

$$S_{out}(T_m) = \int_{0-\infty} S(\tau)h(T_m-\tau)d\tau = \int_{0-\infty} S(\tau)S'(\tau)/S(\tau)d\tau = \int_{0-\infty} S'(\tau)d\tau = 0 \quad (3.19)$$

Thus, the threshold-crossing measurement with the optimum filter is a zero-crossing at time T_m. To collect all the timing information in S(t), we shall simply take T_m larger than the duration of the pulse S(t).

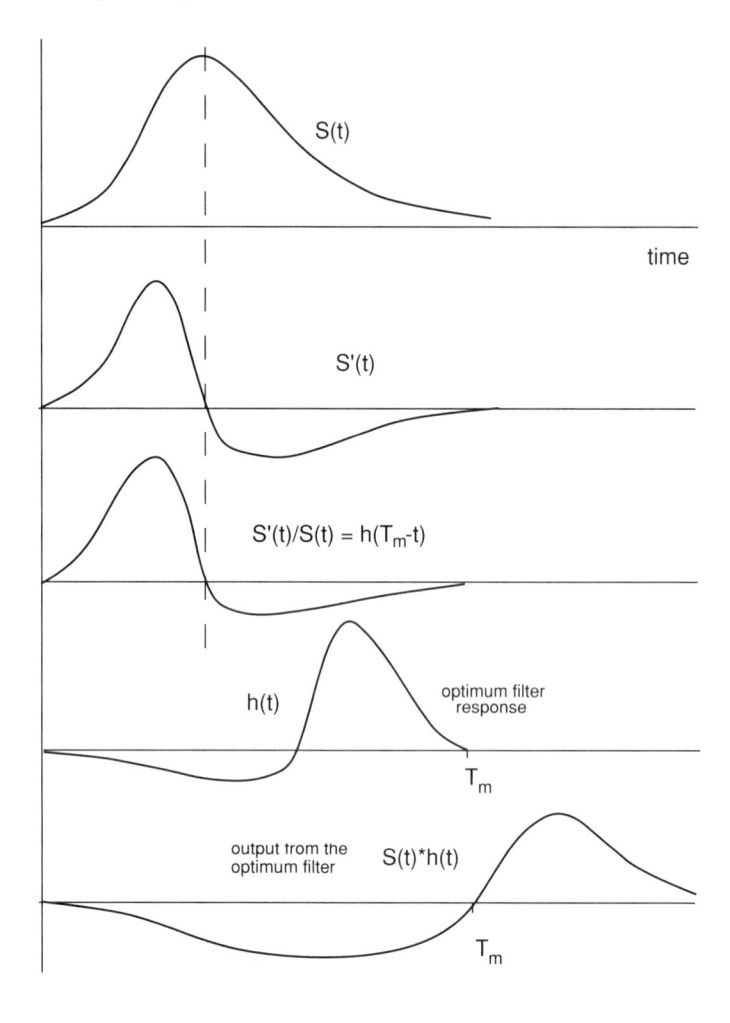

Fig. 3-11 Optimum timing waveforms (top to bottom): mean signal S(t), its derivative S'(t), the filter impulse response h(t), and the output signal S_{out} from the optimum filter. The output exhibits a zero-crossing at the measurement time T_m.

3.2 Time-of-Flight Telemeters

An interesting feature of the optimum filter is that the rigid-shape fluctuations superposed to the pulse are canceled out. In fact, we may write the fluctuations as $\Delta S(t)=\xi S(t)+\zeta s(t)$, that is, as the sum of a rigid-fluctuation $\xi S(t)$ proportional to the mean and a shape fluctuation $\zeta s(t)$, where ξ and ζ are random variables, and $s(t)$ is a random function orthogonal to $S(t)$.

Then, in view of Eq.3.19, just like the mean signal, the rigid fluctuation term gives a zero output at T_m. Only the shape fluctuations remain in the output after the optimum filter and contribute to the timing error.

Using Eqs.3.17 and 3.18, we can calculate the time variance of the signal supplied by the optimum filter as:

$$\sigma^2_{t\,(opt)} = \int_{0-\infty} S(\tau)[S'(\tau)/S(\tau)]^2 \, d\tau \,/\, [\int_{0-\infty} S'(\tau) \cdot S'(\tau)/S(\tau) d\tau]^2$$

$$= 1 / [\int_{0-\infty} S'^2(\tau)/S(\tau) d\tau] \qquad (3.20)$$

Compared with Eq.3.16', which is valid for a fixed-threshold crossing, we see that the multiplying term S/S'^2 contained in it is now replaced by $[\int_{0-\infty} S'^2/S \, d\tau]^{-1}$, the time integral of the timing factor S'^2/S.

In most cases, it is $[\int_{0-\infty} S'^2/S \, d\tau]^{-1} \approx 0.1\text{-}0.3 \, S'^2/S$, or we may get an improvement of the timing variance by a factor 3-10.

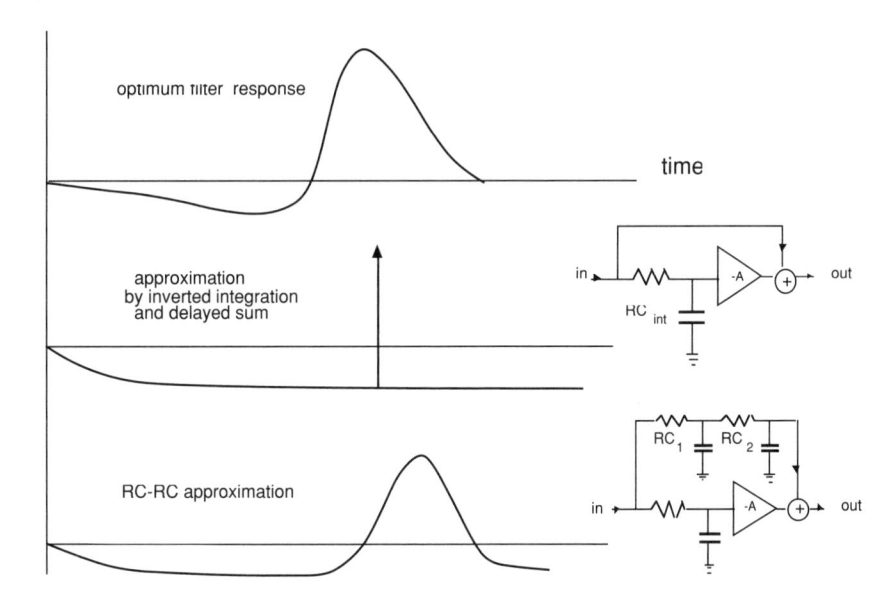

Fig. 3-12 Approximations to the optimum filter: negative integration plus delayed pulse (center) and negative integration plus delayed approximate (RC-RC) integration (bottom).

As numerical example, for a Gaussian pulse waveform of the type $s(t/\tau)=[\sqrt{2\pi}]^{-1}\exp{-(t/\tau)^2/2}$, we can evaluate the timing term $[\int_{0-\infty} S'^2/S \, d\tau]^{-1}$ as $\int_{0-\infty} (2\pi)^{1/2} \tau^2 \exp{-(\tau^2/2)} \, d\tau = 1$, or, the timing variance supplied by the optimum filter is $\sigma^2_{t\,(opt)} = (\tau/\sqrt{N_r})(2\kappa/\eta)$ to be compared with that of the best fixed threshold crossing previously calculated as

$$\sigma_t^2 = (\tau/\sqrt{N_r})(2\kappa/\eta) \, s(t/\tau)/s'^2(t/\tau) = 3.42 \, (\tau/\sqrt{N_r})(2\kappa/\eta).$$

Thus, the optimum filter yields an improvement by a factor 3.42 in variance, or $\sqrt{3.42}=1.85$ in time rms error.

A question is now appropriate; how sensitive is the improvement in time variance with respect to deviations from the theoretical optimum impulse waveform h(t)? Actually, h(t) may be hard to synthesize with conventional filter synthesis techniques because it has a slowly-varying part at short times followed by a sudden jump later in time, which is the opposite of what is usually found in normal impulse responses.

An approximation to h(t) may be constructed as follows. We first integrate the signal S(t) and then sum on the integrated signal an inverted and delayed replica of S(t) (Fig.3-12). A refinement is to smooth the response, by adding two approximate integrators with time constants comparable to the pulse duration.

Summing the pulse with inverted polarity on the integrated level is equivalent to generate a zero crossing dependent on the pulse total area, a concept called Constant-Fraction-Timing (CFT).

A calculation of the resulting timing variance $\sigma_t^2/\sigma^2_{t\,(opt)}$ is shown in Fig.3-13. As it can be seen, values are not far away from the optimum one given by Eq.3.20.

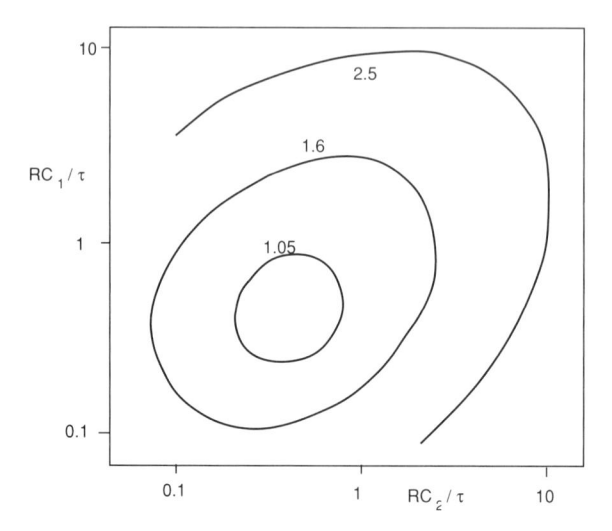

Fig. 3-13 The timing variance obtained by the RC-RC approximation with respect to the optimum value, $\sigma_t^2/\sigma^2_{t\,(opt)}$, as a function of some ratios of RC time constants to pulse duration τ.

3.2 Time-of-Flight Telemeters

3.2.4 Accuracy of the Sine-Wave Telemeter

Sine-wave modulation of power is a good strategy to impress time-localization information on a quasi continuous-wave source. It is especially used with semiconductor lasers that are capable of supplying a substantial dc power and being modulated at high frequency, but are not good for pulsed operation at high peak power level.

Incidentally, the use of a modest power may even be a system requirement in applications, like topography, that call for intrinsic compliance to laser-safety standards.
To analyze the accuracy of the sine-wave telemeter, let us write the transmitted power as:

$$P_s(t) = P_{s0}[1+m \cos 2\pi f_m t], \qquad (3.21)$$

where P_{s0} is the mean transmitted power, f_m is the modulation frequency, and m is the modulation index. The power signal returning from the target at a distance L=2cT is:

$$P_r(t) = P_{r0}[1+m \cos 2\pi f_m (t-T)] \qquad (3.22)$$

In view of the system equation (Eq.3.5), the received mean power P_{r0} is calculated from P_{s0} as $P_{r0}=P_{s0}GD_r^2/4L_{eq}^2$. The photodetected signal then follows as $I_r(t)=\sigma P_r(t)$, with σ being the spectral sensitivity of the photodetector, and is given by:

$$I_r(t) = I_{r0}[1+m \cos 2\pi f_m (t-T)] \qquad (3.22')$$

The distance to be measured is contained in the phase-shift $\varphi=2\pi f_m T$ of the received signal relative to the transmitted one. An error $\Delta\varphi$ in phase reflects itself in a time error $\Delta T= \Delta\varphi/2\pi f_m$. By squaring and averaging, the relation between phase and time accuracy is:

$$\sigma^2_T = \sigma^2_\varphi (1/2\pi f_m)^2 \qquad (3.23)$$

and, of course, the corresponding distance accuracy is $\sigma_L = (c/2) \sigma_T$.

Now, contrary to the pulsed telemeter, we have a long measurement time T_r available for averaging and a large number of sine-wave periods marking the time delay, not a single pulse. Then, to implement the phase-shift measurement, we mix the received signal with a reference local oscillator I_{lo} at the same frequency and with an adjustable phase, that is, with a signal of the form $I_{lo}=I_0 \cos (2\pi f_m t+\varphi_0)$.
Basically, the mixer is a circuit made of a square-law element and a time integrator to average the result, and it produces an output $S_\varphi = \langle I_r I_{lo} \rangle$ of the two inputs I_r and I_{lo} applied to it.
We may also regard the mixer as the circuit performing homodyne (electrical) detection of the signal $I_r(t)$, with the aid of a reference, the local oscillator I_{lo}. The result of the homodyning is a signal carrying the cosine of the phase difference.

Indeed, by inserting the expressions of I_r and I_{lo} in $S_\varphi = \langle I_r I_{lo} \rangle$ and developing the cosine product in sum and difference of the arguments, we get after some easy algebra:

$$S_\varphi = \langle I_0 \cos(2\pi f_m t+\varphi_0) I_{r0} [1+m \cos 2\pi f_m(t-T)] \rangle =$$
$$= (I_0 I_{r0}/2) m \cos [2\pi f_m T - \varphi_0] \quad (3.24)$$

The cos ϕ dependence in Eq.3.24 reveals that the mixer output is sensitive to the *in-phase* component of the received signal, a well-known feature of homodyne detection. For $\phi \approx 0$, the sensitivity of the mixer signal cos ϕ to small variations of ϕ is about zero. To have the maximum sensitivity, we shall work with signals in quadrature, and this can be done by adjusting the local oscillator phase to $\varphi_0 = 2\pi f_m T + \pi/2$.

With this strategy, the mixer signal S_φ is kept dynamically to zero, and the time measurement is obtained as $T=(\varphi_0+\pi/2)/2\pi f_m$.

We can write the total phase in Eq.3.24 as $\phi = \pi/2+\varphi$ to indicate that the mixer works on a small phase signal φ around the π/2 quadrature condition. The mixer output is then $S_\varphi = (I_0 I_{r0}/2)m \cos(\pi/2+\varphi) = -(I_0 I_{r0}/2)m \sin\varphi \approx -(I_0 I_{r0}/2)m\varphi$ for $\varphi \ll 1$.

To take account of the fluctuation ΔI_r superposed to the received signal, we insert $I_r + \Delta I_r$ in $S_\varphi = \langle I_r I_{lo} \rangle$ and change φ in φ+Δφ. Developing the product and averaging, we obtain $S_\varphi = (I_0 I_{r0}/2)m \sin\varphi + \Delta I_r (I_0/2) + (I_0 I_{r0}/2)m\Delta\varphi$.

From this expression, it is easy to find the relation between the phase variance $\sigma_\varphi^2 = \langle\Delta\varphi^2\rangle$, and received signal variance $\sigma^2_{Ir} = \langle\Delta I_r^2\rangle$ as:

$$\sigma_\varphi^2 = \sigma_{Ir}^2 / (m^2 I_{r0}^2) \quad (3.25)$$

The corresponding relation, in terms of equivalent powers at the detector input (see Eqs.3.9 and 3.10) is:

$$\sigma_\varphi^2 = p_n^2 / (m^2 P_{r0}^2) \quad (3.25')$$

Now, combining Eqs.3.25' and 3.23 and using Eq.3.10, we obtain the time variance of the sine-wave modulated telemeter as:

$$\sigma_T^2 = (1/2\pi f_m)^2 [(2h\nu/\eta) B (P_{r0}+ P_{bg}+ P_{ph0})]/(m^2 P_{r0}^2) \quad (3.26)$$

As we can see from Eq.3.26, the primary dependence of accuracy in the sine-wave modulated telemeter is from $1/2\pi f_m$, the inverse of the angular frequency of modulation. By comparing with Eq.3.16', we see that $1/2\pi f_m$ is the counterpart of the pulse duration τ in a pulsed telemeter.

Further, in Eq.3.24 we can let $B=1/2T_r$ for a measurement lasting a time T_r, and can introduce the number of received photons $N_r = mP_{r0}T_r$ contributing to the measurement. With these positions, we can rewrite Eq.3.24 as:

$$\sigma_T = (2\pi f_m \sqrt{N_r})^{-1} [(1/\eta m) (P_{bg}+P_{ph0})/P_r]^{1/2} \quad (3.27)$$

3.2 Time-of-Flight Telemeters

As for the pulsed telemeter, the dependence of timing accuracy is from the inverse square root of the total number N_r of signal photons collected in the measurement time T_r.

The counterpart of Eq.3.16'', expressing the accuracy in terms of the equivalent photon number N_{bg+ph0}, is:

$$\sigma_T = (2\pi f_m \sqrt{N_r})^{-1} [(1/\eta m) (N_{bg+ph0})/N_r]^{1/2} \qquad (3.27')$$

By comparing the two equations, we can see that the performances are theoretically equivalent when we make $1/2\pi f_m$ equal to τ and $2\kappa S_0/s'^2$ equal to $1/m$.

Finally, let us consider the optimum filter treatment of the sine-wave modulated telemeter. If the signal waveform is given by Eq.3.22', from Eq.3.18 the impulse response of the optimum filter is found as:

$$h(T_m-t) = \sin 2\pi f_m t /[1+m \cos 2\pi f_m t] \qquad (3.28)$$

At the measurement time T_m, the optimum filter weights the signal as $s*h = \int_{0-Tm}$ m cos $2\pi f_m t \times \sin 2\pi f_m t /[1+m \cos 2\pi f_m t]$ dt. For m<<1, the weight is $\sin 2\pi f_m t$, exactly that of an in-quadrature homodyne detector.

3.2.5 The Ambiguity Problem

In the preceding sections of this chapter, we have studied the theoretical performance of time of flight telemeters and analyzed the dependence of accuracy from the physical parameters. This is the ultimate limit we can achieve in a well-designed telemeter and, as such, is the first step of design before developing the details.

Now, we shall review the specific problems, if any, that originate from the measurement approach (the time of flight) we are considering. After doing so, we will be ready for the instrumental development. Incidentally, this procedure has a general validity as a good engineering approach to develop instrumentation.

A problem with time of flight telemeters is the measurement ambiguity arising because the signal is inherently periodic (Fig.3-14).
In the sine-wave modulated telemeter, ambiguity shows up when the period $T_m=1/f_m$ of the modulation waveform is shorter than the time of flight T=2L/c (and actually, we want a small T_m to get a good accuracy). Thus, the phase-shift $\varphi=2\pi f_m T$ to be measured will be a multiple $n2\pi$ of the round angle, plus a residual φ', or $\varphi= n2\pi+\varphi'$. The phase meter measures φ' but cannot tell anything about n.
In the pulsed telemeter, we get an ambiguity when the repetition time T_r is shorter than the time of flight T=2L/c, and we send a second pulse before the first pulse has returned to the receiver (Fig.3-14). Thus, for a pulse transmitted at t=0, we receive a correct return at t=T, but also spurious returns at t=T-T_r, T-2T_r, etc., due to the previous pulses still in flight.

A trivial cure to ambiguity is, of course, to keep $T_{r,m} \geq T$ at all time. Using T=2L/c, the condition reads $L \leq cT_{r,m}/2$.

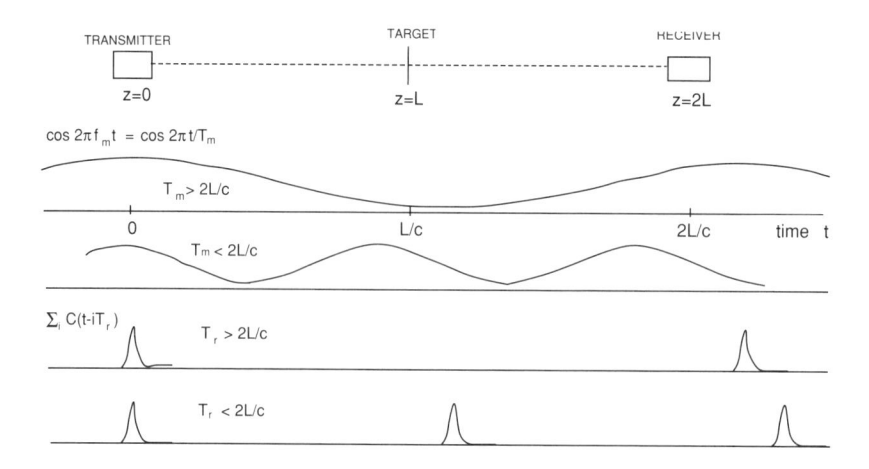

Fig. 3-14 In a time of flight telemeter, here drawn with the propagation path unfolded, the measurements present an ambiguity when more than a single period of modulation (T_m or T_r) is contained in the time-delay interval 2L/c. To avoid this, we shall have $T_{m,r}$ >2L/c or, equivalently, shall limit the useful range of distance to less than $cT_{m,r}/2=L_{NA}$, the nonambiguity distance.

With the equal sign, the distance L becomes the maximum nonambiguity range L_{NA} of the telemeter. This quantity is given by $L_{NA}=cT_{r,m}/2=c/2f_{r,m}$ at a given repetition (or modulation) period T_r (or T_m) or frequency f_m (or f_r).

Now, we can generalize the argument by writing the modulation waveform as a periodic function of time $\Gamma(t/T_{r,m})$, where Γ is a sinusoid for the sine-wave telemeter and is a pulse sequence for the pulsed telemeter. After a propagation on a distance 2L, the waveform is $\Gamma[(t-2L/c)/T_{m,r}] = \Gamma(t/T_{m,r}-L/L_{NA})$, where L_{NA} assumes the meaning of modulation wavelength. The maximum nonambiguity range is then the wavelength of modulation L_{NA}.

In a pulsed telemeter, we usually start with a laser source capable of supplying a short, high peak-power pulse, like a Q-switch solid state laser or a semiconductor laser diode-array. These sources work intrinsically at a low duty cycle, and the typical repetition-rate may be f_r =10Hz...10kHz. The corresponding nonambiguity range is calculated as $L_{NA}=c/2f_r$= 15,000 km...15 km and poses no problem in most cases. Only at the highest repetition-rate may we face ambiguity on very long-distance operation.

In this case, however, we may devise a strategy to manage the superposition of a moderate number N of pulses down the measurement distance. For example, we may use N time-sorters and feed each of them with the correct start-stop pair of pulses, selected by a multiplexer circuit. The multiplexer is made by a fully decoded base-N counter, whose outputs open linear gates connecting the start/stop pulses to the inputs of the time sorters. Operation is initialized at a very low repetition rate ($f_r<c/2L_{max}$, for nonambiguity) with the measurements being distributed cyclically on the N time sorters. Then, the repetition rate is gradually increased up to N times the maximum frequency of nonambiguity, or $f_r=Nc/2L$.

3.2 Time-of-Flight Telemeters

This strategy is equivalent to expanding by a factor N the maximum nonambiguity range.

In a sine-wave modulated telemeter, the ambiguity problem is more severe. Indeed, as we work with a quasi-CW power, we cannot trade duty cycle for nonambiguity because received power and hence accuracy would be sacrificed.

We can only use a modulation frequency f_m low enough to avoid ambiguity on the distance range to be covered. The nonambiguity range is $L_{NA}=c/2f_m$, and for $L_{NA}=0.15$ km...1.5 km we need to keep the modulation-frequency f_m as low as $f_m=1$ MHz...100 kHz.

In contrast, both a semiconductor laser diode and the phase-measuring circuit can easily handle signals with modulation frequency up to several hundred MHz (that is, 3 decade higher). Theoretically, the distance accuracy is $\sigma_L=c\sigma_T \approx c/2\pi f_m \sqrt{N_r}$ from Eq.3.27, and the relative accuracy is $\sigma_L/L_{NA}=1/\pi\sqrt{N_r}$.

Using $N_r \approx 10^7$ detected photons, accuracy would be theoretically limited to $\approx 10^{-4}$ of the maximum range L_{NA}, or we would get a measurement with no more than a four-decade dynamic range.

In practice, the phase-measuring circuit may introduce an additional limit. The phase resolution $\Delta\varphi_r$ is usually 10^{-3} or 10^{-2} of the round-angle 2π. Because the phase 2π corresponds to L_{NA}, the phase $\Delta\varphi_r$ corresponds to one unit of the least-significant digit in our measurement.

This means that the dynamic range is limited to $\approx 1/\Delta\varphi_r$, or only two to three decades in this example. To overcome this limit, we may take advantage that phase resolution does not worsen appreciably as the modulation frequency is increased up to the maximum that can be handled by the phase-measuring circuits.

Because distance resolution is $(c/2f_m)\Delta\varphi_r/2\pi$, we can then recover at least three decades of dynamic range by working at increased frequency.

Of course, to exploit this possibility, we need to combine a high frequency of modulation for resolution and accuracy to a low frequency as required by the nonambiguity regime, which will be shown in the next sections.

3.2.6 Intrinsic Precision and Calibration

In a time of flight telemeter, we measure a time T and convert it to a distance $L=cT/2$. The intervening multiplication factor is the speed of light, c=299,793 km/s in vacuum [7] and c/n in a medium with index of refraction n.

For propagation through the atmosphere, the most common case for a telemeter, n, differs from unity by a modest amount, yet about $300 \cdot 10^{-6}$ or 300 ppm (part-per-million).

Thus, as the time of flight measurement of distance comes to resolve the fourth or fifth decimal digit, we have to care about the index of refraction of the air and apply the n-1 correction to calibrate the instrument. In this way, we can have a calibration up to the fifth decimal digit and eventually be able to reach the sixth.

The problem of air index of refraction correction is shared by interferometers, where we may eventually go to the 7th or 8th decimal digit. Further information on the index of refraction correction can be found in Sect.4.4.4.

3.2.7 Transmitter and Receiver Optics

With regard to the optical system of a time of flight telemeter, we may use either two separate objective lenses for the transmitter and receiver or a single one serving for both transmitter and receiver (Fig.3-15).

The single-objective solution has the advantage of component saving and is the simplest to implement. The field of view (fov) of the receiver and field of illumination (foi) of the transmitter coincide, and there is no dead zone in front of the instrument. A disadvantage of the optical axes of transmitter and receiver being the same is a particularly strong back-scattered power collected by the receiver. This may limit the performance of weak-echo detection, especially if the target in noncooperative. In addition, an insertion loss is incurred because we insert the beamsplitter to deviate the returning beam onto the photodetector (Fig.3-15). In a double-objective system the back-scattered disturbance is much reduced, but field of view and field of illumination become superposed only after a distance L_{dz}, similar to the triangulation distance.

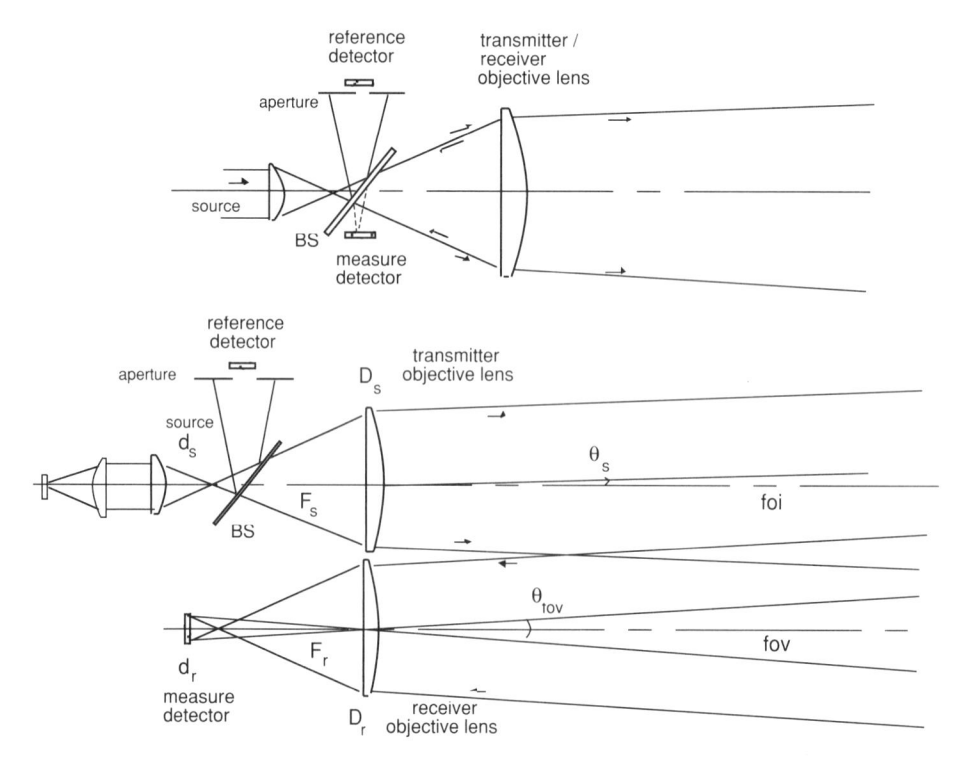

Fig. 3-15 In a time of flight telemeter, we may use either a single objective lens for both transmitter and receiver (top) or a double-objective system (bottom). The beamsplitter allows collecting a fraction of the outgoing optical signal for the reference photodetector to be used as the reference channel.

3.2 Time-of-Flight Telemeters

Thus, unless provisions are taken to move the lens axes or the detector position at short distance, the measurement has a dead zone. A dead zone is, of course, a minor problem with long-distance telemeters for geodesy, whereas it may be of importance in short-range telemeters intended for topography.

In both cases, a beamsplitter (BS in Fig.3-15) is inserted in the transmitter lens to pick up a minute fraction of the outgoing optical signal and to detect it by a reference-channel photodetector. If the same type of photodetector is used in the reference and measurement channels, errors due to residual phase-shifts or delays of them are canceled out.

The beamsplitter reflectivity at the surface hit by the outgoing beam is kept at 1% or less, whereas the other surface is either antireflection coated (double-objective case) or coated to 50% reflection (single-objective case). In the double passage across the beamsplitter of the outgoing and returned beams, the loss is TR=0.25 (or 6 dB) in the single-objective optical system, as opposed to $\approx$ 0 dB in the double-objective system. If we take advantage that the laser output is usually linearly polarized and add polarization-control elements to combine the beams, the loss can be reduced to 3 dB. Last, we may go down to $\approx$0 dB if we replace the beamsplitter with optical circulator, but the cost of this component is usually too high to be afforded. The zero-distance edge of the telemeter yardstick is determined by measurement and reference-channel photodetectors. Considering their virtual position along the transmitter path, as produced by the beamsplitter, it is easy to see that the zero-distance location is the midpoint of the segment connecting these two positions.

Let us now consider the design of receiver and transmitter lenses. The constraint to start with is the maximum allowable size of the objective lenses D_r and, of course, we want to be able to use the smallest detector size d_r for the best noise and speed performance. To minimize d_r, we need to keep the spot size w_t on the target as small as possible.
For a Gaussian beam, the minimum obtainable size as set by the diffraction limit is given by $w_t = \sqrt{(\lambda L)}$, where L is the target distance (see Sect. 1.1). Correspondingly, the transmitter objective lens shall have a diameter $D_t=(2w_t)^{1/2}=(2\lambda L)^{1/2}$ to project the spot w_t at distance L. For L=100 m...10 km and at λ=1µm, we get D_t = 1.4...14 cm, which are quite reasonable values.

At the distance L, the spot w_t is seen under the angle $\theta_{fov} = w_t / L$ (=0.1...0.01 mrad in the previous example). Working with a cooperative target (a corner cube), the retroreflected beam has a size $2w_t$ at the receiver, and this value is the lens-size D_r we require.
When the target is noncooperative (a diffuser), the received signal is proportional to D_r^2 (Eq.3.4) and we will use the largest D_r that is practically allowed by the application.
To collect from a θ_{fov} field of view, we need a detector size $d_r > \theta_{fov} F_r$, where F_r is the focal length of the receiver objective. Letting F_r=100 mm, we get d_r >10µm for θ_{fov} = 0.1mrad.

With respect to the transmitting objective, if the source collimated beam has a waist w_l, we need a magnification $M=w_t/w_l$.
Then, we set $M=F_t/f$, the ratio of the focal lengths of the two lenses constituting the collimating telescope indicated in Fig.3-15, top. If the source provides a near-field spot as an output (for example, is the output facet of a diode laser), we will use a separate collimating lens to enter the telescope.
This lens will include an anamorphic corrector to circularize the beam.

3.3 INSTRUMENTAL DEVELOPMENTS OF TELEMETERS

In the following, we describe some conceptual approaches that have been used to develop pulsed and sine-wave modulated telemeters. As it is well known in instrumentation, the optimum design approach strongly depends on the available components and may become much different as the performance of components undergoes even minor changes. A telemeter is no exception, and the examples reported in the following are illustrative of the design approaches and ingenuity of researchers in the last two decades.

3.3.1 Pulsed Telemeter

A very simple pulsed telemeter can be developed starting from the block scheme shown in Fig.3-16. Two objective lenses are used for the transmitter and receiver, and the start pulse is derived with a beamsplitter from the outgoing optical beam. The stop pulse is detected with the second objective lens aimed to the target.

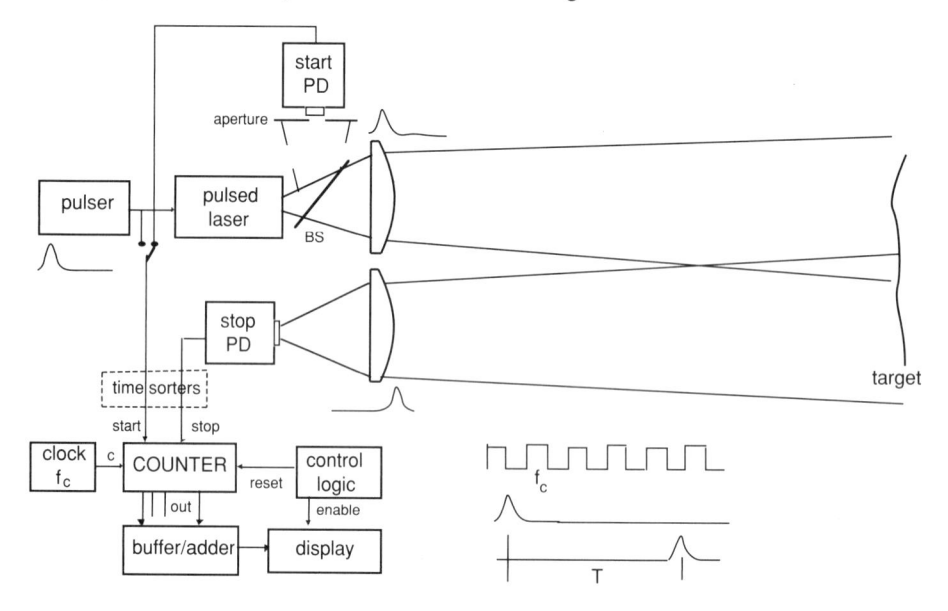

Fig. 3-16 Block scheme of a pulsed telemeter using separate objective lenses for transmitter and receiver. The start and stop pulses are used to gate on and off the counter, for which the input is a clock at frequency f_c. After a time $2L/c$, the counter content is then $N_{count} = f_c 2L/c$, and this is the result of a single measurement. With the aid of the buffer/adder, we may sum the results of N_m successive measurement, getting a total $N_m N_{count}$ as the output of the display. If the laser pulse response has a low time jitter, we may even use the driving signal of the pulser as a start pulse.

3.3.1 Pulsed Telemeter 67

If the light pulse is well correlated to the electrical drive of the laser, that is, the laser response to pulse drive has a low time jitter, we can also use the pulser signal itself as the start pulse. This variant allows us to save a few parts (BS, PD, and preamp in Fig.3-16), and it can be used with semiconductor-diode sources, whereas crystal Q-switched lasers have an excessive jitter.

Start and stop pulses enter time sorters to improve time localization and to standardize the waveform for the subsequent input to the counter circuit. The counter is fed by a clock at frequency f_c, and the periods of f_c are counted in the time interval $T=2L/c$ defined by start and stop pulses. Thus, in a single measurement, the content of the counter we get is:

$$N_{count} = f_c T = f_c\, 2L/c = L/(c/2f_c) \qquad (3.29)$$

[In Eq.3.29, we implicitly assume to take the integer part of the quantities at the right hand side of the equation]. Eq.3.29 tells us that the quantity $c/2f_c = L_1$ is the scale factor representing the length of a single unit of N_{count}, so that the mean value of distance L is:

$$L = N_{count} L_1 \qquad (3.30)$$

The quantity L_1 also represents the unit U of rounding off in the measurement of L. Because the rounding-off error has a variance $U^2/12$ and we make two truncation errors with the start and stop pulses sampling the clock, the total of the rounding-off error is twice as much, or $U^2/6$. In terms of distance, this amounts to a variance $\sigma^2_{L(ro)} = L_1^2/6$.
Adding the timing error σ^2_t (Sect.3.2.3) multiplied by $(c/2)^2$, we get the total distance variance of our telemeter as:

$$\sigma^2_{L(tot,\,1)} = (c^2 \sigma^2_t)/4 + L_1^2/6 \qquad (3.31)$$

We may repeat the measurement N_m times to improve resolution and accuracy, provided the application at hand allows for an increased total measurement time, now given by $N_m T_r$ where T_r is the pulse repetition period. Repeating the measurement N_m times and summing up the results of each measurement, we get as an average of the accumulated counter content:

$$N_{count} = N_m f_c T = N_m L/(c/2f_c) \qquad (3.29')$$

Also, the distance variance decreases by a factor N_m, or:

$$\sigma^2_{L(tot,\,N_m)} = [c^2 \sigma^2_t/4 + L_1^2/6]/N_m \qquad (3.31')$$

Now, we choose the clock frequency f_c according to two criteria: (i) it shall have numerals such that N_{count} represents the target distance L in metric units, and (ii) it shall be high enough to attain the desired resolution.
Because of requirement (i), we need $L_1 = N_m c/2f_c$ to be an exact decimal number, for example 0.1, 1 or 10 m.

As a first approximation, it is $c=3·10^8$ m/s, and hence the frequency should be f_c =$1.5·10^9$, $1.5·10^8$, and $1.5·10^7$ Hz in the previous example for N_m=1.

More precisely, if we take n=1.0003 and c_{vac}= 299,793 km/s (see Sect.3.2.6), we have c= c_{vac}/n=299,6731, and the numerals of f_c are 1.4983655 (in place of 1.5). This value requires a further correction against wavelength and temperature/pressure as we go to resolve the 5th-6th decimal digit (see Sect.3.2.6).

Regarding the clock, we will use a good quartz-oscillator circuit, for which its accuracy in frequency may easily reach the 5th-6th decimal numeral, and even the 6th-7th in special units.

Because of requirement (ii), the order of magnitude of f_c shall be compatible with the optical pulse duration τ, and its value determines the required speed of the counter circuits.

As a rule of thumb, we should choose f_c not much exceeding the inverse of the pulse duration $1/\tau$, so that the two terms in Eq.3.31' contributing to the distance variance have approximately equal weights.

Indeed, if we just take $f_c = 1/\tau$, we have one count per pulse duration. This loosens the speed requirement of circuits, but probably wastes some of the accuracy available in the pulse waveform. On the other hand, taking $f_c = 10/\tau$, we need faster circuits and better time sorters, but are able to optimize the time accuracy down to the limits discussed in Sect.3.2.3.

In general, when writing $f_c = M/\tau$ where M=1..10 in the previous examples, the single count of the telemeter represents a length-equivalent increment $c/2f_c = c\tau/2M$.

To give a few examples, with a pulse duration τ=1, 10, 100 ns we have $c\tau/2$= 0.15, 1.5, 15 m, and $L_1=c\tau/2M$ is a submultiple of these values, for example 0.05, 0.5, 5 m for M=3. Using $f_c = 3/\tau$, the required clock frequency is f_c = 3 GHz, 300 MHz, and 30 MHz respectively.

The case τ=30 ns, M=3, and f_c =100MHz is representative of a pulsed-telemeter based on a semiconductor laser source. Counters and sorters can work easily at this medium speed, and the circuits can be implemented at a relatively low cost. The least significant digit in a single measurement of this hypothetical telemeter is 1.5 m.

By repeating the measurement N_m (say 15...150) times, the resolution is increased by a factor N_m and reaches 10... 1-cm, while accuracy improves by $\sqrt{N_m}$ (=3.9...12). At a typical pulse repetition rate f_r=3 kHz, the total measurement lasts N_m/f_r =5...50 ms.

The case τ =5 ns, M=7.5, and f_c =1.5 GHz is representative of a pulsed telemeter based on a Q-switched solid state laser. The least significant digit is now 0.1 m in a single measurement, a figure already adequate for applications requiring a good accuracy and a single-shot measurement time.

Typical applications calling for the previously mentioned specifications are (i) altimeters for helicopters; (ii) terrain profilers, like the MOLA telemeter mission (Fig.1-5, see [8]); (iii) satellites ranging observatories (see [9] as an example); (iv) military telemeters and (v) long-distance geodesy [10].

In all these applications, as we go to long distances, we have to increase the peak power of the pulse and the collecting aperture of the receiver by using a large-aperture telescope.

Then, it is advisable to use just one optical element for both transmitter and receiver, as shown in Fig.3-15 (top).

3.3.1 Pulsed Telemeter

In this configuration, a beamsplitter inserted in the outgoing optical path collects a fraction of the optical pulse for use as the start signal.

Several different approaches for pulsed telemeters have been proposed in the scientific literature, and an interesting selection of papers on the subject is provided by Ref.[1]. More results were presented recently at specialized conferences [11].

3.3.1.1 Improvement to the basic pulsed setup

By slightly modifying the time interval measurement described in the previous section, we can get rid of the start/stop rounding-off error and improve the distance variance down to the ultimate limit of pulse-timing variance σ_t^2.

The idea consists of combining the coarse measurement of time interval, performed by clock counting, with a fine measurement of the fraction-of-period excess ΔT_{ex} in both start and stop pulses.

As illustrated in Fig.3-17 for the start pulse, we can recover this fraction ΔT_{ex} by using the discriminator output to trigger the start of a Time-to-Amplitude Converter (TAC).

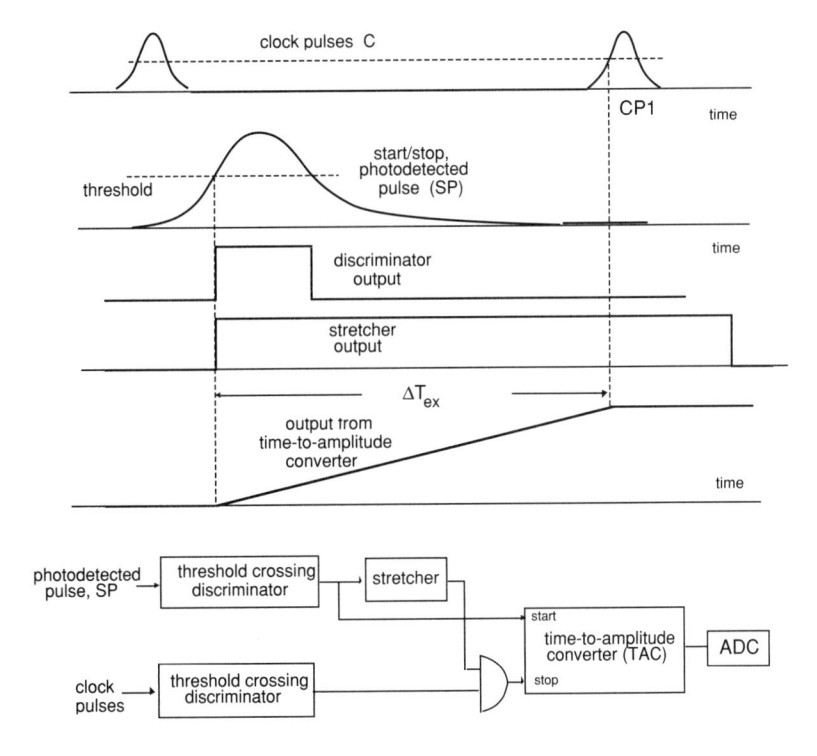

Fig. 3-17 The fraction-of-period time interval ΔT_{ex} between start pulse SP and the first counted clock pulse (CP1) that can be recovered by this circuit to improve accuracy of the telemeter. The scheme is duplicated for the stop pulse.

The stop to the TAC is provided by the first clock pulse arriving after the discriminator switch-on. Using a stretcher to increase the duration of the discriminator output, in combination with an AND gate receiving the clock pulse, we can sort out this first clock pulse (CP1 in Fig.3-17).

Then, the amplitude of the output from the TAC is proportional to the required time ΔT_{ex}. By converting the TAC signal amplitude to a digital output with the aid of an Analog-to-Digital Converter (ADC) we get ΔT_{ex} (and the corresponding L_1) digitized in, say, one or two decades. Of course, the same procedure will be used for the stop pulse to recover the fraction of period excess of it and to correct the result accordingly.

Note that the two more discriminations in Fig.3-17, performed on the clock pulses, do not appreciably degrade the accuracy. In fact, like the start (photodetected) pulse, clock pulses are large in amplitude, and their time variance σ_t^2 is negligible [from Eq.3.16] as compared with the σ_t^2 of the stop (photodetected) pulse.

3.3.1.2 Pulsed telemeters using slow pulses

Two techniques can be used to improve resolution down to nanoseconds when the pulse available from the laser source is rather slow, for example, microsecond-long as for CO_2 laser sources. They are the vernier technique and the pulse-compression technique discussed below.

Vernier Technique. This is a technique well known in nuclear electronics [12] to increase time resolution in measurements with slow waveforms.

As shown in Fig.3-18, the optical start and stop pulses are detected by a photodiode, a nearly ideal current generator. The currents excite two resonant, parallel LC-circuits. The LC circuits are designed to have a high Q-factor, and the voltage V_{LC} across them is taken through high impedance to preserve the initial Q.

With a reasonably high Q, the voltage V_{LC} is a nearly sinusoidal oscillation, and it damps out slowly after excitation, making many periods of oscillation ($N \approx \pi Q$) available for the measurement.

Timing is performed by looking at the zero-crossing of the V_{LC} waveform. Indeed, V_{LC} starts oscillating after the pulse has ended and its charge has been completely collected by the capacitor. It can be shown that V_{LC} carries the timing information associated with the center of mass (or time-centroid) of the pulse, provided the pulse duration is short as compared with the oscillation period.

Now, let us compare the start and stop waveforms V_{LC1} and V_{LC2} (Fig.3-18). The time difference between the start and stop centroids (or, between the negative-going zero-crossings of V_{LC1} and V_{LC2}) can be written as $T=T_0+\Delta T$. Here, T_0 is the coarse delay between the V_{LC1} and V_{LC2} oscillations and is given by an integer multiple of the resonant period $1/f_0$, whereas ΔT is a fine delay, given by a fraction of $1/f_0$ (Fig.3-18).

The coarse measurement is readily obtained by counting the number of periods N_c, which are contained in V_{LC1} after its first zero-crossing and before V_{LC2} makes its own first zero-crossing.

3.3.1 Pulsed Telemeter

To get the fine measurement, the resonant frequencies of the start and stop circuits are made slightly different, let us say f_0 (start) and $f_0/(1-\Delta)$ (stop).
Then, signal V_{LC1} initially leads V_{LC2}, but suffers a small delay, with respect to V_{LC2} at each period. The small delay is $1/f_0 - (1-\Delta)/f_0 = \Delta/f_0$. After a number N_f of periods has elapsed, the stop waveform V_{LC2} finally leads the start waveform V_{LC1}.

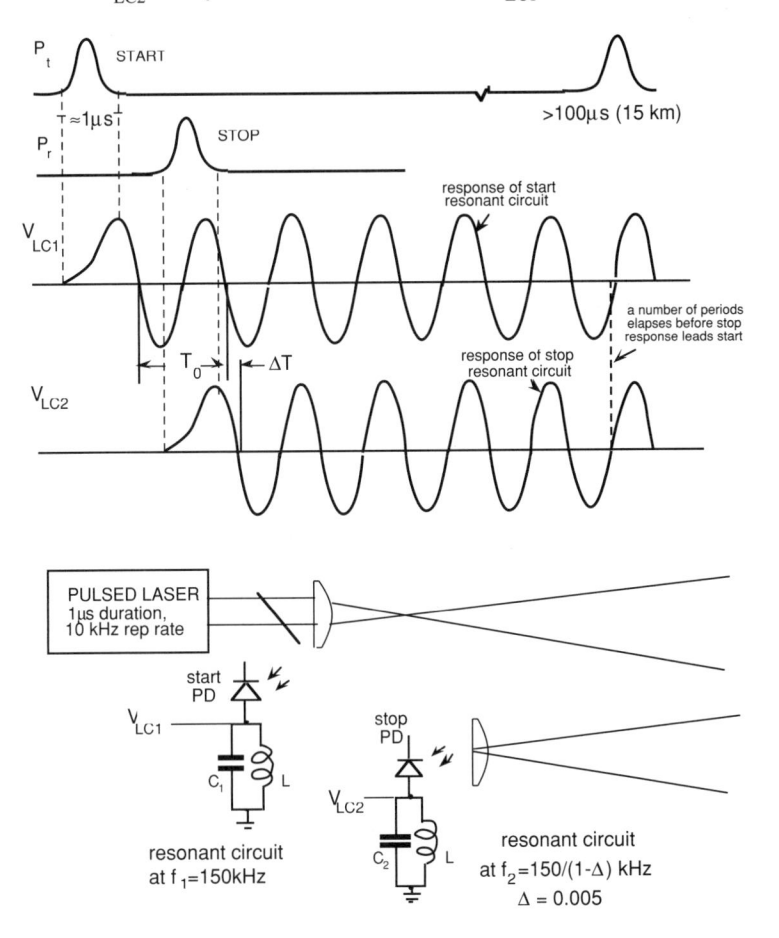

Fig. 3-18 The vernier technique uses two resonant circuits fed by the start and stop photodiodes to perform the time delay measurement. It resolves a small fraction of the resonant period, even with relatively slow (low-noise, however) optical pulses. The resonant circuits have a slight offset in resonant frequency. The coarse measurement of time delay is made by counting the number of periods of the start oscillation before the stop is received (time T_0). The fine (or vernier) measurement of the fraction of period ΔT is made by counting the number of periods contained in the start oscillation before the start oscillation, initially in delay, becomes in the lead (see Fig.3-19 for the electronic block scheme).

Counting the number N_f of periods we find in V_{LC1} after V_{LC2} has done its first zero-crossing, until V_{LC2} leads V_{LC1}, we get the measurement of ΔT in units of Δ/f_0.

Example. With a $\tau=1$-μs pulse (for which the resolution is $c\tau/2=150$m), we may choose $f_0=150$ kHz as the resonant frequency of the LC circuits. The period $1/f_0$ corresponds to a distance increment $c/2f_0=1000$ m and is the least significant digit of the coarse measurement. In addition, by taking $\Delta=0.005$ as the resonance offset, we resolve $\Delta c/2f_0 = 5$ m. This is the least significant digit of the fine measurement. The maximum number of periods $1/f_0$ we have to count in the vernier operation is $1/\Delta = 200$ and, therefore, we need a Q factor of about 200 to keep damping of V_{LC} waveforms reasonably low. Of course, the $\tau\Delta=$ 5ns resolution will be actually obtained if the received pulse carries an adequately large number of photons (Eq.3.16).

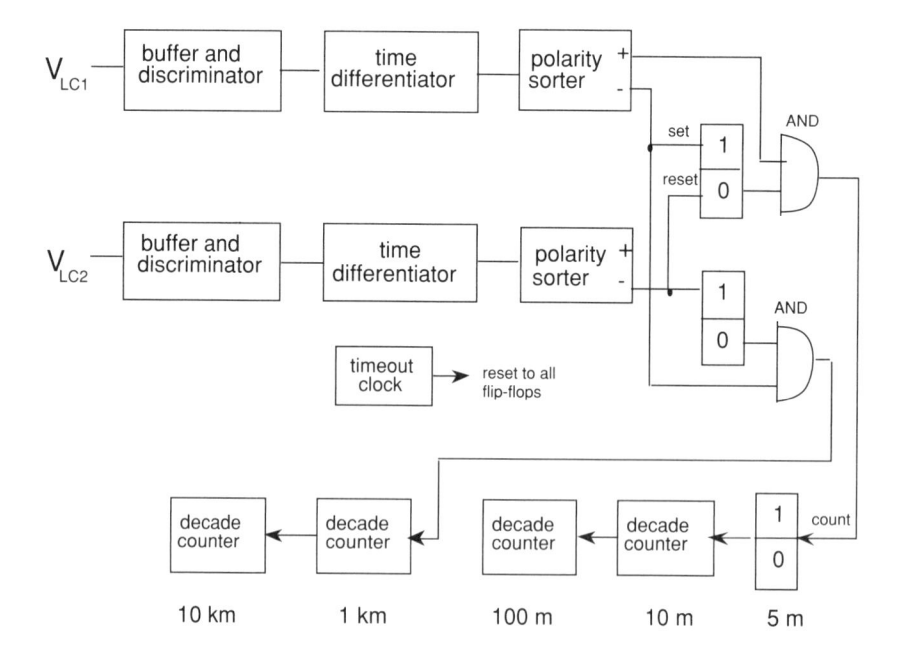

Fig. 3-19 Electronic processing of signals from the vernier telemeter of Fig.3-18 can be made with a variety of approaches. The one shown here is perhaps the easiest to implement. Resonant signals are buffered and passed though a discriminator. By time differentiation, we get positive and negative pulses marking the periods. A logic circuit (middle flip-flop and AND gate) counts the negative transitions of V_{LC1} until V_{LC2} makes the first negative-going oscillation. These counts go to the 1 and 10-km decades. The vernier measurement is made by the top flip-flop and AND gate. Here, we count the positive transitions of V_{LC1} until the negative-going V_{LC2} leads the negative-going V_{LC1}. This gives the two-and-a-half decades with resolution down to 5m.

3.3 Instrumental Developments of Telemeters

The centroid-timing corresponds to weighting the pulse waveform by t-t$_0$. This is actually the operation performed by the optimum filter when the received pulse has a Gaussian waveform (see Sect.3.2.3.1). Because the Gaussian is a very popular approximation, we may conclude that the LC timing and vernier technique are near-to-optimal choices in most practical cases.

Pulse-Compression Technique. This technique has been demonstrated with a CO_2 laser since the early times of laser telemeters [13] as an adaptation to optical frequency of a well-known method used in the microwave radar.

As shown in Fig.3-20, we may start with a Continuous Working (CW) source, such as one like the CO_2 laser that may be difficult to pulse, but that makes a large power available, e.g., a few Watts.

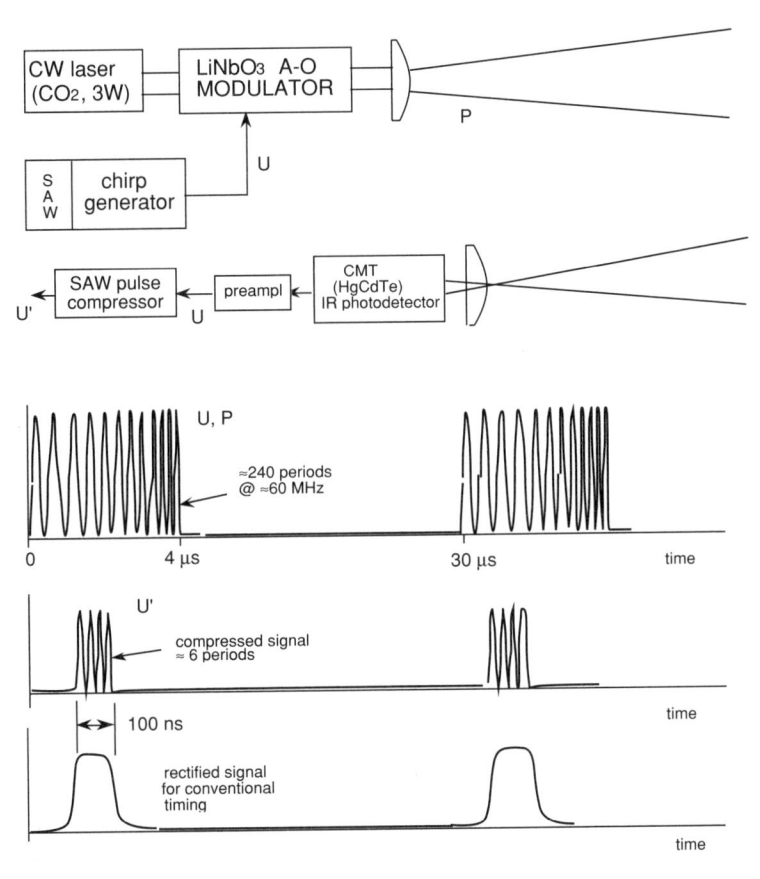

Fig. 3-20 With the pulse compression technique, we can transmit slow (μs) optical pulses and shorten them after photodetection to a width adequate for conventional timing. Compression is achieved by a scheme similar to that used in radar techniques, which involves chirped modulation and demodulation.

A reason to use a CO_2 laser source is that the Far-Infrared (FIR) wavelength of emission, $\lambda=10.6\mu m$, is favorable for a much larger range of operation in turbid media, such as fog and smokes (see App.A3.1 and Fig.A3-7). The relatively large CW power is readily provided even by compact CO_2 sources and is desirable to compensate for the decreased detector sensitivity and increased propagation attenuation.

Using an acousto-optics modulator [14], we are able to impress a modulation on the average power in form of sine-wave bursts, as indicated in Fig.3-20.

The repetition period T_b of the burst is dictated by the desired ambiguity range (Sect.3.2.5), and a typical choice for an airborne telemeter is $T_b=30$ μs ($L_a=4.5$ km). A burst duration τ_b which is relatively short for timing purposes, but long enough to accommodate the chirp modulation, is the value $\tau_b=4\mu s$ shown in Fig.3.20.

With a CO_2 laser, these times are short as compared with the active-level lifetime ($\tau_{21} \approx 400\mu s$), and therefore the average power is the same as the CW value P_{CW}. Then, the peak power of the burst is $P_p=P_{CW}/\eta$, or, it is increased by the inverse of the duty cycle $\eta=\tau_b/T_b$ (a factor of 7.5).

For the pulse compression, we shape the burst waveform as a chirped sinusoid, or a linear-sweep frequency modulation ranging from f_0 at the beginning of the burst to $f_0+\Delta f$ at the end of the burst. Such a chirped sinusoid can be synthesized by a number of circuits, yet the Surface Acoustic Wave (SAW) is the most neat and simple to implement [14]. Typical values compatible with the performance of the lithium-niobate modulator are $f_0=60$ MHz and $\Delta f = 10$ MHz. Thus, we have ≈60MHz×4μs=240 periods of sinusoidal oscillation in the burst, with an instantaneous frequency increasing from 60 to 70 MHz.

The chirp is a sort of (frequency) dispersion of the impulse response of the circuit generating the modulating wave. We may cancel it by cascading, after detection, a second chirp-generating circuit with the same dispersion, but opposite sign. The circuit shall have a resonant response lasting the same time τ_b, but with $f_0+\Delta f$ at the beginning and f_0 at the end of the burst. With a SAW device, this is particularly easy because we need only to interchange input and output of the device.

Theoretically, the compression ratio is found as $\tau_b \Delta f$ [13]. With the previously chosen values, it is a factor 10MHz×4μs=40, and then the compressed duration is 4μs/40=100ns, and rise/fall times may be ≈1/60MHz≈15ns.

In the compression, we gain an increase of (equivalent) peak power of $\tau_b \Delta f$. In addition, a factor $1/\eta=T_b/\tau_b$ comes from the repetition rate. Thus, the total increase respect to the CW value is factor $(T_b/\tau_b)(\tau_b \Delta f)=T_b \Delta f=300$, a remarkably good value indeed.

After rectification of the compressed envelope, we may proceed in the timing operation by conventional processing of signals.

3.3.2 Sine-Wave Telemeter

The sine wave telemeter allows using a low-power laser, provided it can be modulated at reasonably high frequency. Indeed, the theoretical accuracy of the sine-wave telemeter equals that of the pulsed telemeter when the inverse angular frequency $1/2\pi f_m$ of modulation is equal to the pulse duration τ (Sect.3.2.4), and we use the same number of collected photons.

3.3 Instrumental Developments of Telemeters

Low power is desirable and may even be a mandatory requisite in view of laser safety issues (App.A1.5). This is especially true for outdoor applications in the presence of the public, for example the topographic telemeter replacing the theodolite, the laser level and alignment equipment. In such applications, either we use barriers to prevent exposure of humans to a hazardous level, or work with the maximum power allowed by a Class-1 laser.
For semiconductor laser diode operating at a wavelength λ=700-900 nm, this power is P_t =0.2-0.6 mW, a reasonable value readily provided by ternary GaAlAs lasers. These lasers are cheap, reliable, and compact, and can be easily modulated through the injected current up to several hundreds MHz.

The problem of ambiguity (Sect.3.2.5) has been approached by a number of methods [1,11], and in the following we describe two of them, perhaps the most practicable because they have been incorporated in products. They are the *frequency-sweep method* and the *multi-frequency method*.

Frequency-sweep method. We use a sinusoidal waveform for modulating the laser power and vary the modulation frequency f_m in time. We start with a frequency f_{m0} low enough to ensure the desired non-ambiguity distance L_{NA} and sweep it up to a large value f_{M0} that allows us to reach the desired distance resolution ΔL, as indicated in Fig.3-21.

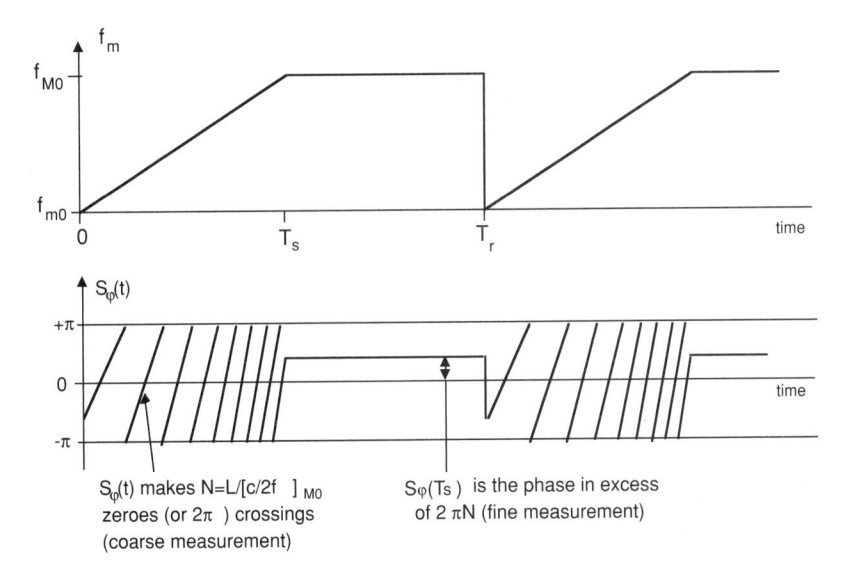

Fig. 3-21 In a sine-wave modulated telemeter, we want to avoid the distance ambiguity and yet be able to use a high-modulation frequency. To this end, frequency is made to sweep from a very low value f_{m0} (typically 15 kHz, avoiding ambiguity up to 15 km) to a fairly high f_{M0} (typically 150 MHz, adequate for a 1-cm resolution). The coarse measurement is recovered from the number of zero-crossings contained in the phase signal S_φ, while frequency is sweeping from f_{m0} to f_{M0}.

Specifically, we will use (Sect.3.2.5) $f_{m0} \leq c/2L_{NA}$ to obtain the nonambiguity range L_{NA}, and $f_{M0} \geq c/2N\Delta L$ to obtain the resolution ΔL in a phase measurement with a 1/N resolution of the round angle.

For example, if our telemeter measures up to 15 km, we choose $f_{m0} = 3 \cdot 10^8/2 \cdot 15 \cdot 10^3 = 10$ kHz. Also, if we want a 1-cm resolution out of a phase sorter with a 1/100 resolution of 2π, we use $f_{M0} = 3 \cdot 10^8/2 \cdot 100 \cdot 0.01 = 150$ MHz.

During the frequency sweep, the instantaneous phase signal that we obtain by mixing the transmitted and received waveforms is written, from Eq.3.24 with $\varphi_0=0$, as:

$$S_\varphi(t) = (I_0 I_{r0}/2) \, m \cos [2\pi f_m(t) T] \qquad (3.32)$$

At the beginning of the sweep, we start from a phase $2\pi f_{m0} T$ less than 2π because of our choice $f_{m0} \leq c/2L_{NA} = 1/T$. Then, as frequency increases, we soon reach a value for which $2\pi f_m T = 2\pi$. At this value, the actual distance L contains exactly one full period of the modulation waveform, or $2\pi f_m(2L/c)=2\pi$, and the phase signal S_φ is zero.

As time elapses and f_m increases further, we find new zero-crossings of the phase S_φ. This happens each time the actual distance L becomes an integer multiple M of the instantaneous modulation period $c/2f_m$, or $L=M(c/2f_m)$. This can also be written as $2\pi f_m(2L/c)= 2\pi f_m T=2\pi M$, so (Eq.3.32) $S_\varphi=0$.

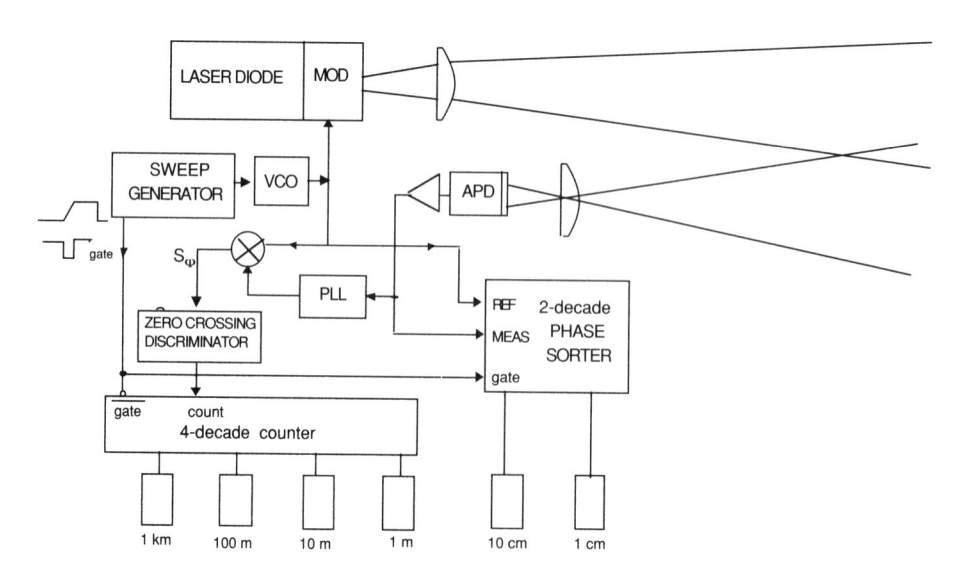

Fig. 3-22 The sweep generator feeds a modulator (or directly) the laser diode current) with the frequency sweep of Fig.3-21, and also gates the coarse and fine-measurements decade counters. The phase signal is generated by mixing the sweep waveform and photodetected signal, properly amplified and filtered by the Phased-Locked-Loop (PLL).

3.3 Instrumental Developments of Telemeters

When the sweep finally reaches its maximum frequency f_{M0}, the number of zero-crossings that the phase signal S_φ has done is exactly the number of times M_0 the distance $c/2f_{M0}$ is contained in the distance L to be measured.

We may count the zero-crossings of S_φ by using the (positive) slope of the sweep signal as a gate input to a 3-decade counter (Fig.3-22). In this way, we obtain the coarse measurement of distance in units of $c/2f_{M0}$ (=$3 \cdot 10^8/2 \cdot 150$ MHz = 1 m).

In the schematic of Fig.3-22, this processing is done by mixing the signal of the modulator and the signal received from the photodetector. Actually, the signal from the photodetector may be noisy and, to filter it, a Phase-Locked-Loop (PLL) block is used.

The PLL is equivalent to a narrow pass-band filter with the central frequency dynamically tracked to the signal frequency. A narrow bandwidth is desirable to effectively filter out noise from the phase signal, to avoid counting spurious zero-crossings. The smallest bandwidth we can use in the PLL filter is determined by the sweep duration T_s as $B \approx 1/T_s$.

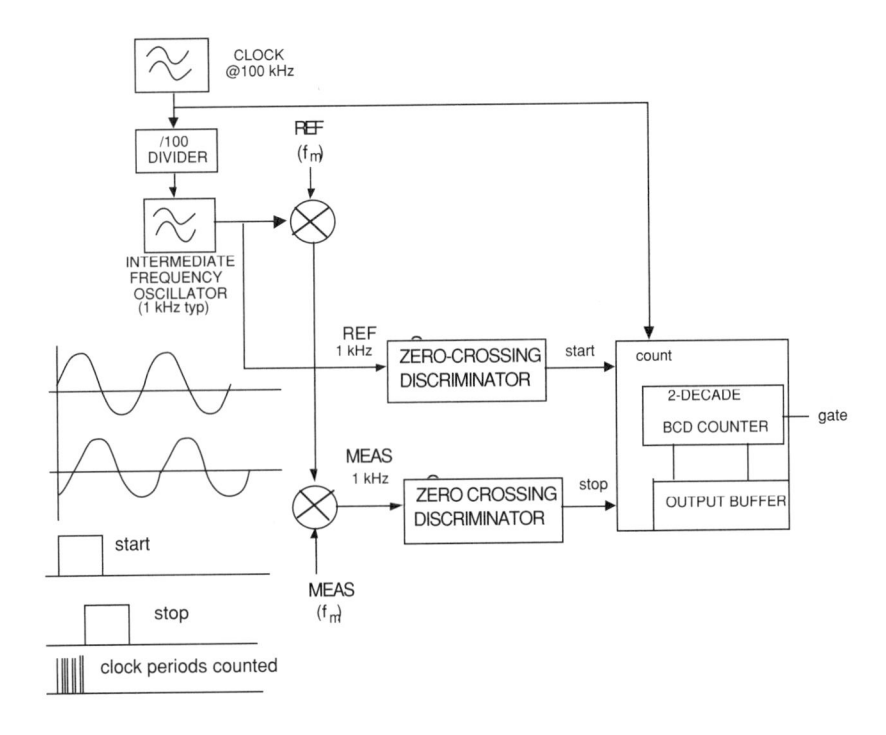

Fig. 3-23 Simplified block scheme of a typical 2-decade phase meter. The measurement is performed at an intermediate frequency (1kHz), that is tied through a :100 divider to a main oscillator (100 kHz), which is also used as a clock of the counter. The MEAS signal is down-converted to 1kHz by mixing with the REF signal shifted by 1kHz. Start and stop to the counter are obtained by zero-crossing discrimination of the MEAS and REF signals.

The fine measurement is performed on the phase-signal S_φ during the period from T_s to T_r (Fig.3-21) when the modulation-frequency is kept constant ($f_m = f_{M0}$). A conventional 2-decade digital phase meter is adequate to sort out the measurement (Fig.3-22).

As detailed in Fig.3-23, a typical design uses a conversion of the signal (MEAS) to an intermediate frequency, usually $f_{IF}=1$ kHz. The periods of a clock running at $f_{ck}=Nf_{IF}$ (100 kHz for N=100) are counted using REF and MEAS signals as start and stop.

Fig.1-4 is an illustration of a commercial sine-wave modulated telemeter developed on the frequency-sweep concept for the application in topography. The resolution of such a telemeter may be 1 cm (for a 10-s measurement time). Distance of operation can go up to 100 to 300 m when using a cooperative target. Usually, the target is a corner-cube array with ≈50 to 200 cm^2 area. The optical source is typically a 0.1 mW GaAlAs laser, and the entire instrument, including the electronic processing unit, fits in a briefcase for use in the field.

Multifrequency method. Another approach to the sine-wave telemeter is using a frequency f_{m1} low enough to avoid distance ambiguity, together with its multiples, $10f_{m1}$, $100f_{m1}$, etc., to achieve the desired resolution. The ingenuity is in properly combining the measurements performed by the different frequencies.

For example, if we choose three frequencies that are a multiple of a factor 100 and use a 1/100 round-angle phase measurement, each frequency supplies two decades, and as a whole we cover 3×2= 6 decades in distance.

As shown in Fig.3-24, the three measurements can be made in sequence by multiplexing the modulation waveforms to the laser and distributing the results to the appropriate counters.

To substantiate the concept, let us take f_{m1}=15 MHz as the highest frequency. The corresponding distance is $L_{m1}=c/2f_{m1}=10$ m, and the 10-cm and 1-m digits come out from the $2\pi/100$ phase measurement.

Then, the second frequency will be f_{m2} =150 kHz to supply the 10-m and the 100-m digits, and the third frequency f_{m3} =1.5 kHz, gives the 1-km and the 10-km digits.

Theoretically, the measurement associated to each frequency is the pair of digits in L remaining from the next lower frequency. In fact, from Eq.3.24, the phase argument of the cosine, the quantity to be measured in $2\pi/100$ intervals, can be written as:

$$\varphi_{m1} = 2\pi f_{m1} 2L/c = 2\pi\, L/L_{m1}$$

where L_{m1} is the scale factor associated to f_{m1}. As the phase is modulo 2π, we measure the remainder of L/L_{m1} from the integer part, or the first two decimals. Considering that $f_{m2}=f_{m1}/10^2$ and $f_{m3}=f_{m1}/10^4$, we have $\varphi_{m2}=2\pi 10^2 L/L_{m1}$ and $\varphi_{m3}=2\pi 10^4 L/L_{m1}$. Thus, the measurements with f_{m2} and f_{m3} are the remainders of $10^2 L/L_{m1}$ and $10^4 L/L_{m1}$ from the integer part, or the second two decimals and the third two decimals, respectively, of L/L_{m1}.

A subtle problem arises, however. We may even tie the three frequencies very precisely, i.e., deriving them from a single oscillator through dividers, but unavoidably measurement errors will prevent the results from being exact multiples of each other.

3.3 Instrumental Developments of Telemeters

Thus, when the 10-cm and the 1-m digits pass from 99 to 00 (or vice-versa), the digit of the 10-m does not increment (or decrements) the content by one. Instead, the 10-m digit may switch erratically, when the 10-cm and the 1-m digits are somewhere, e.g., between 90 and 10 (as they pass through 99 and 00).

This problem is clearly a very basic one, common to all readouts from demultiplied scales and calls for a technique of digital synchronization.

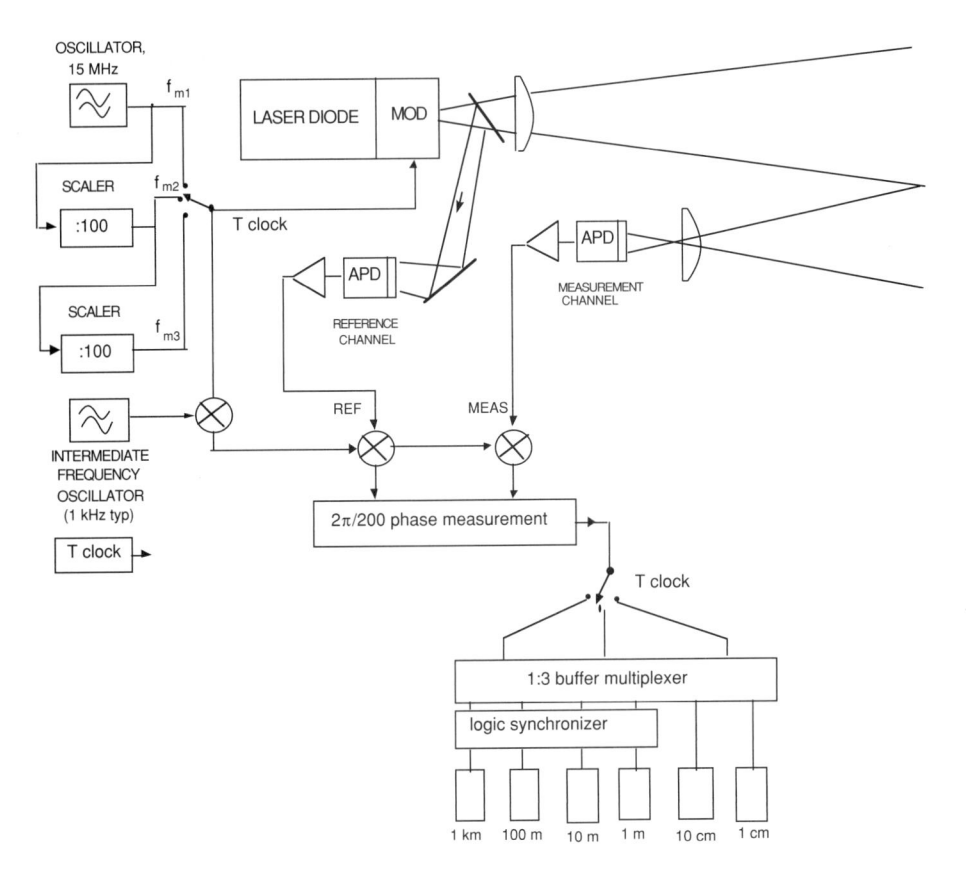

Fig. 3-24 The multifrequency method of the sine-wave modulated telemeter uses three frequencies, multiples of 100, to obtain a large nonambiguity range with the smallest of them ($f_{m3} \approx 1.5$ kHz), yet has a good resolution with the largest available ($f_{m1} \approx 15$ MHz). To try balancing the electronic delays, the REF signal is obtained by a photodetector and amplifier combination identical to that of the MEAS for the pulse returning from the target. The 1-KHz local oscillator shifts all three frequencies so that, after mixing with REF and MEAS, the phase measurement is always performed at the same intermediate frequency. A logic circuit provides synchronization of the three pairs of digits.

To synchronize the digits, we add a redundancy to the coding scheme. Instead of the $2\pi/100$ resolution for the $\times 100$ frequency ratio, we double the resolution to $2\pi/200$ in f_{m2} and f_{m3}.

In this way, we can make available a logic variable, say it $X_{1/2}$, which indicates the state (00 to 49 or 50 to 99) of f_{m2}. Nominally, $X_{1/2}$ replicates the most significant bit of the f_{m1} readout, let us say it Y_{50}. This bit Y_{50} changes from 0 to 1 and from 1 to 0 when the digits of f_{m1} go from 00 to 49 and from 50 to 99, respectively. Because of measurement errors, $X_{1/2}$ deviates from Y_{50}, however.

By comparing the two variables, it is easy to determine if the f_{m2} (and f_{m3}) bit shall be incremented by one or not. Thus, synchronization is obtained at the expense of a modest loss in resolution (a factor of 2).

The multifrequency technique is incorporated in commercially available telemeters, and it yields performance comparable to those of frequency-sweep, sine-wave modulated telemeter.

3.4 IMAGING TELEMETERS

A number of applications, ranging from military to biomedical, call for a spatially resolved measurement of distance, or imaging telemeters. When single-point telemeters reached a technical maturity in the 1990s, research started to deal with the extension of the measurement on a two-dimensional (2-D) array of points (or pixels).

Reason for interest in image telemetry, also called imaging laser radar and three-dimensional (3-D) imaging, stems from the benefit that distance information adds to the image morphology.

In a parallel but separate branch of technical evolution, multispectral imaging has also matured in the years. This is a segment offering a variety of information about the details contained in a scene, viewed at visible (λ=0.4-0.7 µm) as well as at near infrared (λ =0.8-2 µm) and far infrared (λ =3-12 µm) wavelengths.

Adding the distance information to imaging systems greatly helps solve the problem of Automatic Target Recognition (ATR), both in the military and robotic segments of application [15].

Of course, an imaging telemeter can be developed by combining a normal single-point telemeter and a 2-D scanner device that aims, in a raster or TV-like format, to each pixel of the scene in an ordered sequence. We need also an optical switch in the path outward from the laser, because we shall turn off the beam when the scanner reaches the end of a line and the beam shall flyback to the beginning of the next line. As a scanner, we may use a pair of X and Y mirrors, mounted on galvanometer actuators.

This solution enables us to form an image with typically N=100×100 pixel, a 5°×5° angular field of view, and a reasonable frame repetition rate (10 to 30 frames per second).

The switch could either be an electro-optical cell (KDP or LiNbO$_3$ crystals), or an acousto-optical switch (LiNbO$_3$ or CdTe). Both switches are fast enough (settling time of microseconds) to match the speed required by galvanometer-scan systems. Wavelength of operation of the imaging telemeter can go, depending on the application, from the visible, using diode

3.5 The LIDAR

lasers, to the near and far infrared. The benefit of far infrared wavelengths is the mitigation of scattering attenuation and turbulence effects (Ref.[15] and App.A-3), both in a normal atmosphere and in light fogs and hazes. A few imaging laser radars operating at several wavelengths, from 1.0 to 10 µm, have been evaluated in [15] in good and low visibility conditions.

Because the measurement time is subdivided between N pixels, the theoretical accuracy of the imaging telemeter is worse than that of a still staring telemeter by a factor $\sqrt{N}$ (≈ 100 in the previous example). However, this is not a severe hindrance from the practical point of view. It turns out that the extra information gathered from the frame-size image largely alleviates the resolution or precision requirement as compared with the single-point measurement.

About the measurement configuration, imaging telemeters were first demonstrated with pulsed laser sources [15], so they were time-of-flight pulsed telemeters. The sine-wave configuration has been recently used as well [16] in connection to a 3-D imaging application. The advantage of sine-wave modulation is that we can use an image detector and retrieve the desired phase from each pixel without scanning the scene.

In the experiment [16], a wide-angle laser beam from a 900-mW LED source illuminates a 64×25-pixel scene. The source is amplitude-modulated at 20 MHz, and the measurement is limited to the nonambiguity distance L_{na}=7.5 m. A CCD array (Ref.[2], Sect.9.2) receives the light diffused by the scene through an objective lens conjugating each scene pixel to a pixel in the CCD. In the CCD, the electrode commanding the storage of the photogenerated charge is driven by the modulating waveform at 20 MHz.

The efficiency of charge collection depends on the voltage on the storage electrode, so the useful CCD output current is the product of the received radiant power and of the modulating waveform, or the phase $\phi=2ks$. The phase is read with $<10^{-3}$ resolution resulting in a distance error of a few cm.

3.5 THE LIDAR

Actually not a telemeter, but similar to it (see Fig.1-6), the LIDAR (acronym for Laser Identification Detection and Ranging) is an instrument for the remote sensing of the natural media (mainly the atmosphere, but also water masses). In particular, the LIDAR is routinely used to probe pollutants of the atmosphere and to obtain distance-resolved maps of them for ambient protection (Fig.1-6). The LIDAR has also been successfully used in a number of scientific surveys, including ozone and vertical temperature profile in Antarctica, plankton and chlorophyll on the sea surface, etc. [17-20].

In a LIDAR, we use a pulsed source with high peak power, usually a Q-switched laser, and choose the wavelength in correspondence to an absorption peak of the atmospheric component to be measured. The wavelength range of interest spans from the near-UV (300 nm) to the far IR (10 µm) as opposed to millimeter and centimeter wavelength employed by a microwave radar. Plenty of useful lines are found at optical wavelengths [17], virtually for any kind of gaseous species we can find in the atmosphere. These include, for example nitrogen, oxygen, and carbon oxides; water vapor, ozone, and dangerous species like sulphur oxides, hydrocarbons, fuel gases, chlorofluorocarbons (CFCs), etc.

From one side, the abundance of lines means that hardly a species will be found that cannot be measured in the optical range by a LIDAR technique. On the other hand, the crowding of many lines pertaining to different species will make it hard to distinguish between different constituents. Thus, the LIDAR will make measurements and actually work with unparalleled sensitivity on constituents known to be present in the medium under test, whereas the discrimination between similar species will be poor, in general. With regard to the sensitivity, the actual figures will depend on parameters like species, measurement technique, power source, and wavelength range. The typical sensitivity, expressed as the minimum detectable concentration per unit volume, may range from 10 ppb×m (part per billion, or 10^{-9}, per meter of propagation path) to perhaps 10 ppm×m (part per million per meter).

To tune the wavelength on a wide spectral range and cover most species of interest, the laser may be an Optical Parametric Oscillator (OPO) [21]. The OPO is pumped by a Q-switched laser at fixed λ, like Nd:YAG at 1.06µm or at 0.53µm with second harmonic generation (SHG), and can be tuned from about 0.6 to 3.5µm. Other options for tunable sources in the visible and near-infrared are the alexandrite laser (0.8-0.9µm) and several dye lasers (0.4 to 1.0µm) [22]. For the infrared, CO_2 and CO lasers have been widely used, as well as lead-salt semiconductor diodes.

The optical beam is collimated by the transmitting telescope and sent through the atmosphere along the line of sight. There is no need to place a distant target, except for safety issues. Line of sights of successive pulses are arranged in a raster sequence so that a map of measurement can be constructed. For each pulse, power back-scattered by the atmosphere is collected back by the receiving telescope and is converted into an electrical signal by the photodetector. The time dependence of the photodetector signal is easily translated into a distance dependence of the back-scattering strength.

To calculate the signal received by the LIDAR, we start assuming that the transmitted pulse is very short and write it as P(t)=Eδ(t), where δ is the Dirac-impulse function and E is the energy contained in the pulse. Upon propagation to a distance z=ct, the optical signal is delayed and attenuated, and it can be written as:

$$P(t) = E\ \delta(t-z/c)\ \exp\ -\int \alpha\ dz$$

Here, z/c is the time delay, and $\alpha=\alpha(z)$ is the attenuation coefficient of the atmosphere. The back-scattered power going toward the source in the interval z—z+dz is then:

$$dP(t) = E\ \delta(t-z/c)\ [\ \exp\ -\int \alpha dz\ ]\ f(\theta)\ \alpha\ dz\ d\Omega$$

In this expression, θ and $d\Omega$ are the scattering angle and the solid angle of collection, respectively, and $f(\theta)$ is the scattering function (see Appendix A3-1 and Fig.A3-5).

By integrating $d\Omega$ around $\theta=180°$, the direction along which we collect the back-scattered radiation, we get $\Delta\Omega=\pi D^2/4z^2$ as the total solid angle subtended by the receiving lens of diameter D. Letting $\sigma_{bs}(z)= f(180°)\alpha$ for the back-scattering coefficient, and also considering the delay and the attenuation experienced in the backward propagation, the back-scattered power collected at the receiver is obtained as:

3.5 The LIDAR

$$dP_{bs}(t) = E\, \sigma_{bs}(z)\, (\pi D^2/4z^2)\, \delta(t-2z/c)\, [\exp-2\int\alpha dz]\, dz \qquad (3.33)$$

From Eq.3.33 we can see that, by launching a short δ-like pulse at time t=0, we receive at time t=2z/c an elemental power contribution $dP_{bs}(t)$ in an elemental time dt=dz/c. The intensity $dP_{bs}(t)/dt$ is proportional to the back-scattering of the atmosphere $\sigma_{bs}(z)$, to the exponential attenuation exp -2 ∫ αdz, and to a distance-dependent term $(D/z)^2$.

As an illustration, in Fig.3-25, we report the typical $P_{bs}(t)$ waveform received from a LIDAR instrument. Here, a relatively slow pulse (duration ≈150 ns) is used, and a remote beam stop at L≈570 m cuts off the returning signal for t>4μs. The negative-exponential trend of the back-scattering signal (Eq.3.33) is clearly seen in Fig.3-25.

Now, we may take account of the finite response-time T of the detector and associated circuitry. For simplicity, let us suppose that the detector integrates the signal for duration T, so that its output is:

$$P_{out}(t) = \int_{0-T} dP_{bs}(t)$$
$$= E \int_{0-cT/2} \sigma_{bs}(z)\, (\pi D^2/4z^2)\, [\exp-2\int\alpha dz]\, dz \Big|_{z=ct/2} \qquad (3.34)$$

Eq.3.34 tells us that the finite response-time has the consequence of averaging the distance-dependent factors on an interval cT/2. Equivalently, we can say that the distance resolution in the determination of the distance-dependent waveform is limited to cT/2 or c/4B because of the response time T or of the bandwidth B of the detector.

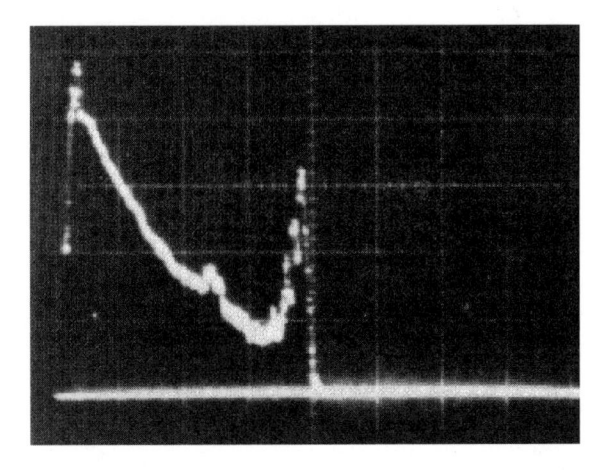

Fig. 3-25 Typical back-scattering waveform $P_{bs}(t)$ received by a LIDAR. The horizontal scale is 1 μs/div, which corresponds to a distance z=150 m/div. The signal is close to a negative-exponential in the first three divisions, and then it is clipped (at about 3.8 div or L≈570 m) because of a remote beam stop. The beam stop gives a strong back-scattering peak followed by a sudden return-to-zero of the signal.

The same argument applies if we take account of the finite duration of the transmitted pulse. In general, we may take the pulse waveform as $\tau^{-1}F(t/\tau)$, where τ is the time duration and $F(t/\tau)$ is the dimensionless pulse-waveform normalized at unit area. Then, it is easy to see that we shall multiply each elemental contribution given by Eq.3.33 by $\tau^{-1}F(t/\tau)$ and integrate it on the time variable $dt=dz/c$ from 0 to ∞ to obtain the received power. This yields:

$$P_{bs}(t) = E\,\tau^{-1} \int_{0-\infty} \sigma_{bs}(z)\,(\pi D^2/4z^2)\,F(t/\tau)\,\exp{-2\int\alpha dz}\;dz \Big|_{z=ct/2}$$

$$\approx E\,\tau^{-1} \int_{0-c\tau/2} \sigma_{bs}(z)\,(\pi D^2/4z^2)\,\exp{-2\int\alpha dz}\;dz \Big|_{z=ct/2} \qquad (3.35)$$

In the second integral of Eq.3.35, we have used the finite duration of $F(t/\tau)$ to reduce the integration limits to 0—$c\tau/2$, the pulse-duration length. Thus, also in this case, the resolution is limited by the value $c\tau/2$ associated with the pulse duration.

We can now focus on the processing of the received signal. Without loss of generality, we can assume $\alpha\approx$const. and write the received signal as:

$$S(t) = E\,\sigma_{bs}(z)\,(\pi D^2/4z^2)\,\exp{-2\alpha z}$$

(in which $z=ct/2$). To extract the physical parameters contained in $S(t)$, we can first remove the inverse square distance dependence by multiplying (electronically) the received signal by a term $(t/t_0)^2$, which increases with the square of time, so that $S(t)(t/t_0)^2=S(t)(z/ct_0)^2$. Then, we take the log of the result and get a signal $L(t)$ given by:

$$L(t) = \ln E\,(\pi D^2/4c^2 t_0^2) + \ln \sigma_{bs}(z) - 2\alpha z$$
$$= \text{const.} + \ln \sigma_{bs}(ct/2) - \alpha c t \qquad (3.36)$$

In this equation, the main dependence of $L(t)$ is in the last term $-\alpha ct$, a negative-going ramp for constant attenuation α along z. The second term $\ln\sigma_{bs}(ct/2)$ is significant at those locations where the scattering σ_{bs} is unusually high and is revealed by a (positive) spike at time $t=2z/c$ (see Fig.3-26). The spike is followed by a negative-going step in $L(t)$ (Δ in Fig.3-26) because, where the local attenuation $\alpha(z)$ is higher than the normal average value, the ramp $-\alpha ct$ has a larger slope.

A sample of LIDAR waveforms is provided in Fig.3-26 to illustrate the previous considerations. The received power $P_{bs}(t)$ waveform is smoothed by the distance correction (Fig.3-26, middle). After taking the logarithm, the final result (bottom diagram in Fig.3-26) is a nearly linear trend with a few superposed peaks, which is indicative of local high-scatter centers.

A very useful way to represent distance-dependent scattering maps is that of using the data of Fig.3-26 (bottom) to intensify the trace (z-axis input) of an oscilloscope with a signal proportional to the scattering at the given distance z.

Maps of scattering intensity versus distance are reported in Fig.3-27. These maps reveal the state of pollution, or of the scattering concentration, as a function of distance or of the elevation angle.

3.5 The LIDAR

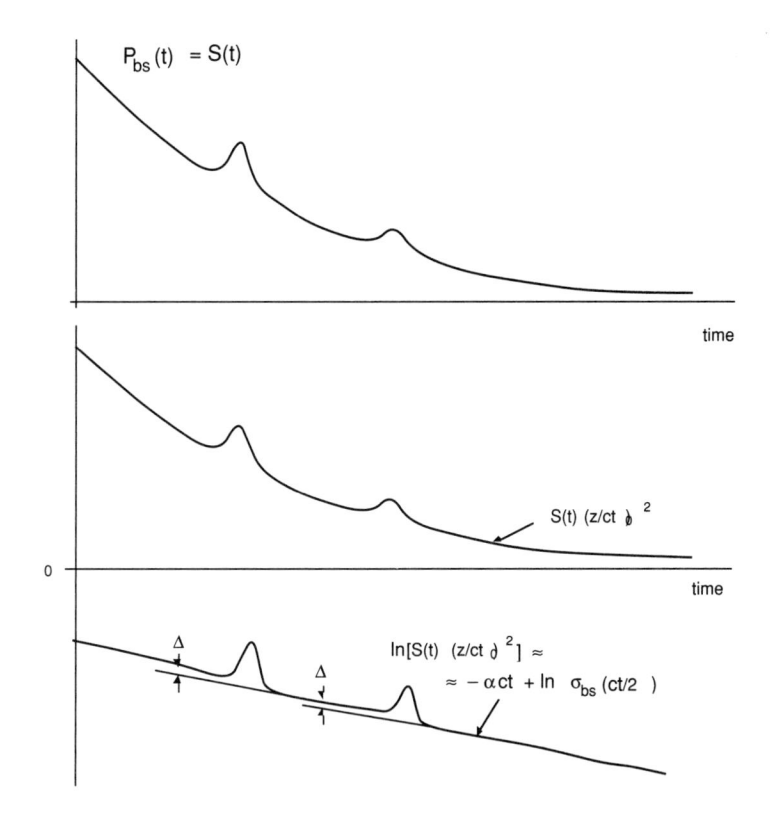

Fig. 3-26 By multiplying the back-scattering power $P_{bs}(t)$ (top) by the square-inverse distance dependence (middle), and taking the logarithm of the result, a signal is generated (bottom) that unveils (i) the attenuation $\alpha(z)$ of the atmosphere (given by the negative slope of the diagram) and (ii) the local back-scattering $\sigma_{bs}(z)$ coefficient of the medium along the propagation path (positive-going spikes).

The LIDAR can be operated in a number of ways [17-20]. The one considered so far, with the optical axis of the illuminating beam parallel (or coincident) to the optical axis of the viewing telescope, is classified as *monostatic* (or *coaxial monostatic*). When the two axes are separated by a significant angle, the LIDAR is called *bistatic*.

Regarding the scattering signal we shall look at, we may use a single wavelength for the transmitter and receiver.
In this case, we have an *elastic-scattering* LIDAR, which supplies information on the distance-resolved concentration of the atmospheric scattering, but basically is unable to sort out the nature of the scatterer.

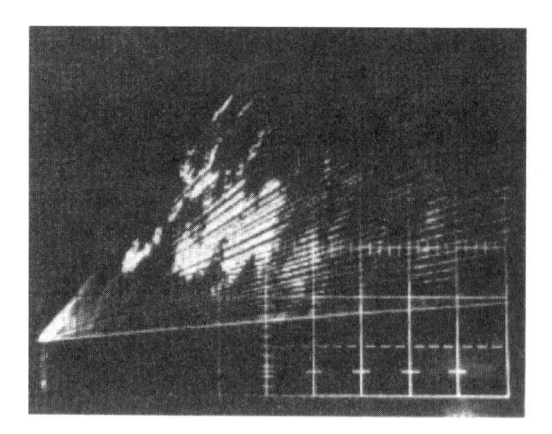

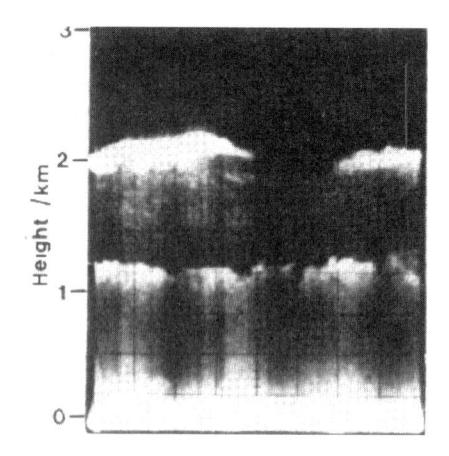

Fig. 3-27 Maps of back-scattering signals detected by a LIDAR. Left: signal as a function of the elevation angle at which the beam is aimed (10 to 60 deg; horizontal scale: 200 meter/div). Right: signal as a function of height above ground when the beam is aimed vertical and time elapses (x-scale: 10 minute/div).

With the advent of tunable laser sources, another mode of operation of the LIDAR has become feasible and is very frequently implemented: the *Differential-Absorption-Lidar* (DIAL). In a DIAL, the scattering measurement is carried out at two wavelengths: one centered at the absorption peak of the species or pollutants we want to monitor and the other just off the peak. Off the peak, absorption is contributed only by the normal atmosphere. Therefore, subtracting the two measurements gives the differential (or extra) absorption caused by the species, a quantity proportional to the concentration (see Eq.A3-3) of it.

The wavelength most suited for the DIAL monitoring of the atmosphere depends, of course, on the considered species, their targeted concentration, and coexisting lines of normal atmospheric constituents. Normal constituents of the atmosphere strongly attenuate the probe beam outside the transmittance windows, so an obvious choice is to operate inside or at the boarder of one of these windows.

For example, to measure methane, we can operate in the near–IR window around 1.6 µm as shown in Fig.3-28. With an OPO pulsed-source pumped by a Q-switched Nd:YAG laser, a sensitivity of 50ppm×m has been reported on a 50 to 300-m range of operation [23]. The unit is mounted on a medium-size van for outdoor operation and can also sense NO, NO_2, O_3, and SO_2 with a 0.25-1 ppm×m resolution on a range of 15-250 m.

Another mode of operation is that of using fluorescence. In the targeted molecules, fluorescence can be induced by an intense beam illuminating the atmosphere with a suitable wavelength, usually in the blue or UV range, typically λ_{ex}<400nm. This source can be obtained by second- or third-harmonic generation (SHG or THG) of a Q-switched solid state laser. Following the excitation, the molecules decay back to the ground state and emit a fluorescence photon, at a specific wavelength $\lambda_{fl} > \lambda_{ex}$, usually in the red-to-near infrared range.

3.5 The LIDAR

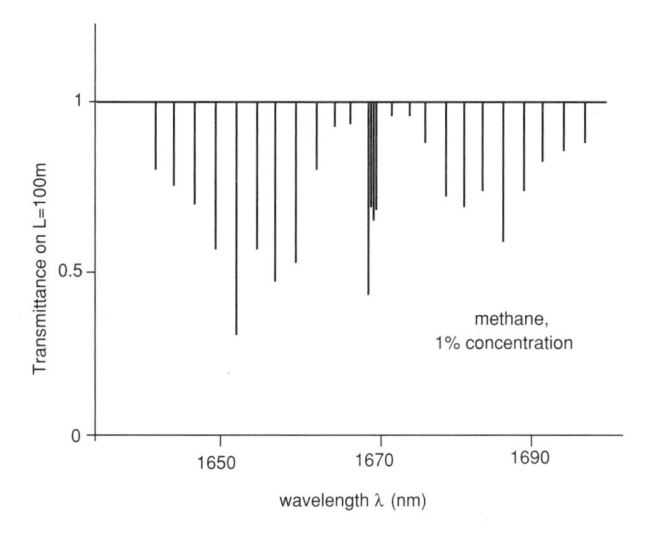

Fig. 3-28 Transmittance of the atmosphere with diluted methane (C=1% in volume) for a 100-m path. Several lines of absorption are shown, for which amplitudes are proportional to concentration C. Other pollutants (hydrocarbons, nitrogen oxides, etc.) have similar lines in the near-infrared spectral range.

It is easy to see that Eq.3.35 can be modified to describe the *fluorescence-LIDAR* by changing $\sigma_{bs}(z)$ in $\sigma_{fl}(z)$, the fluorescence cross-section, and $\exp-2\int\alpha dz$ in $\exp-\int(\alpha_{fl}+\alpha_{ex})dz$ to account for the different wavelengths λ_{fl}, λ_{ex} of the go-and-return propagation.
Thus, a signal of the form L(t) as in Eq.3-36) can still be obtained, which gives the concentration of the targeted species.

Because the fluorescence signal is at a wavelength different from the illuminating wavelength, the contribution of a given species or pollutant is much better distinguished as compared to a scattering signal where a lot of species do contribute to absorption.

A variant of the two-wavelength LIDAR is the *Raman LIDAR*, where we look to the Raman, wavelength-shifted, back-scattering signal. In water, the Raman signal is strong and nearly constant with wavelength so that its variations can be used to monitor pollutants (especially hydrocarbons at the surface) and particulate. In the atmosphere, the Raman signal is weak, but we can take advantage of its dependence from temperature to make a height-resolved air-temperature measurement.

REFERENCES

[1] T. Bosch and M. Lescure (editors), "*Selected Papers on Laser Distance Measurement*", SPIE Milestone Series, MS115: Bellingham, WA, 1995.
[2] S. Donati, "*Photodetectors*", Prentice Hall: Upper Saddle River, 2000.
[3] R.G. Dorsch, G. Hausler, and J.M. Hermann, "*Laser Triangulation: Fundamental Uncertainty in Distance Measurements*", Applied Optics, vol.33 (1994), pp.1306-1314.

[4] S. Donati and A. Gilardini, *"Advanced Techniques of Laser Telemetry"*, Selenia Technical Review, vol.8 (1982), pp.1-12.
[5] E. Gatti and S.Donati, *"Optimum Signal Processing for Distance Measurements with Lasers"*, Applied Optics, vol.10 (1971), pp.2446-2451.
[6] S. Donati, E. Gatti, and V. Svelto, *"Statistical Behaviour of the Scintillation Counter: Theory and Experiments"*, in Adv. in Electronics and Electron Physics, vol.26, 1969, edited by L. Marton, pp.251-307.
[7] K.P. Birch and M.J. Downs, *"An Updated Edlen Equation for the Refractive Index of Air"*, Metrologia, vol.30 (1993), pp.155-162.
[8] D.E. Smith and M. Zuber, *"The Mars Observer Laser Altimeter Investigation"*, J. Geophys. Res., vol.97 (1992), pp.7781-7797; also Proceedings ODIMAP III, edited by S. Donati, Pavia, 20-22 September 2001, pp.1-4; also: Physics Today (Nov.1999) pp.33-35.
[9] G. Bianco and M.D. Selden, *"The Matera Ranging Observatory"*, in Proceedings ODIMAP II, edited by S. Donati, Pavia, 20-22 May 1999, pp.253-260; *"The Matera Ranging Observatory: Observational Results"*, in Proceedings ODIMAP III, edited by S. Donati, Pavia, 20-22 September 2001, pp.90-98.
[10] B. Querzola, *"High Accuracy Distance Measurement by Two-Wavelength Pulsed-Laser Source"*, Appl. Optics, vol.18 (1979), pp.3035-3047.
[11] S. Donati (editor), *"Proceedings 'ODIMAP II, 2nd Int. Conf. on Distance Measurements and Applications"*, Pavia, 20-22 May 1999, LEOS Italian Chapter, 1999; and*"Proceedings 'ODIMAP III, 3rd Int. Conf. on Distance Measurements and Applications"*, Pavia, 20-22 September 2001, LEOS Italian Chapter, see leos.unipv.it/odimap.
[12] K. Kowalski, *"Nuclear Electronics"*, J. Wiley and Sons: Chichester, 1975.
[13] C.J. Oliver, *"Pulse Compression in Optical Radar"*, IEEE Trans. on Aerospace Systems, vol.AES-15 (1979), pp.306-324.
[14] F.K. Hulme, B.S. Collins, and G.D. Constant, *"A CO_2 Rangefinder using Heterodyne Detection and Chirp Pulse Compression"*, Opt. Quant. Electr., vol.13 (1981), pp. 35-45.
[15] G.R. Osche and D.S. Young, *"Imaging Laser Radar in the Near and Far Infrared"* , Proc. of the IEEE, vol.84 (1996), pp.103-125.
[16] R. Lange and P. Seitz, *"Solid-State Time-of-Flight Range Camera"*, IEEE Journal of Quantum Electronics, vol.37 (2001), pp.390-397.
[17] W.B. Grant, *"LIDAR for Atmospheric and Hydrospheric Studies"*, in Tunable Laser Applications, edited by F.J. Duarte, M. Dekker, Inc.: New York, 1995, pp.213-306.
[18] R.M. Meausers, *"Laser Remote Sensing"*, J.Wiley: New York, 1984.
[19] L.Pantani and G.Cecchi, *"Fluorescence LIDAR in Environmental Remote Sensing"*, in Optoelectronics for Environmental Science, edited by A.N. Chester and S. Martellucci, vol.131, Plenum Press: New York 1990.
[20] E. Galletti, R. Barbini, A. Ferrario, et al. *"Development of a CO_2 Pulsed Laser for Spaceborn Coherent Doppler LIDAR"*, 5th Conf. on Coherent Laser Radar, edited by C. Werner and J.W. Bilbo, SPIE Proceedings vol.1181 (1989), pp.113-124.
[21] R.L. Sutherland, *"Handbook of Nonlinear Optics"*, M. Dekker, Inc.: New York, 1996, see Chapter 3: Optical Parametric Generation, Amplification and Oscillation.
[22] J. Hecht, *"The Laser Guidebook'"*, McGraw-Hill: New York, 1986.
[23] R. Barbini, *"Measurement of Physical Parameters of the Atmosphere with a LIDAR'"*, Proceedings Elettroottica'90, Milan (Italy), 16-18 Oct.1990, pp.335-342.

CHAPTER **4**

Laser Interferometry

*I*nterferometry is one of the most powerful tools in measurement science. It provides a very high sensitivity, unequalled by any other techniques, and is widespread in virtually every segment of engineering and physics. Interference of light was one of the phenomena marking the birth of optics and challenging the seventeenth-century scientists to explain the nature of light, notably Newton and Huygens [1]. In the nineteenth century, though a good understanding of interference was reached, interferometry had few notable applications because of the limited coherence length of the sources then available [1]. With the advent of lasers in the 1960s, the full capability of this powerful technique was finally unleashed. Many impressive sensors and measurement devices were developed: submicrometer displacement interferometers, laser Doppler velocimeters, and gyroscopes, just to name the big engineering achievements of the electro-optical science.

In this chapter, we adopt a bottom-up presentation, first describing the operation of the instrument with a minimum of basics. Then, we expand the coverage by considering the many facets of modern interferometry. Among these, we will treat optical configurations, fundamental limits of sensitivity and accuracy, different schemes of superposition, and all the

90   Laser Interferometry   Chapter 4

topics of general relevance. Other chapters in this book will deal specifically with applications of interferometry to velocimetry, gyroscopes, and fiberoptic sensors.

As a primer, let us now briefly outline the basic paradigm of interferometry. A suitable source, usually a laser, is used to direct an optical beam into the experiment or the physical ambient to be sensed (Fig.4-1).

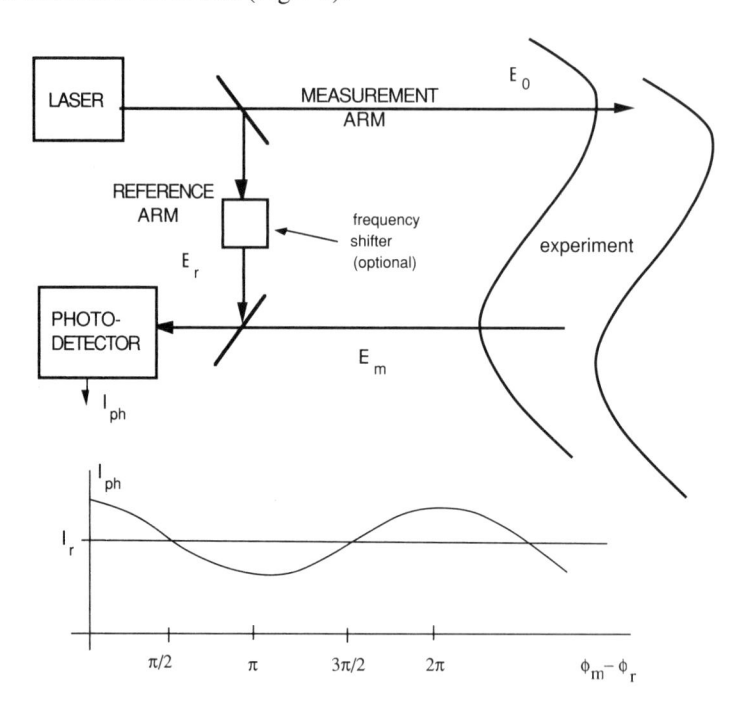

Fig.4-1   The conceptual scheme of interferometry (top) and the signal out from the photodetector (bottom)

A fraction of the beam is kept in the instrument to be used as a reference. Upon returning back from the experiment, the measurement beam is superposed to the reference beam onto a photodetector.

If E_m and E_r are the optical fields of the two beams, the photogenerated current I_{ph} obtained as a signal out from the photodetector is:

$$I_{ph} = \sigma \ |E_m + E_r|^2 = \sigma \ |E_m \exp i\phi_m + E_r \exp i\phi_r|^2 \qquad (4.1)$$

where the fields $E_{m,r} = E_{m,r} \exp i\phi_{m,r}$ have been represented as rotating vectors with amplitude $E_{m,r}$ and phase $\phi_{m,r}$, the bars (|..|) denote modulus, and σ is a conversion factor between current and squared field (or power). Developing Eq.4.1 yields:

$$I_{ph} = \sigma [E_m^2 + E_r^2 + 2E_m E_r \, \text{Re}\{\exp i(\phi_m - \phi_r)\}]$$

$$= I_m + I_r + 2(\sqrt{I_m I_r}) \cos(\phi_m - \phi_r) \qquad (4.2)$$

Here, $I_{m,r} = \sigma E_{m,r}^2$ are the currents that the measurement and reference fields would provide individually, and the last term is the interference signal.

As we can see, the interference signal is proportional to the cosine of the optical phase shift $\phi_m - \phi_r$ between measurement and reference beams. If we keep the reference beam fixed and unaffected by any disturbance (ϕ_r =const.), while the measurement beam carries the phase shift collected in the propagation, the signal from the photodetector is a quantity measuring the optical phase ϕ_m.

Usually, the beam propagates to a distant reflector and back, totaling a length 2s which amounts to a phase shift $(2s/\lambda)2\pi = 2ks$, where λ is the wavelength, 2π is the number of radians per wavelength and $k = 2\pi/\lambda$ is the wave-number.

Apart from a constant term $I_m + I_r$, Eq.4.2 tells us that the interferometric signal is of the type cos 2ks (the cosine-signal), and its amplitude is proportional to $\sqrt{(I_m I_r)}$, the geometric mean between reference and measurement powers.
This amplitude is $2\sqrt{(I_m I_r)}/I_m$ times larger that of the signal I_m being detected alone. Thus, in the interferometer, there is a sort of internal gain, $G = 2\sqrt{(I_r/I_m)}$, of photodetection. This is a consequence of the beam superposition, resulting in a coherent scheme [3] of detection, a scheme further possessing the very welcome property of working at the quantum-noise limit of the received signal. Because of that, the sensitivity of the interferometer is very good, which is detailed later in this chapter.

4.1 OVERVIEW OF INTERFEROMETRY APPLICATIONS

Fig.4-2 presents a pictorial overview of the big tree of interferometry. Applications may be divided in the two branches of industrial/avionics and scientific measurements.
In the first, we find the well-established applications to the following:
- Mechanical metrology (positioning of tool machines, mechanical workshop calibrations, and measurements)
- Fluid anemometry (the so-called laser-Doppler velocimetry)
- Vibrometry (rotating machinery diagnostics and control)

In avionics, a very wide segment of application is covered by the gyroscope. This is an interferometer and senses the Sagnac phase shift induced by rotation. The gyroscope is the heart of inertial sensors like Inertial Navigation Units (INU), Heading Reference Systems (HRS), and attitude and spin control systems.

About scientific uses, a number of amazing examples of application have been successfully reported.

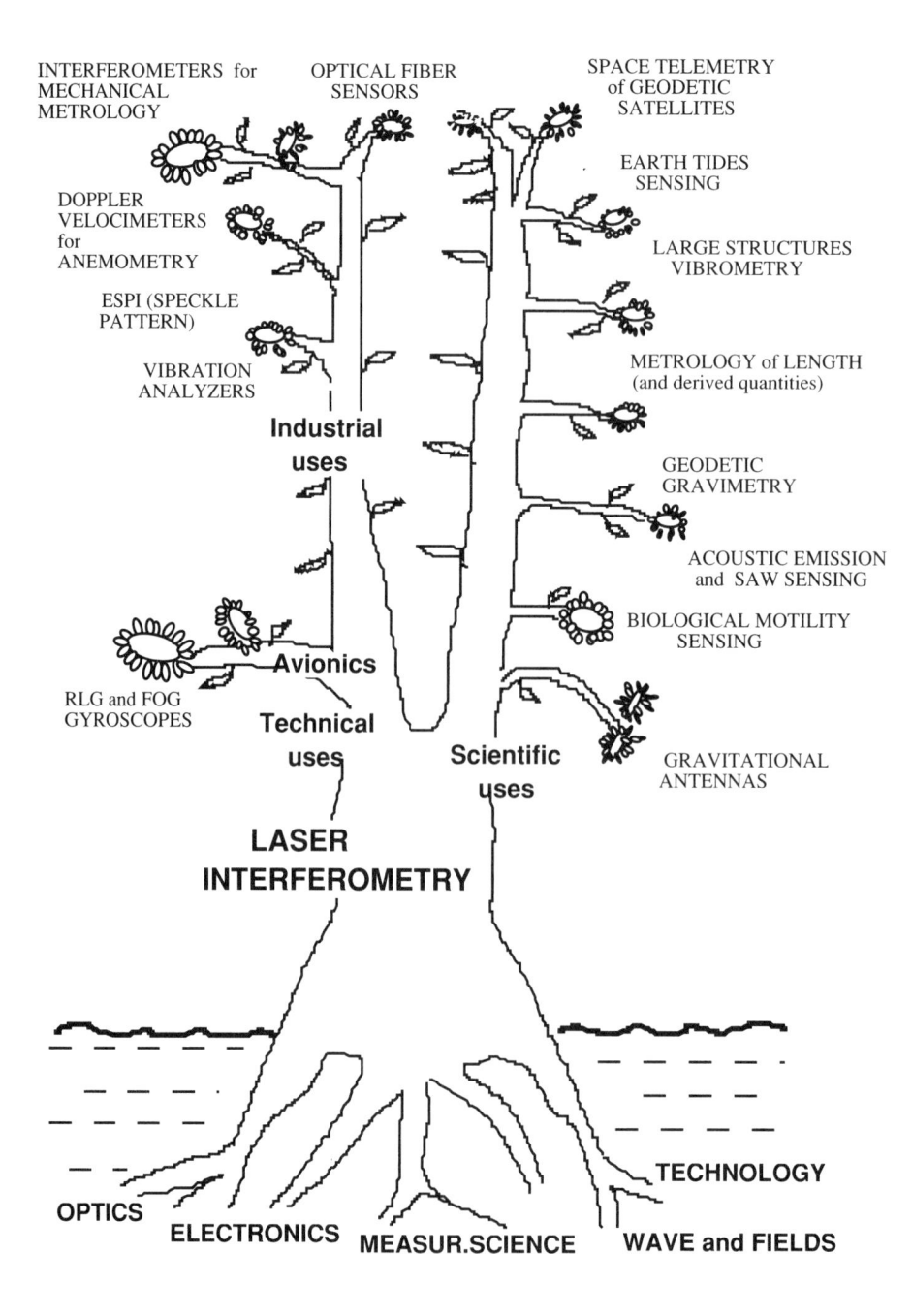

Fig.4-2   The tree of interferometry applications

4.2 The Basic Laser Interferometers

Classifying them as in Fig.4-2 according to the distance or the baseline on which the measurement is performed, we find the following:
- Terrestrial tide monitoring (for geodesy)
- Satellite-to-subsatellite displacement pickup (to unveil *mascons* – mass concentrations useful to oil-field discovery)
- Long-distance vibrometry (for testing the structural integrity of towers, bridge and dams) and the metrology of length and derived quantity (gravimeter, thermal expansion of materials)
- Pickup of biological motility (respiration sounds and cells motility for fertility)
- Pickup of surface acoustic waves (study, design, and testing of SAW devices and AOM modulators).

Most recent on the long baseline end, we find the biggest interferometers ever built (or, near to completion), the Virgo and LIGO (Fig.1-11). These experiments are gravitational antennas based on an interferometer. They are aimed to detect very minute (10^{-18}m) bumps on a target mass acting as an antenna, coming from very remote (Megaparsec) sources of gravitational waves – giant-star collapses.

4.2 THE BASIC LASER INTERFEROMETERS

The term *laser interferometer* is commonly used to indicate an electro-optical instrument capable of performing the measurement of displacement of a corner cube target, with a fraction-of-wavelength resolution and a 10^{-6} accuracy or precision. The target is usually mounted on the carriage of a tool machine or the like, and the distance range covered by the measurement is up to the scale of meters.

Soon developed after the invention of the He-Ne laser in 1961, the laser interferometer has become a well-established instrument for measurements and especially *calibration*. The precision is exceptionally good because the scale factor is directly connected to the wavelength of the laser. This quantity can be easily stabilized to a relative accuracy better than $\Delta\lambda/\lambda \approx 10^{-8}$ in commercial, cheap He-Ne frequency-stabilized units (see Appendix A1).

Diode lasers are rapidly catching up in this application, but, to obtain at least $\Delta\lambda/\lambda \approx 10^{-6}$, we should have either a DBR laser diode, presently not yet available as standard products at visible wavelengths (610-680nm) where operation is preferable because of safety, or an external-grating stabilized diode, which is rather expensive.

Therefore, in the following, we refer to interferometers based only on the He-Ne laser.

In next sections, we describe the two main approaches considered the best for the development of practical laser interferometers: the *dual beam* and the *two frequency*.
Both approaches are used in commercial products, and give rise also to several variants in respect to frequency stabilization of the laser and/or electronic processing of the interferometric signals.

4.2.1 The Two-Beam Laser Interferometer

It may be instructive to follow the evolution of the Michelson interferometer, see Fig.4-3a, up to the modern two-beam laser interferometer of Fig.4-3c. As soon as the laser was available for use as the source in a Michelson interferometer, displacement measurements on a mirror target at a distance s_m have been performed with $\lambda/2$-resolution. This is done simply by counting the transitions (e.g., the upgoing semiperiods) of the cosine signal out of the photodetector, as shown in Fig.4-1. However, to avoid losing counts while moving the mirror target in a displacement measurement, angular alignment has to be ensured, which is a critical point that makes the setup impractical.

The solution was to adopt the Twyman-Green interferometer using corner cubes in place of mirrors. In a glass cube corner with dihedral angles of 90°, an impinging beam is returned parallel to the incidence, irrespective of the incidence angle (or of corner cube tilt). The lateral displacement of the returning beam is beneficial because the returning beam does not fall in the laser any more, like with mirrors. In fact, back-injection disturbs the laser oscillation, and may spoil the coherence length, generally. With the Twyman-Green interferometer, both back injection and alignment criticality are eliminated. It suffices that the measurement corner cube is moved with a small transversal error, less than the beam size, for a proper superposition with the reference to take place on the photodetector.

Also, a glass-cube beamsplitter is better than a thin beamsplitter because it is easier to mount in a wavelength-stable holder, and the reference corner cube can be cemented to one of its faces for increased stability of the reference-arm length s_r. The cube beamsplitter has the semitransparent surface fabricated by multilayer dielectric film, giving the desired splitting ratio at the wavelength of operation and has antireflection coating on the entrance/exit surfaces.

Last, to work also with large arm mismatch s_m-s_r (that is, at sizeable distance) without losing the beating signal (Fig.4-1), we shall use a long-coherence length source, such as a frequency-stabilized laser, which also brings about the benefit of wavelength calibration.

Thus, we come to the configuration of Fig.4-3b, which is actually working and satisfactory in some applications. In it, the source is normally a small-power ($P_{laser} \approx 1mW$), compact He-Ne unit, which is frequency-stabilized by means of one of the several possible schemes (see App.A1.2).

Entering the optical interferometer, the beam is divided by the cube beamsplitter BS into two equal parts with powers P_r and P_m, one reflected and one transmitted toward the reference and the measurement corner cubes. The corner cubes reflect the beams back to the beamsplitter, where the beams are recombined after propagation on a distance $2s_r$ and $2s_m$, respectively.

After recombination, we get two equal beams, one of which is collected by the photodiode PD. The signal out from the photodiode is given by:

$$I_{ph} = {\textstyle\frac{1}{2}} I_m + {\textstyle\frac{1}{2}} I_r + \sqrt{I_m I_r} \cos 2k(s_m - s_r) = I_m [1 + \cos 2k(s_m - s_r)] \qquad (4.3)$$

where $I_m = \sigma P_m$ and $I_r = \sigma P_r$ are the currents separately given by measurement and reference

4.2 The Basic Laser Interferometers

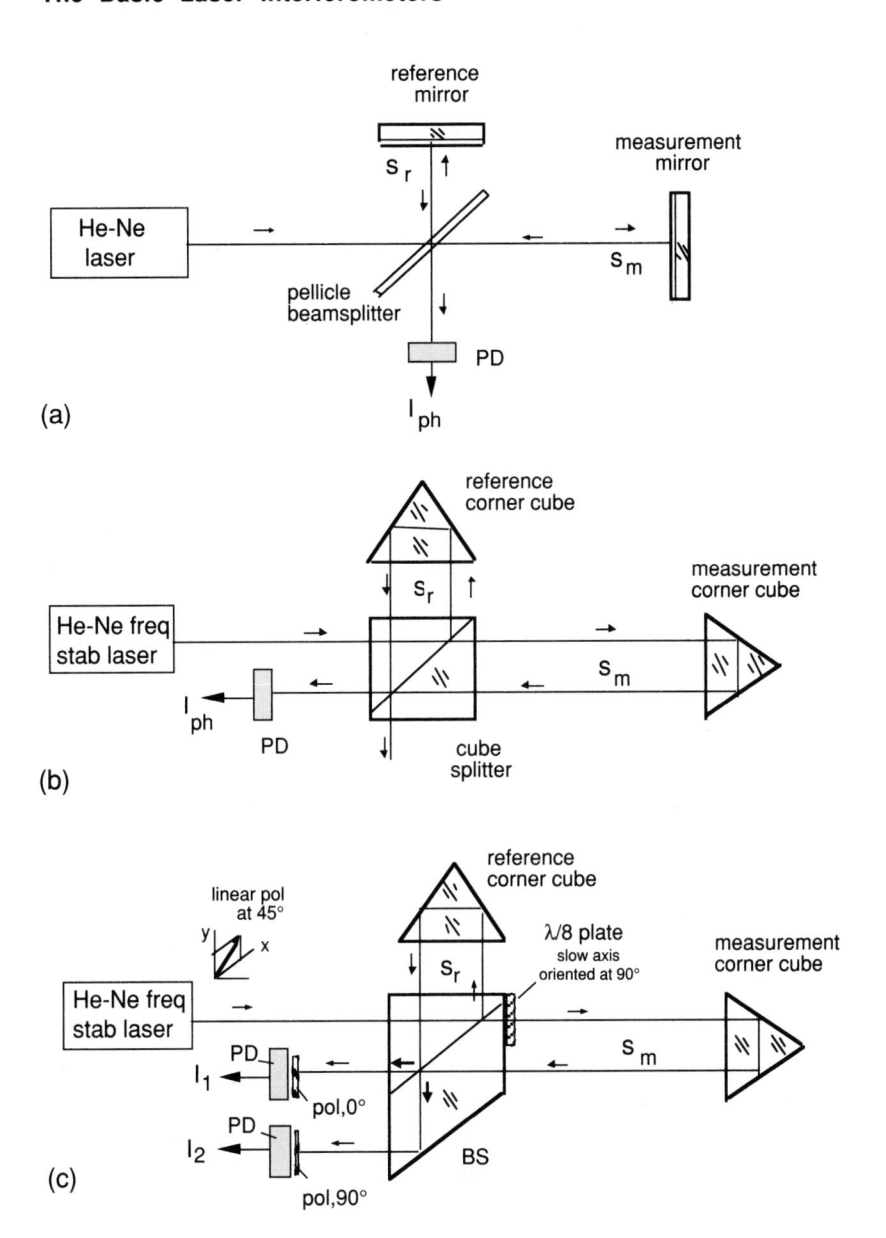

Fig. 4-3 Evolution from the Michelson interferometer (a) to the Twyman-Green interferometer (b) using angular-alignment tolerant corner cubes, and to (c), the modern two-beam laser interferometer measuring the displacement with its sign.

beams, taken equal in power, $P_m = P_r = P_{laser}/2$.
The interferometric signal $\cos 2k(s_m-s_r)$ is superposed to a dc term (of amplitude unity) and can be processed in several ways, analogue and digital.

For example, if we want to measure small displacements with subwavelength amplitudes, we can try stabilizing the quiescent working point of the interferometer at the so-called *half-fringe* point, that is midway between the maximum and minimum of the cosine function. To do so, assume $2ks_m << 1$ (or, $s_m << \lambda/4\pi$) and take $2ks_r = \pi/2$. Then, Eq.4.3 becomes:

$$I_{ph}/I_m = 1 + \cos(2ks_m - \pi/2) = 1 + \sin 2ks_m \approx 1 + 2ks_m \qquad (4.4)$$

or, we get a linear replica of the displacement s_m waveform directly from the photodetector current. This is the analogue vibrometer-regime of operation, discussed in detail later in this chapter.

On the other hand, if we want to measure fairly large displacements, we can count the interferometric signal transitions. Looking at the signal $\cos 2k(s_m-s_r)$, which can be written as $\cos(2ks_m - \phi)$ with $\phi = $const., we get a positive-going transition at each period of $2ks_m$ or for each variation $\Delta 2ks_m = 2\pi$, that is, for $\Delta s_m = \lambda/2$ being $k = 2\pi/\lambda$.
Thus, displacement s_m is measured in steps of $\lambda/2 = 316$ nm, which is a digital readout without limit in dynamic range other than the accuracy of the wavelength yardstick.

However, a serious drawback of the configuration in Fig.4-3b is that it cannot distinguish increasing from decreasing displacement because the cosine function is even. The other unused output from the beamsplitter [seeFig.4-3b] cannot help either because the signal it supplies is $1 - \cos 2k(s_m - \pi/2)$, i.e., has the same ambiguity. Therefore, the single-channel scheme of Fig.4-3b only works correctly with monotonic displacements $s_m(t)$.

To recover the sign of displacement, we double the measurement channel and take advantage of polarization diversity, as in the *two-beam interferometer* of Fig.4-3c. The linear-polarization of the laser mode is adjusted to enter the beamsplitter at 45° with respect to the incidence plane. Thus, two independent components, linearly polarized at 0° and 90°, are provided. Both components share the same physical path, down the corner cubes and back.

By inserting a $\lambda/8$-retardance plate at the beamsplitter output, as in Fig.4-3c, we add an extra path in one component (e.g., the one at 90°, parallel to plate slow axis) with respect to the other component (at 0°, the plate fast axis). In the go-and-return path, the total path length is $\lambda/4$, or $\pi/2$ in phase. Then, after the (polarization-independent) recombination at the beamsplitter, we have two beams feeding each photodetector PD1 and PD2. The beatings of 0° and 90° components are obtained by polarizers which are oriented at 0° and 90°. Taking into account the 1/2-attenuation introduced by the polarizer, the photodetected currents are written as:

$$I_{ph1} = \tfrac{1}{4}I_m + \tfrac{1}{4}I_r + \tfrac{1}{2}\sqrt{I_m I_r}\cos 2k(s_m - s_r) = \tfrac{1}{2}I_m[1 + \cos 2k(s_m - s_r)]$$

$$I_{ph2} = \tfrac{1}{4}I_m + \tfrac{1}{4}I_r + \tfrac{1}{2}\sqrt{I_m I_r}\cos 2k(s_m - s_r + \lambda/8) = \tfrac{1}{2}I_m[1 - \sin 2k(s_m - s_r)] \qquad (4.5)$$

4.2 The Basic Laser Interferometers

Now, with the pair of *cosine* and *sine* interferometric signals, the argument $s_m - s_r$ of the trigonometric functions can be recovered without ambiguity. One possible processing strategy is illustrated in Fig.4-4.

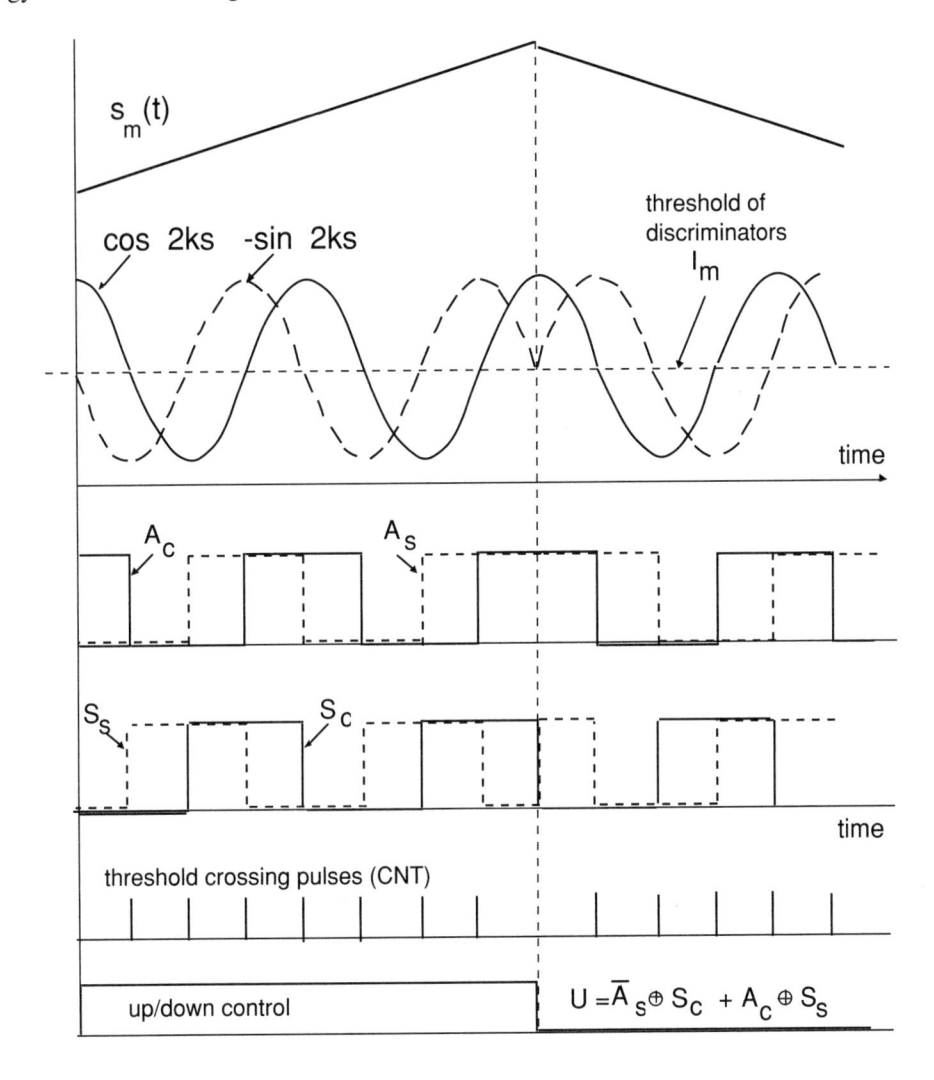

Fig. 4-4 Signal handling in a two-beam interferometer. Cosine and sine signals are passed through discriminators, logic signals of amplitude and slope are obtained, and an appropriate logic combination of them acts as the up/down command at the counter input.

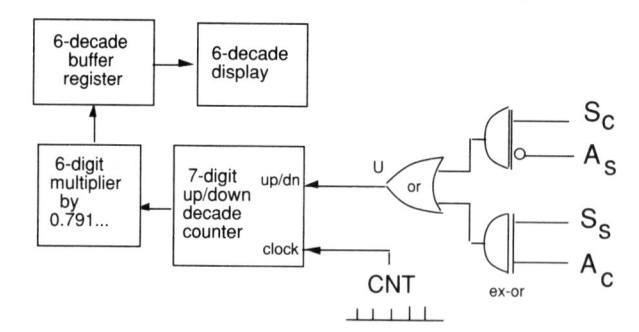

Fig. 4-5 The up/down 7-decade decimal counter stores the displacement under measurement in λ/8 units. A 6-digit multiplier brings the reading to metric units for the buffer register and the display.

Each signal out from the photodetector is passed through a comparator, with the threshold placed at the average ($1/2\ I_m$) photocurrent so that *amplitude* logic signals A_C and A_S are generated (A=1 for high amplitude, A=0 for low amplitude).
Next, *slope* logical signals S_C and S_S are generated for each channel, according to the sign of the signal slope (S=1 for positive slope, S=0 for negative).

Count pulses are generated for the switching of both sine and cosine signals, for example, by differentiating the discriminator's outputs (either upgoing or downgoing) and by rectifying the obtained pulses so that they are all positive. Because the sine and cosine signals provide four switchings per λ/2-period, one count represents a λ/8 (=0.079 µm for a 633-nm He-Ne laser) increment of displacement s_m.

Count pulses are sent to an up/down counter as clock (or count) pulses, and a logic combination of amplitude and slope enables the up/down input in the correct direction. With reference to Fig.4-4, let us suppose that the displacement first increases and then decreases. At the first switching, it is A_S=0 and S_C=0, at the second A_C=0 and S_S=1, at the third A_S=1 and S_C=1, and at the fourth A_C=1 and S_S=0, whereas in the decreasing portion the reverse is true.

The appropriate logic for the up-count command is therefore written as:

$$U = S_C{}^*A_S{}^* + A_C{}^*S_S + S_C A_S + A_C S_S{}^* = S_C{}^* \oplus A_S + A_C \oplus S_S \qquad (4.6)$$

where * stands for the negation (or complement) and $\oplus$ for the exclusive-or logic operation. The calculation of U is easily implemented by using a few logic gates, as shown in Fig.4-5. The counter is normally a 7-decades counter, so that it can nominally accommodate readings of displacements from 0.079 µm up to 0.79 m.

Now, after the counter, we shall bring the counter content to decimal. This requires a 7-digit multiplier, with the other number entering the multiplier being the length corresponding to one count, e.g., 791....., usually settable in a register for calibration.

4.2 The Basic Laser Interferometers

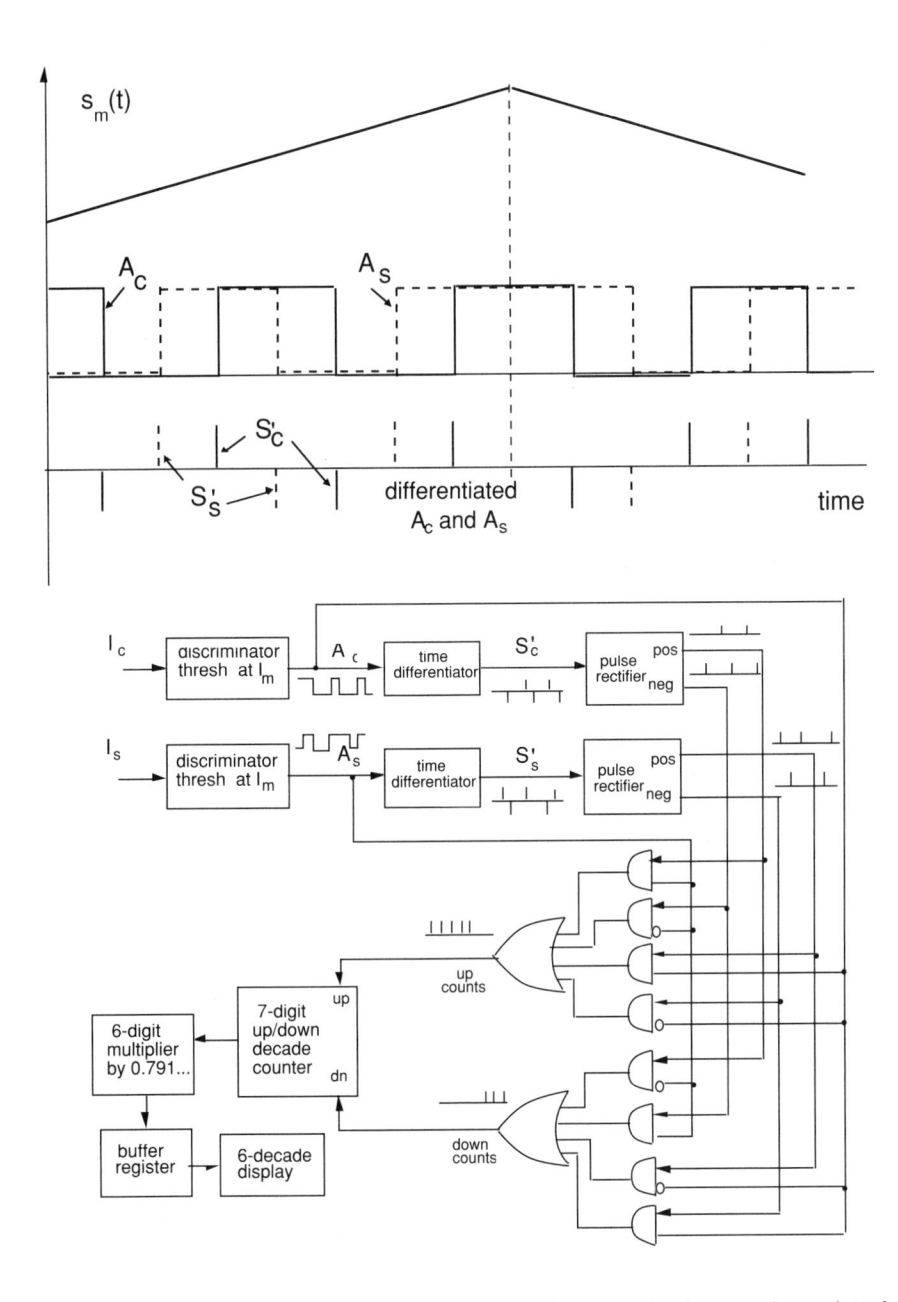

Fig. 4-6 Another more robust strategy for pulse counting in a two-beam interferometer. Pulses S_c' and S_s' from the switching edges of the discriminator outputs are processed directly by the AND/OR gates, and the results are counted in a 7-digit up/down counter.

The result is stored in a 6-digit buffer register, which has 0.1-µm as the least-significant digit and 0.1-m as the most-significant digit. Therefore, a displacement up to ±0.999999 m can be measured. The buffer is connected to the numeric display. This allows the display to be refreshed periodically (e.g., every 0.1 s) and to avoid the last digits from changing fast and becoming unreadable during count buildup. It is not advisable to go beyond 6 digits because of several errors intervening at the ppm (or 10^{-6}) resolution level.

The processing scheme described previously works well with well-behaved displacement waveforms, but has a weak point. The slope logical signals $S_{S,C}$ are obtained necessarily by a high-pass circuit that rejects quasi-dc components of $I_{ph1,2}$. We can indeed design it with a very low frequency cutoff covering the practical range of $s_m(t)$ expected frequency content, but can never reach zero. Therefore, an objectionable possible loss of counts for very-slow waveforms $s_m(t)$ remains.

This drawback can be overcome with the more robust processing scheme of Fig.4-6. Here, we perform the up/down logic operation directly on the switching signal edges, derived from the amplitude comparator outputs, which make the same logic function as described by Eq.4.6, but without the need for a slope logic signal. Accordingly, the scheme of Fig.4-6 removes the low-frequency constraints on $s_m(t)$.

The position of the discriminator threshold, nominally set at $1/2\, I_m$ in electronic processing, is a critical factor to the performance of the two-beam interferometer. Because the photodetected signals I_{ph1} and I_{ph2} (see Eq.4.5 and Fig.4-4) can go well below the average value simultaneously, the strategy to derive the threshold is not trivial, unless an assumption on the expected behavior of $s_m(t)$ is made. Perhaps, the less stringent assumption is that, because of ambient-related vibrations, $s_m(t)$ undergoes several λ-cycles in a medium-term period. Then, a good practical choice is to take the threshold as the semisum $1/2\, (I_{ph1}+I_{ph2})$ of the interferometric signal amplitudes, integrated on say, a 1-s period.

Regarding the high-frequency cutoff of the circuits, if we take all of them having at least (a specified) bandwidth B, the interferometer can correctly count λ/8 steps of displacement at a rate of one per period 1/B. The maximum velocity allowed to target displacement is then: $v_{target}= (λ/8)B$. For B=10 MHz and λ/8=0.079 µm, we get the quite satisfactory value $v_{target}= 0.8$ m/s.

When going to the field, the two-beam interferometer is recognized as satisfactory by users. However, it reveals the following drawbacks that need to be eventually corrected:
- If the optical beam is interrupted, counts are lost and the measurement run shall be repeated
- High-frequency (EMI) disturbances sometimes leads to counting errors.
- Ambient-induced vibrations can occasionally lead to incorrect counts.

For the first point, a strategy is to monitor the amplitudes of signal I_{ph1} and I_{ph2} and give a warning (with a panel-mounted LED) when both fall below, say 5% of the time-averaged threshold.

The second and third points call for a good shielding from electromagnetic, as well as mechanical disturbances, but are difficult to be eliminated intrinsically. The basic reason is

4.2 The Basic Laser Interferometers

that the two-beam interferometer works on threshold crossing by *baseband signals*. Thus, all the disturbances falling in a spectral range 0-B are indistinguishable from signals and, even if small in amplitude, may lead to incorrect switching of the discriminators.

4.2.2 The Two-Frequency Laser Interferometer

In this approach, we take advantage of the frequency difference between the two-orthogonal polarization modes emitted by a Zeeman-stabilized He-Ne laser (see App.A1.1.3) to make the interferometric signal available on a *carrier* frequency instead in baseband.
The basic schematic of the two-frequency interferometer is reported in Fig.4-7. The Zeeman laser is one with an axial magnetic field and generates two counter-rotating, circularly polarized modes, that are spaced in frequency by f_1-f_2 =5 MHz approximately. The circular polarizations are transformed into linearly polarized waves by a quarter-wave retarder inserted at the laser output.

Before entering the optical interferometer, we first get a reference signal I_{phR} at the frequency difference f_1-f_2. To do so, a small fraction η of the beam power (for example, $\eta \approx 5\%$) is deviated to the reference photodiode PD_R, in front of which a polarizer oriented at 45° is placed. The photocurrent is therefore written as (cf. with Eq.4-2):

$$I_{phR} = \tfrac{1}{2}\,\eta\,\{I_1+I_2+2\sqrt{I_1 I_2}\cos[2\pi(f_1-f_2)t+\varphi]\} = \eta\,I_{av}\,\{1+\cos[2\pi(f_1-f_2)t+\varphi]\} \quad (4.7)$$

where $\varphi = \varphi_1 - \varphi_2$ is the relative phase of the two modes, and we let $I_1=I_2=I_{av}$ for simplicity.

In the optical intereferometer, we use the usual corner cubes and a Glan-cube beamsplitter (or polarization splitter), one with the property of dividing linear polarizations. The Glan is usually made by calcite, a well-known birefringent crystal. The two halves of the Glan are cut along different orientations of the crystal, to have a large difference of index of refraction for the two polarizations. In this way, at the separation surface, incidence is beyond total reflection angle for one polarization (the perpendicular to incidence plane), which is accordingly reflected, while it is below the total reflection for the other polarization (the parallel to incidence plane), which is transmitted.

Thus, by virtue of the Glan-cube splitter, we are able to direct one polarization (the perpendicular, at frequency f_1) along the reference path, and the other (parallel, f_2) to the measurement path. On the return path, the two polarizations recombine in a beam directed on the photodetector (Fig.4-7).

In the propagation to the corner cubes and back, the two waves have cumulated a phase shift $2ks_m$ and $2ks_r$ and, accordingly, the signal I_{phM} from the photodetector PD_M is:

$$I_{phM} = \tfrac{1}{2}(1-\eta)\,\{I_1+I_2+2\sqrt{I_1 I_2}\cos[2\pi(f_1-f_2)t+\varphi+2k(s_m-s_r)]\,\}$$

$$\cong \tfrac{1}{2} I_{av}\,\{1+\cos[2\pi(f_1-f_2)t+\varphi_t+2ks_m]\} \quad (4.8)$$

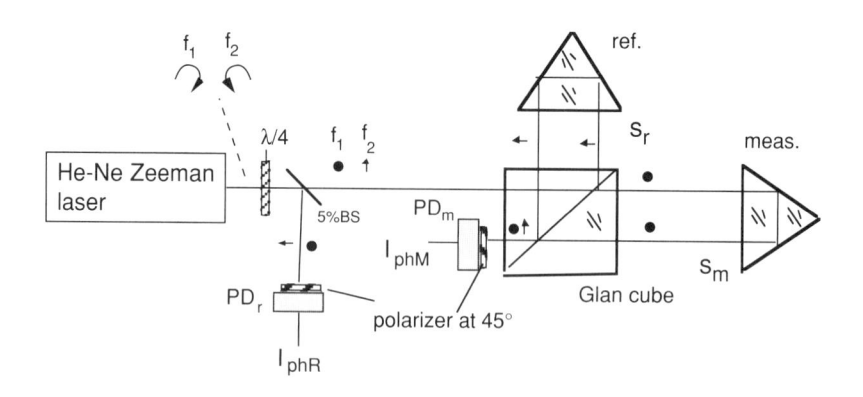

Fig. 4-7 The two-frequency interferometer uses the two modes of a Zeeman laser, separated by $f_1-f_2 \approx 5$ MHz and with orthogonal polarizations. The polarization Glan beamsplitter sends one mode to the reference path and the other mode to the measure path. Upon recombination, the phase difference $2k(s_m-s_r)$ is found on the carrier frequency f_1-f_2.

where we have taken $I_1 = I_2 = I_{av}$, neglected the small η, and let $\varphi_t = \varphi + 2ks_r$ for the phase term, which is a constant if $2ks_r$ is kept constant.

Eq.4.8 shows that the quantity to be measured, $2ks_m$, is now a phase superposed on a carrier at frequency f_1-f_2. To recover it, we have several possible choices.

For example, we may want to go back to analog interferometric signals. Then, we can electrically mix the photodetected signals I_{phM} and I_{phR} and generate sum-and-difference frequencies. The difference frequency term is $\cos 2k(s_m-s_r)$ and is recovered with a low-pass filtering of the mixer output. Also, by phase-shifting the reference I_{phR} by $\pi/2$ and mixing it again to I_{phM}, we obtain $\sin 2k(s_m-s_r)$. This approach works, but does not fully exploit the advantages of the two-frequency arrangement, which are made clear later.

A better signal processing is illustrated in Fig.4-8. The photodetected signals, I_{phM} and I_{phR}, are squared by comparators and, from the transitions (e.g., positive going), one pulse per period is obtained. We use two counters, one for the measurement and the other for the reference channel. The counters are allowed to count freely, and their content is transferred to buffers registers by a gate pulse G1. This pulse has a period T, which is the renewal time of measurement (typically 0.1 s). The reason for the buffer registers is that pulses may occur simultaneously in the two channels, both running close at $f_1-f_2 \approx 5$ MHz, and they cannot be handled directly by an up/down counter (unless we use a complicate logic circuit).

To find the content of the counters at the end of period T, we use Eqs.4.7 and 4.8 with $(d/dt)\varphi_t = 0$, i.e., assume a still reference-arm. Also, we recall that frequency is the time derivative of the phase, $f = (1/2\pi)d\varphi/dt$, and that counters perform an integration operation, which yields the integer part of the quantity $C = \int_{0-T} f \, dt$.

Thus, we obtain:

4.2 The Basic Laser Interferometers

$$C_R = \int_{0-T} (1/2\pi)(d\varphi/dt) \; dt = \int_{0-T} (f_1-f_2) \; dt = (f_1-f_2) \; T,$$

$$C_M = \int_{0-T} [(f_1-f_2)+(1/2\pi) \; d/dt(2ks_m)] \; dt = \int_{0-T} [(f_1-f_2)+2v_m/\lambda] dt$$

$$= (f_1-f_2) \; T + 2\Delta s_m/\lambda \qquad (4.9)$$

In this equation, $v_m=(d/dt)s_m$ is the velocity of the measurement-arm corner cube. The term v_m/λ in the second line of Eq.4.9 can be recognized as the change in frequency induced by the Doppler effect, usually written as $\Delta f=(v/c)f$. The term Δs_m in the last line of Eq.4.9 is the displacement occurred in the period T, in units of $\lambda/2$.

The subtractor (Fig.4-8) makes the difference of the buffer contents, and the result is:

$$S = C_M - C_R = 2\Delta s_m/\lambda \qquad (4.10)$$

Last, the content of the subtractor is transferred to the main output register by the gate pulse G2 (delayed respect to G1, see Fig.4-8, to allow for subtraction time). In the output register, the $2\Delta s/\lambda$ counts from the subtractor are added (with sign) to the content already present there. Thus, the content at time t represents the counts in $\lambda/2$ units of $s=\int_{0-t}\Delta s$, which is the total displacement from time t=0 (the general reset) to current time t.

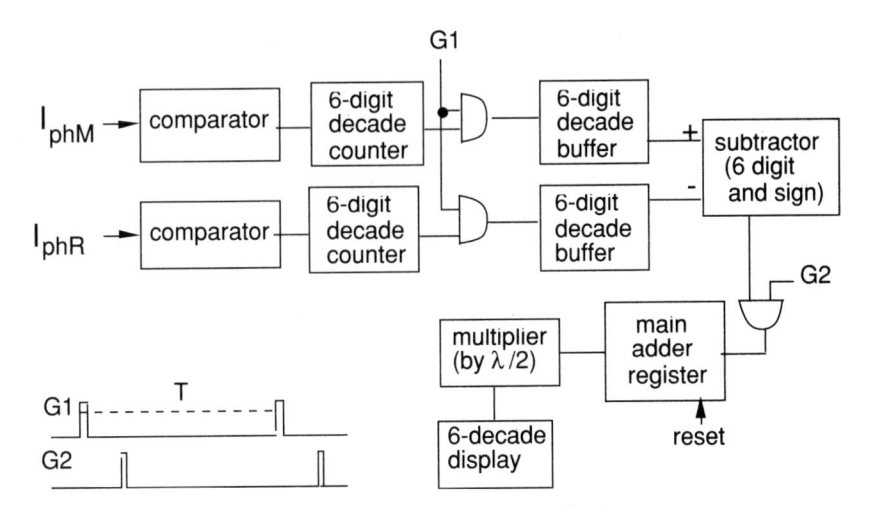

Fig. 4-8 Signal processing for the two-frequency interferometer. Measurement and reference signals are counted for a period T, and then are transferred to buffers. The results are subtracted to yield 2kΔs. A main adder gives 2ks, and a multiplier converts the counts in decimal units for the display.

The measurement resolution is now $\lambda/2$, but we can bring it readily to $\lambda/4$ by doubling the pulses obtained from the comparators (that is, by adding the negative-going transitions in both channels, properly rectified).

In a two-frequency interferometer, the maximum speed of the corner cube we can follow with correct counts is again $v=(\lambda/4)B$, where B is the bandwidth of processing circuits and, also, of the signal I_{phM} and I_{phR} filtering around the carrier $f_1-f_2=5$ MHz.

Allowing for a reasonable fraction of the carrier frequency, i.e., B=1MHz, we get a speed limit typically about $v=0.15\mu m \cdot 10^6 s^{-1}=0.15$ m/s.

The advantages of the two-frequency method over the two-beam method can be summarized as follows:

- The threshold of discrimination can be placed at zero because the dc components of I_{phM} and I_{phR} can be removed without loss of information.
- The rejection of electromagnetic disturbances is much better, at the 5±1 MHz frequency, as compared to the baseband.
- Any beam loss or interruption is readily detected, looking at the I_{phM} signal amplitude.

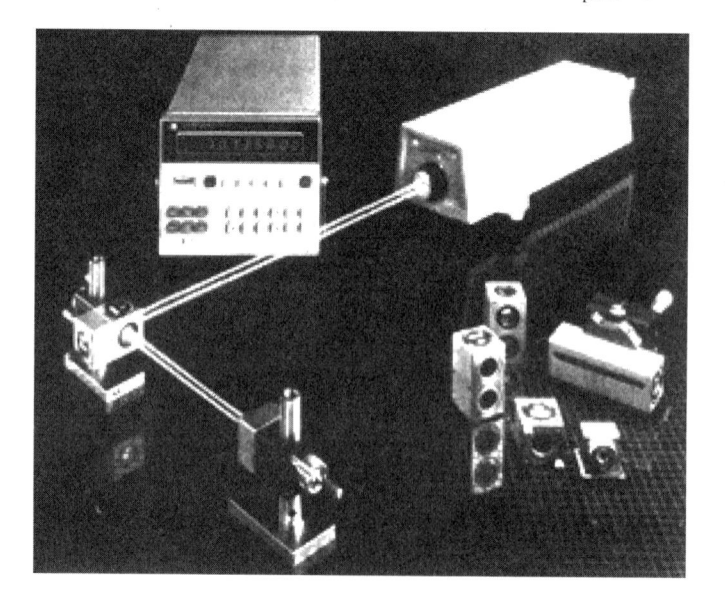

Fig. 4-9 A typical laser-interferometer instrument (from [2], by courtesy of Hewlett Packard)

Based on the two-frequency approach, a number of commercial laser interferometers have been developed and are available from several vendors worldwide. Perhaps the most popular has been the 5525 Hewlett-Packard *"Laser Interferometer"* (Fig.4-9). This instrument was first released in 1967 and can nowadays been considered a great commercial success of electro-optical instrumentation.

4.2 The Basic Laser Interferometers

4.2.2.1 Extending the displacement measurements to nanometers

Also, we can go well beyond the $\lambda/4$ resolution and down to the ultimate limits of sensitivity by means of an electrical phase-shift measurement performed directly on the I_{phM} and I_{phR} waveforms, as illustrated in Fig.4-10.

To do so, it is customary to electrically mix both signals with a local oscillator at a frequency $f_{LO}=(f_1-f_2)-f_{IM}$, where f_{IM} is the intermediate frequency (usually $f_{IM} \approx 10\text{-}100$ kHz) at which we perform the electrical phase-shift measurement. A feature interesting to be noted is that the phase shift $2ks$ is transferred from the f_1-f_2 carrier to a lower intermediate frequency f_{IM} with no error.

By measuring the phase shift with a 100-interval subdivision of the 2π angle, as provided by a counter at 100 times the intermediate frequency, we obtain a $\lambda/200$ resolution from the interpolator.

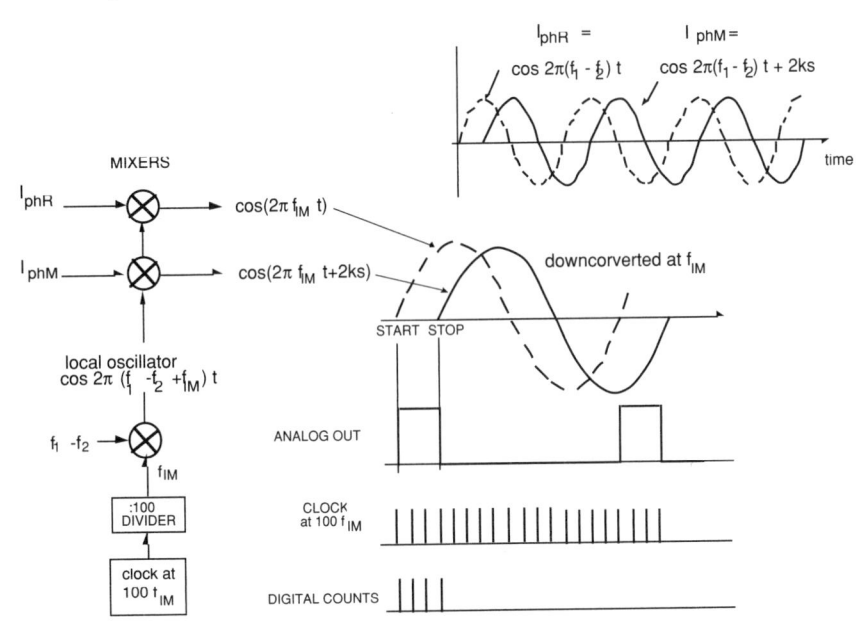

Fig. 4-10 The concept of interpolation to extend the resolution to $\lambda/200$ in a two-frequency interferometer.

The output of the interpolator is in digital format and is suitable for supplying the user with two additional decades of counts, that represent, say 0.01 and 0.001 µm.

Alternatively, we may also amplify and low-pass filter the analog output of Fig.4-10 to make available an analog-format signal that supplements the digital readout for small displacements.

It is interesting to remark that a nm-resolution of a displacement measurement is seldom required in machine-tool applications and, if we actually want to attain it, operation of the instrument on an antivibration table is mandatory.

On the other hand, measurement of vibrations (that is, periodic phenomena) with amplitudes down to nanometers and picometers really makes sense and has been actually developed (see next sections). The oscillating character of the phenomenon can be used to develop much simpler and effective approaches.

We may also wonder if the two-beam interferometer can be extended to operation in the nanometer range. Trying to resolve a submicrometer displacement Δs from the two signals given by Eq.4.5 leads to an ambiguity. Indeed, the signals X=1+cos2ks and Y=1-sin2ks lead to a second-degree equation in the argument 2ks.

To remove the ambiguity, we need at least to triplicate the measurement channels of the interferometer. One of the possible methods [6] is illustrated in Fig.4-11. It consists of segmenting the propagating beam aperture in three sectors, each detected by a separate photodetector. On the propagation path, a segmented retardance plate is inserted, with 0, 120, and 240 degrees of retardance. Thus, the signals are:

$S_1 = I_{ph1}/I_{ph0} = 1 + A\cos 2ks$, $S_2 = I_{ph2}/I_{ph0} = 1 + A\cos(2ks + \pi/3)$, $S_3 = I_{ph3}/I_{ph0} = 1 + A\cos(2ks + 2\pi/3)$

where A allows for nonunitary fringe contrast. By combining the signals, we have:

$$S_2 + S_3 - 2S_1 = -(2+\sqrt{3})A\cos 2ks, \quad S_2 - S_3 = A\sin 2ks_1 \qquad (4.11)$$

Now, the term 2ks can be solved for as: $2ks = -\text{atan}(2+\sqrt{3})(S_2-S_3)/(S_2+S_3-2S_1)$. This calculation requires a small microcomputer receiving the signals $S_{1,2,3}$ through an A/D converter interface. In laboratory experiments [6], the approach has demonstrated to reach a 1-nm resolution easily.

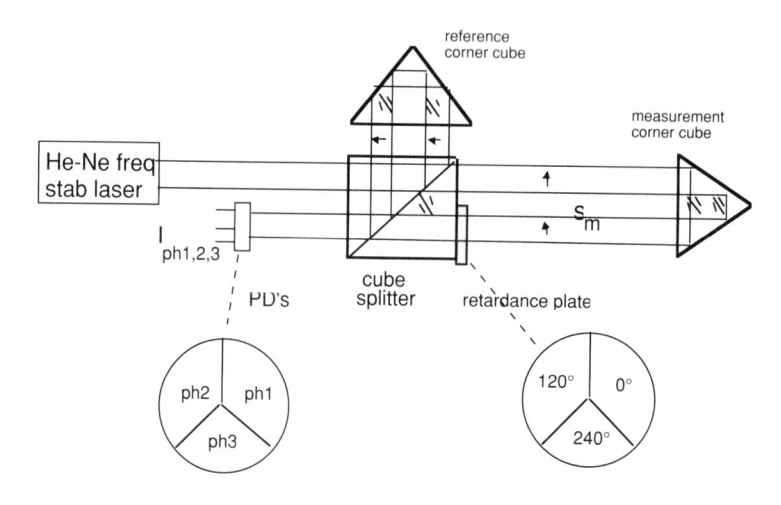

Fig. 4-11 Extension of the two-beam interferometer to nm-resolution.

4.2 The Basic Laser Interferometers

4.2.3 Measuring with the Laser Interferometer

The laser interferometer, in any of the previously described implementations, measures the *incremental displacements* of a target (the retroreflector). As such, it requires an initial reset (zeroing) of the counters with the target positioned on the mechanical zero of the system under control.

The target distance is dynamically measured by counts and requires that regular counts are developed and that no beam interruption occurs at any time during target movement. Interruption of the beam while the target is moving results in a loss of counts, and thus a wrong measurement that cannot be corrected anymore. In this case, we shall go back to the initial reset and restart operation.

For the same reason, immunity of the instrument to Electro-Magnetic Interference (EMI) is of the utmost importance, because we have no way to discriminate spurious EMI pulses from true displacement pulses. Interference may result in a wrong measurement with no warning to the user.

Another source of spurious counts comes from microphonics. This term means sensitivity to ambient-related mechanical disturbances that induce vibrations into the measurement path. Vibrations generate up/down counts that have nominally zero average and are harmless, but may introduce an error when the content is sampled instantaneously.

In interferometers with fraction-of-λ resolution, vibration-induced spurious counts are kept low or negligible by mounting the measuring setup on a suitable antivibration table. On the other hand, if we deal with instruments reaching the nanometer resolution, mechanical isolation is more demanding and may require a special design or arrangement of the experimental layout.

Another specific requirement of the laser interferometer is the need for a corner cube reflector.

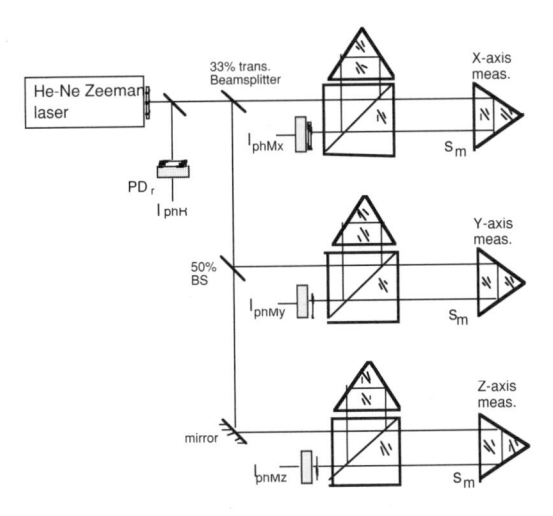

Fig. 4-12 Three-axis extension of the interferometer measurement

The corner cube is mounted on the moving carrier of the tool-machine under measurement to serve as the mechanical reference. The device is generally compact (typically, 1-2 cm in diameter), yet mounting it in the experiment means that the measurement is invasive. Additionally, we need to keep it reasonably clean in the surrounding environment.

In the practical operation of the instrument, several systematic errors may occur.
One is the *cosine* error, arising because the beam wave vector $\underline{k}$ and the motion vector $\underline{s}$ are not strictly parallel, but form an angle α_{ks}. As the target is moved, the displacement s is measured as if it was s cos α_{ks}. To adjust the parallelism of $\underline{k}$ and $\underline{s}$, despite vector $\underline{k}$ being immaterial, we must check alignment of an optical component (usually the front surface of the corner cube) at the beginning and at the end of a displacement stroke [3].

Another systematic error comes from minute spurious reflections of the reference and/or the measurement beams on a parasitic optical path. If ε is the fraction of optical power leaking to the unwanted path, either reference or measurement, a cyclic error $\sigma_{cyc}= (\lambda/16)\sqrt{\varepsilon}$ is generated. This *cyclic error* is a ripple error, affecting the true value of measurement with an amplitude σ_{cyc} and a periodicity $\lambda/2$ versus displacement [4].

Using good engineering practice, the previous errors can be minimized and the laser interferometer can be used in a number of circumstances, as detailed in the following sections. Frequently, the ultimate limits of performance discussed in Sect.4.4 are actually reached.

4.2.3.1 Multiaxis extension

Tool machines with numerical control usually require *three-axis* positioning. We do not need to triplicate the interferometer to perform a three-axis measurement, however.

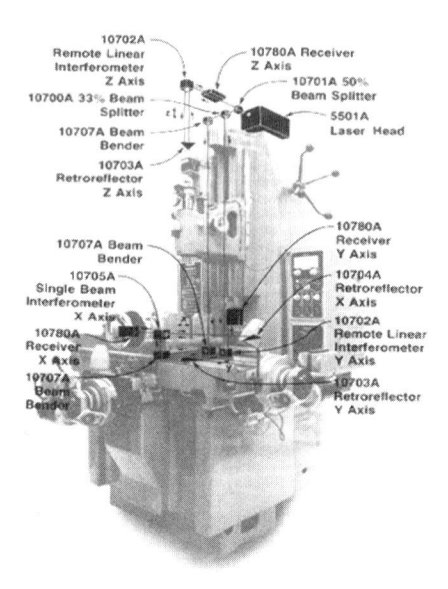

Fig. 4-13 A universal tool machine equipped with a three-axis laser interferometer (from Ref. [3], by courtesy of Hewlett-Packard)

4.2 The Basic Laser Interferometers

To save parts, we may share the laser, taking advantage of the power being still adequate when reduced to 1/3, and use for each axis the two-frequency scheme, with the same reference for all axes, as illustrated in Fig.4-12.

A universal tool machine equipped with the three-axis laser interferometer is shown for illustration in Fig.4-13.

4.2.3.2 Planarity measurement

Another measurement commonly performed in mechanical workshops is *planarity*.

We may use the scheme shown in Fig.4-14 where, by aid of a beamsplitter and mirror combination, we direct two beams to a couple of corner cubes mounted on a square.

The square with the corner cubes is moved along the plane to be tested, whereas the reference mirror and beamsplitter assembly is kept fixed.

The measurement photodiode receives the recombined beams returned from the corner cubes, and it supplies a signal containing the interferometric phase difference $2k(s_{upper}-s_{lower})$, where s_{upper} and s_{lower} are the path lengths of the upper and lower beams.

When the corner cube square is moved and it remains parallel to itself, s_{upper} and s_{lower} increase by the same amount, and no count is developed. However, if a tilt α is suffered, the path length difference $2\alpha L$ is counted in units of $\lambda/4$, with L being the distance between the corner cubes (Fig.4-14). Thus, the angular resolution corresponding to a single count is $\alpha_{1c} = \lambda/8L$. Taking L=100 mm and being λ=0.633 µm, we get α_{1c} =0.8 10^{-6} rad or 0.16 arc-sec, a very good resolution indeed.

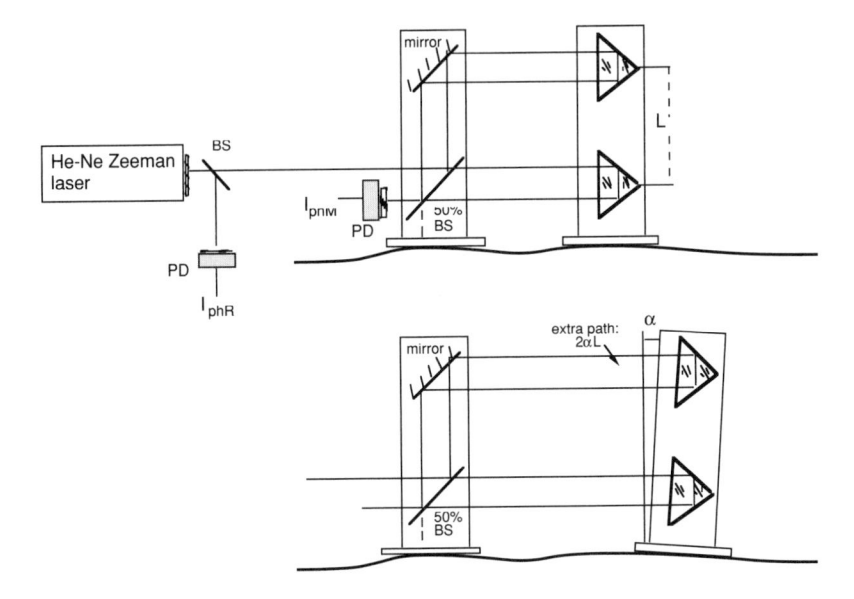

Fig. 4-14 Planarity measurement scheme

As a further step, we may divide the surface under test in individual cells, typically of the size of the corner cubes square basement, let's say $L_B=100$ mm, and collect the measurement counts N_c developed along the surface.

From these data, the profile $z(x,y)$ of the surface under test can be easily computed as a function of spatial coordinates x and y simply by adding step by step the vertical displacement $\Delta z = \alpha L_B = N_c L_B \lambda/8L$ from one cell to the next.

The vertical resolution we obtain in the planarity deviation of the surface under test is $\Delta z = N_c \lambda/8$ (for $L_B = L$), or, $\lambda/8 = 0.08$ μm using a He-Ne laser interferometer.

4.2.3.3 Rectangularity measurement

A variant to the planarity scheme is obtained by adding a 90° deviation of the beams by means of a pentaprism, as indicated in Fig.4-15.

The pentaprism has a dihedral angle of $\gamma=45°$, and it is easily seen to deviate the incident beam by $2\gamma=90°$ irrespective of the incidence angle. With respect to using a mirror oriented at 45°, the pentaprism eases the alignment and introduces no error associated to incidence angle.

The counts developed in the arrangement of Fig.4-15 are related to the planarity errors, plus the deviation error from rectangularity of the surface under test on which the mobile corner cube square slides on.

Again, the resolution in rectangularity is the same as for planarity, $\alpha_{1c} = \lambda/8L$, while the precision, is of course, affected by the pentaprism dihedral angle error $\varepsilon = \gamma - 45°$ (typically <1 arcsec in best units).

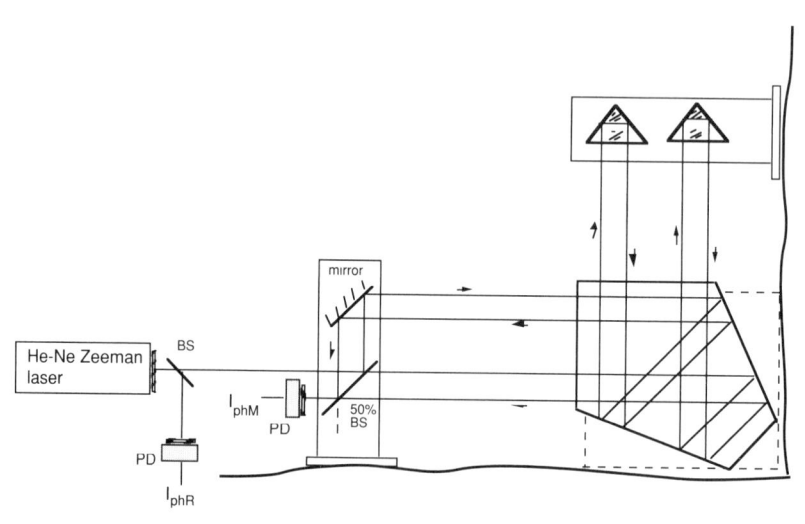

Fig. 4-15 Rectangularity measurement scheme

4.2 The Basic Laser Interferometers

4.2.3.4 Extending the measurement on diffusing targets

Still another extension of the basic interferometer, either dual-beam or two-frequency, is that of attempting operation on *diffusing surfaces*. To improve collection of light returning to the measuring section, we need a focusing lens in front of the diffusing target that is used in place of the corner cube, as shown in Fig.4-16.

Though the scheme in Fig.4-16 is probably not the optimal approach, it is instructive to analyze it to point out the problems we face using a diffuser. On the target, the spot focused by the lens ideally has a (radial) dimension $w_t = \lambda/\pi NA$ (see Eq.A2-10 in Ref.[5]), where NA is the numerical aperture of the lens used by the incoming beam, given by NA= w_1/F, where F is the lens focal length and w_1 the input beam spot size.

Radiation rediffused by the target follows Lambert's law (see [5], Appendix A2) and the amount of it superposed to the reference path spot of area πw_1^2 on the photodiode PD_m is readily found as $P_m = (1/\pi) P_0 (\pi w_1^2)/F^2 = P_0 (w_1/F)^2$ where P_0 is the power from the laser leaving the beamsplitter. Because it is usually $(w_1/F)^2 \ll 1$, first we shall expect a weak signal back to the readout section of the interferometer.

This is not a serious problem, however, because the interferometer detection is a truly coherent detection scheme (see [5], Sect.8.1) and, accordingly, it works at the quantum limit of the received power. When the target moves, a signal of the form $I_{phM} = I_m + I_r + 2\sqrt{I_m I_r} \cos 2ks$ is developed, that can be processed like in a conventional interferometer, even though its amplitude $2\sqrt{I_m I_r} = 2\sqrt{[\sigma P_0 \, \sigma P_0 (w_1/F)^2]} = 2 \, \sigma P_0 w_1/F$ is attenuated by w_1/F with respect to the signal from a corner cube target.

Of more concern is the range of displacement allowed for the measurement before a large speckle error is suffered.
When the diffuser moves along the z-axis appreciably, it runs out of focus with respect to the (fixed) lens.

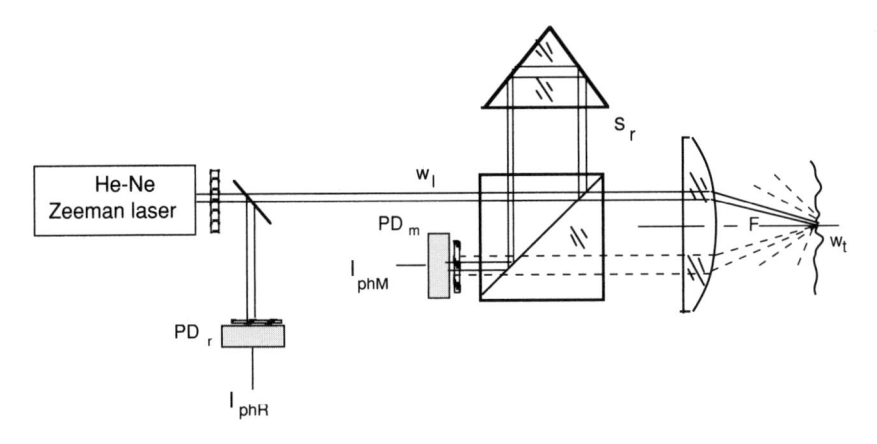

Fig. 4-16 Modification of the optical setup for operation on a diffusing target

The size of the spot w_1 then increases and, much worse, the sample of random elemental areas contributing to the returned field also changes, adding a random phase error ϕ_{sp} to the expected phase shift 2ks.

This error is called the speckle-pattern phase error, and it will be analyzed in detail in Section 5.1. For the moment, let us just mention that ϕ_{sp} becomes comparable with 2π as the displacement Δz brings the target out of focus of the lens. As a consequence, the diffuser statistical sample generating the speckle is changed over, and the phase error is $\approx 2\pi$.

For this to happen, we need $\Delta z = w_t/(w_1/F)$ where $w_1/F = NA$ is the numerical aperture used by the beam of spot size w_1. Using $w_t = \lambda/\pi NA$ in this expression, we get $\Delta z = \lambda/\pi NA^2$ as the dynamic range of maximum displacement for a λ-error. Even at a small NA, the resulting range is clearly much less than the tens of cm to meters we require in applications of the interferometer to mechanical metrology.

On the contrary, the dynamic range allowed by the speckle statistics is adequate for vibration measurements, where the amplitude of the periodic displacement is much smaller (typically ranging from nm to hundreds μm).

4.3 PERFORMANCE PARAMETERS

Irrespective of the working principle, interferometers can be characterized by a set of parameters describing their ability to detect target movement in a variety of situations.

The first is the minimum detectable amplitude of displacement that is set either by a threshold of discrimination or by the noise inherent to detected radiation or to circuits.

Second, the maximum amplitude of displacement that is accommodated is limited by the dynamic range of circuits.

Third, we find a maximum (and eventually, a minimum) frequency of detectable displacement due to the frequency cutoff B of the circuits handling the interferometric signal.

Last, we shall consider a fourth limitation—the speed or Doppler limit—due to the fact that even a slow displacement can develop a high-frequency interferometric signal. Indeed, if the target moves at a speed v, the interferometric signal is cos2ks=cos2kvt, or its frequency content is $f=2kv/2\pi=2v/\lambda$. This quantity shall be less than the circuit bandwidth B, indicating that large and fast displacements have a $v < \lambda B/2$ limitation in performance.

We can represent these four parameters in a diagram of performances (Fig.4-17) which is also called Wegel's or lemon diagram because of its shape. Each square in the diagram is a 10 by 10 factor of covered performance. The more decades found in the diagram, means the more the instrument performs.

Of course, several other parameters are necessary to complete the interferometer performance description. The most common are the following:

(i) The accuracy of the readings (depending on the calibration and stability of the wavelength).
(ii) The dependence on temperature and atmospheric pressure and composition
(iii) The dependence on quasistatic magnetic fields.
(iv) The immunity to EMI disturbances.

4.4 Ultimate Limits of Performance

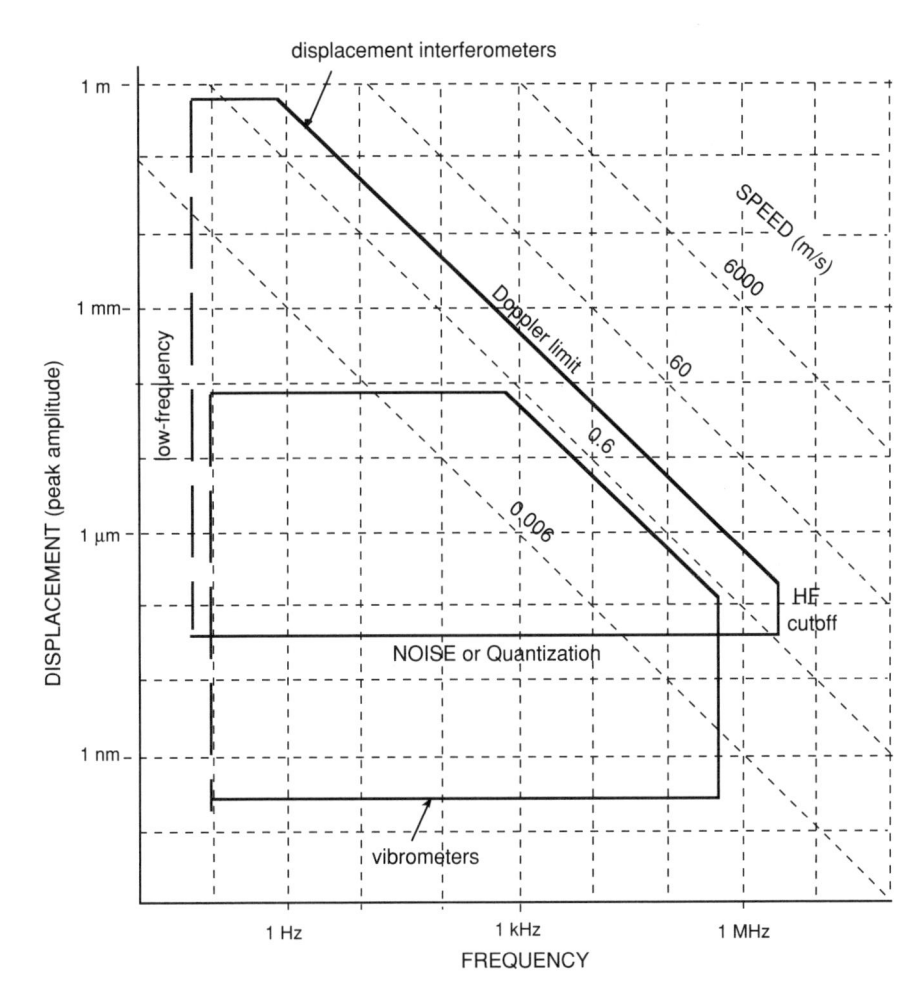

Fig. 4-17 Wegel's diagram for interferometer performances. Operation is inside the thick-line perimeter. Typical performances of a displacement-measuring interferometer and of a vibrometer are indicated.

4.4 ULTIMATE LIMITS OF PERFORMANCE

We have seen in the preceding sections that laser interferometers easily reach several digits (e.g., six) of dynamic range and submicrometer resolution. To ascertain the fundamental limits of operation posed by physical laws, we analyze in this section several factors that affect the instrument performance.

4.4.1 Quantum Noise Limit

The ultimate limitation to the sensitivity of a laser interferometer is the quantum limit associated with the detection of the signal returning from the optical interferometer.

Let us consider it in detail, starting with the photocurrent written in the generalized form as:

$$I_{ph} = I_0[1+V\cos Rk(s_m-s_r)] \qquad (4.12)$$

where R is the responsivity and V the fringe visibility (see App.A2). For simplicity, the reader may start taking V=1 and R=2, as in previous sections.

We assume that we can adjust the reference path so that the interferometer works in a quiescent point of maximum sensitivity, that is at $Rk(s_m-s_r)=\phi-\pi/2$ (or, at half fringe). In this case we have from Eq.4.12:

$$I_{ph}/I_0 = 1+V\sin\phi \approx 1+V\phi \quad \text{(for small } \phi\text{)}.$$

Explicitly, the phase ϕ will then be given by the small displacement to be measured, $\phi=Rk\Delta s$. For a small deviation around the quiescent point, the photocurrent deviation ΔI_{ph} is:

$$\Delta I_{ph}/I_0 = V\phi \qquad (4.13)$$

Superposed to the useful signal I_{ph}, we find a fluctuation ΔI_{ph} associated with the shot (or quantum) noise (see [5], App.A4) of the quiescent point current I_0. The rms value i_n of such noise is proportional to the square root of the average current I_0 and of the observation bandwidth B:

$$i_n = (2eI_0B)^{1/2} \qquad (4.14)$$

The associated signal-to-noise ratio of the current amplitude measurement around I_0 is S/N= $i_n/I_0 = (2eB/I_0)^{1/2}$. Equating $\phi=\phi_n$ and $\Delta I_{ph}=i_n$ in Eqs.4.13 and 4.14, we get:

$$\phi_n = i_n/VI_0 = (2eB/I_0)^{1/2}/V = 1/(S/N)V \qquad (4.15)$$

Now, we may introduce the Noise-Equivalent-Displacement (NED) which is defined as the value of displacement Δs giving the same effect as the intrinsic noise ϕ_n. By definition and Eq.4.12, it is $Rk\Delta s = \phi_n$ or:

$$NED = \phi_n/kR = (\lambda/2\pi)/(S/N) \; RV$$

$$= (\lambda/2\pi)(2eB/I_0)^{1/2}/RV \qquad (4.16)$$

4.4 Ultimate Limits of Performance

The diagram of Eq.4.16 is plotted in Fig.4.18 for the representative case of λ=633nm, η=0.9, and R=V=1.

As we can see from the figure, the quantum limit is indeed very low and theoretically allows us to reach very high sensitivities (e.g., $\phi_n \approx 10^{-6}$-10^{-8} rad in most cases).

For example, a displacement sensor working with an instrumental bandwidth of, say, B=1 Hz, and with 1mW of detected power, we can go down to a resolution of just a few femtometers (10^{-15} m). For a vibrometer aimed at detecting oscillations up to a few hundred kHz, the quantum noise limit is around a few picometers (10^{-12} m).

These very challenging figures are indeed obtained in practice, for example, in the gyroscope. However, we shall stress that generally they require a very careful design to first get rid of several perturbing effects, usually much larger than the quantum limit.

For RV ≠1, the values of ϕ_n read in Fig.4.16 shall be divided by V, the fringe visibility, and those of NED by RV.

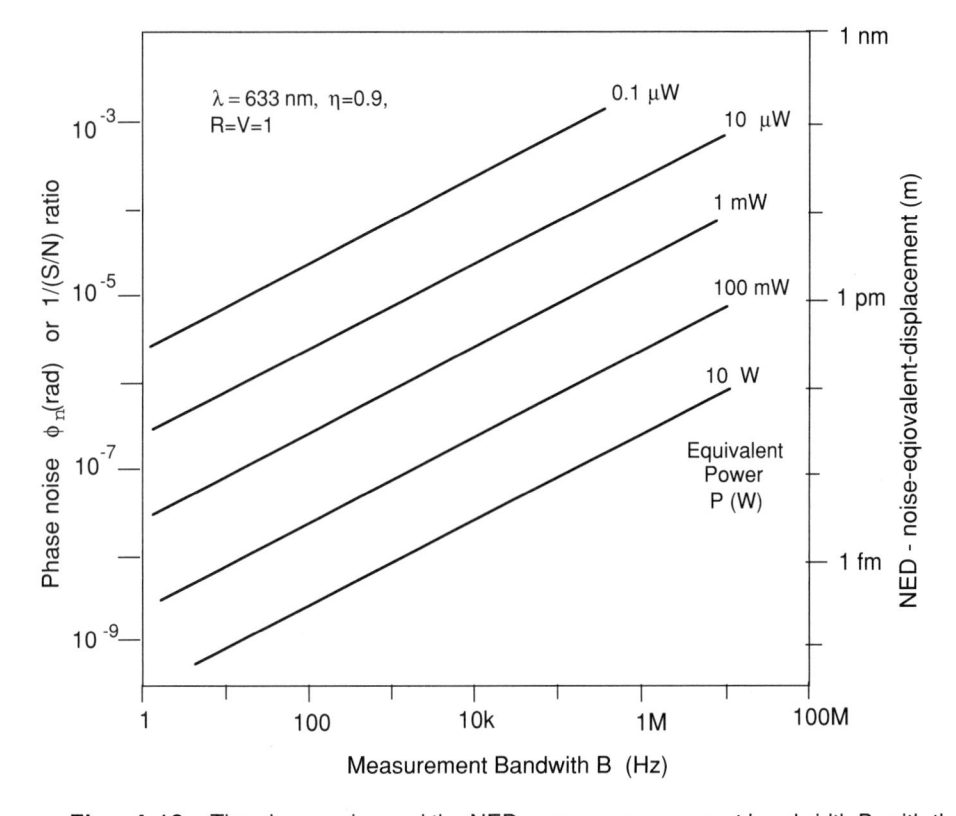

Fig. 4-18 The phase noise and the NED versus measurement bandwidth B, with the detected power P as a parameter, at the quantum noise limit of operation in an interferometer

4.4.2 Temporal Coherence

When recombining two beams propagated on different lengths along the measurement s_m and reference s_r paths, we actually superpose two field contributions delayed of a time $T=(s_m-s_r)/c$. If the optical frequency ν is not truly constant, but fluctuates in time, a decrease from the ideal unity visibility of fringes is incurred, and additionally, a phase error in the measurement of $k(s_m-s_r)$ is generated.

An easy way to describe the decrease of fringe visibility is to consider the finite line width $\Delta\nu$ of the source. If frequency ν_0 is defined not better than $\Delta\nu$ in the time-dependent term $\exp i2\pi\nu t+i\varphi = \exp i2\pi(\nu_0+\Delta\nu)t+i\varphi$, after a time $T_{coh}= 1/\Delta\nu$ the initial phase φ has changed by 2π, and coherence is lost (see also Appendix A1.1.2).

Time T_{coh} is therefore called the coherence time of the source. Alternatively, we talk of coherence length $L_{coh}=cT_{coh}$ as the distance traveled by light in the coherence time T_{coh}. The quantity L_{coh} represents the maximum value allowed to the path length difference s_m-s_r of the interferometer if a beating is to be obtained.

Usually, if the shape of the line is a Lorentzian, $p(\nu)=[1+4(\nu-\nu_0)^2/\Delta\nu^2]^{-1}$, the decrease of fringe visibility is a negative exponential: $A=\exp-T/T_{coh}$.

To study the phase error due to incomplete temporal coherence, let us assume that the frequency is a constant ν_0 and ascribe all the fluctuations to the phase term $\varphi(t)$. Writing the superposed fields as:

$$E_m = E_0 \exp i2\pi\nu_0 t +i\varphi(t)-iks_m, \quad E_r = E_0 \exp i2\pi\nu_0 t +i\varphi(t+T)-iks_r$$

The beating on the photodetector becomes:

$$I_{ph}/I_0 = 1+ \cos[k(s_m-s_r) +\varphi(t)-\varphi(t+T)] \qquad (4.17)$$

Now, let us express the phase term $\varphi(t)= \varphi_0+\varphi_n(t)$ as the sum of a mean value φ_0 and a random fluctuation $\varphi_n(t)$ with zero mean value, or $\langle\varphi_n(t)\rangle=0$. Then, the phase difference in Eq.4.17 becomes $\varphi(t)-\varphi(t+T)=\varphi_n(t)-\varphi_n(t+T)= \phi_n$ and we get by inserting in Eq.4.17:

$$I_{ph}/I_0 = 1+ \cos[k(s_m-s_r)+\phi_n] = 1+ \cos k(s_m-s_r) \cos\phi_n+ \sin k(s_m-s_r) \sin\phi_n \qquad (4.18)$$

The mean photodetected current is obtained by taking the average at both sides of Eq.4.18. Because $\langle\phi_n\rangle=0$, it is also $\langle\sin\phi_n\rangle=0$ because $\sin\phi_n$ is an odd function of a zero mean argument, and therefore we have:

$$\langle I_{ph}\rangle/I_0 = 1+ \mu_{tc} \cos k(s_m-s_r)$$

where $\mu_{tc}=\langle\cos\phi_n\rangle$ takes the meaning of coherence factor (see also [5], Sect.8.1.2). Developing the cosine in Taylor's series, $\cos\phi_n=1-\phi_n^2/2+...$, reveals that $\mu_{tc}\approx 1-\langle\phi_n^2\rangle/2$ is related to the variance $\sigma_\phi^2=\langle\phi_n^2\rangle$ of the random-phase difference $\varphi_n(t)-\varphi_n(t+T)$ at times t and t+T.

4.4 Ultimate Limits of Performance

If T is short, we may expect that ϕ_n keeps much less than unity and $\mu_{tc} \approx 1$. The fringe visibility then decreases as $A = \mu_{tc}$ in Eq.4.12.

Moreover, the fluctuation i_n around the mean value $\langle I_{ph} \rangle$ adds a phase error. To evaluate it, let us note first that the $\cos\phi_n$ term in Eq.4.18 is close to unity, in the case of practical interest of not so bad temporal coherence ($\mu_{tc} \approx 1$), and therefore it can be assumed as a constant not contributing to fluctuations. By taking the differentials at both sides of Eq.4.18, and then squaring and averaging, we get for the variance $\langle i_n^2 \rangle$:

$$\langle i_n^2 \rangle / I_0^2 = \sin^2 k(s_m - s_r) \langle \sin^2 \phi_n \rangle$$

Again assuming the case of $\mu_{tc} \approx 1$, the term $\langle \sin^2 \phi_n \rangle$ can be approximated to $\langle \phi_n^2 \rangle = \sigma_\phi^2$. Now, we recall that the phase error is related to the amplitude error by Eq.4.15, and assume that the maximum sensitivity condition $k(s_m - s_r) = -\pi/2$ to obtain the result:

$$\phi_n^2 = \sigma_\phi^2 \qquad (4.19)$$

The phase error σ_ϕ^2 can be traced to the two-sample Allan's variance $\sigma_v^2(2, T, \tau)$ describing the frequency stability of a generic oscillator. This quantity is defined as:

$$\sigma_v^2(2, T, \tau) = (1/\tau) \langle \int_{0-\tau} [v(t) - v(t+T)]^2 dt \rangle \qquad (4.20)$$

or, it is the mean square deviation of two frequency samples separated by a delay T and averaged on a time interval τ. The phase error is found to be related to the frequency variance by $\sigma_\phi = (2\pi\tau) \sigma_v(2, T, \tau)$, and in its turn σ_v is related to a two-consecutive sample variance by $\sigma_v(2, \tau, \tau) = (\tau/T) \sigma_v(2, T, \tau)$ for $T \ll \tau$.

Using Eqs.4.16, 4.19 and 4.20, we then get the NED_{tc} due to temporal coherence effects as:

$$NED_{tc} = (\lambda/2\pi)\phi_n = (\lambda/2\pi)(2\pi T) \sigma_v(2, \tau, \tau) = cT \sigma_v(2, \tau, \tau)/v$$

Recalling that $T = (s_m - s_r)/c$, we may write the result in the expressive form:

$$NED_{tc} = (s_m - s_r)(\sigma_v/v) \qquad (4.21)$$

This result tells us that incomplete temporal coherence, beyond reducing the fringe visibility, produces a noise-equivalent random error equal to a fraction of the arm mismatch $(s_m - s_r)$. The fraction is given by the relative frequency stability σ_v/v of the source and is evaluated in the integration time τ (where $\tau = 1/2B$ in terms of observation bandwidth).

With practical frequency stabilized lasers, frequency stability of 10^{-9} to 10^{-11} are obtained, so a 1-count or $\lambda/4 = 0.15$ µm error is generated for an arm mismatch $s_m - s_r = 0.15$ µm $/(10^{-9}..10^{-11}) \approx 0.15..15$ km.

4.4.3 Spatial Coherence and Polarization State

In the superposition at the photodetector, the reference E_r and measurement E_m fields shall have the same spatial mode distribution. If not, their interference term, and according the fringe visibility, will be reduced from the full value to a factor:

$$\mu_{sp} = \int_A E_m(x,y) E_r^*(x,y) dx dy / [\int\int_A |E_m(x,y)|^2 dx dy \int_A |E_r(x,y)|^2 dx dy]^{1/2} \quad (4.22)$$

Eq.4.22 explains why interferometers shall use single-mode beam spatial distributions. If the beam contains a mixture of N modes sharing the total power, because the integral product of different modes is zero, only homologous modes will contribute to μ_{sp}. The net result is that μ_{sp} cannot be larger than 1/N. More commonly, it will be $\mu_{sp} \ll 1/N$ because, in addition, the different modes may have a slightly different propagation constant and smear out the interferometric phase difference.

Similarly, we shall use the same State of Polarization (SOP) for both modes, either linear or circular, or generically elliptic. If not, the signal will be reduced by a factor:

$$\mu_{pol} = \mathbf{E_m \cdot E_r} / |\mathbf{E_m}| |\mathbf{E_r}| \quad (4.23)$$

where $\mathbf{E_m \cdot E_r}$ is the product of Jones matrixes of signal $\mathbf{E_m}$ and local oscillator $\mathbf{E_r}$, and $|..|$ indicates the modulus of vectors.

Thus, the combined effect of spatial, temporal, and polarization matching is to reduce the fringe visibility to:

$$A = \mu_{tc} \, \mu_{sp} \, \mu_{pol} \quad (4.24)$$

No extra noise is generated by the loss of visibility due to spatial and polarization effects, however, because of the deterministic nature.

4.4.4 Dispersion of the Medium

A nice feature of interferometric measurements is the very precise yardstick on which they are inherently calibrated: the wavelength. However, we usually propagate the beams in air, and therefore the unit of measure is λ/n_{air}. Though not too different from unity, the index of refraction can indeed introduce a calibration error in measurements with several significant digits.

In standard condition (T=15°C and p=760 mbar) and with its standard composition, air has an index of refraction that is well approximated by the following expression (see Ref.[7] of Ch.3):

$$(n_{air} - 1)|_{st} = 272.6 + 4.608/\lambda_{(\mu m)} + 0.061/\lambda_{(\mu m)}^2 \quad (ppm) \quad (4.25)$$

4.4 Ultimate Limits of Performance

We can see from this expression that in standard conditions the interferometric measurement requires a correction of about 280 ppm (ppm = part-per-million) at λ=633 nm. In addition, the correction is also dependent on the wavelength of operation, and the amount is a few ppm along the visible to near infrared regions.

At temperatures and pressures other than those of the standard conditions, the excess to 1 of the index of refraction is easily computed by noting that this quantity is proportional to the mole number per unit volume n/V, and therefore is equal to P/RT for the law of perfect gases. Therefore we can write:

$$n_{air} - 1 = (n_{air} - 1)|_{st} (P/760) (288/T) \qquad (4.26)$$

From Eq.4.26 we can calculate the temperature and pressure coefficients of the wavelength and of the correction to the corresponding interferometric measurement:

$$d(n_{air} - 1)/dT = -(n_{air} - 1)|_{st} (288/T^2) \approx -1 \text{ ppm/°C}$$

$$d(n_{air} - 1)/dP = -(n_{air} - 1)|_{st} (1/760) \approx +0.36 \text{ ppm/mbar} \qquad (4.27)$$

If we take reasonable values for ΔT and ΔP, for example 10°C and 10 mbar, we can see that temperature and pressure variations affect the 5th and 6th decimal digits of the displacement measurement. As already noted in Sect.4.2.1, the correction is performed by two sensors with outputs that change the multiplication factor (as shown in Figs.4-6 and 4-8) used to convert the fraction-of-λ counts to a metric scale.

Ideally, the sensors should average T and P on the actual optical path, but this is clearly awkward in practical operation.
With point sensors, located internal to the instrument, or close to the measurement optical path covered by the instrument, the correction is effective to the 6th decimal digit in the normal laboratory conditions.

4.4.5 Thermodynamic Phase Noise

In optical fiber sensors with interferometric readout (see App.A2 and Fig.8-28)), the thermodynamic fluctuation of index of refraction of the fiber introduces a phase noise ϕ_{th} that may become comparable or even larger than quantum noise. By an analysis of the phenomenon [7], the NED$_{th}$ is found as:

$$\text{NED}_{th} = \phi_{th}/k = 0.37 \, [k_B T^2 LB/\kappa]^{1/2} \qquad (4.28)$$

where k_B is the Boltzmann constant, T the absolute temperature, L the fiber length, and κ the thermal conductivity. Because of the interference of counter-rotating waves sharing the same medium, the Sagnac configuration (App.A2) provides a partial cancellation of ϕ_{th} and the best immunity to thermodynamic phase noise [7].

4.4.6 Brownian Motion

When aiming to small-mass targets, Brownian motion may add a fluctuation that has to be taken into account in interferometric measurements of vibrations or displacements, despite not being a fault of the instrument.

On the line of sight along which we are performing the measurement, we find a thermodynamic degree of freedom and hence an energy $(1/2)kT$. This energy shall be equated to kinematic energy $(1/2)m\langle v^2\rangle$, thus obtaining:

$$\langle v^2\rangle = kT/m \tag{4.29}$$

In the same way, for a rotating target for which we want to measure the angular speed of rotation Ω, we have a kinematic energy $(1/2)I\langle\Omega^2\rangle$, where I is the inertia momentum of the target. Equating to $(1/2)kT$ gives a variance of the angular speed:

$$\langle\Omega^2\rangle = kT/I \tag{4.30}$$

Letting numbers in Eqs.4.29 and 4.30 reveals that even for not so small masses (e.g., 1 to 10 g), the Brownian-induced speed or displacement may become comparable to the quantum noise NED.

Example. If we let T=300 K and being k= $1.38 \cdot 10^{-23}$ J/K the Boltzmann constant, we get for a 1mg mass from Eq.4.29: $\langle v^2\rangle = 4 \cdot 10^{-21}/10^{-3} = 2 \cdot 10^{-18}$, or $\sqrt{\langle v^2\rangle} = 1.4 \cdot 10^{-9}$ m/s, which is a value well in the reach of a typical interferometer (it corresponds to s=1nm and f=1Hz in the diagram of Fig.4-17). Also, for a disk of mass 0.01-g and radius 1-mm, the inertia momentum is $I=(1/2)mr^2 = 5 \cdot 10^{-3} \cdot 10^{-6} = 5 \cdot 10^{-9}$, and we get from Eq.4.30: $\sqrt{\langle\Omega^2\rangle} = \sqrt{(4 \cdot 10^{-21} \cdot 0.2 \cdot 10^9)} = 0.9 \cdot 10^{-6}$ r/s = 0.18deg/h, again a value within the readout sensitivity of a gyroscope (see Ch.7).

4.4.7 Speckle-Related Errors

When the interferometer works on a diffusing surface, a random phase error originates form the speckle-pattern statistics.
In addition, we find fading of field amplitude, as well as a deterministic field-curvature error. When these effects are cured, we are left with the speckle-induced phase error discussed in this section.
Of course, this situation is relevant only for the detection of vibrations on nonreflective surfaces, whereas in a corner cube interferometer the error is absent.

The statistical properties of the speckle field will be treated in Chapter 5. To discuss the ultimate error of an interferometric measurement on a diffuser, we refer to the conceptual scheme of Fig.4-19 (similar to the practical setup of Fig.4-16).
The three basic contributions to speckle errors, according to the displacement considered, are

4.4 Ultimate Limits of Performance

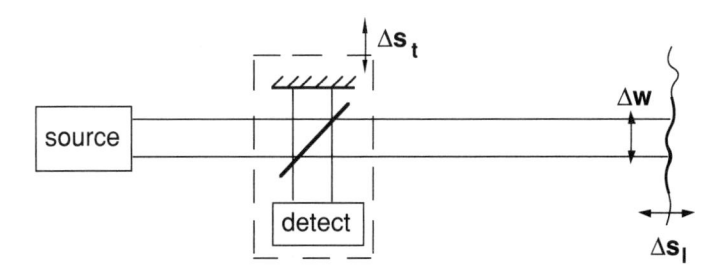

Fig. 4-19 Conceptual scheme to evaluate the NED of an interferometric measurement performed on a diffusing surface

the following:
- (i) On-axis (along the line of sight)
- (ii) Transversal (perpendicular to the line of sight)
- (iii) Projection (change of the diffuser portion illuminated by the beam)

For a *longitudinal* displacement Δs_l of the target along the line of sight, the noise-equivalent displacement is found as (see also Sect.5.1):

$$\text{NED}_{ax} = (C/8\sqrt{2})\, \lambda \Delta s_l / S_l \qquad (4.31)$$

where $S_l = \lambda s^2/\Delta w^2$ is the longitudinal speckle size, and C is a factor depending on the intensity of the particular speckle on which we are making the measurement.

If $I(0)$ and $I(\Delta s_l)$ are the intensities (or powers collected) at the extremities of the displacement Δs_l, and $\langle I \rangle$ is the mean intensity, we have the *conditioned* value of C:

$$C = \{\langle I \rangle / \sqrt{[I(0)I(\Delta s_l)]}\}^{-1/2} \qquad (4.32)$$

On the other hand, if we disregard the intensity or are not able to choose a particular speckle, the average (in quadratic sense) of the error given by Eq.(4.31) corresponds to the *free* value of C:

$$C = [\,3 - 2\ln(1+\Delta s_l/4S_l)^2\,]^{1/2} \qquad (4.33)$$

As we can see from Eq.4.31, the error is of the order of the wavelength multiplied by the dynamic-range to speckle size ratio. The factor C is of the order of unity (up to $\Delta s_l \approx S_l$) in the free measurement (Eq.4.33), but, if we choose a bright speckle with $I(0)$ and $I(\Delta s_l) \approx I(0)$ say K times larger than the mean intensity $\langle I \rangle$, the NED is decreased with respect to the free value of a factor $\approx \sqrt{K}$ (Eq.4.32).

For a *transversal* displacement Δs_t of the interferometer in the speckle field, one that should theoretically give a zero output, we get an error:

$$\text{NED}_{tr} = (C/8\sqrt{2})\, \lambda \Delta s_t /S_t \qquad (4.31a)$$

Here, $S_t = \lambda s/\Delta w$ is the transversal speckle size, and C is again given by Eqs.4-32 and 4.33 for the free and conditioned measurements, with Δs_l and S_l changed to Δs_t and S_t.

Last, a *projection* error is generated when the diffusing target moves transversally of a quantity Δw across the spot size w, a situation that should theoretically give no output. Because the random sample of the diffuser is changed, a random error is generated, and the corresponding NED is:

$$\text{NED}_{tp} = (C/4\pi\sqrt{2})\, \lambda \Delta w /w \qquad (4.31b)$$

All the errors considered previously can be kept to a small value by narrowing the allowed range of displacement as compared with a characteristic length. This length is the speckle-sizes S_l and S_t for longitudinal/transversal displacements, and the spot-size w for a diffuser change.

Because both transversal and projection errors are likely to occur simultaneously in a practical setup, it may be useful to note that their composition is quadratic, that is $\text{NED} = (\text{NED}_{tr}^2 + \text{NED}_{tp}^2)^{1/2}$.

4.5 READ-OUT CONFIGURATIONS OF INTERFEROMETRY

The laser interferometer we have considered so far can be classified as an *external configuration*, and in applications it is by far the most commonly used configuration. In it, the laser feeds an optical interferometer external to the source, and from the recombination of the propagated beams, a signal $I_0 \cos\Phi$ is provided, that is, an intensity signal carrying the *phase* information $\Phi = 2ks$ (see Fig.4-20).

This is not the only possibility we have available to make an interferometric read-out, however.

We may also think of using the mirrors of the laser itself as the optical interferometer, as shown in Fig.4-20. This is the *internal configuration*, which generates an output signal of the form $I_0 \cos \Omega t$, i.e., a *frequency* signal Ω proportional to the optical phase shift, $\Omega = \chi ks$, χ being a suitable constant. The internal configuration is put to advantage in the Ring-Laser-Gyro (RLG) version of the gyroscope (Ch.7).

A third configuration is the *injection-modulation*, or self-mixing, interferometer, (Fig.4-20, bottom). Here, the optical interferometer external to the source is absent, and we rely just on the interaction of the returned field from the remote target into the laser cavity field to produce a modulation of the emitted field related to Φ.

The self-mixing configuration generates interferometric signals in form of Amplitude Modulation (AM) and Frequency Modulation (FM) of the oscillation (or, laser field), which carry the driving terms $\cos\Phi$ and $\sin\Phi$, respectively.

4.5 Read-Out Configurations of Interferometry

EXTERNAL INTERFEROMETRY

$I = I_0 \cos 2ks$

optical pathlength is read on INTENSITY

laser interferometers, Doppler velocimeters, etc.

INTERNAL INTERFEROMETRY

$I = I_0 \cos \Omega t,$
$\Omega = (c/2L) \, 2ks$

optical pathlength is read on FREQUENCY

RLG gyroscope

SELF-MIXING INTERFEROMETRY

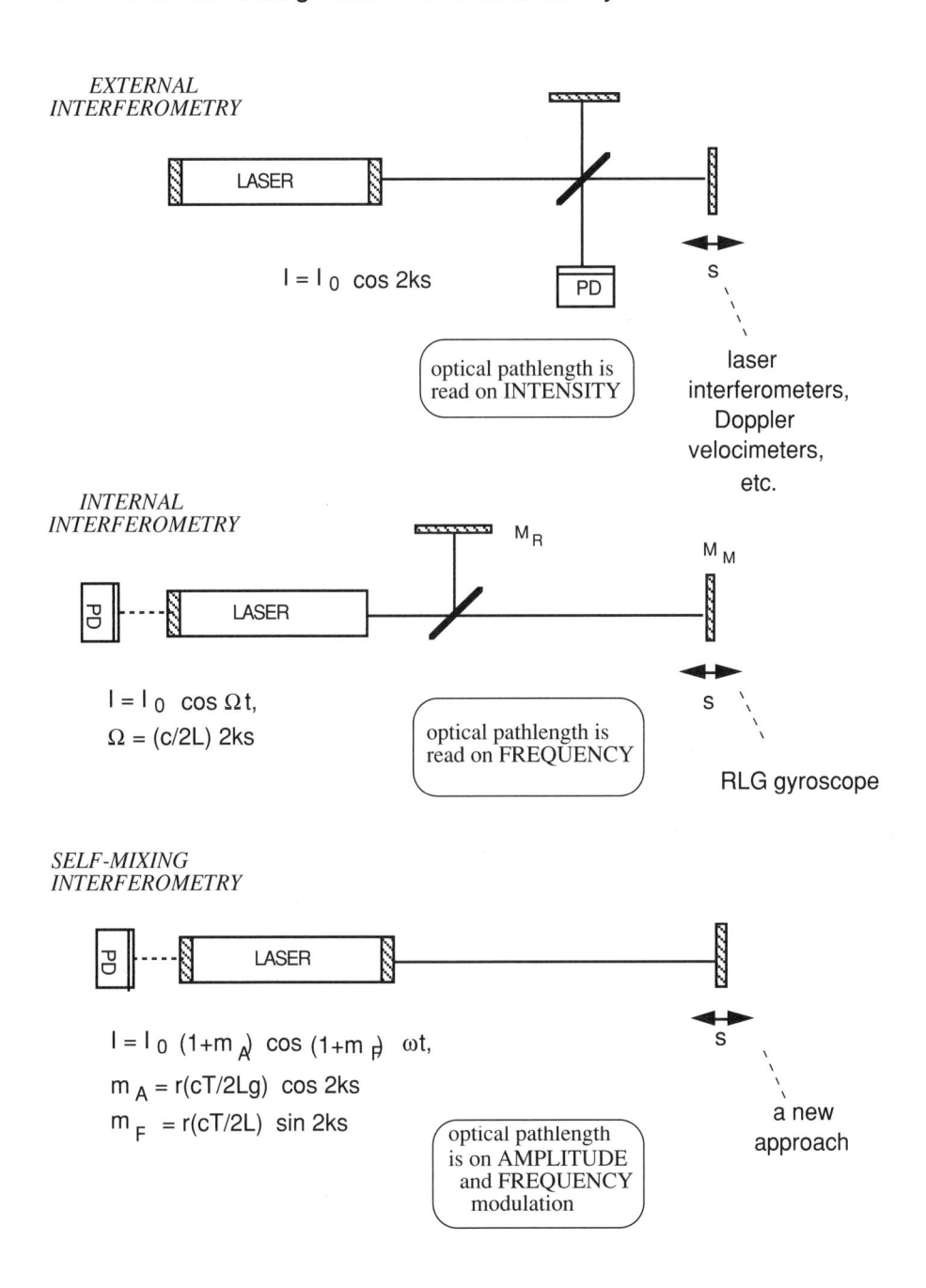

$I = I_0 (1+m_A) \cos (1+m_F) \omega t,$
$m_A = r(cT/2L_g) \cos 2ks$
$m_F = r(cT/2L) \sin 2ks$

optical pathlength is on AMPLITUDE and FREQUENCY modulation

a new approach

Fig. 4-20 Configurations of interferometry: external (top), internal (middle), and injection or self-mixing (bottom)

4.5.1 Internal Configuration

Let us consider a laser oscillating in a single spatial and longitudinal mode, for example, a 20-cm He-Ne tube with external mirrors (see App.A1-1). As we move one mirror by a small displacement Δs along the cavity axis, the oscillation frequency will change by:

$$\Delta f = (c/2L) \, \Delta s/(\lambda/2) \qquad (4.33)$$

This result is explained because the spacing of longitudinal modes is $c/2L$, where L is the mirror distance (see Fig.A1-1) and, for each $\lambda/2$-increase of the laser cavity length L, frequency increases by $c/2L$ (and the order of the mode increases by 1). Therefore, the ratio R_f of frequency variation Δf to displacement Δs, called responsivity R_f, is given by:

$$R_f = \Delta f/\Delta s = (c/\lambda L) \qquad (4.34)$$

By letting $\lambda=0.633$ μm and L=20 cm so that c/2L=750MHz, we get from the internal configuration the remarkable responsivity to mirror displacement R_f = 750 MHz/316 nm = 2.4 MHz/nm, which is a very high value.

However, we cannot measure optical frequency directly and, to exploit this result, we need a second mode to be used as a reference frequency so that the frequency shift Δf can be converted down to an electrical frequency.

Thus, we have the arrangement of Fig.4-20, where a beamsplitter is used to allow the laser sustaining the oscillation of two modes, defined by the mirrors M_M and M_R. When M_R is moved, the frequency difference f_2-f_1 carries the interferometric signal and has a frequency variation proportional to Δs as given by Eq.4.34, a value that we will measure from the photodetector output. The maximum range of displacement is $\Delta s=\lambda/2=316$ nm, and the corresponding frequency is $\Delta f= f_2-f_1=c/2L=750$MHz. As we now have a Michelson interferometer brought inside the active cavity, the name internal configuration is justified.

In an actual experiment, however, the linear response expected from the configuration is obtained only when the frequency difference f_2-f_1 is not too small (Fig.4-21). When f_2 approaches f_1, first a decrease in response is found, and then, for $f_2 \approx f_1$, the signal suddenly disappears [8,9].

The dead band is due to the frequency *locking* of the two oscillations. Locking is a very general phenomenon in coupled oscillators, which comes from even very minute coupling of power from one oscillation to the other. A residual very small interchange of power is unavoidable (for example, because of mirror scattering and gain-medium coupling), and therefore the dead band will never be zero (see also Ch.7).

However, we first try to reduce it as much as possible, and an improved scheme is shown in Fig.4-21. Here, because we use a Glan beamsplitter, two modes oscillate with linear orthogonal polarizations. At the rear mirror output, we can insert a polarizer in front of the photodetector, and orient it at 45° for the modes to beat on the photodetector.

4.5 Read-Out Configurations of Interferometry

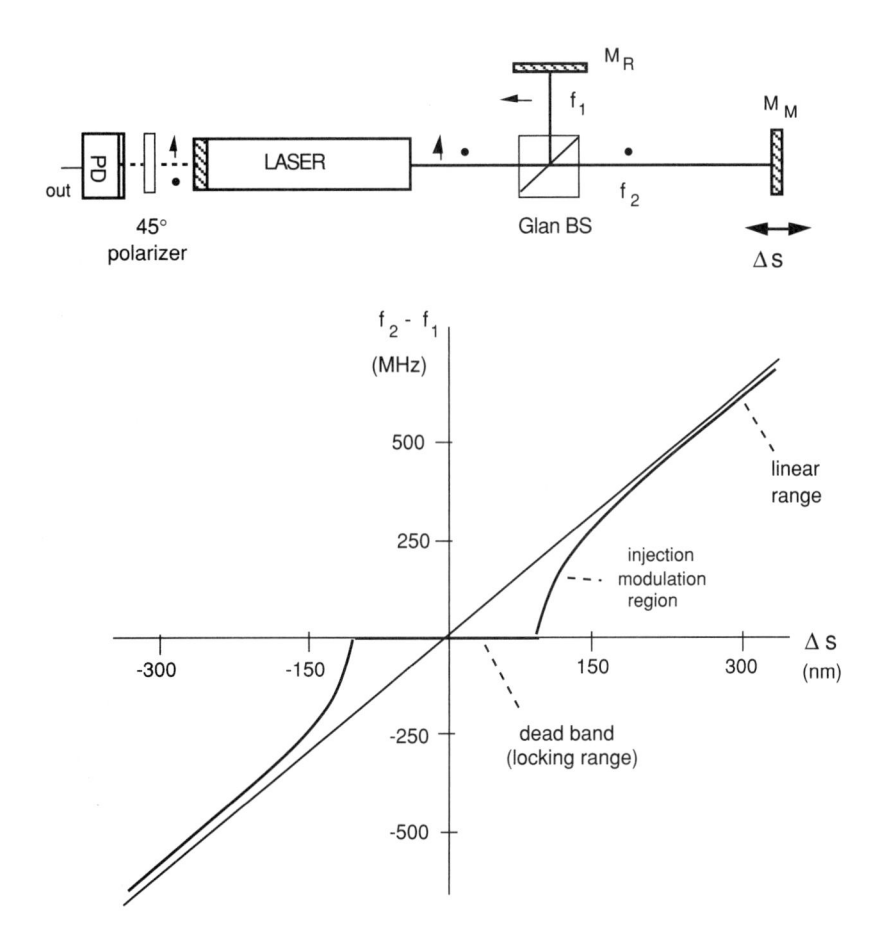

Fig. 4-21 Internal interferometry. Top: scheme with polarization-split modes. Bottom: the theoretical response of frequency difference f_1-f_2 versus displacement Δs is linear, with a range of c/2L=750 MHz for a Δs =λ/2=316 nm displacement (values for a 30-cm He-Ne). In practice, a dead band around $f_1-f_2\approx0$ is found because of locking.

Because of the orthogonal polarizations, the gain coupling in the active medium is nearly suppressed. In addition, we get rid of the 50% loss of a normal beamsplitter used in Fig.4-20, which is hard to be compensated in a normal He-Ne laser.

Using low-scatter mirrors and a high-quality Glan cube to realize a He-Ne internal interferometer, we may go down to a few MHz of locking range as compared to several hundreds MHz of the basic scheme. Then, we can bias the interferometer to operate far away from $f_2-f_1=0$, e.g., at f_{bias}=100 MHz, so we can also detect the sign of Δs.

This can be done by finely adjusting the position of mirror M_R. A further refinement is to mount M_R on a piezo actuator and make a servo loop on the reference mirror position to stabilize the interferometer at $f_2-f_1=f_{bias}$=const. The Δs signal is then obtained from the error signal of the feedback loop.

The internal configuration is critical to operate because mirror M_M shall be kept aligned to the cavity during motion, and the tolerable error is much less than the diffraction limit, <1 arc-second in practice. Though the displacement we are aiming at might be very small, alignment is still very critical.

On the other hand, if the internal configuration concept is useful in Sagnac interferometers like the gyroscope (Ch.7). In this case, optical path length variations are induced in a balanced cavity from outside, and the very high responsivity of the readout is put to great advantage.

4.5.2 Injection (or Self-Mixing) Configuration

Even without using any optical interferometer external to the laser source, we can make a laser interferometer by using the interaction, produced in the laser cavity, by a small fraction of the field emitted from a laser and returned back after propagation from the remote target.

First reported in 1978 [10], the resulting configuration is variously referred to in the literature as *induced-modulation*, *injection*, *self-mixing*, or *feedback* interferometry.

A straightforward explanation of the phenomenon comes from considering the field oscillating in the laser cavity as a rotating vector (Fig.4-22). The cavity field E rotates in the phase plane at optical frequency ω. If a fraction αE of it is allowed out for propagation to a remote target at distance s, the optical phase shift accumulated in the go-and-return path is $\Phi=2ks$.

Thus, upon re-entering in the laser cavity through the mirror with transmittance t, we find a rotating vector of complex amplitude $aE \exp i\Phi$, where $a=t\alpha\eta$, and η is an eventual propagation loss. This contribution adds as a vector to the existing field E to give the instantaneous new cavity field. The addition of rotating vectors is a well-known result of communication theory.

This result can be stated as follows. The in-phase component $aE \cos \Phi$ produces amplitude modulation of the pre-existing field E, and its modulation index (or depth) is $a \cdot \cos \Phi$. The in-quadrature component $aE \sin \Phi$ produces frequency modulation of the pre-existing field E, and its modulation index (or depth) is $a \cdot \sin \Phi$.

Thus, the laser cavity field acts as the optical carrier of AM and FM modulations induced by the perturbation returning from the remote target. The modulation indexes are exactly the $\sin \Phi$ and $\cos \Phi$ signals of the optical path length $\Phi=2ks$, which we were attempting to procure in conventional configurations based on an optical interferometer.

In other applications of lasers, injection phenomena are usually undesired, and we want to get rid of them. For example, in fiber-optics communications, it is common practice to protect with an optical isolator the narrow-line laser transmitter, from the unwanted back-scattering that spoils the laser line and adds amplitude noise.

4.5 Read-Out Configurations of Interferometry

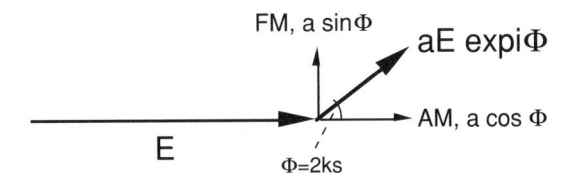

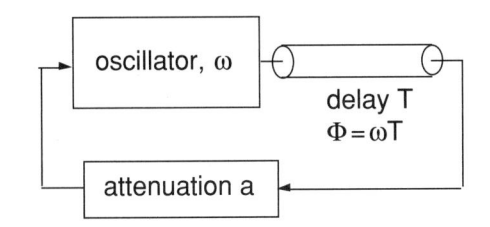

Fig. 4-22 Rotating vector description of injection modulation (top), and the circuit analogy of the injection phenomenon (bottom)

The injection interferometer, on the other hand, uses injection in a well-controlled way to measure the returning field. Of course, we need a narrow-line, single-mode laser to make clean modulation waveforms available.

It is interesting to observe that the injection phenomena are quite general, and the discussion here reported for a laser source actually applies with minor changes to virtually all kinds of oscillators.

For example, echo detection by feedback modulation has been reported in microwaves as well as in ultrasonic measurements. In addition, if we make an Op-Amps version of the circuit model of Fig.4-22, we will readily obtain for voltage signals the same induced-modulation waveforms that will be presented later in this section.

It is also interesting to note that we focus here on the interferometric measurements performed with the injection configuration, but the scheme of Fig.4-22 is conceptually a coherent detection scheme [5] that allows applications based on the phase, as well as the amplitude of a returning weak signal [11].

To describe the injection phenomena, we have to distinguish three levels of injection according to the fraction of power returned to the cavity, and to the ratio of cavity length L to target distance s.

As it will be shown later, the dependence is summarized by a feedback parameter C defined as [12]:

$$C = a \; n_1 \; L\sqrt{(1+\alpha_{en}^2)}/s \qquad (4.35)$$

where a is the field-amplitude feedback parameter (or, the square root of returning power fraction), α_{en} is the linewidth enhancement factor [13] ($\approx$1 in He-Ne and $\approx$2..6 in diode lasers), and n_1 is the refractive index of the active medium. Then, we have the following cases:

- *Weak feedback*, for C<<1. If the output mirror has a high reflectivity (R$\approx$0.99, as in a He-Ne laser) and distance is L<10m, we are probably in this case. The simple picture of rotating vectors is applicable, the laser has nearly the same properties as in the unperturbed state, and the interferometric waveforms are sinusoidal.
- *Moderate feedback*, for C$\approx$1. This is easily found in semiconductor laser diodes (R$\approx$ 0.05...0.3). The interferometric signal becomes a distorted-sinusoid waveform, and switching between levels in each 2ks=2π period may occur. Both spectral line and coherence exhibit significant deviation from the unperturbed state.
- *Strong feedback*, for C>>1. The interferometric waveform exhibits multiple switching in each 2ks=2π period, coherence length and line width are strongly affected and the laser starts oscillating on the external cavity [13-15,20].

4.5.2.1 Analysis of injection at weak-feedback level

In weak feedback conditions, we can apply the description based on the well-known Lamb's equations [12] used in the original derivation [10] to analyze injection in He-Ne lasers. In the standard treatment, we start writing the laser cavity field **E** and the returned signal field **E**$_s$ as rotating vectors, i.e.:

$$\mathbf{E} = E \exp i\varphi \quad \text{and} \quad \mathbf{E}_s = a E \exp i(2ks+\varphi),$$

$$\text{with} \quad \varphi = \Omega t+\psi = 2\pi\nu t+\psi$$

Amplitude E and phase φ of the oscillating field are assumed to be slowly varying quantities, and a balance is written of their time derivatives dE/dt and dφ/dt. Taking into account the external field re-entering the cavity after propagation to a remote target at distance s with *field* attenuation a, the Lamb's equations are written as:

$$dE/dt = [(\alpha-\beta E^2)c - \Gamma] E + (c/2L) aE \cos(\varphi+2ks)$$

$$d\varphi/dt = \Omega_c + \zeta(\alpha-\beta E^2)c + (c/2L) a \sin(\varphi+2ks) \tag{4.36}$$

where:

- $a = At_1^2\eta$ is the total *field* loss suffered by the reinjected field because of: (i) attenuation A in the go-and-return propagation to the target, (ii) double passage through the output mirror with a field transmission t_1, and (iii) mismatch η of the mode (cavity vs. returning) spatial distributions;
- $t_1 = \sqrt{(1-r_1^2)}$ is the *field* transmission of the input mirror of field reflectance r_1 (explicitly, the *power* transmittance and reflection are T= t_1^2 and R=1-T= r_1^2);
- $\alpha = \lambda^2(n_2-n_1)/8\pi\tau_{21}\Delta\nu_{at}$ is the active medium gain rate per unit length (of the field)
- n_2-n_1 is the population inversion concentration (cm^{-3});
- $\Delta\nu_{at}$ is the (atomic) gain line width and τ_{21} is the active level lifetime;

4.5 Read-Out Configurations of Interferometry

β is the gain saturation coefficient;
$\Gamma = \Omega_c/2Q$ is the cavity *field* loss-rate (per unit time), including scattering in the medium and mirror transmission loss, explicitly $\Gamma = t_1 t_2 c/2L + \Gamma_{sc} c$;
Ω_c is the cavity resonant frequency;
L is the cavity length, and c/2L is the longitudinal mode spacing;
$\zeta = (\nu_0 - \nu_{at})/\Delta\nu_{at}$ is the frequency detuning respect to the gain line center ν_{at}.

Letting $(d/dt)E=0$ in Eqs.4.36, we get the steady-state solution.

For the solitary laser (a=0), the quiescent values E_0, Ω_0 are readily found as $E=E_0 = \sqrt{[(\alpha - \Gamma/c)/\beta]}$ and $\Omega = \Omega_0 = \Omega_c + \zeta\Gamma/c$.

For a≠0, we look for a small-perturbation solution of Eq.4.36, by letting $E=E_0+\Delta E$ and considering $\Delta E << E_0$ so that only first-order terms in ΔE are retained. By neglecting the pulling term $\zeta << 1$ and dropping the unessential constant phase φ_0, after some algebra, we obtain the solution as:

$$E = E_0 [1+ (a/2L\gamma_0) \cos 2ks] = E_0 [1+m_A] \qquad (4.37a)$$

$$d\varphi/dt = \Omega_0 + a (c/2L) \sin 2ks = \Omega_0 + \Delta \qquad (4.37b)$$

where $\gamma_0 = \alpha - \Gamma/c$ is the net gain per unit length available in the medium when oscillation is not yet started (it equals βE^2 because, in the permanent regime of oscillation, $\alpha + \beta E^2 - \Gamma/c = 0$).

Thus, at the first order of perturbation, the cavity field has an AM with modulation index $m_A = (a/2L\gamma_0) \cos 2ks$ proportional to the cosine of the external pathlength, and an FM with frequency deviation $\Delta/2\pi = a(c/4\pi L) \sin 2ks$ proportional to the sine of the external pathlength.

Numerical example. The proportionality coefficient of AM is primarily the total attenuation $a=A\eta t_1^2$, while the term $2L\gamma_0$ at the denominator has an order of magnitude not far away from unity in practical cases. Working on a diffusive target with a Gaussian beam with waist w_0, we find $A=\lambda/\pi w_0$ in the near field and $A=w_0/s$ in the far field. For a retroreflector, it is A=1.

The mode superposition efficiency is $\eta=1$ for the diffuser and $\eta^2=2/[(w/w_0)^2+(w_0/w)^2]$ for the retroreflector, where $w^2=w_0^2+(\lambda 2s/\pi w_0)^2$ is the spot size at distance 2s. In the far field, the product $A\eta$ is in both cases $A\eta \propto 1/s$, for which we get an inverse-distance dependence of the signal level, which is typical of a coherent detection scheme (see [5], Ch.8) and well confirmed by experiments. Last, the output mirror (power) transmission is usually $t_1^2 \approx 0.01$ in a He-Ne laser. About the AM modulation index, $\alpha - \Gamma/c$ can be estimated from the amount of extra gain allowed for the startup of oscillation. In a He-Ne laser, usually $\alpha \approx 0.0005$ cm^{-1} and with L=20 cm, we have $2L\alpha \approx 0.02$; with $t_1^2 \approx 0.01$, it would be $\Gamma/c = t_1^2/2L \approx 0.00025$ cm^{-1} and $\gamma_0 = 0.00025$ cm^{-1}.

To conclude our example, let us take a 20-cm He-Ne laser with a 0.2-mm waist w_0 and a 1-mr divergence. Working at s=500 mm distance on a diffuser target (far-field condition), we have: $m_A = A\eta t_1^2/2L\gamma_0 \approx (0.2/500)0.01/0.00025 \cdot 40 = 4 \cdot 10^{-4}$, while, on a retroreflector, it would be $m_A \approx 0.2 \cdot 0.01/0.01 = 0.2$. Correspondingly, the frequency deviation $\Delta/2\pi$ in the two cases is 750 MHz(0.2/500)0.01/6.28=500 Hz and 750MHz·0.2·0.01/6.28 =250kHz, respectively.

Eqs.4.37a and 4.37b are an adequate description of the weak-injection regime (a<<1) in media with a population inversion n_2-n_1 only dependent on the pumping rate ($\alpha,\beta \approx$ constant). These equations are well suited to He-Ne lasers and other media with small gain per unit length, for which they provide a good fit of theory to experimental data. They also hold at first approximation for semiconductor lasers at weak injection.

4.5.2.2 Bandwidth and noise of the injection interferometer

The frequency response of the interferometric signal (E, dφ/dt) to a variation of the distance s is made up by two contributions. One comes from the finite time that the laser medium takes to reach a new equilibrium condition established by the external distance, and the other is the delay associated with the propagation time of the field reflected by the target.

A third, usually dominant, limit is the frequency response of the photodetector translating the laser field into an electrical current signal. Because this limit is not inherent to the injection mechanism and can be separately evaluated, we will not deal with it furtherly.

To estimate the laser medium response time, we may take account that the cold laser cavity has a decay time $\tau_d=1/\Gamma$. In addition, the laser oscillator is a feedback system with loop-gain G=1/2Lγ_0, as can be seen from Eq.4.37a where the effect of an input aE is transferred to an induced field aE/2Lγ_0. As a result of feedback theory, the resulting time constant of response is then $\tau=\tau_d/G=2L\gamma_0/\Gamma$, and the corresponding bandwidth is B= 1/2$\pi\tau$=Γ/4πLγ_0.

The external pathlength introduces a delay T=s/c in the effective field perturbing the cavity. Should the change in (E, dφ/dt) further interact with the laser cavity in the next round trip to the distance s, this delay would actually affect the frequency response. However, if the returned field in the first roundtrip is aE, it will become a^2E in the second, and if a<<1, we may safely neglect this contribution, and state that the external distance effect is simply a delay T in response. This conclusion also applies to the medium-level of injection (Sect.4.5.2.3) that is applicable to semiconductor lasers.

With the usual values of parameters, B is in the range of some hundreds MHz in He-Ne lasers and of gigahertz (GHz) in semiconductor lasers.

To evaluate the noise, let us recall that the signal returned to the cavity is aE. If no amplification mechanism were present, the detected power would be $a^2E^2=a^2P$. Because the detection is coherent, the quantum noise limit is attained. Thus, we may conclude that the minimum detectable displacement (or NED) of the injection interferometer is given by the diagram of Fig.4-18, in which we shall use a^2P as the value of power, P being the power output from the unperturbed laser.

This result can be used as a default estimate of the injection interferometer noise. To improve the evaluation, however, we may account for the amplification undergone by the injected field. From Eqs.4.37, we can see that the laser supplies an AM signal 1/2Lγ_0 larger than the cosine component and a frequency FM signal equal to the sine component. In power, we have thus a gain $G=1+(1/2L\gamma_0)^2$. From optical amplification theory (see [5], Sect.8.3) the noise figure is given by NF=1+2n_{sp}(G-1)/G, where $n_{sp}=n_2/(n_2-n_1)$ is the inversion factor ($\approx$1 for $n_2>>n_1$). Therefore, we may conclude that, at most, noise is increased by a modest 3 dB (or a factor 2) with respect to the a^2P level when G>>1 and $n_{sp}\approx$1.

4.5 Read-Out Configurations of Interferometry

4.5.2.3 The He-Ne injection interferometer

The experiment is carried out with a frequency-stabilized He-Ne laser (Appendix A1.2). In the practical implementation, we can use a Zeeman-split two-frequency source actuated by thermal expansion through a resistive wire wound on the He-Ne capillary tube, as originally reported in [10].

The two frequency split modes oscillate with linear, orthogonal polarized states. With a polarizer P placed in front of the main output mirror (Fig.4-23), one mode (say f_1) is selected for propagation to the remote target. The other mode (at f_2) is prevented from exiting the laser and is kept in cavity to act as a reference local oscillator for detection.

The propagated mode undergoes the AM and FM injection modulations and carries the optical path length 2ks information. When both modes are superposed on the photodetector placed on the rear mirror (Fig.4-23), they will beat and downconvert the AM and FM modulations to an electrical carrier at f_1-f_2.

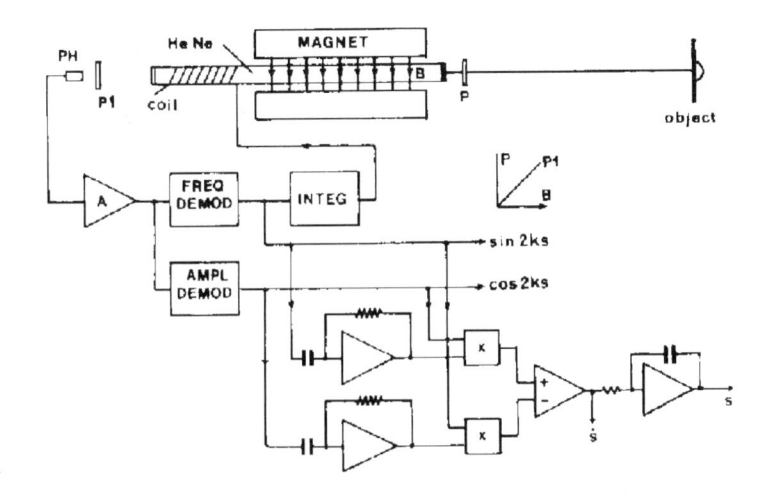

Fig. 4-23 Injection interferometry with a He-Ne laser: two modes at slightly different frequencies (f_1-$f_2 \approx$ 100kHz) are generated by Zeeman-splitting due to a transversal magnetic field applied to the capillary tube. The two modes have linear orthogonal polarizations. One mode is allowed to propagate through the polarizer P, and the other mode is kept in the cavity. Polarizer P1 in front of the photodetector PH is oriented at 45° to allow beating of the propagated mode with the unperturbed mode kept in cavity. After frequency and amplitude demodulation, the signals sin 2ks and cos 2ks are obtained. The frequency signal is also used for frequency stabilization through the thermal actuator. Through differentiation and cross-multiplying of sin 2ks and cos 2ks, the v and s are recovered (from Ref.[10]).

The experimental waveforms of the photodetector current are shown in Fig.4-24. These are obtained with a 20-cm long tube, 0.5-mW commercial He-Ne laser aimed at a diffuser-surface target (a loudspeaker driven by a triangular waveform at audio frequency) placed at a s=40-cm distance. After AM and FM demodulation of the photodiode electrical signal, the waveforms of the interferometric signals S=sin 2ks and C=cos 2ks are obtained, as shown in Fig.4-25. Starting from these signals, several schemes can be devised to recover the displacement s.

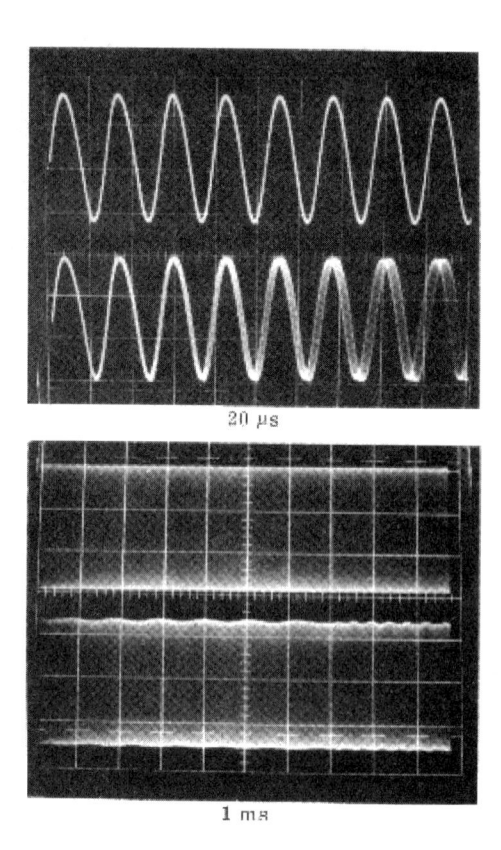

Fig. 4-24 Waveforms of the electrical signal out from the photodiode when the target is a diffuser (white paper on a loudspeaker driven at audio frequency). To ease comparison, bottom traces with injection are shown along with top traces with the beam blocked off. On a 20 µs/div scale, the waveform reveals the FM signal in form of a time jitter; on a 1ms/div scale, the AM signal shows up in form of a ripple. Using a corner cube target, the signals become much larger and require an attenuation inserted on the beam (from Ref.[10]).

4.5 Read-Out Configurations of Interferometry

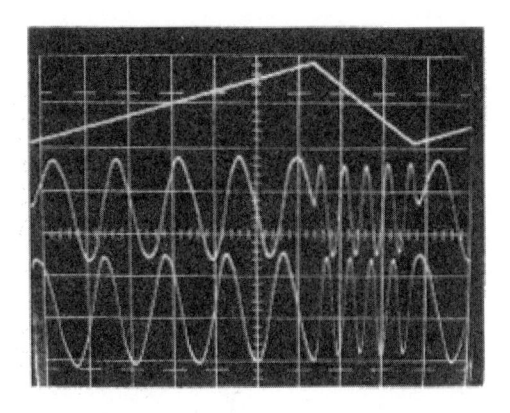

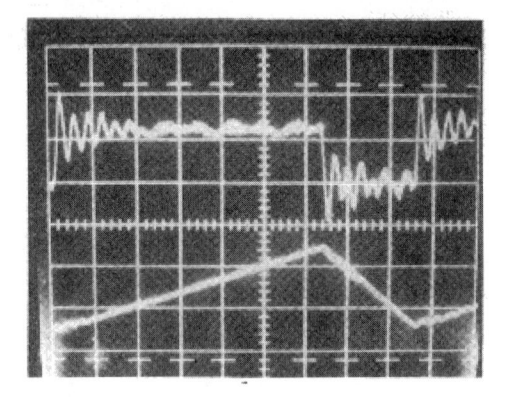

Fig. 4-25 Top: waveforms of target driving signal and of the interferometric signals S=sin 2ks and C=cos 2ks demodulated from the photodiode output; bottom: the reconstructed velocity signal v=ds/dt and its integral, the displacement s. Time scale: 1ms/div (from Ref.[10]).

For example, we can acquire S and C through an A/D interface and use a small dedicated computer to calculate s. For small displacements $\Delta s < \lambda/2$, Δs is simply computed as $\Delta s = (1/2k)$ atan S/C.

For large displacements, because atan is a multivalued function, we shall compute s looking at the evolution of Δs. Using a digital processing, we may add to (1/2k) atanS/C an increment $\pm \lambda/4$ for each zero-crossing of C, with the sign (+) for U=1 and (–) for U=0, where U is given by Eq.4.6.

Alternatively, we may prefer to use a straightforward analogue processing performed by a few electronic analogue circuit blocks. Taking the time derivatives of S and C, we get:

$$dS/dt = 2kv \cos2ks, \quad dC/dt = -2kv \sin2ks$$

where v=ds/dt is the velocity. By cross-multiplying the two derivatives to C and S and subtracting, we have:

$$C \, dS/dt - S \, dC/dt = 2kv \cos^2 2ks + 2kv \sin^2 2ks = 2kv \quad (4.39)$$

and, with a time integration, we obtain the displacement $s = \int v \, dt$ [10].

The circuit performing this analog processing is shown in Fig.4-23. With this circuit, the reconstructed velocity and displacement for a triangular waveform excitation of the target are those reported in Fig.4-25. Note that the s(t) and v(t) waveforms deviate somehow from those expected from the drive waveform. However, this is simply a consequence of the loudspeaker real response and is actually a measurement of the response itself.

Here, the target was a piece of plain white paper glued on the loudspeaker. Accordingly, the back-scattered field received by the laser obeys the speckle-pattern statistics. The dynamic range is then limited by the longitudinal speckle dimension $S_l = \lambda s^2/w^2$ (where w is the spot size on the target and s is the target-laser distance). A peculiar feature of the injection interferometer is the ability to self-filter the returned field spatially.
The transversal speckle size is $S_t = \lambda s/w$, and the far-field spot size is $w=(\lambda/w_0)s$. By substitution, we get $S_t=w_0$, that is, the returning speckle (or coherence) size S_t exactly matches the laser mode size w_0, and no waste of power occurs.

Another field of application is the pickup of biological motility signals, like respiratory sounds and blood pulsation [16]. In Fig.4-26, we can see the experimental trace of the blood pulsation detected on a finger, revealing the cardiac pulsation, detected remotely without contact with the tissues under measurement.

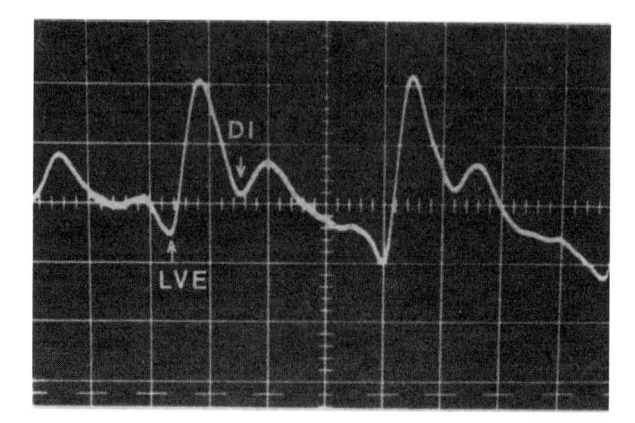

Fig. 4-26 Pickup of biological motility with the He-Ne laser injection interferometer. The blood pulsation waveform is measured on a finger and reveals the left ventricular ejection (LVE) as well as the dicrotic incisure (DI). The vertical scale is 0.1 μm/div, and the horizontal scale is 0.2 s/div.

4.5 Read-Out Configurations of Interferometry

Of course, the injection interferometer of Fig.4-23 can operate on a corner cube target as well [17]. In this case, the returned signal is much larger than with the diffusing target, and an optical attenuator is inserted on the beam to avoid high-level injection effects. When counts are drawn out of signals C and S, the performance is much the same as the two-frequency interferometer.

The He-Ne laser is well suited for the weak-level injection interferometer, because it readily provides two frequency-split modes, one for propagation and the other for internal reference and down-conversion to electrical frequency. However, because the He-Ne is rather bulky and its reliability is questionable, we may think of replacing it with a semiconductor laser.

In diode lasers, unfortunately, we are not able to make the equivalent of a Zeeman dual-frequency oscillation. Thus, just the AM signal is available as a modulation in emitted power, whereas the FM signal cannot be recovered. This brings about the ambiguity of the cos 2ks function, which is not easily circumvented for the application to displacement measurements, at least in the weak-injection regime. On the other hand, the single interferometric AM signal cos 2ks is sufficient for small-vibration measurements, when amplitude Δs is less than or comparable to $\lambda/2$ (Sect.4.6).

4.5.2.4 Analysis of injection at medium-feedback level

The regime of injection at moderate levels of a back-reflected field is important in applications because it permits collecting, with a single AM signal, the same information supplied by the signal-pair cos 2ks and sin 2ks. In this regime, we can perform an unambiguous measurement of displacement, in increments of a fraction of wavelength with the correct sign (e.g., + for increasing distance, – for decreasing), yet with a single signal.

Experimentally, we have the setup of Fig. 4-27, in which a laser (usually, a semiconductor single-mode laser) is aimed through a collimating objective lens to a mirror target. A variable attenuator is used for adjusting the level of power returned into the laser cavity. Detection can be performed by a photodiode placed everywhere on the beam (because the AM

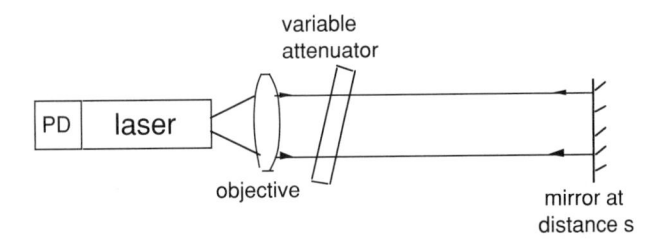

Fig. 4-27 Conceptual arrangement for experiments with an injection interferometer: using a variable optical attenuator, we adjust the level of injection and look at the AM waveform using the rear-mirror photodiode PD.

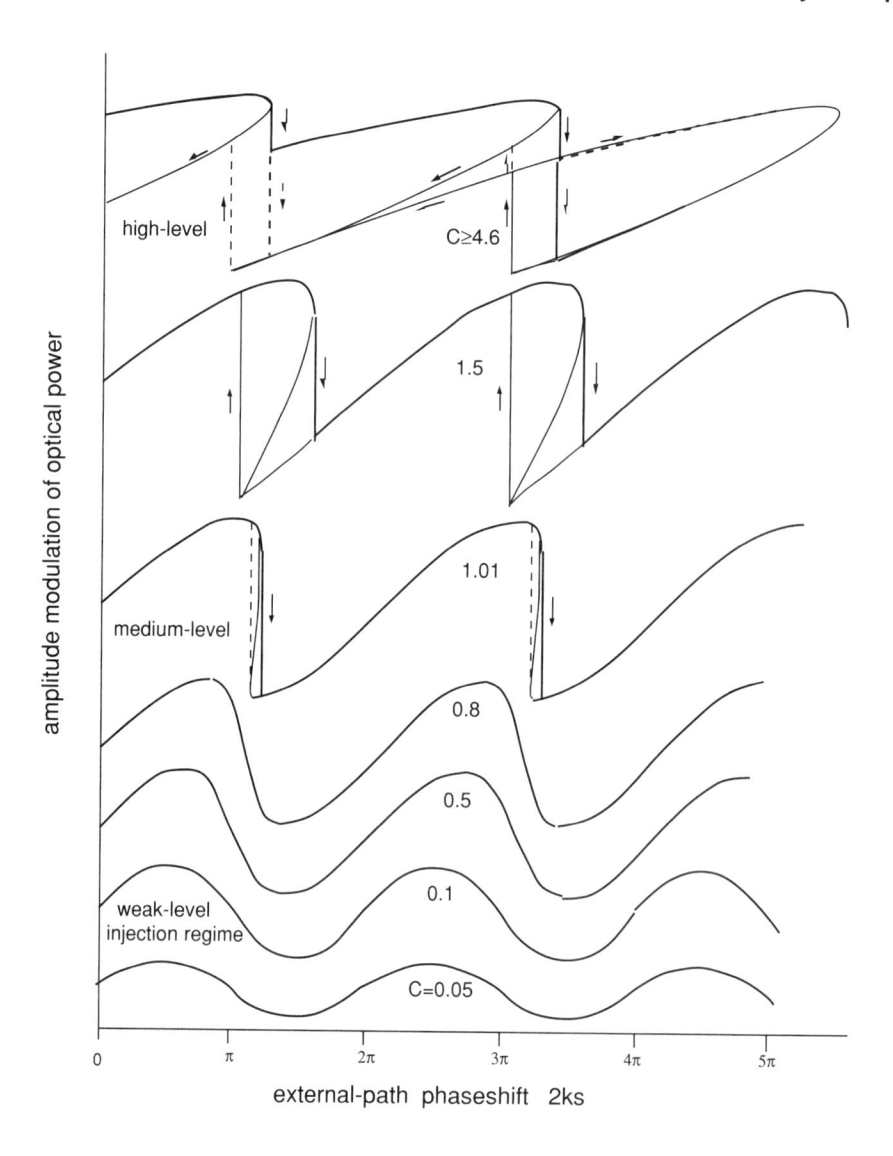

Fig. 4-28 Waveforms of optical power in an injection interferometer. At weak levels of injection (C<<1), the normal dependence cos2ks from external phase shift is found. Increasing the injection, the waveform becomes progressively distorted and asymmetric. As C reaches unity, a switching occurs with period 2π. If the target reverses its motion, the phase shift decreases, and the waveform becomes time-inverted, except at the switching point where now it follows the up arrow. Beyond C≥4.6, more than one switching per period occurs, and operation becomes affected by errors.

4.5 Read-Out Configurations of Interferometry

signal is carried by the beam itself, a distinctive feature compared to conventional interferometers. Most conveniently, however, we may use the monitor PD usually incorporated in the diode-laser package at the rear mirror of the laser.

Thus, a signal I=σP is obtained, which is proportional to the power P emitted by the laser and that carries the AM modulation waveform, that is, cos2ks at low-injection level.

We enter the moderate or medium level of injection when, at increasing back-reflected power, the interferometric signal cos 2ks becomes appreciably distorted (Fig.4-28). First, the waveform exhibits a leading semiperiod slower than the trailing semiperiod. Then, when injection is increased further and C (Eq.4.35) reaches unity, the waveform becomes sawtooth like, with a sharp switching in amplitude. Then, we find one switching per period of the optical path length 2ks, that is, one for each $\Delta s=\lambda/2$ increment of displacement and, important to note, the switching is negative-going when 2ks is increasing (Fig.4-28).

Now, if the motion of the target is reversed, the diagram of Fig.4-28 still applies, but with 2ks runs from right to left. The waveform has now a leading semiperiod faster than the trailing semiperiod and, most important, the switching is now positive-going (and occurs at a slightly changed value of 2ks because of hysteresis).

The strategy to make a displacement measurement without ambiguity easily follows. By operating the injection interferometer at C>1, we count each switching corresponding to $\lambda/2$. Counts are with the positive sign if the transition is up-going and with the negative sign if it is down-going. Operation is limited to the range 1<C<4.6 where there is just one switching per period (Fig.4-28) [15]. The experiment will be developed in more detail in next section. In the basic setup of Fig.4-27, the FM signal cannot be detected. However, we should expect that the sin2ks signal, similar to the cos2ks, is asymmetrical and has switching on its period [15].

4.5.2.5 Analysis by the three-mirror model

Let us consider the injection interferometer as a three-mirror cavity laser, as shown in Fig.4-29. Analyzing the round trip loop, we can write the field returning to the initial point as the sum of two contributions, from the paths internal and external to the laser cavity, as:

$$E\ r_1 r_2 \exp 2\alpha^* L \exp i2kL + E\ a \exp i2ks \qquad (4.40)$$

where r_1 and r_2 are the mirrors' field reflectivity, $\alpha^* = \alpha - \beta E^2 - \Gamma/c$ is the net gain (or, gain less loss) per unit length, L is the laser cavity length, and $a = A(1-r_1^2)\eta$ is total field loss in the external propagation to the third mirror. Thus, the loop gain is:

$$G_{loop} = r_1 r_2 \exp 2\alpha^* L \exp i2kL + a \exp i2ks = 1 \qquad (4.41)$$

Because of the well-known Barkhausen criterion of oscillators, G_{loop} shall be exactly unity in the permanent regime of oscillation, and phase ϕ_{loop} exactly zero, or:

$$G_{loop} = |G_{loop}| \exp i\phi_{loop} = 1, \quad \text{or also:} \quad |G_{loop}|=1, \quad \phi_{loop}=0 \qquad (4.42)$$

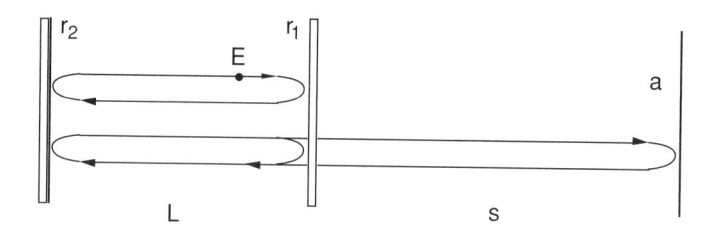

Fig. 4-29 Schematization of the injection interferometer as a three-mirror laser

When a=0, Eq.4.42 requires $|G_{loop}|=r_1 r_2 \exp 2\alpha^* L=1$ (net gain equal to mirror losses) and $\phi_{loop}=2kL=0_{[mod 2\pi]}$ (cavity resonance condition), or that 2ks is an integer multiple of 2π. Writing the condition as $2 \cdot 2\pi n_1 L v_0/c = N2\pi$, the unperturbed frequency is found as $v_0=Nc/2n_1 L$, with N being the order of the mode and n_1 the index of refraction of the active medium. Near to resonance, if the actual frequency v deviates from v_0, the part of 2kL in excess of 2π can be expressed as $2kL=4\pi n_1 L(v-v_0)/c$ [as $4\pi n_1 L v_0/c=N2\pi$].

In presence of feedback (a≠0), the phase condition is $\phi_{loop}=\mathrm{atan}[\mathrm{Im}G_{loop}/\mathrm{Re}G_{loop}]=0$ or, more simply, $\mathrm{Im}\{G_{loop}\}=0$, and explicitly reads:

$$r_1 r_2 \exp 2\alpha^* L \sin 4\pi L n_1 (v-v_0)/c + a \sin 2ks = 0 \qquad (4.43)$$

Using $r_1 r_2 \exp 2\alpha^* L=1$ in Eq.4.43, assuming in the first term that $v-v_0$ is small (so that $\sin x \approx x$) and noting that $2ks=4\pi s v/c \approx 4\pi s v_0/c$, we obtain:

$$v-v_0 + (c/4\pi L n_1) \, a \sin 4\pi s v_0/c = 0 \qquad (4.44)$$

This equation tells us that, in presence of injection, the actual frequency v deviates from the frequency v_0 of the solitary laser by an amount $(c/4\pi L n_1)$ a sin2ks, which is easily recognized to coincide with the factor Δ that is derived by Lamb's equations (Eq.4.37) treatment for small a.

Now, if we plot v versus v_0 as given by Eq.4.44, we get a sinusoid curve superposed to a straight line (Fig.4-30). When the distance s is changed by $\lambda/2$, the argument of the sine changes by 2π, and v covers the full swing of the sine curve displayed in Fig.4-30. Thus, the actual frequency obtained depends on the residue Δs of distance $s= N\lambda/2+\Delta s$ to the nearest integer multiple of $\lambda/2$. We can draw a horizontal line (the dotted line in Fig.4-30) to represent the working point determined by Δs, which moves vertically along a full period of the sinusoid as Δs varies of $\lambda/2$.

As long as the amplitude of the sinusoid is minute, only one intersection is found, and we are in the case of weak-feedback injection. However, at increased injection, we may have a curve with three intersections to a horizontal line. The frequency then stops following the sine curve smoothly and will flip suddenly to another portion of the curve (Fig.4-30).

We may take the boundary from the weak- to the moderate-feedback regime as that of the

4.5 Read-Out Configuration of Interferometry

curve with a horizontal flex. Mathematically, the condition of a horizontal flex in the y-x diagram, for a function $y = x + A \sin Bx$ is $AB=1$. Comparing with Eq.4.44, we find the condition $(c/4\pi L n_1) a 4\pi s/c = 1$, or also: $as/L n_1 = 1$.

We may now define a factor describing the strength of injection as $C = as/L n_1$ and state that, for $C>1$, we enter in the regime of frequency switching. This factor coincides with that already quoted in Eq.4.35, except for the term $\sqrt{(1+\alpha_{en}^2)}$ specific of the semiconductor lasers.

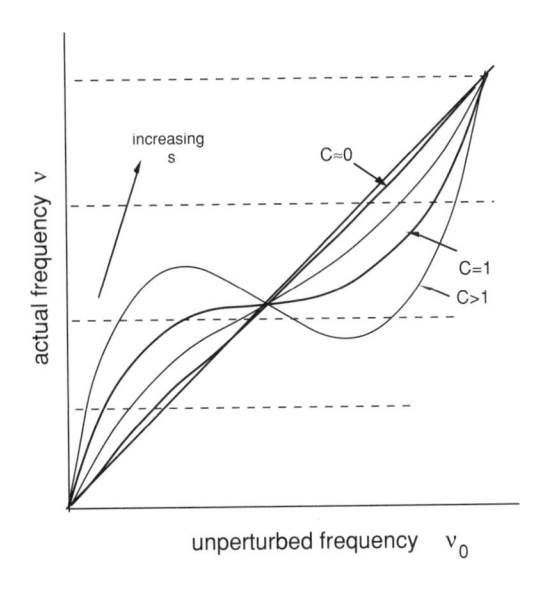

Fig. 4-30 Frequency of the injection-perturbed oscillation ν as a function of the solitary laser frequency ν_0. At increasing feedback level (C values), the curve has three intersects with the dotted line representing the distance s and exhibits switching.

About the AM amplitude signal, from the amplitude Barkhausen criterion, it is $|G_{loop}|=1$, or $Re^2(G_{loop}) + Im^2(G_{loop}) = 1$. Letting $E = E_0 + \Delta E$ in Eq.4.41, developing the mosulus $|G_{loop}|$ and considering $\Delta E << E_0$ in it, after some lengthy calculation we find:

$$\Delta E = E_0 (a \cos 4\pi s\nu/c)/[2L\gamma_0(1 + a \cos 4\pi s\nu/c)] \qquad (4.45)$$

This expression is coincident with the weak-feedback result (Eq.4.37a) for $a<<1$. Moreover, when the factor a increases, the waveform becomes distorted because of the cosine dependence at the denominator. At C=1, when frequency switches, ΔE and the AM signal will, too, because of the ν-dependence of the cosine term at the numerator.

By solving numerically Eq.4.44 (or using the diagram of Fig.4-30), we can find a function $\nu = F(4\pi s\nu_0/c)$ that replaces ν in $\sin 4\pi s\nu/c$ at the moderate feedback level.

Using ν=F(s) in Eq.4.45, the diagram of Fig.4-28 can be obtained.

As a conclusion of the previous analysis, by applying the Barkhausen's criteria to the three-mirror cavity model [15,18,22], we have been able to derive the basic results describing the moderate-level feedback regime. In particular, switching of frequency and amplitude signals are found to occur at a critical value C≥1. Using them, we are able to perform an unambiguous measurement of displacement using the amplitude waveform only.

4.5.2.6 Analysis by the Lang and Kobayashi equations

A more rigorous treatment of moderate-level feedback, including the effects specific to a semiconductor laser, is provided by the description through the Lang and Kobayashi equations [13,14]. These equations rephrase Lamb's equations for amplitude E and phase φ in a perhaps more detailed way, and also add a third equation for the active-level population and its dependence from pumping rate and cavity field. In presence of feedback, the Lang and Kobayashi equations read [14,15]:

$$dE(t)/dt = [g_N(N-N_{tr}) - 1/\tau_p]E(t) + (ac/2n_lL) E(t-\tau) \cos[\omega_0\tau + \varphi(t) - \varphi(t-\tau)] \quad (4.46a)$$

$$d\varphi/dt = \alpha_{en}g_N(N-N_T) - (ac/2n_lL) E(t-\tau)/E(t) \sin[\omega_0\tau + \varphi(t) - \varphi(t-\tau)] \quad (4.46b)$$

$$(d/dt)N = J/ed - N/\tau_s - g_N (N-N_{tr}) E^2(t) \quad (4.46c)$$

where: g_N is the gain coefficient [typically $8 \cdot 10^{-7} cm^3 s^{-1}$]; in terms of the Lamb's coefficient α, (Eq.4.36) it is $g_N = \alpha c/(n_2-n_1) = \lambda^2 c/8\pi\tau_{21}\Delta\nu_{at}$.

N=N(t) is the carrier (electron-hole pairs) density (cm^{-3}) in the active region.

N_{tr} and N_T are the carrier density (cm^{-3}) at transparency and at threshold of the solitary laser [typically a few $10^{18} cm^{-3}$], and it is $g_N(N_T-N_{tr}) = 1/\tau_p$.

τ_p is the photon lifetime [typically a few ns] related to absorption and mirror losses
$a = At_1^2\eta$ is the total *field* loss suffered by the reinjected field in the roundtrip path.
L is the laser cavity length, and n_l is the active-medium index of refraction.
τ =2s/c is the external roundtrip time.
ω_0 is the (angular) frequency of the unperturbed laser.
α_{en} is the linewidth enhancement factor [13] [typically 3 to 6 in diode lasers].
J is the pump current density (A/cm^{-3}).
τ_s is the charge-carrier lifetime [typically a few ps].

The analysis of the moderate-level injection [15] starts with letting dE(t)/dt=0 and (d/dt)N =0 in Eqs.4.46a and 4.46c, and solving for the stationary values E_{0f} and N_{0f} in presence of feedback. From Eq.4.46a we get:

$$N_{0f} = N_T - (ac/2n_lLg_N) \cos \omega_{0f}\tau \quad (4.47)$$

The term on the right-hand side shows that feedback induces a modulation of the carrier density N. As a consequence, a variation of voltage developed across the laser-diode junction is

4.5 Read-Out Configuration of Interferometry

induced [15, 21]. This is a signal that can indeed be recovered and used as the interferometer output in place of the normal photodiode output.

By inserting Eq.4.47 in Eq.4.46b, we find the perturbed oscillation frequency ω_{0f} as:

$$\omega_{0f} = \omega_0 - (ac/2n_1L) [\alpha_{en} \cos \omega_{0f}\tau + \sin \omega_{0f}\tau] \quad (4.48)$$

This equation can be brought to coincide with the Barkhausen phase-condition (Eq.4.44). Now, we can use the feedback parameter C, already defined by Eq.4.35 and repeated here for convenience:

$$C = a\, s\, (\sqrt{1+\alpha_{en}^2}) / n_1 L$$

By multiplying both members of Eq.4.48 by $\tau=2s/c$, it becomes:

$$\omega_{0f}\tau = \omega_0\tau - C/\sqrt{(1+\alpha_{en}^2)}[\alpha_{en}\cos\omega_{0f}\tau + \sin\omega_{0f}\tau] \quad (4.49)$$

or also, using an identity of trigonometry:

$$\omega_{0f}\tau = \omega_0\tau - C \sin(\omega_{0f}\tau + \operatorname{atan}\alpha_{en}) \quad (4.50)$$

In a 2π-period of $\omega_0\tau$, Eq.4.50 has just one solution for C<1 (weak-level injection), and three solutions for 1<C<4.6 (moderate-level injection). One solution is unstable, however, and implies a switching between the other two states, the stable solutions. Last, for C>4.6, multiple switching is found in a single 2π-period of $\omega_{0f}\tau$ (strong-level injection). From Eq.4.46c, letting $(d/dt)N=0$, the electric field with feedback E_{0f}^2 is:

$$E_{0f}^2 = [J/ed - N/\tau_s]/g_N(N_{0f} - N_{tr}) \quad (4.51)$$

whereas, from Eqs.4.46a and 4.46c, with no feedback, we have the quiescent point $E_0^2 = \tau_p g_N(N_0 - N_{tr}) = \tau_p(J/ed - N_0/\tau_s)$, and $P_0 \propto E_0^2$.

Using this result and N_{0f} as given by Eq.4.47, we obtain, after some algebra, the optical power variation $\Delta P \propto E_{0f}^2 - E_0^2$ induced by feedback as:

$$\Delta P = P_0 (ac\tau_p/n_1L) \cos\omega_{0f}\tau / [1+(ac\tau_p/n_1L) \cos\omega_{0f}\tau] \quad (4.52)$$

which is similar to Eq.4.45 and becomes the familiar $\Delta P_0 m_A \cos 2ks$ expression for a<<1.

A last result worth mentioning about the moderate-level injection [15,22] is the reconstruction of the displacement waveform s(t) from the AM interferometric signal given by Eq.4.52. By combining Eqs.4.50 and 4.52, we can obtain the following expression relating $\omega_{0f}\tau=2ks$ to the instantaneous $\Delta P/\Delta P_0 = F(t)$ modulation:

$$\omega_{0f}\tau = \pm \arccos F(t) - C\{-\alpha_{en}F(t) \pm \sqrt{[1-F(t)^2]}\}/\sqrt{(1+\alpha_{en}^2)} + 2m\pi \quad (4.53)$$

where the sign shall be taken '+' for $dF(t)/dt \times (ds/dt) < 0$, and '−' for $dF(t)/dt \times (ds/dt) > 0$, and m shall be increased by 1 every two zero-crossings of F(t) [15,22].

4.5.2.7 The laser-diode injection interferometer

Starting from the basic setup of Fig.4-27, a number of researchers [11,15,18-27] have reported about the development of laser-diode injection interferometers for measurements of displacements, velocity and for the detection of remote targets. The typical waveforms of the current detected under moderate feedback conditions with single-mode laser diodes emitting either in the visible (λ=680 nm) or in the near infrared (λ=800 nm) are shown in Fig.4-31.

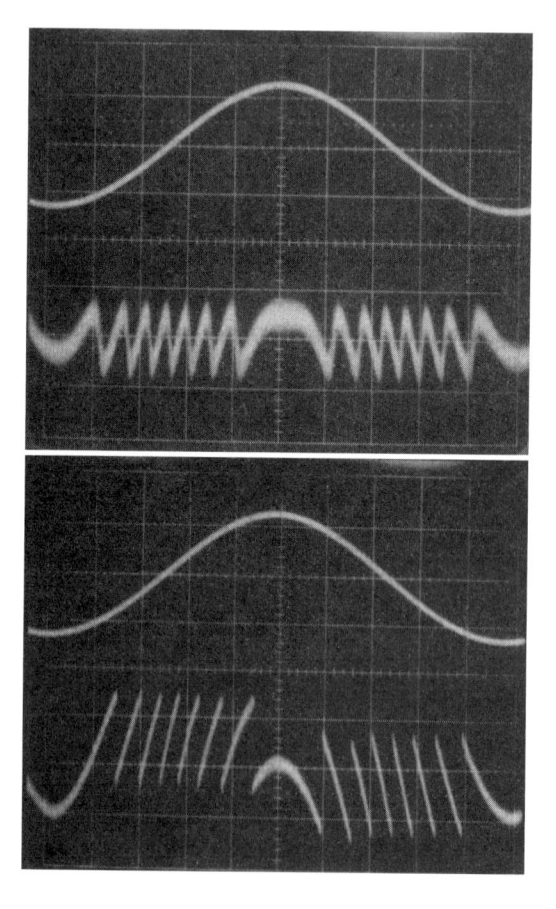

Fig. 4-31 Injection interferometry experimental signals, in response to sinusoidal drive s(t) signal (top trace in both pictures). Top picture: at C=0.6, the cos 2ks waveform is slightly distorted, with falling semiperiods somewhat faster than the rising ones; at C larger than 1, a switching occurs in the fall semiperiod (bottom picture is for C=2.2). Wavelength: λ=850 nm. Time scale: 2 ms/div. (from Ref.[15], by courtesy of IEEE).

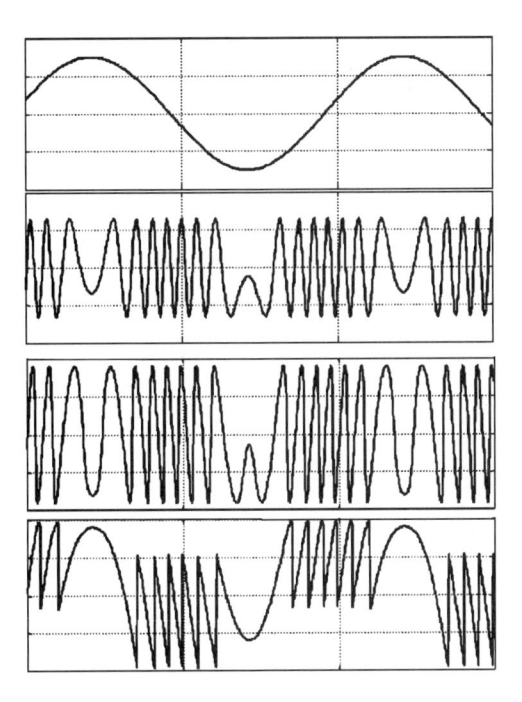

4.5 Read-Out Configuration of Interferometry 143

Fig. 4-32 Drive s(t) (uppermost) and theoretical ΔP/P waveforms corresponding to the experimental data of Fig.4-31, for a few values of the feedback parameter C, ranging from weak to moderate level of injection (C=0.01, 0.7, and 3.3 from top to bottom). The relative vertical scale of ΔP/ P signal is 0.5, 3, and 10 (top to bottom). Horizontal scale: 1ms/div.

Experimental results nicely match the analytical expressions found in the previous section, as exemplified by the curves reported in Fig.4-32 for a few values of C comparable with those of Fig.4-31.

The block scheme of a displacement interferometer working on λ/2-step counting is shown in Fig.4-33. The target is aimed through a collimating objective lens (typically a 5-mm focal length) or by a telescope (with typical ×20 magnification). An optical attenuator is inserted in the target path to adjust the feedback parameter to be in the range 1<C<4.6, the desired regime of a single switching per period.

The signal out from photodiode (in most practical schemes, the monitor photodiode of the laser chip is adequate for use) is passed to a transimpedance preamplifier and then to a differentiator with a short time constant τ_d=RC. Out of the differentiator, the up/down switchings are a sequence of positive/negative short ($\approx\tau_d$) pulses. We can easily separate the pulses in two channels, rectify the negative ones, and send one channel to the up-count input of a counter, and the other to the down-count input. Last, multiplying by the scale factor λ/2 brings the readout of the display to decimal metric units.

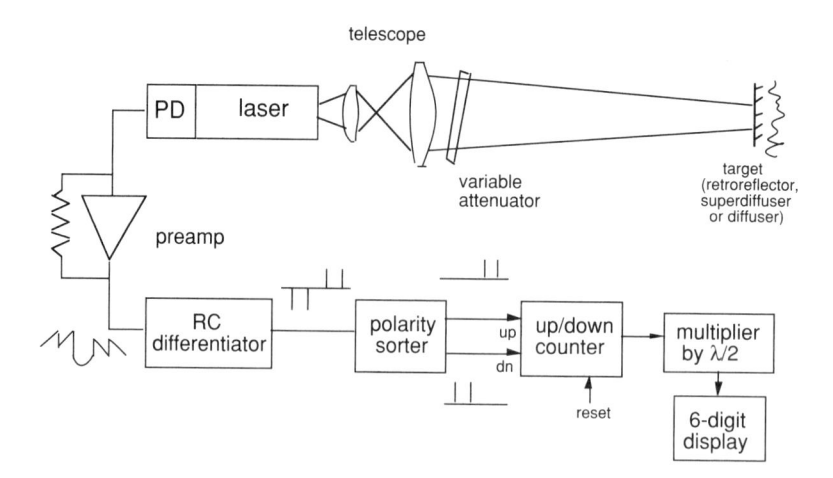

Fig. 4-33 Block scheme of the laser-diode injection interferometer working on the waveform switching to count displacement in steps of $\lambda/2$

Thus, with a minimum part count, we are able to build an interferometer measuring target displacement in steps of $\lambda/2$, or 425-nm per least-significant digit (at $\lambda = 850$-nm).

As in all interferometers, the maximum distance that can be covered in the displacement measurement is limited by the coherence length of the laser diode. This can be up to a several meters in selected narrow-line plain (Fabry-Perot) lasers.

The target for the measurement may be a corner cube, to allow for a sizeable attenuation, or a Scotchlite tape or varnish. This provides a gain ≈ 40 with respect to the normal Lambertian diffuser. Also, we may use a simple plain diffuser, that is, white paper or the like. In all cases, we shall keep the signal in the single switching (1<C<4.6) regime. This requires a careful design of the objective optics when using the plain diffuser because the signal from it is barely sufficient. With the corner cube or the superdiffuser, the signal is even too large, and we shall adjust it by the variable attenuator, eventually changing its setting with the working distance. It may then be advisable to add a servo-loop control of the attenuation to get a self-adjusting operation.

A strategy to stabilize the signal amplitude in the 1<C<4.6 range is the following. Superposed to the dc laser diode current, we add a small modulation signal i(t) at audio frequency. The emitted wavelength then varies as $\Delta\lambda(t) = \alpha_I \Delta i(t)$, where α_I is the current coefficient (see App.A1.3.1). We choose $2\Delta ks = 4\pi s \Delta\lambda/\lambda^2 \geq 2\pi$ so that at least one switching occurs in a period of modulation, even for a target at rest. Then, we look at the differentiated signal and, if it is missing, we decrease the optical attenuation until the signal shows up and allow for a moderate (e.g., x2) overdrive. To have a simple and cheap controlled-voltage attenuator, we can use a Liquid Crystal (LC) sandwiched between two polarizers. The birefringence of the LC is controlled by the applied voltage, resulting in an adjustable attenuation. The up and down switching of the modulation has a zero-mean value and does not interfere with the displacement measurement.

4.5 Read-Out Configuration of Interferometry

About precision, the laser-diode interferometer is different from the He-Ne laser because it does not possess an inherent calibration of wavelength. In a Fabry-Perot semiconductor laser, wavelength is markedly dependent on drive current and temperature, mode-hopping can occur, and the actual $\lambda=\lambda(I,T)$ curve is also dependent on the previous story of I,T changes. In the most common and cheaper Fabry-Perot cavity lasers, the typical temperature and current coefficients are $d\lambda/\lambda dT = 2-5 \ 10^{-6} (°C^{-1})$ and $d\lambda/\lambda dI = 1-3 \ 10^{-6} (mA^{-1})$ (see also App.A1.3.1).

Thus, we may expect an error in the displacement measurement of several ppm for a 1-°C or a 1-mA variation of chip temperature or diode drive current.

The situation is much better in DFB or DBR lasers, where the incorporation of a grating reflector alleviates the temperature dependence and nearly eliminates the current dependence. However, these lasers are not easily available so far for wavelengths other than the fiber third-window ($\lambda \approx 1500$ nm) and are much more expensive than Fabry-Perot types.

Altogether, the previous points call for a thermal stabilization of the laser chip (with a TEC or Peltier cell module) and of the drive current (by a regulated and low-ripple supply). With Fabry-Perot lasers, using a stabilization of 0.01 °C in temperature and ±10 μA in current, the instrumental accuracy of the injection interferometer has been evaluated to be ±1 count ($\approx 0.4 \mu m$) at 50 cm distance, or ±1 ppm [15]. In addition, DFB lasers with a long-term stability of $\Delta\lambda/\lambda = 0.1$ ppm have been reported in the laboratory [28].

To extend the operation of the interferometer to sub-λ resolution, the photodetected signals can be acquired by a computer, and the calculation of 2ks is then performed with the aid of Eq.4.53. This scheme of analogue processing has been shown [10,22] to be able to reconstruct the displacement waveform s(t) with a ±5nm accuracy, with no constraint imposed on the signal s(t) waveform.

In conclusion, the advantages and disadvantages of the laser-diode injection interferometer, as compared to conventional types, can be summarized as follows:
- It has a minimum optical part-count (no external optical interferometer) and relatively simple electronics.
- The optical head (laser, lens, attenuator) is very compact with a distinct advantage for ease of mounting in experiments short of space.
- No critical alignment on the target is required.
- No spatial, wavelength or stray-light filtering are necessary; the laser cavity itself provides these functions.
- Operation on a normal diffusing target surface is readily achieved, without modifications of the setup.
- The interferometric signal can be detected everywhere on the beam, and also from the target side.
- $\lambda/2$ resolution with digital processing and sub-λ sensitivity with analogue are easily achieved up to several hundreds kHz or MHz bandwidth.
- Performances are comparable to those of conventional interferometers.
- Wavelength accuracy and long-term stability may be of concern.
- Flexibility of reconfiguration is modest.

4.6 LASER VIBROMETRY

A laser vibrometer is an interferometer especially designed for use as an instrument capable of detecting and measuring small vibrations in a variety of applications related to mechanical, automotive, avionics, and construction engineering, as well as in biological structures. Usually, vibration means a periodic displacement waveform, with typical amplitudes of interest ranging from less than 1 nm to tens of µm, and frequency ranging from less than 0.1 Hz to perhaps 1 MHz.

Vibrometry is an important diagnostic tool and is commonly used to measure vibrations of a structure under test or to determine its frequency response to a periodic excitation applied externally. Both modes of operations allow collecting important information about the structure for design, verification, and testing purposes. With respect to other measurement techniques based on electronic sensors, the laser vibrometer has the key advantage of noncontact operation and does not disturb the structure under test.

The best configuration of the vibrometer is different according to the allowed or required distance of operation of it from the target under test. In the following sections, we describe vibrometers aimed to a performance optimized on three ranges of interest: (i) short distance (say less than 10 cm), (ii) medium distance (a fraction to several meters), and (iii) long distance (up to hundreds meters).

Basically, we may also use a displacement interferometer as a vibrometer (Sect.4.2.3.4), but three specific features shall be considered for design optimization.
First, superposed to the small displacement, we may find a relatively large (up to hundreds of µm) ambient-induced microphonics disturbance. This can be strongly reduced if we are permitted to work on an antivibration laboratory table, but cannot be avoided if we aim to in-the-field operation of our instrument. Second, we are forbidden to use anything that disturbs the target and its vibration state, like a corner cube, and the vibrometer is to work directly on the natural target surface, normally a gray ($\delta<1$) diffuser. Thus, the returned field is much attenuated and subjected to the speckle-pattern statistics (Sect.4.4.7 and Ch.5). Last, the range of displacement to be measured is very small and it may permit relaxing the coherence length needed when operation is on short distance, or when we can balance the interferometer arm path lengths.

4.6.1 Short-Distance Vibrometry

Early experiments in short-distance interferometric vibrometry were developed in the 1970s around a Michelson configuration, which was modified by adding lenses focusing the reference and measurement beams and by adding a provision for dynamical tracking of the reference path to compensate for the microphonics-induced ambient disturbances (Fig.4-34). This is accomplished by the piezo actuator in the reference path, which is either a diffuser (as shown in Fig.4-34) or a reflective surface. Light scattered by the diffusing target in the measurement arm is collected back by the lens and recombined on the photodetector.

4.6 Laser Vibrometry

The photodetected signal, $I_{ph}=I_0[1+\cos 2k(s_m-s_r)]$, is preamplified by the front-end giving a signal $V_{ph}=RI_0[1+\cos 2k(s_m-s_r)]$ and is low-pass filtered with corner frequency f_1. A constant voltage V_{ref} is then subtracted from V_{ph}, and the result, amplified by A, constitutes the error-signal supplied to the piezo.

Because of the feedback loop, the system adjusts itself at a quiescent point where the error signal is nearly zero, or $V_{ph}-V_{ref} \approx 0$. Setting $V_{ref}=RI_0$, we get the condition $0= V_{ph}-RI_0 = RI_0 \cos 2k(s_m-s_r)$. For s_m very small, the quiescent point is $2ks_r= \pi/2$, and we get:

$$V_{out} \approx RI_0\, 2ks_m = V_{ref}\, 2ks_m \qquad (4.54)$$

The best choice for the low-pass frequency f_1 depends on the actual level and frequency content of the disturbance coming from the ambient. In most cases, the disturbance spectrum is found to damp off quickly between, say $f_{dist}= 100\text{-}1000$ Hz.

Choosing $f_1 \approx f_{dist}$, the low frequency components ($f< f_1$) are suppressed because of the feedback effect. Instead, signals $s_m(t)$ with frequency content $f>f_1$ find the loop ineffective (interrupted by the filter) and are passed on to the output. From a well-known result of feedback theory, the actual suppression factor is given by the loop-gain G.

By inspection of Fig.4-34, the dc loop gain is easily found as $G_0=R\, I_0 A\kappa$, where $\kappa =\Delta s/V$ (μm/Volt) is the conversion factor of the piezo. We can readily make $G_0=10^3$ and nearly eliminate the disturbance. In frequency, the loop gain is $G(f)= G_0/[1+(f/f_1)^2]^{n/2}$, where n is the order of the filter (n=1 for a simple RC). Signals with $f>f_1$ find a small $G(f)$ and are not reduced, so we can take them either from the transimpedance output or from the filter high-pass output as indicated in Fig.4-34.

Of course, signals with frequency content $f<f_1$ are suppressed, like the disturbances. To minimize loss of useful spectrum, we will make the filter frequency f_1 adjustable and trim it to optimize the outcome of the measurement by inspecting the result.

The speckle-pattern regime of operation is of little problem for the vibrometer. Should eventually the reference or measurement fields fall on a dark (weak amplitude) speckle, it will suffice to move the target surface transversally just a little bit (by a few μm) to find another, bright high-amplitude speckle.

The same result can be obtained by a fine position adjustment of the lenses, too.

Regarding the phase error generated by the speckle-pattern statistics (Sect.4.4.7), the re-radiated field received through the lens is in the far-field of the target [large S_1 in Eq.(4.31)] and generates a negligible error, whereas changes in the spot projected on the target due to longitudinal or transversal displacements may be of concern.

The longitudinal displacement introduces a negligible error as far as the amplitude of vibration (signal plus disturbance) is small as compared to the depth of focus of the lens Δf. This is given by $\Delta f =\lambda/\pi NA^2$, where NA is the lens numerical aperture and typically is $\Delta f\approx$ 0.1-1 mm.

On the other hand, transversal displacements of the target are to be more tightly controlled. They are required to be small with respect to the focused spot size $\Delta w= \lambda/\pi NA$ (Ref.[5], App.A2), which may have a typical value of ≈5 to 30 μm.

Thus, it may be important to align the specimen under test to get the vibration to be parallel to the impinging beam axis.

In addition to the mechanical segment, a short-distance vibrometer is useful for noncontact measurements of otherwise undetectable minute vibrations in a variety of devices, including electronic and piezoelectric components. As an illustration, we report in Fig.4-35 the measurement of sub-nm amplitude waves in a Surface Acoustic-Wave (SAW) filter, obtained with a setup similar to that reported in Fig.4-34. The result has pointed out the difference between the Stoneley regime and the normal SAW regime [29] of surface waves.

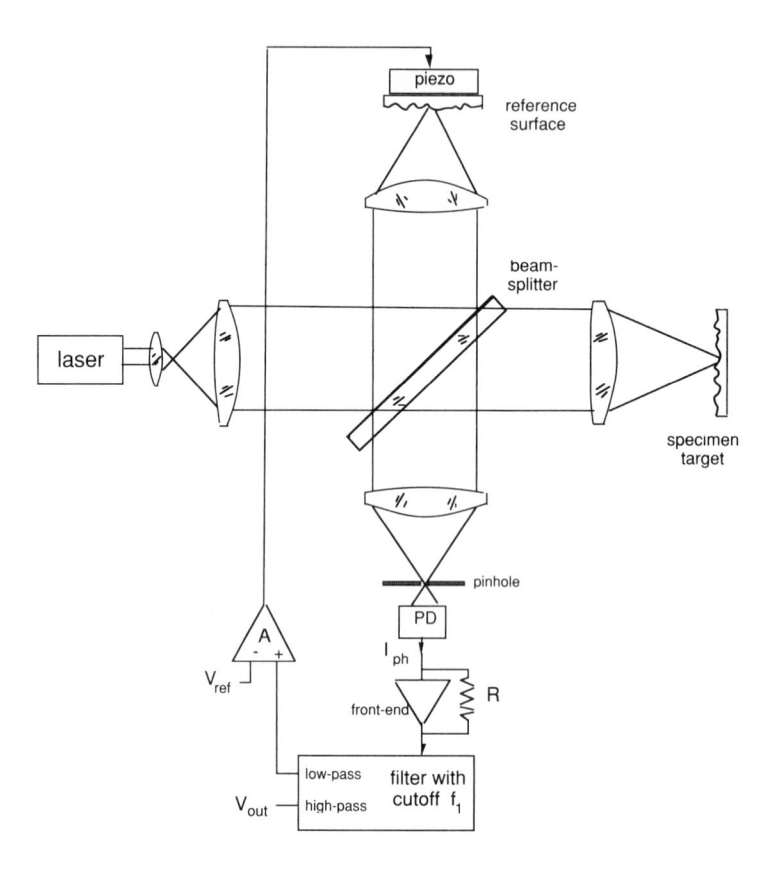

Fig. 4-34 A short-distance vibrometer for measurements on small size specimen. The reference beam is focused on the diffuser surface driven by a piezo actuator. The interferometric signal is low-pass filtered, amplified, and sent to the piezo. Because of the feedback loop, the working point is locked to $I_{ph}=V_{ref}/R$, the half-fringe point of the interferometric signal $I_{ph}=I_0[1+\cos 2k(s_m-s_r)]$. The vibration signal, at $f>f_1$, is taken out from the high-pass output V_{out} of the filter as $RI_0\, 2ks_m$.

4.6 Laser Vibrometry

To evaluate the ultimate sensitivity (or NED) of the vibrometer, we may start considering that the optical power P_0 supplied by the laser is sent half to the measurement and half to the reference path. All this power reaches the target, but, in the rediffusion process, only a fraction η of it is sent back into a single spatial mode for coherent superposition at the photodetector.

If w_1 is the spot size of the Gaussian beam and $\theta=\lambda/\pi w_1$ is the associated diffraction angle, the power fraction is $\eta=(\lambda/\pi w_1)^2$. With typical values for a He-Ne laser, $P_0=1$mW, $\lambda=0.633$ μm and $w_1=0.5$ mm, we have $\eta=1.6 \cdot 10^{-7}$. The useful power is then $\eta P_0/2=0.8 \cdot 10^{-10}$ W, and, from the diagram of Fig.4-18, we can find the minimum measurable amplitude of vibration as NED=$(0.6 \cdot 10^{-6}$m$)\sqrt{B/P}$= 6.7 pm/$\sqrt{B_{(Hz)}}$, where B is the bandwidth of the measurement in Hz.

This very appealing result is to be regarded as a conservative estimate. Indeed, a real target is not an ideal diffuser, but may even have a minor (e.g.<0.1%) residual specular reflection. The reflection helps because it makes the statistics more regular. This will readily overwhelm the very small η and make the signal increase even of a few order of magnitude.

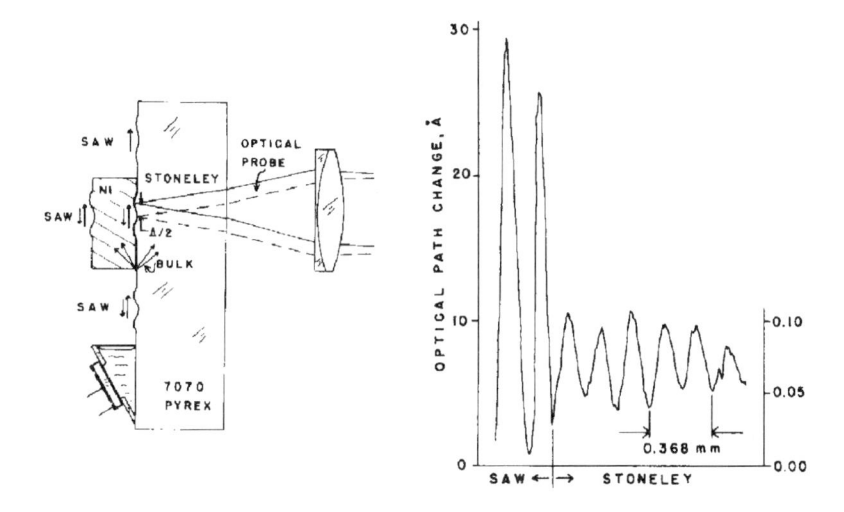

Fig. 4-35 Detection of sub-nm (1Å=0.1 nm) vibrations in a SAW device. Bulk Stoneley waves and ordinary surface waves are discriminated by their amplitude. Central frequency of the SAW is 8 MHz (from Ref.[29], by courtesy of the American Physical Society).

With regard to the spatial resolution of the vibration measurement, this is given by the spot size Δw projected by the lens on the target in the focal plane.
In general we have (Eq.A.2.10, Ref.[5]) Δw=λ/πNA, where NA=arcsin D/2F is the numerical aperture of the lens. However, if the lens diameter D is not filled completely by the beam, NA is determined by the spot size radius as NA= arcsin Δw/F.

Using $\lambda=0.633$ µm, $w_1=0.5$ mm, and F=20 mm, we obtain a spot size $\Delta w=16$ µm. This figure represents the elemental pixel of spatial resolution.

Last, the bandwidth of response of the vibrometer is primarily determined by the high frequency cutoff of the photodiode transimpedance amplifier and associated electronic processing circuits. With an available optical power in the $\eta P_0 \approx 100$ pW range (or, $I_{ph} \approx 100$ pA), we may use a 10 to 100kΩ feedback resistance and get, from state-of-the-art design (see Sect.3.5, Ref.[5]), bandwidth of 1 to 10 MHz.

In conclusion, the approach described previously is a good technical choice to develop a short-distance vibrometer with nm-sensitivity, ≈20 µm sampling size, and MHz bandwidth.

We can classify this vibrometer as homodyne with low-frequency half-fringe stabilization as compared to other types described in the following sections. Homodyne is because the reference and measurement beams share the same frequency, half-fringe stabilization is because we work at the midpoint of the sine-signal dynamic range and, by doing that with a feedback loop, we get rid of slow undesired low-frequency microphonics induced by the external ambient.

4.6.1.1   High-frequency fringe tracking

Another approach, evolved from the basic scheme of Fig.4-34, is that of homodyne with *high-frequency* fringe tracking. If we remove the low-pass filter in Fig.4-34 and use a small, high frequency piezo (or any other path-length actuator) with say, a $f_p \approx 1$ MHz passband, we are able to lock the interferometric signal always at half-fringe, irrespective of the signal spectrum, up to f_p.

Because the error signal is always null-balanced, or $RI_0 \sin 2k(s_m-s_r) \approx 0$, the photocurrent I_{ph} shows no appreciable signal. However, because it is also $s_m - s_r \approx 0$, the reference s_r tracks the measurement displacement s_m. Thus, if the piezo response is linear, its drive voltage $V_{dr} = \Delta s_r/\kappa$ is just proportional to the desired signal s_m to be measured.

One advantage of the high-frequency tracking is the much increased dynamic range and improved linearity. Because the working point is always at half fringe, the responsivity $R = \Delta s_m/V_{dr}$ is a constant at a value found from $V_{dr} = ARI_0 \sin 2k(s_m-s_r)$ as $R = 1/(ARI_0 2k)$. This is true even for a large ($\gg \lambda/2$) amplitude of vibration, which suffer from the nonlinearity of the sine function in the low-pass scheme. Also the dynamic range is increased to a value $G_0 \lambda/2$, because of feedback, and can easily attain several hundred micrometers.

On the other hand, a drawback of the high-frequency scheme is that spurious ambient-related disturbances are not eliminated, and we need to look for an alternative strategy to get rid of them.

As the superposed disturbance is physically undistinguishable from the signal, we need supplementary information about it to succeed. One is about the spectral content of the signal. If we look to signals with frequency well in excess of the typical disturbance frequency ($f_{dist} < \approx 100$Hz), we can obviously simply filter out the disturbance from the output V_{dr} signal. Second and less demanding, frequently it happens that the experiment allows us to switch on and off the signal, and the perturbation is statistically well behaved and stationary (that is, has an average power spectrum stable in time).

4.6 Laser Vibrometry

Then, we may take advantage of the linear regime of response. Because of this, the output $V_{dr}=V_m+V_{arv}$ is the sum of the effects separately produced by measurement s_m and ambient-related disturbance s_{arv}. We cannot hope the disturbance will replicate itself in different runs, i.e., computing the signal as $V_m=V_{dr}-V_{arv}$ is not acceptable, whereas the corresponding power spectrum $S_{arv}(\omega)$ is much likely to be constant, at least in a short period of measurement time. Therefore, we acquire the signal V_m during two equal time slots, one with signal on and one with the signal off and compute its power spectrum (e.g., by a FFT routine). The corresponding power spectra are $S_{dr1}=S_m+S_{arv}$ and $S_{dr2}=S_{arv}$. By subtraction, $S_{dr1}-S_{dr2}=S_m$ is the required signal power spectrum that is obtained.

The price for such a power-spectrum subtraction scheme is that the phase relationship of different frequency components in the signal is lost because the power spectrum is the square modulus of the Fourier transform of the s(t) signal in the time domain.

Instead of a high-frequency feedback loop, we may also think of a quasi-dc, very-low frequency compensation of the reference path, again looking at the photocurrent half-fringe. This is easily implemented with nearly any kind of actuator (motorized micropositioning stage, loudspeaker cone, or even a manual control) and has been experimented with since the early time of vibrometry. Of course, we don't have, in this case, any benefit from the loop-gain on the dynamic range (s_m-s_r is limited to $\lambda/2$) nor on linearity (the straight sin2ks dependence applies). However, the fringe tracking avoids fringe hopping and the related erratic functioning of the vibrometer, and we can again apply the spectrum subtraction scheme, though with a much more limited performance.

4.6.2 Medium-Distance Vibrometry

In the range of medium distances (a fraction to several meters) the concept of injection interferometry is readily applied to develop a homodyne, high-frequency fringe-stabilized vibrometer with high performances.

4.6.2.1 An Injection Vibrometer, fringe stabilized

As it will be clear in the following, the best choice is to operate at the medium-level of injection (Sect.4.5.2.3), that is, with 1<C<4.6.

Because we have C≈as/L from Eq.4-35, the desired level C≈1 requires a distance s≈L/a. Recalling the result of the efficiency calculation in last section, $a=\sqrt{\eta}=\lambda/\pi w_1$, we may have typically a≈4 10^{-4}. For a normal L≈200-μm diode-laser cavity, this requires a distance s≈ 0.2mm/ 4 10^{-4}= 50 cm, which is a value classified as medium-distance range (s≈0.3...3 m) of operation. This range is still covered by the coherence length of plain Fabry-Perot laser diodes, the ideal choice for low cost and simple set-up.

In the medium-level injection regime, the photocurrent I_{ph} is a distorted sine wave response resembling a saw-tooth waveform (Figs.4-28 and 4-31 or 4-32). We will use the quasilinear ramp semiperiod for operation.

Thus, linearity of response (before feedback is applied) is improved with the additional advantage of expanding the dynamic range of control.

The injection interferometer does not have a reference path, and this may appear a serious hindrance. However, we can take advantage of the wavelength dependence $\lambda=\lambda(I_{dc})$ from the laser drive current to adjust 2ks exactly to $\pi/2$. As shown in the schematic of Fig.4-36, we look at the interferometric signal from the front-end preamplifier, $V_{ph}=RI_0(1+\cos 2ks)$, and subtract the half fringe dc term $V_{ph0} =RI_0$. The loop ends up driving the laser current, with the error signal superposed to the fixed dc level $I_{dc} =V_{bias}/R_c$ chosen to set the laser power.

To lock the signal at half-fringe from any starting phase, we have to change the optical phaseshift $\phi=2ks$ by up to 2π at a distance s from the target.
We do that by impressing a wavelength variation $\Delta\lambda$ to the initial λ_0. This produces a phase variation $\Delta\phi=2s\Delta k =-2s(2\pi/\lambda_0^2)\Delta\lambda$, having used $k=2\pi/\lambda$.
By letting $2s(2\pi/\lambda_0^2)\Delta\lambda=2\pi$ and solving for $\Delta\lambda$, we obtain the necessary $\Delta\lambda$ as $\Delta\lambda= \lambda_0^2/2s$. Because the wavelength coefficient of the current dependence is $\alpha_1=\Delta\lambda/(\lambda_0\Delta I)$, the current variation required for a 2π-regulation of ϕ is found as $\Delta I=(1/\alpha_1)\Delta\lambda/\lambda_0 =(1/\alpha_1)\lambda_0/2s$.
Using the values of App.A1.3.1, we get for $\lambda=800$ nm and s=200 mm the typical value $\Delta I=(2\ 10^{-6})^{-1}0.8\ 10^{-3}/400=$ 0.16 mA.

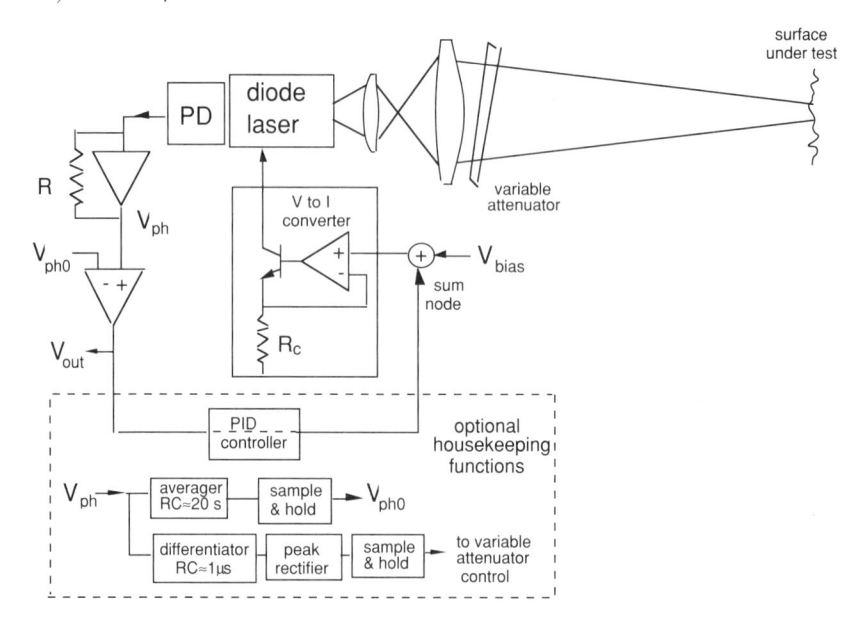

Fig. 4-36 An injection vibrometer for medium distance (simplified schematic) is designed around a feedback loop ending on the laser drive current to keep 2ks stabilized to $\pi/2$. The output signal $V_{out}=V_0 2ks$ is taken from the amplified error signal. The high-frequency loop ensures a very wide dynamic range and an improved linearity.

4.6 Laser Vibrometry

153

This is a small value as compared to perhaps I_{dc}=30-70 mA of normal rating current of our laser diode. We may take advantage of this circumstance by allowing a much larger current regulation range, e.g., I_{reg}=8 mA. Therefore, we may recover a large number $I_{reg}/\Delta I$=50 of 2π-phase periods, or, equivalently a $\lambda(I_{reg}/\Delta I)$=40µm drift, still keeping the linearity at the best half-fringe value.

The upper limit to I_{reg} is the incipient mode hopping, depending on the specific laser and ≈10 mA in best units with a TEC control of chip temperature (App.A1.3.1).

About the frequency response of the current-actuated half-fringe feedback loop, the theoretical limitations are: (i) the 2s/c delay (typ. a few ns), (ii) the laser response, involving the τ_p and τ_s time constants (ps to ns, see Sect.4.5.2.6), and (iii) the frequency response of the electronics, including front-end and laser drive-circuit response.

This response can be readily designed to exceed 10 MHz.

The output signal can be taken from the output of the transimpedance, proportional to the current error signal ΔI. Indeed, from $\Delta(2ks)=-2sk\Delta\lambda/\lambda+2k\Delta s=-2ks\ \alpha_I\Delta I+2k\Delta s$, we can solve for Δs as $\Delta s=s\alpha_I\Delta I$. This equation reveals that the instrument shall be calibrated once for all against the current coefficient α_I and in the actual use against the distance s. Experimentally, the scale factor $1/s\alpha_I = \Delta I/\Delta s$ fairly well matches the 1/s dependence (see Fig.4-37) and is well reproducible (with a <1% error) from measurement to measurement.

In the practical implementation of the injection vibrometer, a few refinements are worth addition, as outlined in Fig.4-36.

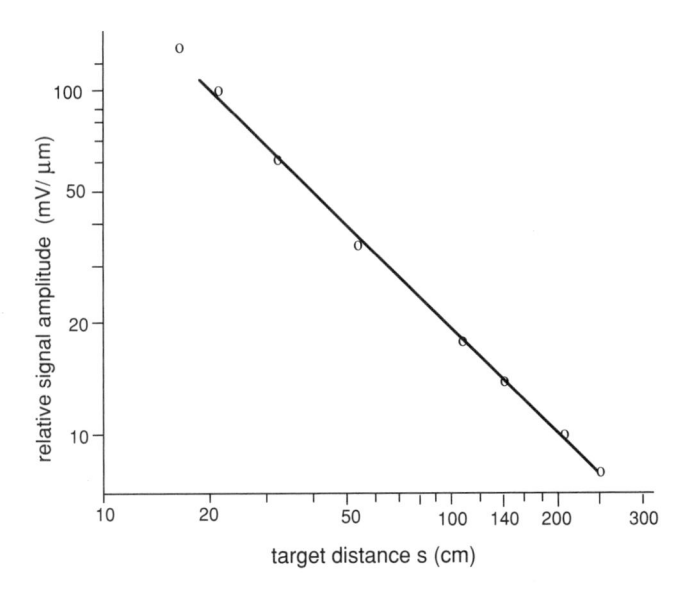

Fig. 4-37 The scale factor of the output signal from the injection vibrometer. Line: theoretical 1/s dependence; points: experimental values.

A PID controller (a network, well known in control technique, including a gain trimmer, a low-pass, and a high-pass filter) can be inserted to optimize the frequency response of the regulated loop. Also, an automatic reference circuit, made by a long time-constant averager and a sample-and-hold can be used to avoid readjusting the reference V_{ph0} when optical power changes.

Last, to track the 1<C<4.6 condition against distance and target diffusivity coefficient, a control of the variable attenuator may be envisaged.

The performance of an instrument developed [30] with the half-fringe stabilization concept and operating in the injection regime described previously, are shown in the diagrams of Fig.4-38. Limit sensitivity is down to ≈100 pm/√Hz, and the maximum amplitude is about hundred μm (in total, six decades in amplitude dynamic range). A frequency response of up to hundreds of kHz is covered.

The linearity of the measurement is better than 1% as a deviation measured from true s(t). Also, the instrument operates on untreated target surfaces with diffusivity as low as 0.3 [30].

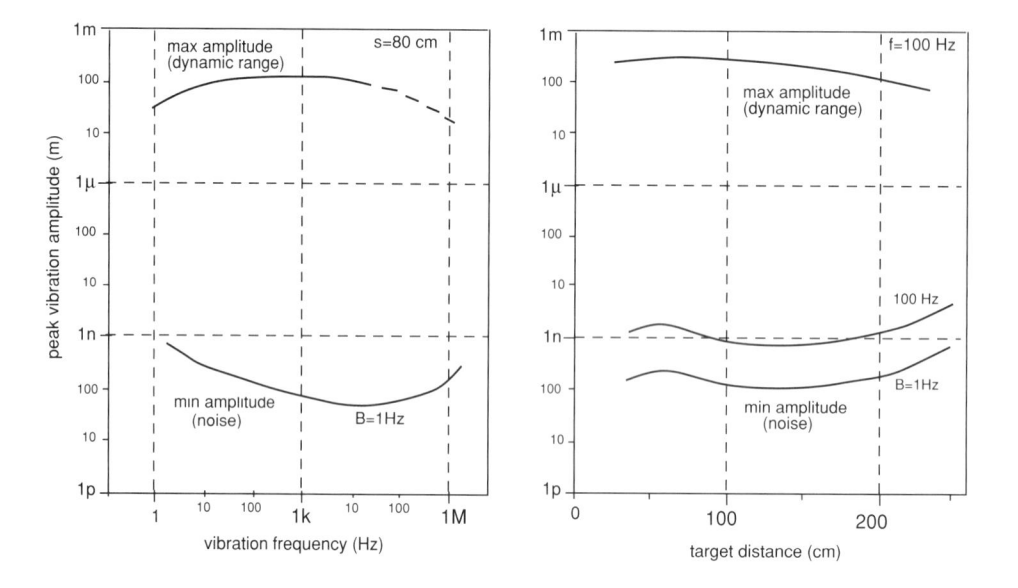

Fig. 4-38 Performance diagram of an injection vibrometer, stabilized to half fringe, operating on a diffusing target surface [30]. The laser source is a 25 mW near infrared (λ=780 nm) commercial laser diode. Left: at a representative s= 80 cm distance, amplitude of vibration is measured up to >100kHz with 100 nm (or better) resolution (bandwidth is B=1Hz) and up to 100 μm as limited by the dynamic range of the fringe tracking circuit. Right: at a representative 100 Hz signal frequency, resolution is 0.1 to 1 nm (B=1 to 100 Hz) from s=30 to 250 cm, still with up to 100 μm maximum amplitude.

4.6 Laser Vibrometry

4.6.2.2 A plain injection vibrometer

In some circumstances, fringe stabilization at C>1 is unwieldy, either because the target diffusivity is very low or the laser diode has a long cavity and small spot size. Another reason is that we may want to trade performance for low cost and layout simplicity while retaining the basic advantages of the injection concept (self-alignment, self-filtering, etc).

Then, we can simply arrange for a minimum-count injection vibrometer made by an optical section including a diode laser, an objective lens, and an attenuator, and an electronic section made by the front-end preamplified connected to the laser monitor photodiode.

With this arrangement, we get the plain interferometric signal $I_{ph}=I_0(1+\cos 2ks)$, where $s=s_m(t)+s_{micr}(t)$, $s_m(t)$ is the useful vibration signal and $s_{micr}(t)$ is the ambient-induced microphonics. As previously noted, we can separate these two addends if they lie in different frequency bands. Also, the cosine ambiguity is no problem if we look to a sinusoidal vibration $s_m(t)$, especially if the $s_m(t)$ swing has a multi-λ amplitude. The $s_{micr}(t)$ disturbance then helps avoid the signal to be pinned at the $2ks=n\pi$ worst condition.

An example of application of the approach is provided by a prototype instrument intended for MEMS testing [31], shown in Fig.4-39. The beam is focused on a MEMS Si-chip, typically 500µm by side, incorporating a comb-like spring structure and a mass, to diagnose the mechanical properties of the structure. The mass is aimed at by the focused laser spot (typ. 100 µm), but, even if some nonmoving parts are illuminated, by virtue of the injection process, they do not disturb the detection of the useful signal.

In the measurement setup (Fig.4-39), the MEMS is mounted inside a vacuum chamber to be able to test it at different ambient pressures. The window of the glass chamber does not need to have a good optical quality, because the wave front distortion is just a small additional attenuation, tolerated by the injection vibrometer. The out-of-plane vibration of the MEM mass is viewed at an angle Φ ($\approx 20°$ in Fig.4-39), and the appropriate cos Φ correction on s(t) is applied to the measurement.

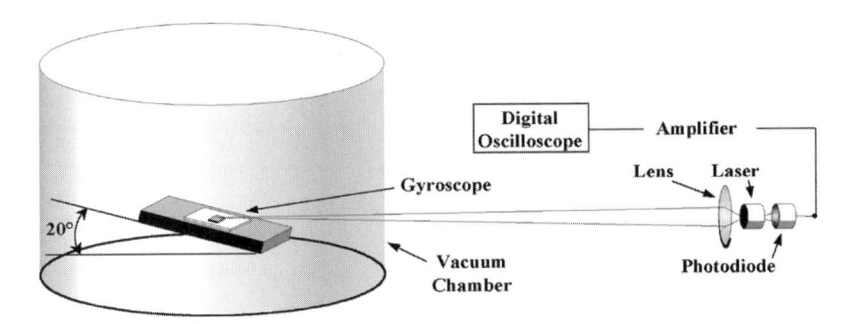

Fig. 4-39 Setup for measuring the mechanical properties of a Si-MEMS by means of an injection vibrometer. Light from the laser is focused on the vibrating mass of the MEMS chip through the glass wall of the vacuum chamber (from [31], by courtesy of IEEE LEOS).

156   Laser Interferometry   Chapter 4

The signal out from the preamplifier is acquired by a digital oscilloscope, and the desired s(t) waveform (Fig.4-40) is computed from the $I_{ph}=I_0[1+\cos(2kA_0\cos\omega_0 t)]$, by means of a routine minimizing the rms errors of the desired quantities, amplitude A_0 and frequency ω_0.

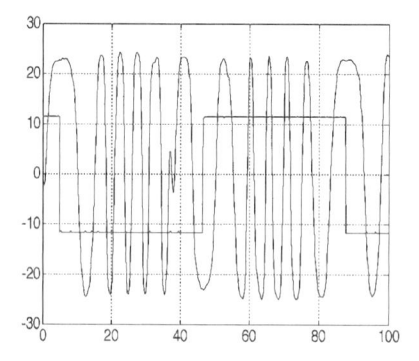

Fig.4-40 The MEMS drive signal (square wave) and the interferometric signal output $I_{ph}=I_0[1+\cos(2kA_0\cos\omega_0 t)]$, as obtained from the vibrometer of Fig.4-39.

By sweeping the drive signal frequency, we can measure the frequency response of the MEMS in several conditions of interest and test the dependence on drive voltage and ambient pressure. Experimental results in Fig.4-41 show the effect of increasing drive voltage (and vibration amplitude). The response is linear for low V, but shows incipient hysteresis at V= 8-9 Volt, indicating the risk of creep in the structure. Also, the effect of damping of the mechanical Q-factor because of insufficiently low ambient pressure is clearly evidenced.

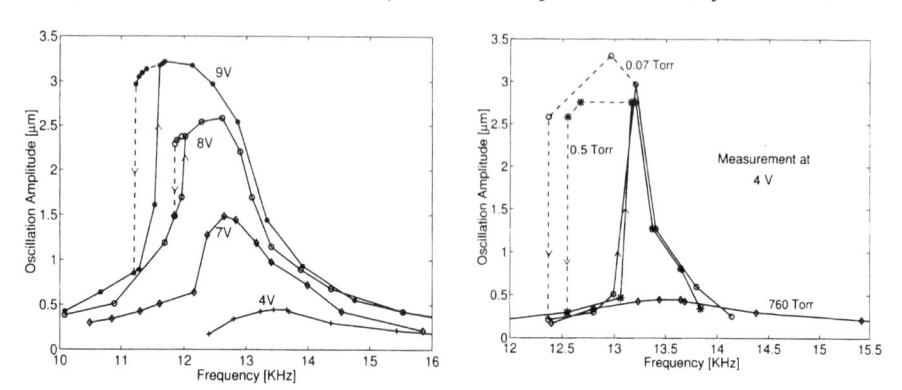

Fig.4-41 The MEMS frequency response, as measured by the injection vibrometer (horizontal scale in kHz, vertical scale in relative units). Left: at increasing drive voltage up to incipient hysteresis; right: at different ambient pressures up to Q-factor damping (from [31], by courtesy of IEEE LEOS)

4.6 Laser Vibrometry

4.6.3 Long-Distance Vibrometry

Early experiments in long-distance vibrometry were aimed to remote, noncontact diagnostics of large structures, like towers, bridges, and other construction hardware of industrial, civil, or historical interest. Since the 1970s, several instruments have been developed internationally by several research centers [32].

To circumvent the ambiguity, around 2ks=0 or π, of the baseband cos2ks signal, the approach selected has been that of a *heterodyne* detection of the remote target echo. This is equivalent in performance, with other parameters being equal, to the homodyne with the half-fringe tracking approach discussed previously, and is much akin the two-frequency interferometer (Sect.4.2.2).

Of course, as a source adequate for long distance operation, we need a frequency-stabilized laser with a coherence length in excess of the distance 2s to be covered. For s≈300 m, this requires a laser linewidth $\Delta\nu$=c/2s=500 kHz or better. Also, the power should be adequate to allow for an increased attenuation, which calls for units with power typically in the range 20 to 50 mW (He-Ne) or ≈1 W (CO_2).

The basic configuration to start with is illustrated in Fig.4-42. The source is a single-mode frequency-stabilized laser (He-Ne or a CO_2 for larger distances). The beam is sent to a beamsplitter, and one output is directed to a telescope aimed to the target, while the other is used as the local oscillator for heterodyning. To get a frequency-shifted reference, an acoustooptical modulator (or Bragg cell) is used and provides an output shifted by $\Delta\omega$ =10-50 MHz typically.

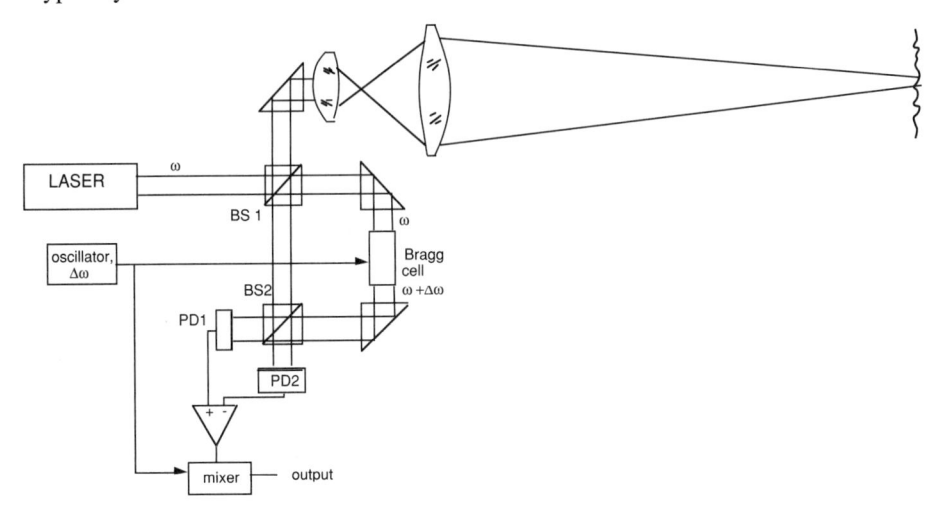

Fig.4-42 Scheme of a long distance vibrometer. The frequency-stabilized laser is frequency-shifted by a Bragg acoustooptic cell to generate a reference local oscillator. A beamsplitter (BS2) is used for beam recombination on the balanced detector. The mixer brings the signal to the base band.

After propagation to the target and back, the optical signal is collected by the telescope and superposed to the local oscillator with the aid of a second beamsplitter BS2, ending on a balanced detector (see Ref.[5], Ch.8.2). Here, by beating of the fields, we obtain the photodetected signals as $I_{ph1}=(I_0/2)[1+\cos(\Delta\omega t+2ks)]$ and $I_{ph2}=(I_0/2)[1-\cos(\Delta\omega t+2ks)]$ and, after subtraction, we get $S= I_0 \cos(\Delta\omega t+2ks)$. By mixing it with the original frequency difference $\Delta\omega$ or a fraction of it, the vibration-signal $2ks=A_0\cos\omega_0 t$ is brought to the baseband (or to an intermediate frequency) for further processing.

Several improvements can be introduced in the basic scheme of Fig.4-42. First, the beamsplitter BS1 halves the optical signal returning from the telescope and directs half of it back into the laser. We can eliminate this return by using a Glan-polarizing beamsplitter for BS1 and by adding a polarizer before entering BS1.

Second, superposed to the useful signal coming back from the target, we find a back-scattering contribution, generated by the atmosphere along the propagation path. This contribution can even be larger than the useful echo when distance increases. To get rid of it, we may take advantage of the polarization properties of scattering, by which a propagating circularly polarized (CP) beam (say right-handed, RH) scatters back a left-handed (LH) contribution. In contrast, the target depolarizes the echo for any impinging state of polarization, and the useful returning signal contains as much RH as LH power. We may use a CP beam in transmission and cut off the back-scattered power by adding a polarizer and a quarter-wave plate at the telescope entrance. If the laser beam has a linear polarization and we align it to the polarizer, no loss in transmission will be experienced, while the back-scatter power will find the polarizer in the crossed state and be eliminated. The outgoing CP signal is then combined at the balanced detector, again using a Glan beamsplitter for BS2, with the reference beam either in a CP or in a linear, 45°-oriented state.

To protect the frequency-stabilizer laser source from even minute, unwanted reflections eventually originated in the optical devices internal to the instrument, we may consider adding an Optical Isolator (OI) at the laser beam output. The OI is well-known device, based on magnetooptical nonreciprocal rotation and is widely used for protecting lasers in fiber-optics communications. It has ideally a T=1 transmission in one direction, and T=0 in the reverse direction. Versions of OI without fiber pigtails (called free-space OIs) are readily available in the wavelength range λ=850-1600 nm of interest for fibers, while materials for the visible range are rather exotic and may not be easily available (at reasonable price).

Still another magnetooptical device may be useful, the Optical Circulator (OC). As an evolution of the OI, the circulator has four access ports for the beam. A beam entering in the i-th input exits from the (i+1)-th port with T=1 transmission, while it is T=0 for all other ports. Placed at the entrance of the telescope, the OC provides us with an input and output port spatially (or physically) separated without introducing any loss, which is different from normal beamsplitters. The OC also eases the layout of transmitted and received beams and cuts out unwanted reflections and scattering.

The typical sensitivity performances (or, NED) of a vibrometer built around the heterodyne concept are reported in Fig.4-43. As it can be seen, sensitivity is dominated at low frequency (<100 Hz) by an excess noise. This is due to turbulence effects, overwhelming the fundamental, flat-spectrum shot noise and electronics noise, which is finally reached at higher

4.6 Laser Vibrometry   159

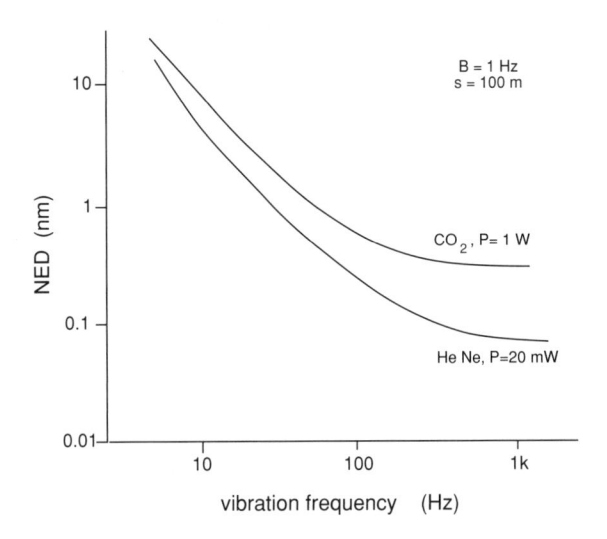

Fig.4-43 Typical NED of a long-distance vibrometer, as a function of frequency, for the detection of small remote vibrations on a diffusing target with He-Ne and CO_2 lasers. Operation is at s=100m distance, B=1Hz, in a clear atmosphere.

frequency (≈1kHz).

As outlined in App. A3, turbulence introduces a random fluctuation of the index of refraction and hence of the optical path length summed to the useful vibration signal. In addition, we find beam wandering around the aiming direction and distortion of the wave front in form of scintillation or bright spots. Both effects combine with the speckle regime of the remote-diffusing surface to add an error.

Thus, operation of the vibrometer on substantial distance (say, in excess of a few tens of meters) requires taking care of propagation condition to keep turbulence effects low.

We can try to mitigate or almost compensate for these effects of turbulence by electronic and optical, as well as measurement strategies.

Electronically, we may subtract the spectrum of the fluctuations, in a measurement with the vibrodyne off, from the useful spectrum with the vibrodyne on. This helps reduce the noise floor at low frequency, perhaps by a factor of 10, but not better because turbulence is not strictly stationary, statistically speaking, in separate measurement sessions. By optical means, we can attempt to compensate turbulence with adaptive optics, theoretically up to a perfect cancellation, but practical schemes are rather involved and have not yet demonstrated a valuable improvement, so far. Last, we can take advantage of the variability of turbulence from day to day and in daily hours. By waiting for a calm day and the best hours to carry out the measurement, turbulence effects are greatly reduced. This strategy is indeed acceptable when evaluating the structural integrity of buildings, which requires an outcome be determined within, say, a few days rather than instantly.

In performing the measurement, we should prefer operating on the shortest distance that is practicable. As a rule of thumb, if the article we are going to test has an out-of-ground height H, we will position our instrument at a distance d≈H to be able to test the bottom and top parts of it in a single session (Fig.4-44).

To generate a test vibration, a vibrodyne is used, which is fastened securely atop the structure. The vibrodyne is an electromechanical device consisting of an electrical motor (typ. ≈50 W in power) with an eccentric-mounted mass (typ. ≈1 kg) rotating at a frequency f_{vib} (typ f_{vib}≈0.1 to 10 Hz). Because of the mechanical torque generated by the vibrodyne, even a modest mass like 1 kg readily generates amplitude of displacements 10 to 100 nm in the tower top surface.

Of course, this amplitude is well within safety limits, and corresponds to the natural stress imparted by a moderate wind (say, with a speed 0.1 to 1 m/s).

The test session consists of measuring the vibration amplitude on a number of points selected along the structure and repeating the measurements for a set of frequencies covering the natural vibration frequencies of the structure. For the simplest article, a tower, we may take, for example, 10 sampling points in H/10 steps, and 10 frequencies, from a fraction (e.g., 1/4) of the lowest-mode frequency f_0 to 10 times as much, in equally log-spaced steps.

In this way, we are able to obtain the frequency response F(h,ω) to a mechanical excitation, a response which is resolved spatially. Now, the measured frequency response can be compared with what is expected from a solid structure, as calculated by standard methods well known in structural engineering. Also, we may introduce modifications in the calculation to simulate the effect of localized flaws and yields for ease of comparison.

Let us consider an illustrative example. A tower or chimney has bending modes at frequencies $f_n = n\eta\sqrt{k/M}$, where n=1,3,5... is the order of the mode, k is the elastic constant (N/m), M is the mass, and η is a dimensionless factor depending on height, diameter, wall thickness, etc.

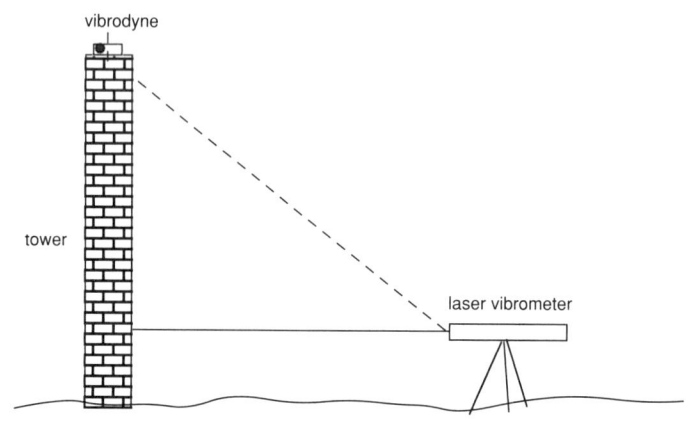

Fig. 4-44 Arrangement for testing a tower by means of a laser vibrometer and a vibrodyne

4.6 Laser Vibrometry

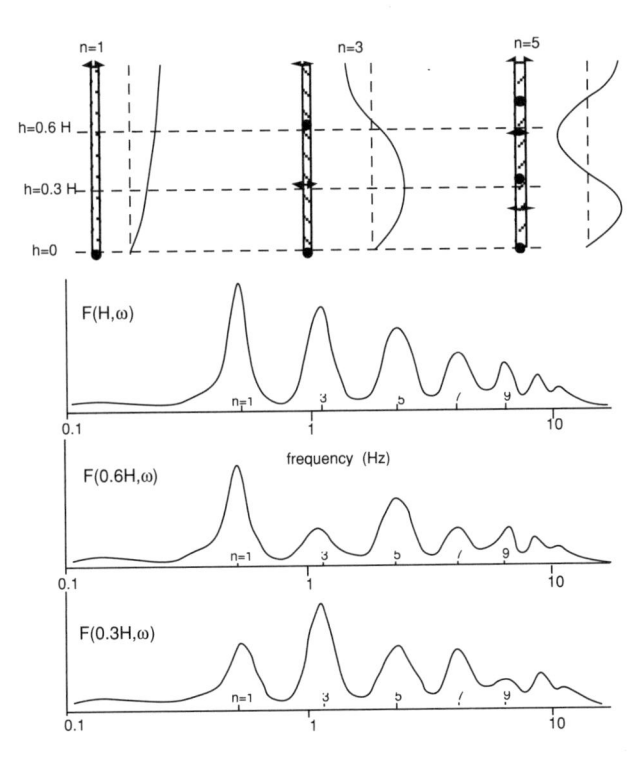

Fig. 4-45 Top: a tower or chimney has bending vibration modes with nodes (indicated by points) and antinodes (arrows) at integer fractions of the height H. The modes are revealed by the frequency response (bottom) taken at different heights h.

The lowest mode has a node of vibration at h=0 and an antinode (or maximum swing) of vibration at h=H, or, the standing wave of vibration has half-period in the height H. Higher order modes have n/2 periods in H (see Fig.4-45). Additional modes can be found in a structure lacking radial symmetry or having finer structural details, but here we ignore them for simplicity.

The frequency of the n=1 mode is found at f_1=0.02 to 1 Hz, typically (Fig.4-45), in most practical cases of interest (10 to 50 m towers). Higher modes have resonant frequencies that are odd-multiples of the fundamental one.

As a function of height, the amplitude of vibration changes with the mode order (Fig. 4-45), and we can therefore recognize them by looking at the relative obtained at different heights. The sharpness of the resonance decreases with the order of resonance and, more important, with the deviations from the structural integrity of the article. A solid structure has a well-defined resonance spectrum, whereas a near-to-collapse article usually exhibits a broad spectrum of frequency response $F(h,\omega)$, with a barely recognizable resonance pattern.

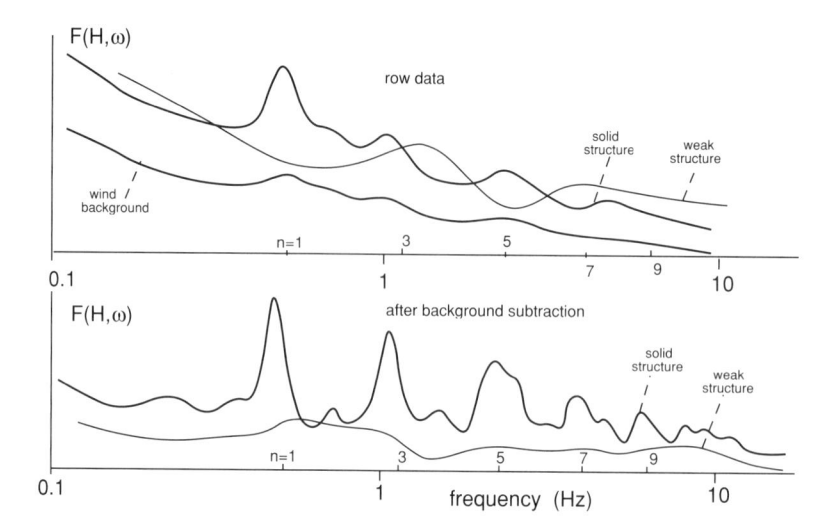

Fig. 4-46 Vibration measurement (top) and the result after background subtraction (bottom), revealing the state of conservation by means of the resonance spectrum

The raw interferometric signal of the vibrometer has a frequency spectrum with a strong background due to turbulence (Fig.4-46). After background subtraction through a separate measurement with the vibrodyne off, the resonance peaks of a solid structure are clearly highlighted. For a weak structure, the spectrum is rather flat without the expected resonances, and the measurement of the frequency response $F(h,\omega)$ at various heights adds further evidence.

At a closer look, in the background spectrum due to wind and turbulence we find, for a solid structure, modest peaks at resonances (Fig.4-46, top). This can be explained because the wind is a sort of white-noise excitation of the structure, and the mechanical response is enhanced at resonance. Thus, under favorable circumstances — a wind strong enough to impart a mechanical excitation to the structure, but not so strong to overwhelm the useful signal by turbulence — we can dispense with using the vibrodyne and need not climb the article to mount it.

Subtraction of the background cannot be performed as it was previously, in this case. However, from the set of measurements at different heights, and knowing the relative amplitudes of the expected resonances, we can process the data as to subtract the common background with a least-mean-square error routine, for example.

In conclusion, despite the limitation of turbulence, long-distance vibrometry offers a powerful diagnostic tool for noncontact testing of buildings. This is exemplified by examination of masterpieces like the historical Pisa tower [33] and similar articles, as well as in the diagnostic of structural integrity of industrial articles, notably cooling towers and chimneys of cement factories, electrical generating plants, and the like.

4.7 Other Applications of Injection Interferometry

About the choice of the wavelength best suited for long-distance operation of the vibrometer, let us consider the factors involved, by comparing He-Ne (λ=0.63 µm) and CO_2 (λ=10.6 µm) lasers. The turbulence-induced Δn fluctuation and the phase noise (App.A3.2) is proportional to $\sigma_\chi \propto \lambda^{-7/12}$, thus noise decreases of a factor $(\lambda_{He-Ne}/\lambda_{CO2})^{7/12}$ =0.24 going to the infrared.

The NED is proportional to wavelength and to $(S/N)^{-1}$ ratio, on its turn proportional to λ: so we get an increase of it by $\sqrt{(\lambda_{CO2}/\lambda_{He-Ne})}$ = 4.1 in the IR.

Optical device transmission and detector quantum efficiency η are worse in IR as compared to visible, and we may take a factor $\approx$3 of increase in the IR.

Last, power available from a small CO_2 laser is readily $\approx$1W, instead of 50 mW of a He-Ne laser, and this gives a factor $\sqrt{0.05/1}$=0.22 of NED decrease in IR.

Collecting the contribution, the NED we should expect in IR is 0.24·0.4·16.8· 3·0.22=1.06 times that of the He-Ne, or, substantially the same. However, the effects of turbulence are smaller in the IR, as indicated by Fig.4-43 where the corner frequency of noise for the CO_2 wavelength is smaller.

4.7 OTHER APPLICATIONS OF INJECTION INTERFEROMETRY

Recently, a few additional applications of injection interferometry have been reported [34], which are particularly attractive, though they are still at the development stage and not yet incorporated in commercial products, like those found for the other topics discussed in this chapter.

4.7.1 Absolute Distance Measurements

Laser interferometry is intrinsically a phase-difference tool, and is definitely the best for high-resolution measurement of displacements with sub-λ increments, but it requires a moving target sliding along the displacement to develop the counts. On the other hand, telemeters based on beam modulation (Chapter 3) can operate on still targets and are truly absolute distance meters.

However, because interferometry works at the quantum limit of sensitivity and the injection configuration is especially very attractive for its simplicity, we may try to devise a scheme circumventing the basic limitation of incremental measurement.

The starting point is to sweep the wavelength of the source, a feature readily available in diode lasers through modulation of the drive current (App.A1.3). Sweeping the wavelength changes the number of wavelengths N=2s/λ contained in the go-and-return path to the target. Each time one more wavelength fits in 2s, the interferometric signal undergoes a 2π phase shift (and a switch in the C>1 injection regime). By detecting it, we may be able to trace back the absolute distance s.

As a wavelength modulation, let us now consider a linear sweep on a swing $\Delta\lambda$.

This modulation is obtained by superposing a linear sweep on the drive current, with an amplitude $\Delta I = \Delta\lambda/\alpha_I$ (App.A1.3) superposed to the quiescent dc current I_{dc}. The interferometric signal $I_{ph}/I_0 = 1 + \cos 2ks$ at the start and at the end of the sweep period is, respectively:

$$I_{ph}/I_0 = 1 + \cos 4\pi s/\lambda_0, \quad \text{and}$$

$$I_{ph}/I_0 = 1 + \cos 4\pi s/(\lambda_0+\Delta\lambda) \approx 1 + \cos 4\pi s(1-\Delta\lambda/\lambda_0)/\lambda_0$$

Thus, the phase variation $\Delta\Phi$ produced by the sweep is:

$$\Delta\Phi = 4\pi s(1-\Delta\lambda/\lambda_0)/\lambda_0 - 4\pi s/\lambda_0 = -4\pi s \Delta\lambda/\lambda_0^2$$

By dividing $\Delta\Phi$ by 2π, we get the number of interferometric waveform periods found in the photodetected current as $N = 2\Delta\lambda/\lambda_0^2 s$. For $C>1$, N is the number of switchings to be counted. From N, the distance is obtained as:

$$s = N \lambda_0^2/2\Delta\lambda \tag{4.55}$$

Each count (N=1 in Eq.4.55) corresponds to a distance $\Delta s = \lambda_0^2/2\Delta\lambda$, and this represents the resolution of the measurement performed in the injection regime (Sect.4.5.2), with $C>1$. In a normal Fabry-Perot laser, the practical limit to the wavelength swing we may impose with a reasonable excursion of the drive current is $\Delta\lambda \approx 4$nm (Fig.A1-13), and thus the theoretically attainable resolution at $\lambda=1000$ nm can be estimated as $\Delta s = (1000)^2/2\cdot 4 = 125$ μm, a very attractive value indeed.

Unfortunately, Fabry-Perot laser diodes are affected by mode hopping (Fig.A1-13) at large $\Delta\lambda$ swings. Though not a really fundamental limitation, mode hopping is detrimental to signal processing and, to avoid it, reported experiments [35] limit the useable wavelength swing to about $\Delta\lambda \approx 0.2$nm, so that the resolution presently attained is about $\Delta s \approx (1000)^2/(2\cdot 0.2) \approx 2.5$ mm, up to a distance s of a few meters.

An experimental arrangement is shown in Fig.4-47. A triangular modulation is applied to the laser drive current. This produces a wavelength modulation and, in addition, also a power amplitude modulation of the same waveform. We can subtract the power-dependent term, however, using a fraction of the modulating signal, eventually slightly distorted so as to correct for the P=P(I) non-linearity of the laser diode characteristics.

After the subtraction, we obtain the interferometric signal S(t) (Fig.4-47) and one full period for each $\Delta\Phi=2\pi$ change. In the injection regime with C not too small, the leading edge is faster than the trailing edge. Thus, by time-differentiation, we will get pulses with the positive part larger than the negative one, easily discriminated. To count the positive pulses during the $\Delta\lambda$ sweep time, we may simply differentiate the modulating waveform (Fig.4-47) and use it as a gate signal for the counter.

The number of counts N is used in Eq.4.55 to determine the distance s, with a scale factor given by $\lambda_0^2/2\Delta\lambda$.

Two sources of error shall be considered. First, because of the discrete counting process, we incur into a ±1 count as round off error at the beginning and end of the sweep. As an average, the systematic error is +1-1=0. Recalling that the variance of a uniform distri-

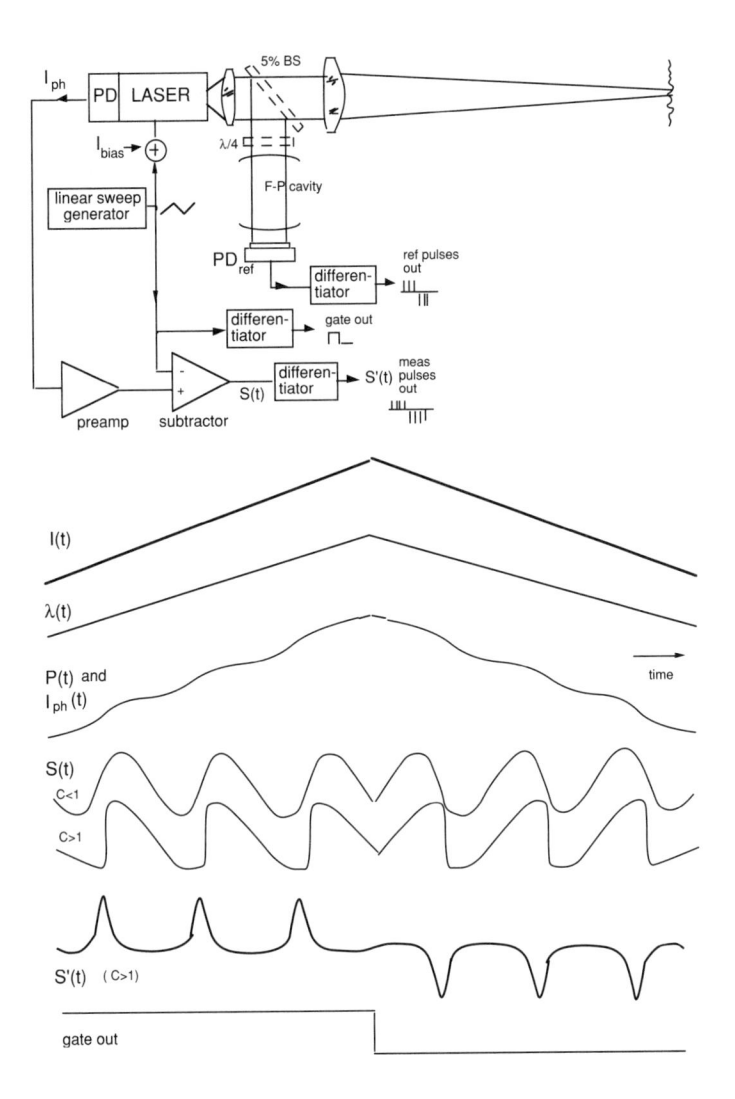

4.7 Other Applications of Injection Interferometry 165

Fig.4.-47 Absolute distance meter based on injection interferometry. The laser wavelength is swept with a triangular waveform, using a current modulation superposed to the bias I_{bias}. The photodiode signal, after subtraction of the power-dependent term duplicating the modulation, exhibits spikes corresponding to $\Delta\Phi = 2\pi$ variations. Counting them, the distance is found as $s = N \lambda^2/2\Delta\lambda$. To improve the measurement accuracy, we may add the section with the beamsplitter and the Fabry-Perot to count $N_{ref} = 2L_{FP}\Delta\lambda/\lambda^2$ with a separate circuit (not shown in figure) and get $\Delta\lambda = N_{ref} \lambda^2/2L_{FP}$, and accordingly $s = (N/N_{ref}) L_{FP}$.

bution is 1/12, we have a random rms error $\sigma_{dis} =\sqrt{(1+|-1|)}/\sqrt{12}=1/\sqrt{6}$ counts. This error is reduced by a factor $1/\sqrt{N_T}$ if we repeat the measurement N_T times, provided the initial phase is randomly distributed between 0 and 2π [ambient vibration superposed on s usually suffices for that].

Second, the accuracy of the scale factor depends on how precisely we know λ_0 and $\Delta\lambda$. Usually, wavelength is known to better than one part in 10^3, whereas $\Delta\lambda$ may scatter by as much as a few percent around the nominal value $\alpha_I \Delta I$.

An elegant method to circumvent the accuracy problem is illustrated in Fig.4-47. We pick out a fraction (say, 5%) of the power projected on the target by a beamsplitter and send it to an interferometer. The interferometer may be a Mach-Zehnder as first reported in [36], or a Fabry-Perot for a simpler implementation as shown in Fig.4-47. Feeding the interferometer under the wavelength sweep, we get a periodic sequence of signal peaks at the output (see Fig.A2-2), with a period $c/2L_{FP}$ in frequency and $\lambda_0^2/2L_{FP}$ in wavelength. Now, we need only to duplicate the circuit of the main section to read these peaks. Counting them, we get a total number $N_{ref}=2L_{FP}\Delta\lambda/\lambda^2$ in a sweep, where L_{F-P} is the cavity length of the Fabry-Perot (which is known to a very good accuracy).

Using the scale factor $2\Delta\lambda/\lambda^2=N_{ref}/L_{FP}$ in Eq.4-55, we now can obtain the desired distance as $s= L_{FP}(N/N_{ref})$, that is, we are completely released from the λ and $\Delta\lambda$. Note that the interferometer needs a length L_{FP} comparable to the distance s if the count discrete error of it is to remain low.

4.7.2 Angle Measurements

Alignment of mirrors along the line of sight defined by a laser-beam wave vector is one of the early applications reported [37] for injection interferometry.

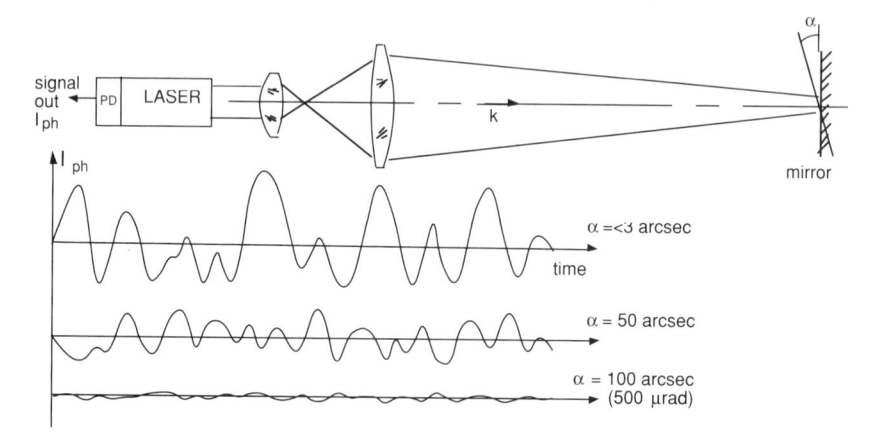

Fig. 4-48 Alignment of a remote mirror to the propagation wave vector k of the laser: ambient vibrations provide an interferometric signal, whose maximum is found when the error angle α is at a minimum (typ. <3 arcsec).

4.7 Other Applications of Injection Interferometry

Fig.4-48 shows the straight arrangement for such an autocollimator measurement. The laser beam is expanded by a telescope and sent to the mirror target. On a substantial distance (say, >30 cm) the normal ambient-induced vibrations readily provide an interferometric signal, for which the amplitude is maximum when the alignment is best.

More specifically, the maximum is attained when the angle error α is fairly less than the diffraction limit of the transmitted beam, with a typical resolution <3 arcsec as indicated in Fig.4-48.

Like working with a normal optical autocollimator, the alignment of several reflecting surfaces is achieved by separately aligning each surface to the k vector, used as an angular yardstick.

The injection-interferometer autocollimator has been found useful in alignment with IR lasers (originally, with a 3.39μm He-Ne laser). However, it does not provide a true angle measurement, but just a sensitive null detection.

By adding a line-of-sight modulation to the basic scheme, we can work out a true angle measurement. The beam can be steered by an actuator, e.g., an x-y translation of the laser chip or of the first lens of the telescope, or by means of a prism inserted in the transmitted beam. To substantiate the design, let us assume a deflection $\Delta x = \Delta x_0 \cos\omega_0 t$ of the first lens. Then, the angle of k is modulated with a deflection $\alpha = \Delta\alpha_0 \cos\omega_0 t$, where $\Delta\alpha_0 = \Delta x_0 / F$ and F is the focal length of the first lens. As illustrated in Fig.4-49, with no modulation the response versus α is parabolic, whereas adding the modulation and using phase detection, the response is linear up to the amplitude $\Delta\alpha_0$ of the angle swing.

Typical performance of the angle meter implemented with a laser diode in the injection-interferometry configuration [38] is a noise-limited resolution of ≈ 0.2 arcsec on a ≈ 5 arcmin dynamic range.

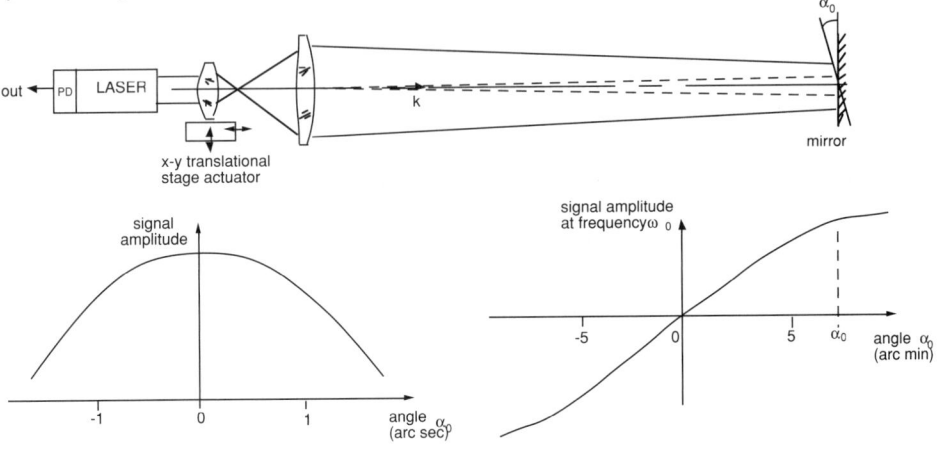

Fig. 4-49 Angle measurement by injection interferometry: the aiming direction is modulated so that the parabolic amplitude response is transformed into a linear dependence on the angle α_0 to be measured.

4.7.3 Detection of Weak Echoes

Besides the phase difference $\varphi_m - \varphi_r$, the generic interferometric signal (Eq.4.2) contains information on the amplitude of the field E_m returning to the measuring port. Actually, any interferometer can be regarded as a coherent detection scheme (see [5], Ch.8) in which E_m is the signal and $E_r \gg E_m$ is the local oscillator. By beating of the fields E_r and E_m at the photodetector, an internal gain is generated, and the quantum limit of detection is attained. Thus, as an amplitude detector, the interferometer has the advantage of a very good sensitivity to small signal and echo returned after a strong attenuation.

The coherent detection scheme is classified as homodyne or heterodyne according to whether $f_m = f_r$ or $f_m \neq f_r$ and as an injection scheme if the returning signal is fed into the source. Classical homodyne or heterodyne detection is used in fiber optics communication, whereas injection detection is attractive in instrumentation because of several features (Sect.4.5.2.5).

Using Eq.4.52, we can write the photocurrent signal $i_{ph} = \sigma \Delta P$ of the self-mixing configuration in the weak-injection regime as:

$$i_{ph} = I_0 \kappa a \cos 2ks / (1 + \kappa a \cos 2ks) \qquad (4.56)$$

Here, $I_0 = \sigma P_0$ is the dc photocurrent, $\kappa = c\tau_p / n_l L$ is a constant of the order of unity, and a is the total (field) loss in the path to the target at distance s and back, explicitly $a = \sqrt{A}$ in terms of the total power attenuation A.

Eq.4.56 tells us that, at weak injection levels (a<<1), the signal has an amplitude proportional to the square root of power attenuation, $a = \sqrt{A}$, as expected from a coherent detection process. When $\kappa a \approx 1$ or $C > 1$ (moderate feedback), the amplitude does not increase any more and we get a saturation of the signal versus attenuation (Fig.4-50).

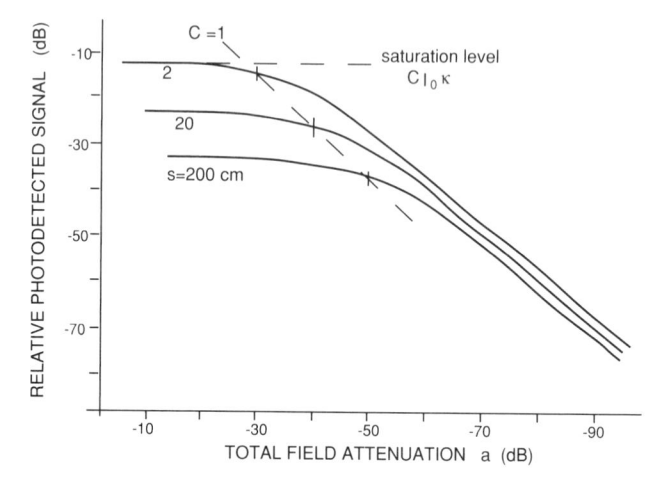

Fig. 4-50 The theoretical dependence of signal amplitude from external attenuation (or return loss) in an injection interferometer, for some values of distance.

4.7 Other Applications of Injection Interferometry

To be able to measure the amplitude $I_0\kappa a$ without being disturbed by the phase term $\cos 2ks$, we need either the distance to be a constant or to be varying in a known way. If we are operating on a substantial distance (say s >50 cm), ambient-related microphonics usually contributes with a random jitter $s_j(t) \gg \lambda$ added to the mean $\langle s \rangle$, and therefore $\cos 2ks$ is a random waveform with zero mean value and a rms value $1/\sqrt{2}$. Then, we will measure the rms i_{ph} value of the signal and get the attenuation as $a = \sqrt{2} i_{ph(rms)}/I_0\kappa$.

If distance is short or we want to move the signal off the dc, we may add in the optical path a phase modulation $\Phi = \Phi_0 \cos\omega_m t$, with a deviation $\Phi_0 > 2\pi$ large enough to have a phase term $\cos(2ks+\Phi)$ swinging from -1 to $+1$. In this way, the signal is modulated on a carrier frequency ω_m, at which the measurement of amplitude will be performed.

Typical examples of echo attenuation measurement [39-41] are shown in Fig.4-51. The first setup [40,41] is for the measurement of the return loss from a fiber device (DUT, device under test). The optical path length is modulated with the aid of a piezoceramic (PZT) phase modulator driven at frequency ω_m, and the output signal at frequency ω_m is proportional to the square root of the ratio $P_{back}/P_0 = A_{RL}$. In this scheme, we may also add a second photodetector PD2 to measure the ratio P_{tr}/P_0, of the DUT, that is, its insertion loss.

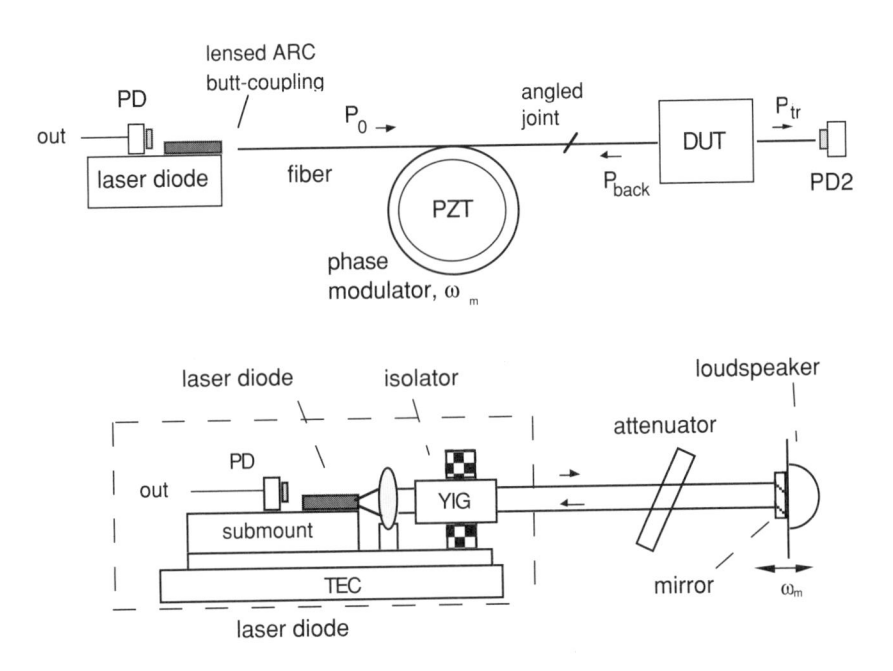

Fig. 4-51 Typical arrangements for the measurement of weak echoes by self-mixing interferometry. Top: for measuring the return loss of an all-fiber device (DUT), we can use an in-line PZT phase modulator. Bottom: for testing the isolation of an optical isolator mounted in the laser package, a remote vibrating mirror mounted on a loudspeaker supplies the path length modulation.

In the second example, aimed to test the isolation factor of an optical isolator mounted in front of the laser chip [39], we shall use a path-length modulation external to the device. This can be a mirror aligned to the transmitted beam and mounted on a loudspeaker, driven at the desired frequency ω_m. In this case, the signal is proportional to the square root of the isolation factor P_{back}/P_0.

To make the point clear, 10 dB (or a decade) of attenuation corresponds to a half a decade of variation in the current or voltage signal, that is, to a 10 dB change in it.

Typical performances of the echo detector based on the injection interferometer are illustrated in Fig.4-52. Several single-mode laser diodes operating at different wavelengths have been experimented, with either plain Fabry-Perot or DFB (grating reflector) structure. For all of them, the range of measurable attenuation spans from −25 dB to −80 dB. By adding an attenuator in the optical path (Fig.4-51, bottom), we can extend the upper limit to about 0 dB.

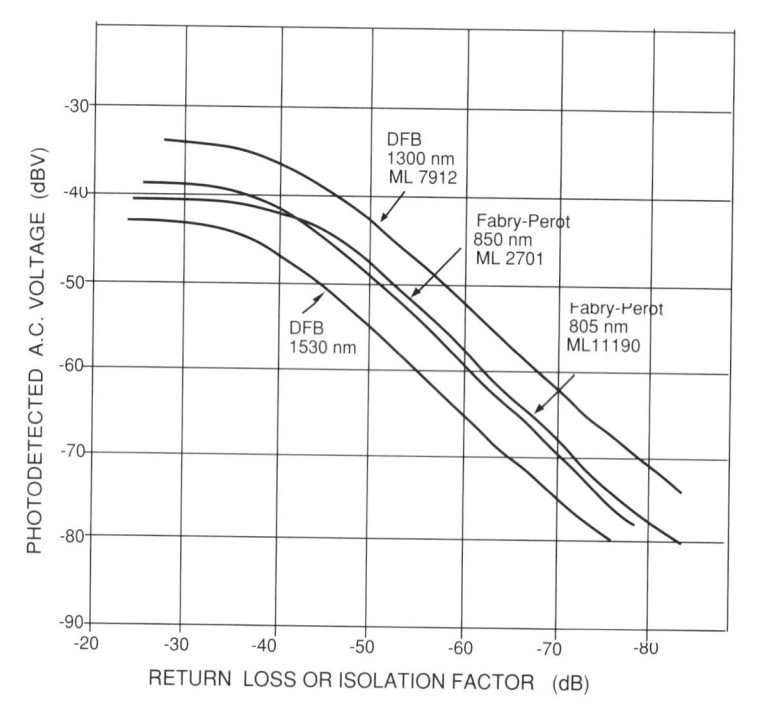

Fig. 4-52 Typical attenuation or optical isolation that can be measured by a diode laser operating in the injection interferometer regime of echo detection. At about -80 dB the limitation is S/N ratio, whereas at ≈-30 dB is saturation.

Other interesting applications of the echo detection by injection-interferometer schemes have been reported in the fields of optical testing [42], compact-disk readout [43], and scroll sensors [44].

4.7 Other Applications of Injection Interferometry

In CDs, optical pits correspond to the bits of recorded information. In conventional design, to read the bits we need a laser and photodiode combination, a beamsplitter to divide the input and output beam paths, and a conjugating lens to focus on the spot.

Using an injection detector, we can dispense for the beamsplitter and the lens (Fig. 4-53) by placing the laser diode close to the disk (typically at 10-20 µm off the surface). The rear photodiode will supply the readout signal in form of spikes corresponding to the bits superposed to the dc quiescent current [43].

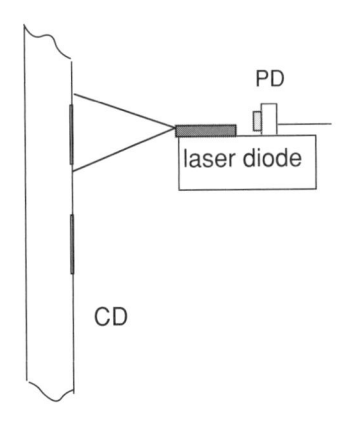

Fig. 4-53 Readout of optical bits in a CD by injection interferometry. The laser and photodiode combination provide a readout signal. Unwritten portions of the CD surface reflect light and give a large signal, whereas pits diffuse light and give a small injection signal detected by the rear photodiode.

Another recently reported example of an application of the readout of echoes by injection interferometry is the scroll sensor [44]. The scroll sensor is used in mobile telephone sets and is one of the two axes of the computer mouse. It supplies a signal indicating the up/down direction of a finger or another common surface presented at the viewing window of the sensor, and operates with no physical contact or a particularly clean window.

As shown in Fig.4-54, we can design a scroll sensor by means of two low-cost single mode lasers in a package with folding mirrors. The mirrors direct the beams on the output window. The incidence angles of the two beams are ≈45° and the sign opposite, so that the phase term $2\underline{k}\cdot\underline{s}$ driving the self-mixing modulations (AM and FM) has an opposite sign in the two laser cavities.

As usual, we get the AM signal directly from the photocurrent of the monitor photodiode mounted in front of the rear mirror of each laser. Using a difference amplifier to subtract the two signals and time-differentiating the result, we obtain a signal with a polarity which indicates the up/down direction of the scroll, and with an amplitude which is proportional to the speed $2\underline{k}\cdot\underline{s}'$. The external surface we sense does not need to be treated or be in contact with the window.

As wavelength of operation, a visible (typically 620 to 680 nm) wavelength is preferred because of safety issues (App.A1.5), and because the laser power requirement is modest (≈0.2 mW or less).

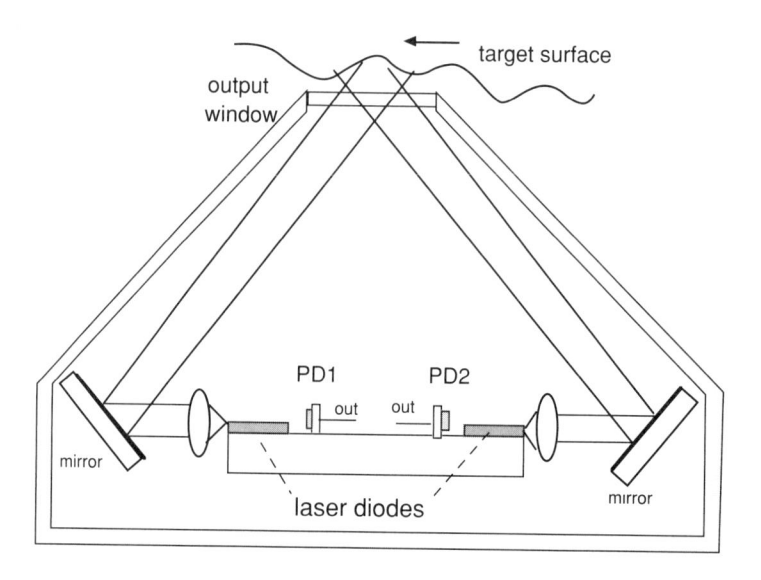

Fig. 4-54 In a self-mixing scroll sensor, two low-cost lasers shine on the sensing window, with beams impinging there at an angle (typically 45°) from opposite sides. The self-mixing signal 2**k**·**s** returned to each laser cavity is then opposite in sign. After detection on the monitor photodiode, subtraction of currents yields the direction of surface movement.

4.8 WHITE LIGHT INTERFEROMETRY

In all applications of laser interferometry described so far, we have called for sources with a long coherence length. Indeed, operation is intended in the coherent regime of superposition of reference and measurement fields so that we can resolve sub-λ path-length differences. Indeed, if we have to accommodate an arm-imbalance 2(s_m-s_r) and yet get a nonvanishing interferometric signal, we need a coherence length l_c>2(s_m-s_r).

In white-light interferometry, this concept is completely overturned.

We use a low-coherence length source, like an LED, a Superluminescent LED (SLED), or a filtered incandescent lamp with a line width so large (Δλ≈20-100 nm around λ≈1μm) that the coherence length is limited to a very small value like l_c=$λ^2$/Δλ≈10-50μm.

4.8 White Light Interferometry

Now, we extend the strategy of operation by looking at both the incoherent and coherent regimes of superposition. For $2(s_m-s_r)>l_c$, the superposition is incoherent and the photodetector yields a dc current which is just the sum of the intensities provided by the reference and measurement beams.

On the other hand, when the arms are nearly balanced, $2(s_m-s_r)<l_c$, the photodetector yields the interferometric beat signal added to the previous dc level.

Thus, by waiting for a spike to appear in the photodetector output signal, we are able to distinguish the balance condition $s_m=s_r$ with a $l_c \approx$ 10-50µm resolution.

Recalling Eq.4.12, we may write the photodetected signal from a Fabry-Perot interferometer in white light (Fig.4-55) as:

$$I_{ph} = I_0 [1+V\cos 2k(s_m-s_r)] \quad (4.57)$$

Recalling the result of Sect.4.4.2 and 4.4.3 and assuming a Lorentzian line, the visibility V can be expressed as: as:

$$V = \mu_{sp} \mu_{pol} \exp-2(s_m-s_r)/l_c \quad (4.58)$$

To illustrate the practical implementation of the concept, let us refer to Fig.4-55.

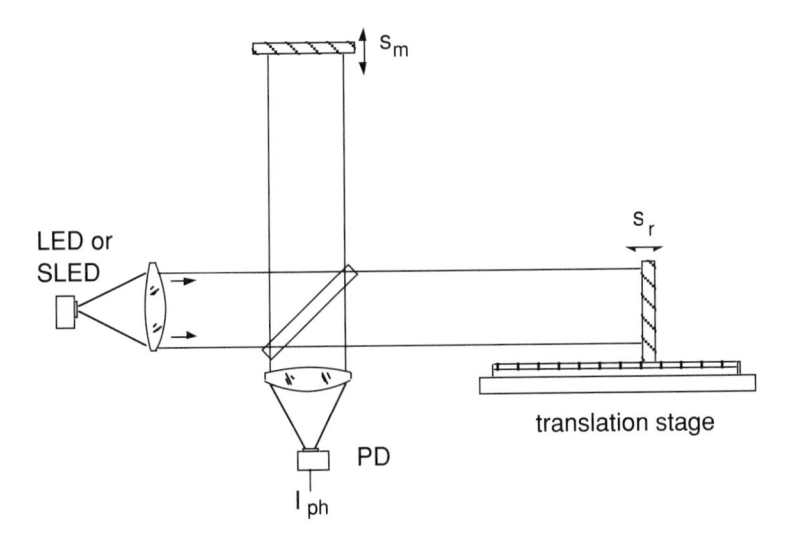

Fig. 4-55 In a white-light interferometer, the interference signal only appears when arms are very close to balance ($s_m=s_r \pm l_c$) because the coherence length l_c is very short ($\approx$20 to 50µm). The measurement condition is searched by scanning the reference arm length s_r, with the aid of a motorized translation stage and is revealed by a spikelike signal in photocurrent.

We may use either an LED or a SLED as the source of the interferometer. The basic SLED is a diode with a laser structure, but with the output mirror missing (the output facet is antireflection coated). Thus, laser oscillation is prevented and emission from the device is in form of spontaneous emission generated in the bulk and amplified down the path to reach the output facet, hence the name of superluminescent light.

The SLED has an emitted power (a few mW) and spectral width (typically $\Delta\lambda \approx 20$ to 50 nm) comparable to a normal LED, but emission is from a small spot size, nearly single-mode area, like that of a laser. Thus the SLED brilliance is high, which is a very useful feature. In view of Eq.4.58, the SLED is better because it has $\mu_{sp} \approx 1$, whereas it may be $\mu_{sp} \approx 10^{-3}$ for an LED. For the same reason, though in principle useable, an incandescent lamp source would provide a very low $\mu_{sp} \approx 1/N$ and useful signal, because of the large number of modes emitted ($N \approx A\Omega/\lambda^2$, see [5], App.A2).

The other factor in Eq.4.58 is usually $\mu_{pol} \approx 0.5$ because either the source is unpolarized or polarization is not preserved in rediffusion from a rough target.

Considering a Fabry-Perot interferometer (Fig.4-55), one mirror is secured to the target being measured, and the other mirror is mounted on a precision translation stage. In this way, the reference arm can be scanned to search for the $s_r = s_m$ condition while keeping the interferometer aligned. Of course, we can replace the mirrors with corner cubes to relax the angular alignment criticality and have much larger signal amplitude.

Now, as we pass through the $s_r = s_m$ condition, oscillations corresponding to fringes show up in the photocurrent signal I_{ph}, as illustrated in Fig.4.56, in accordance with Eq.4.57. If the translation stage is moved with a constant speed, the waveform of I_{ph} versus time is a replica of the dependence versus s_r.

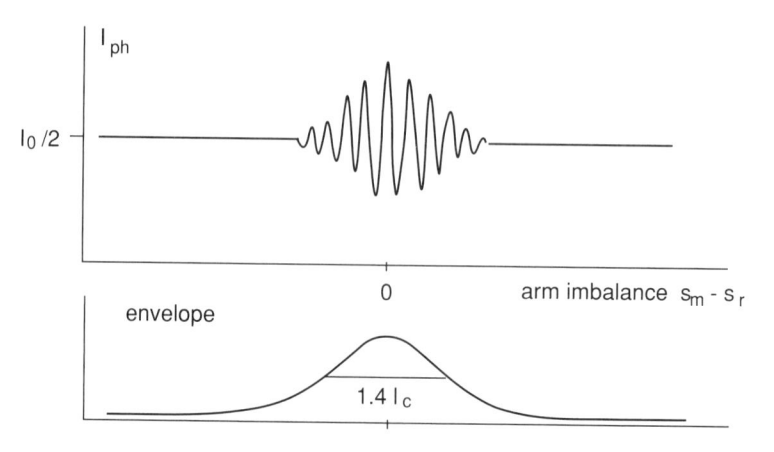

Fig. 4-56 The photodiode signal at the output of a white-light interferometer is an oscillation with a pulse envelope, peaking in correspondence to the balance condition $s_r = s_m$. The envelope width is about equal to the coherence length l_c and determines the resolution of the measurement.

4.8 White Light Interferometry

To accurately localize the spike position, we may implement the following electronic processing. The photocurrent signal I_{ph} is first rectified, thus obtaining its envelope. By time differentiation of the envelope, we get a zero crossing in correspondence of its peak. Finally, a zero crossing comparator provides the logic signal to sample the s_m value, and is then read by an encoder mounted on the translation stage. About the measurement accuracy, several systematic, as well as random sources, are to be considered. By analyzing them, it is found that the accuracy σ_m is a fraction of the response Full-Width at Half Maximum (FWHM), equal to $2 \ln 2 \, l_c = 1.4 \, l_c$ for a Lorentzian line. In practice, we can achieve an accuracy of 10 to 50µm, not as good as in a conventional interferometer, yet a respectable value that can be adequate in a number of applications.

As a comment, the main drawback of the white-light technique is that we need a moveable stage inside the instrument, covering a distance D equal to the range 0-D to be measured. The practical range of measurement is accordingly limited to, say a few tens of cm, at least in the basic configuration, and this makes the technique unsuitable for long-distance, general-purpose measurements.

Of course, we can add a fixed delay in the reference arm to extend the range. For example, a coil of single-mode fiber with objective lenses to focus the input beam and to collimate the output beam readily provides a compact device, which increases the path length by nL and the distance range to nL-nL+D.

This feature coupled to the inherent single-mode propagation in fibers, makes white-light interferometry very attractive [45] for optical fiber sensors (Chapter 9). Also, when the distance range to be covered is rather small (say <1 cm), the technique is easily implemented and can be extended to the image format, as described in next section.

4.8.1 Application to Profilometry

When incorporated in a microscope, the white-light interferometer provides a measurement of the surface height of the specimen, with a spatial definition determined by the size of the illuminating spot. By moving the spot across the specimen in a x-y raster arrangement, we can obtain a map of surface height or a profilometry of the surface [46-48].

Specifically, the surface under test can be either a polished or a rough surface. The working principle is illustrated in Fig.4-57. A light source (normally an incandescent lamp) is filtered in wavelength by a colored filter to choose the desired λ_0 and $\Delta\lambda$, and spatially by a pinhole to work with a nearly single-mode spatial distribution.

With a collimating lens, the beam is sent along the axis of the microscope with the aid of a beamsplitter (BS1). After passing through the objective lens focused on the specimen surface, the illuminating beam encounters a pair of optical flats acting as an interferometer (called a Mirau interferometer).

The first flat F has antireflection coatings on both surfaces and a small central obstruction, which is black (absorbing) on the upper side and metallized (reflecting) on the bottom side. The second flat is partially reflecting and acts as a beamsplitter (BS2) on one of its surfaces.

At BS2, a part of the beam is directed to the specimen and constitutes the measurement beam, whereas the other part is reflected toward F to be used as the reference beam. The portion of the reference beam falling on the central metallization of F is reflected back to BS2 where it recombines with the measurement beam returning from the specimen. The two superposed beams are collected by the objective lens in front of the photodetector PD.

The interferometer arms are balanced when the distances of the beamsplitter BS2 to the specimen and to the flat F are equal, or differ by less than l_c. This condition is checked by the signal I_{ph} of the photodetector (Fig.4-57).

To obtain the condition of balance, the flat F is moved by a micrometer motorized stage. With a phase detector, we are able to lock the stage position at the I_{ph} maximum, with an accuracy of a few hundredths of l_c. The corresponding resolution in surface height is well below the micrometer.

About the stage drive, we do not actually need a precise z-axis movement, but rather a precise measurement of it, as can be supplied by a sensor secured on it. The sensor may be one of the several, well-known linear-motion transducers (e.g., a differential transformer, a linear potentiometer, or even a low-cost interferometer), capable of sub-μm accuracy or better [49].

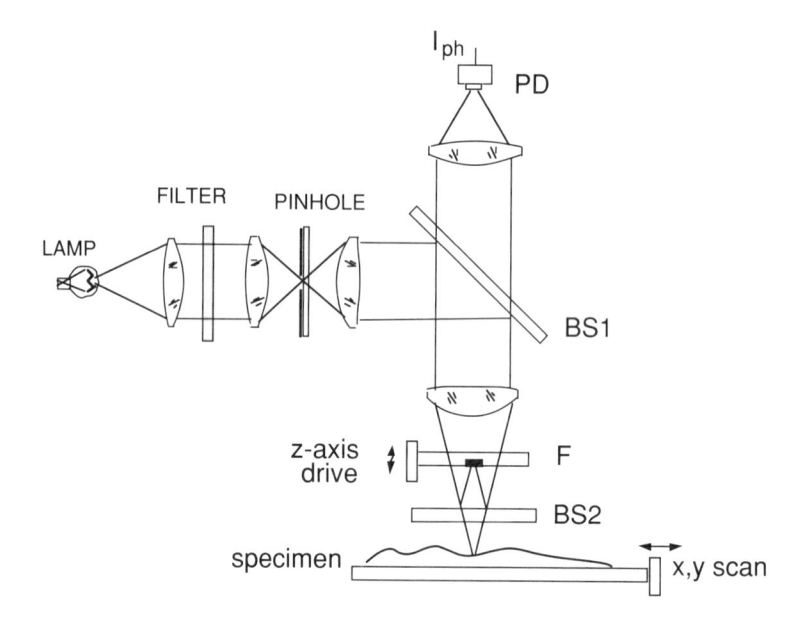

Fig. 4-57 Basic setup of a profilometer based on white-light interferometry. A first beamsplitter BS1 combines the illuminating and returned beams. The flat F and the beamsplitter BS2 act as a Mirau interferometer (F has a small central portion mirrored). The path balance is obtained when the distances of BS2 to F and to specimen are equal. This condition is searched by moving the z-axis drive until the I_{ph} has the spike shown in Fig.4-56.

4.8 White Light Interferometry

As the profile variations from point to point are usually small, and the moving part is of little weight, the time response of the system is adequately short, and acquisition of a single point may require $t_r \approx 1\text{-}10\text{ms}$.

The spot size is determined by the numerical aperture NA of the objective lens as $s=\lambda/\pi NA$ (see [5], App.A2). This is the sampling dimension (or pixel minimum size) of the x-y profile measurement. As a typical value, for $\lambda=0.5\mu m$ and $NA=0.01\text{-}0.05$, we get $s=1\text{-}15\mu m$.

To obtain a full-image profile measurement, a simple way is to move the specimen by a micromotorized stage in an x-y raster fashion. Again, precision encoders will provide the x-y position readout to which the z-axis measurement is assigned. The total scan time of a 100×100 pixel image may be $10^4 \times (1\text{-}10\text{ms}) = 10\text{-}100$ s, a fairly long but acceptable time.

Even better, we can use a CCD array as the photodetector to cover the full specimen image, and simply scan the z-axis. When an individual point in the specimen reaches the balance condition, the corresponding pixel will exhibit a current spike, and the z-measurement is assigned to the pixel coordinates.

A recent refinement to white-light profilometry has been that of improving the vertical resolution down to nanometers [50]. This is achieved by adaptation of the phase-shifting technique discussed in Sect.4.2.2.1 to the white-light interferometer scheme.

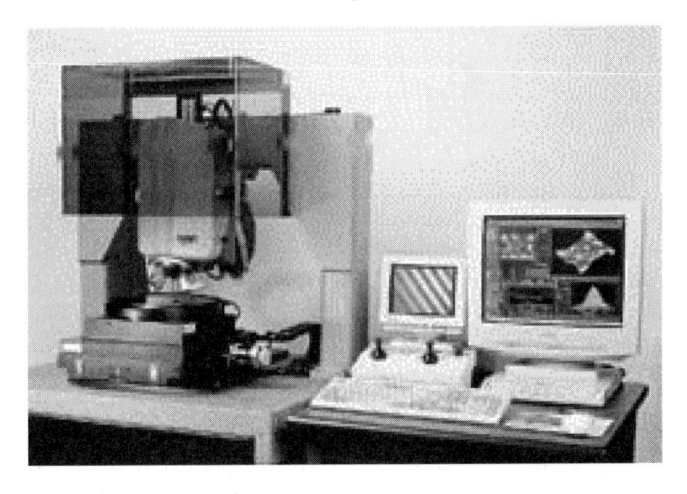

Fig. 4-58 A typical profilometer based on white-light interferometry (top) and two examples of images taken with submicrometer height resolution (courtesy of Zygo, Inc.).

For points in the x-y plane close to the point under test, and deviating vertically by less than $\Delta z < l_c$, a fringe system is generated in the detector plane. Using a CCD, we can sample this fringe pattern and process the data in a computer. By successive acquisitions on overlapping spatial portions, the measurement is extended to the entire specimen.

A commercial profilometer is shown in Fig.4-58, along with typical examples illustrating the results of x-y profile measurements.

Another recent development in the field of white-light interferometry is Optical Coherence Tomography (OCT). In this technique, we produce the images of thin slices of the object under study, a biological tissue, or another semitransparent object [50], just by looking at it through a white-light interferometer. The reference length s_r (Fig.4-55) determines the depth, and the coherence length l_c determines the thickness of the slice.

REFERENCES

[1] I. Newton, "*Lectiones Opticae*", 1678 manuscript published in Cambridge, 1727; C. Huygens "*Traité de la Lumiere*", Cambridge 1690; A. Schuster and J.W. Nicholson "*Theory of Optics*", E. Arnold: London, 1904.
[2] A.F. Rude' and M.J. Ward, "*Laser Transducer System for High Accuracy Machine Positioning*", Hewlett Packard Journal (Feb.1976), pp.2-6.
[3] R.C. Quenelle and L.J. Wuertz, "*A New Microcomputer Controlled Laser Dimensional Measurement and Analysis*", Hewlett Packard Journal (Apr.1983), pp.3-13.
[4] P.G. Halverson and R.E. Spero "*Signal processing and Testing of Displacement Metrology Gauges with pm-Scale Cyclyc Nonlinearity*", J. of Optics A, vol.4 (2002), pp.S304-10.
[5] S. Donati, "*Photodetectors*", Prentice Hall: Upper Saddle River, 2000.
[6] G. Molesini et al., "*Multiphase Homodyne Interferometry*", Proc. Elettroottica'94, edited by S. Donati, pp.53-57, AEI: Milan, Italy, 1994.
[7] V. Annovazzi Lodi, S. Donati, and S. Merlo, "*Thermodynamic Phase Noise in Fiber Interferometers*", Journal Optical and Quantum Electronics, vol.28 (1996), pp.43-49.
[8] V. Annovazzi Lodi and S. Donati, "*Injection Modulation in Coupled Laser Oscillators*", IEEE J. Quant. Electr., vol.QE-16 (1980), pp.859-865
[9] F. Aronowitz, "*The Laser Gyro*", in Laser Applications, edited by M. Ross, Academic Press: New York, 1971, pp.134-199.
[10] S. Donati, "*Laser Interferometry by Induced Modulation of the Cavity Field*", J. Appl. Physics, vol.49 (1978), pp.495-497.
[11] T. Bosch, N. Servagent, and S. Donati, "*Optical Feedback Interferometry for Sensing Applications*", J. of Optical Engineering, vol.40 (2001), pp.20-27.
[12] A. Siegman, "*Lasers*", University Science Books: Mill Valley, CA, 1986, pp.947-953.
[13] Y. Suematsu and A.R. Adams, "*Handbook of Semiconductor Lasers and Photonic Integrated Circuits*", Chapman & Hall: London, 1994.
[14] R. Lang and K. Kobayashi, "*External Optical Feedback Effects on Semiconductor Injection Laser Properties*", IEEE J. Quant. Electr., vol.QE-16 (1980), pp.347-355.
[15] S. Donati, G. Giuliani, and S. Merlo, "*Laser Diode Feedback Interferometer for the Measurements of Displacements without Ambiguity*", IEEE J. Quant. Electr., vol.QE-31 (1995), pp.113-119; also Italian Patent filed 18/10/94 n° PV-94 A00012.

[16] S. Donati and V. Speziali, *"Laser Interferometry for Sensing of Respiratory Sounds"*, Proc. CLEA Conf. on Laser and Electrooptics Appl., Washington, June 1977, Digest in IEEE J. Quant. Electr.,vol.QE-13 (1977), pp.798-87D; also *"Interferometric Sensing of Respiratory Sounds"*, Laser & Elektro-Optik, Munchen, vol.12, 1980, pp.34-35.

[17] S. Donati, *"A Novel Laser Interferometer for Distance Measurements"*, Proc. Conf. on Precision Electromagnetic Measurements, Ottawa (1978), pp.75-77.

[18] N. Servagent, T. Bosch, and M. Lescure, *"A Laser Displacement Sensor using the Self-Mixing for Modal Analysis and Defect Detection"*, IEEE Trans. Instrument. Measur., vol.46 (1997), pp.847-850.

[19] S. Donati, L. Falzoni, and S. Merlo, *"A PC-Interfaced, Compact Laser-Diode Feedback Interferometer for Displacement Measurements"*, IEEE Trans. Instrument. Measur., vol.45 (1996), pp.942-947.

[20] K. Petermann, *"Laser Diode Modulation and Noise"*, Kluwer Academic: Dordrecht (Holland), 1988.

[21] G. Giuliani, P. Cinguino, and V. Seano, *"Multifunctional Characteristics of a 1.5 μm Two-Sections Amplifier-Modulator-Detector"*, IEEE Phot. Techn. Lett., vol.PTL-8 (1996), pp.367-369.

[22] S. Merlo and S. Donati, *"Reconstruction of Displacement Waveform with a Single-Channel Laser-Diode Feedback Interferometer"*, IEEE J. Quant. Electr., vol.QE-131 (1995), pp.113-119.

[23] P.J. de Groot, G. Gallatin, and S.H. Macomber, *"Ranging and Velocimetry Signal in a Backscattering Modulated Laser Diode"*, Applied Optics vol.27 (1988) pp.4475-4480; also *Optics Lett.,* vol.14 (1989) pp.165-167.

[24] S. Shinohara et al., *"Compact and Versatile Self-Mixing Type Semiconductor Laser Doppler Velocimeter with Direction Discrimination Circuit"*, IEEE Trans. Instr. Measur. vol.38 (1989), pp.574-577; also IEEE Trans. Instr. and Measur., vol.41 (1992), pp.40-44.

[25] T. Bosch and M. Lescure (editors), *"Selected Papers on Laser Distance Measurement"*, SPIE Milestone Series, vol. MS115: Bellingham, WA, 1995.

[26] N. Servagent, F. Gouaux, and T. Bosch, *"Measurement of Displacement using the Self-Mixing Interference in a Laser Diode"*, Journal of Optics, vol.29 (1998), pp.168-173.

[27] W.M. Wang, K.T.W. Grattan, A.W. Palmer, and W.J. Boyle, *"Self-Mixing Interference inside a Single-Mode Laser Diode for Optical Sensing Applications"*, IEEE J. Lightwave Techn. vol.LT-12 (1994), pp.1577-1587.

[28] R.S. Vodhanel, M. Krain, and R.E. Wagner, *"Long-Term Wavelength Drift of 0.01nm/y for 15 Free-Running DFB Lasers"*, Proc. OFC-94 Optical Fiber Confer., San Jose, 20-25 Feb.1994, paper WG5, pp.103-104.

[29] R.O. Claus and C.H. Palmer, *"Direct Measurement of Ultrasonic Stoneley Waves"*, Appl. Phys. Lett., vol.31 (1977), pp.547-548.

[30] G. Giuliani, S. Bozzi-Pietra, and S. Donati, *"Self-mixing Laser diode Vibrometer"*, Meas. Sci. Technol., vol.14 (2003), pp.24-32.

[31] S. Donati, V. Annovazzi Lodi, S. Merlo, and M. Norgia, *"Measurements of MEMS Mechanical parameters by Injection Interferometry"*, Proc. IEEE-LEOS Conference on Optical MEMS, Kawai, HI, 21-24 Aug.2000, pp.89-90, see also IEEE J. Microelectromechanical Syst., vol.12 (2003), pp.540-549.

[32] M. Corti, F. Parmigiani, and S. Botcherby, *"Description of a Coherent Light Technique to Detect Vibration of an Arch Dam"*, J. Sound and Vibration, vol.84 (1981) pp.35-45.

[33] G. Macchi and A.Pavese, *"Diagnosis of Masonry Towers Using Interferometric Transducers and System Identification"*, in Proc. ODIMAP III, Pavia 20-22 Sept.2002, pp.51-56; see also A. Miks and J. Novak,*"Non-contact Measurement of Static Deformation in Civil Engineering"*, in Proc. ODIMAP III, Pavia 20-22 Sept.2002, pp.57-62.

[34] G. Giuliani, M. Norgia, S. Donati, and T. Bosch, *"Laser Diode Self-Mixing Techniques for Sensing Applications"*, J. Optics A Pure Appl. Opt., vol.4 (2002), p.S283-S294.

[35] F. Gouaux, N. Sarvagent, and T. Bosch, *"Absolute Distance Measurement with an Optical Feedback Interferometer"*, Applied Optics, vol.37 (1998), pp. 6684-6689.

[36] T. Bosch, S. Pavageau, S. Donati, et al., *"Low-cost Optical Feedback Laser Range-Finder with Chirp Control"*, IEEE Instrum. and Measur. Technol. Conf., Budapest 2001.

[37] H. Matsumoto, *"Alignment of Length-Measuring IR Laser Interferometer using Laser Feedback"*, Applied Optics, vol.19 (1980), pp.1-2.

[38] G. Giuliani, S. Donati, M. Passerini and T. Bosch, *"Angle Measurement by Injection Detection in a Laser Diode"*, Optical Engineering, vol.40 (2001), pp.95-99.

[39] S. Donati and M. Sorel, *"A Phase-Modulated Feedback Method for Testing Isolators Assembled into the Laser Package"*, Phot. Technol. Letters, vol.PTL-8 (1996), pp.405-407.

[40] S. Donati and M. Sorel, *"High-Sensitivity Measurement of Return Loss by Self-Hetero-dyning in a Laser Diode"*, Proc. OFC'97, Opt. Fiber Conf., Dallas, 1997, p.161.

[41] S. Donati and G. Giuliani, *"Return Loss Measurement by Feedback Interferometry"*, Proc. WFOPC'98, Workshop on Fiber Opt. Comp. Pavia, 1998, pp.103-106.

[42] P. de Groot, G. Gallatin, and G. Gardopee, *"Optical Testing using Laser Feedback Metrology"*, Conf. on Laser Interferometry, San Diego, 1989, SPIE vol.1162, pp.435-442.

[43] H. Ukita, Y. Uenishi, and Y. Katagiri, *"Application of an Extremely Short Strong-Feedback External-Cavity Laser Diode System Fabricated with GaAs-based Integration Technology"*, Applied Optics, vol.33 (1994), pp.5557-5563.

[44] J. Hewett, *"Optical Interface to Give Mobile Phone a New Look"*, Optics and Laser Europe, 97 (Jul./Aug.2002), p.9; see also IoP Journal of Meas. Science and Techn. vol.13 (2002), pp.2001-2006.

[45] D.A. Jackson et al., *"An Interferometric Fibre Optic Sensor Using a Short Coherence Length Source"*, Electronics Letters, vol.23 (1987), pp.1110-1112.

[46] M. Davidson, K. Kaufman, I. Mazor, and F. Cohen, *"Application of Interference Microscopy to IC Inspection and Metrology"*, SPIE Proceedings, vol.775 (1987), pp.233-247.

[47] T. Dresel, G. Hausler, and H. Venzke, *"3-D Sensing of Rough Surfaces by Coherence Radar"*, Applied Optics, vol.31 (1992), pp.919-925.

[48] J.C. Wyant and K. Creath, *"Advances in Interferometric Optical Profiling"*, Intl. Journal Machine Tools Manufact., vol.32 (1992), pp.5-10.

[49] P.J. Caber *"An Interferometric Profiler for Rough Surfaces"*, Applied Optics, vol.32 (1993), pp.3438-3441.

[50] A.G. Podoleanu et al., *"OCT En-face Images from the Retina with Adjustable Depth Resolution in Real Time"*, IEEE Sel. Top. Quant. El., vol.STQE-5 (1999), pp.1176-1184.

CHAPTER 5

Speckle-Pattern Instruments

When light from a laser source is shed on a plain surface like a sheet of paper or a white wall, the spot appears granular (Fig.5-1). This appearance is termed speckle pattern and came as a surprise to the first researchers experimenting with lasers.

Indeed, spatial coherence was a predicted feature of the laser beam, and common sense was that a laser spot should look smooth. Actually, the spatial distribution of a mode is smooth, and as such is preserved when the beam impinges on a smooth surface. However, a smooth surface is one with roughness much smaller than wavelength, and only a mirror satisfies this condition.

In a normal diffuser or diffusing surface, one with a brightness that looks the same from any line of sight, the vertical scale of roughness $\Delta z=f(x,y)$, see Fig.5-2, is large compared to wavelength. Each elemental area of the diffuser is then a source of secondary wave leaving the surface with an added random phase $\phi=2k\Delta z$, replica of the random profile Δz.

The random phase is responsible for destroying the spatial coherence of the near field distribution at the diffuser and for the grainy appearance of the far field.

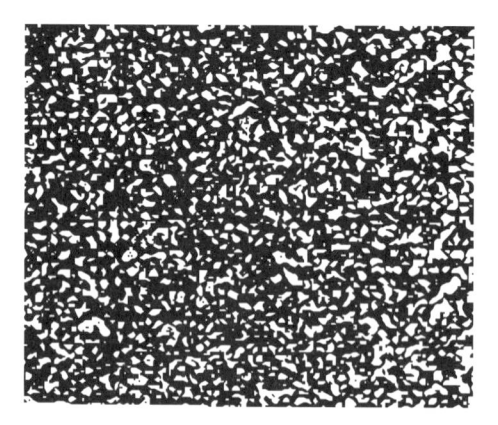

Fig.5-1 Coherent light projected on a diffusing surface appears granular, and this phenomenon is termed speckle pattern.

In this chapter, we start by considering the statistical properties of the speckle pattern field, first described on a qualitative base and then through a more rigorous analytical approach. We proceed by analyzing the errors of speckle in measurements with interferometers and will conclude with the description of an interferometer imaging technique known as Electronic Speckle Pattern Interferometry (ESPI).

5.1 SPECKLE PROPERTIES

In treating speckle properties, it is customary to assume [1] an ideal diffuser to schematize the statistical properties of surface roughness. As it will be readily appreciated in the following, the ideal diffuser is the model of extreme randomness and gives rise to the so-called fully developed speckle statistics [1]. A real diffusing surface may approach the ideal diffuser rather well, or have a limited randomness of surface roughness that translates itself in a reduced variance. Thus, the ideal diffuser case is the limit case of statistics.

5.1.1 Basic Description

In an ideal diffuser, the height profile $\Delta z = f(x,y)$ is described by a random function $f(x,y)$. The function is invariant to translations (i.e., properties are the same for all points of the diffuser), has a zero average deviation, $\langle \Delta z \rangle = 0$, and a quadratic mean value or variance $\sigma_z^2 = \langle \Delta z^2 \rangle$ much larger than λ^2. As a consequence, the phase shift ϕ introduced by the profile corrugation on diffusion at any point x,y is larger than 2π, or $\phi = k\sigma_z >> 2\pi$.

In an ideal diffuser, the profile has no spatial correlation on short-scale, or $f(x,y)$ is uncorrelated to $f(x',y')$ as soon as $|x'-x|$ and $|y'-y| > \lambda$.

This condition can be expressed in terms of the profile correlation function. With the usual

5.1 Speckle Properties

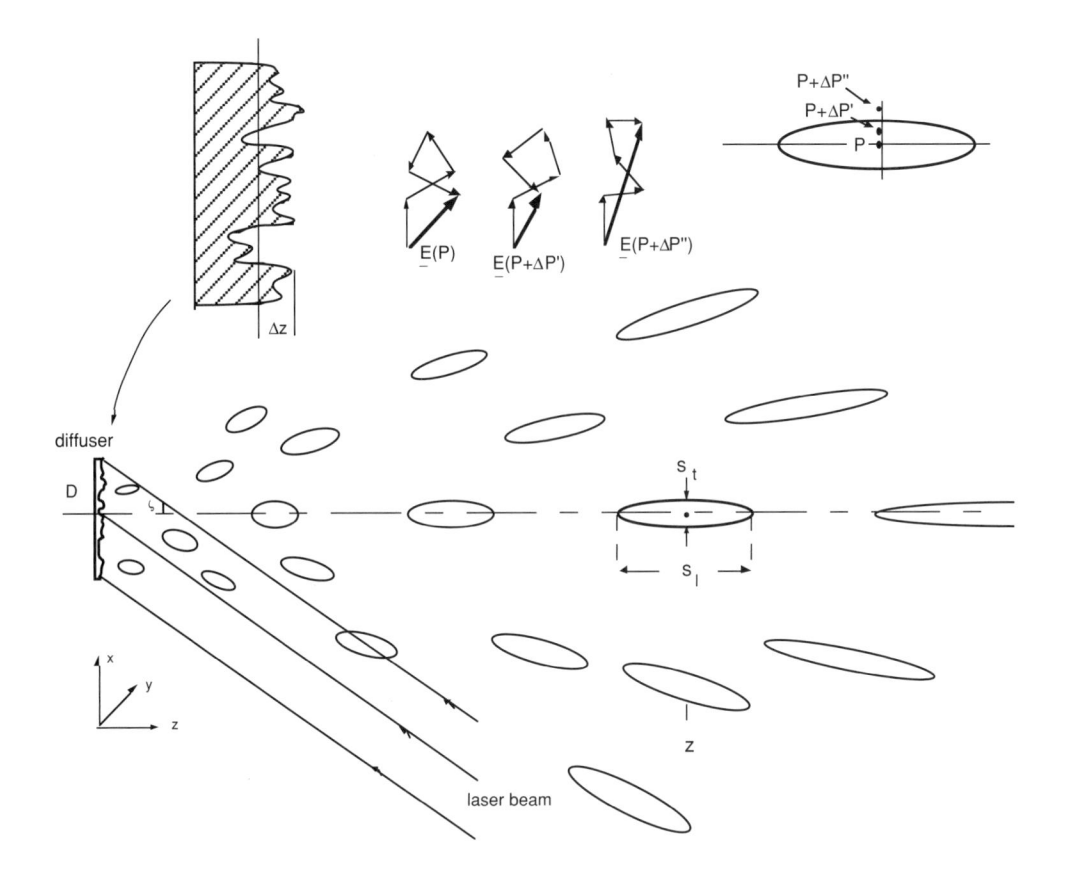

Fig.5-2 Speckle pattern is the field radiated from a diffuser, that is, a surface with roughness larger than wavelength, $\Delta z \gg \lambda$ (top left). The field at a generic point P is the summation of many individual vectors with a random phase (top center). Moving away from P to P+ΔP' and P+ΔP", the field gradually loses correlation. A speckle grain is a region within which correlation is C≥0.5. Grains are cigar shaped and point to the diffuser center.

definition of the correlation function, that is $C_z(\Delta x, \Delta y) = \iint \Delta z(x,y) \, \Delta z(x+\Delta x, y+\Delta y) dx dy$, the result expressing spatial uncorrelation is of the type: $C_z(x-x', y-y') = \sigma_z^2 \, \delta(x-x', y-y')$, where δ is the Dirac's impulse function and σ_z^2 is the variance defined previously.

Now, let us consider the electrical field $\underline{E}(x,y)$ leaving the diffuser at $z \approx 0$, after illumination from the coherent beam $\underline{E}_1$. If we move just a little away from the diffuser, the amplitude $E(x,y)$ is substantially equal to the illuminating one, E_1.

The total phase of E is the sum of a deterministic phase contribution $\phi_1(x,y) = kx \sin\zeta$, due to the incidence angle ζ of the illuminating beam (Fig.5-2) and a random phase term $\phi(x,y)$ due to the diffuser roughness.

Because the random phase shift is large, $k\sigma_z \gg 2\pi$, and has no spatial correlation, the deterministic term is negligible. Therefore, field correlation $C_E(x-x',y-y')$ duplicates the statistics of $C_z(x-x',y-y')$ or, it is δ-like.

Near the diffuser ($z \approx 0$), coherence areas are delta-like regions, and the spatial coherence of the original laser beam is lost, destroyed by the diffuser's roughness.

Because the δ-like condition holds irrespective of the actual amplitude and phase dependence of the illuminating field E_1 from x,y, we may take for simplicity E_1=const. in the following.

The field $\underline{E}(x,y,z)$ at any point P in space (Fig.5-2) is evaluated as the vector summation of contributions emitted by all dxdy elemental areas in the source. Each contribution has constant amplitude, but random phase. Such a vector sum is a statistical process known as random walk. The process is the same in all points P of the space, and only the descriptive parameters change with P, not the distributions.

Now, in a generic point P, we write the field as $\underline{E} = E_R + iE_I$ and recall the results known from statistics about the properties of random-walk process [1-3]. Because of the law of large numbers relative to the addition of many individual contributions, both the real E_R and the imaginary E_I component of the field are normal distributions with zero mean value, $\langle E_R \rangle = \langle E_I \rangle = 0$. Moreover, the intensity $I = E^2$ distribution is found to be a negative exponential. Explicitly, the probability density of intensity is written as $p(I)=[1/\langle I \rangle] \exp -I/\langle I \rangle$, where $\langle I \rangle$ is the average intensity of the speckle field and coincides with the mean square value $\langle E^2 \rangle$ of the field components.

When we move at a substantial distance from the diffuser, the statistical properties remain unchanged, but the δ-like correlation is smeared to a finite-width distribution. In fact, let us go back to point P and consider the field resulting from the addition of elemental vectors with a random phase. Let $\underline{E}(P)$ be the vector sum, as indicated in Fig.5-2. Now we move to point P'=P+ΔP'. If ΔP' is small, it is reasonable to expect that the phase of the individual contributions is not so much changed or that $\underline{E}(P+\Delta P')$ is correlated to $\underline{E}(P)$. This shall certainly happen, because *ab absurdo*, if $\Delta P' \to 0$, $\underline{E}(P+\Delta P')$ cannot but converge to E(P).

Second, if we move substantially away from P to P''= P+ΔP'' and let the deviation ΔP'' increase, we shall ultimately collect the field contributions with new phase terms, different from the initial ones, whence the resulting vector addition $\underline{E}''=E(P+\Delta P'')$ becomes different from the initial result E(P), or, is not correlated to it any more.

Now, the substance is about how large ΔP' shall be before correlation is lost (or, quantitatively, it becomes less than a specified value, e.g., C<0.5).

This quantity defines the so-called speckle size. The speckles then represent the coherence areas we find at a distance z from an ideal diffuser illuminated with a spatially coherent beam.

In next section, we will calculate the speckle size, both longitudinal (s_l along z) and transversal (s_t along x or y). For a circular diffuser of diameter D, uniformly illuminated, the results are:

$$s_t = \lambda \, z/D, \qquad s_l = \lambda \, (2z/D)^2 \qquad (5.1)$$

5.1 Speckle Properties 185

As can be seen in Fig.5-2 and from Eq.5-1, speckles are cigar shaped, with the longitudinal size larger than the transversal, and become more elongated as we move away from the diffuser. Speckles point to the diffuser center, and the projection of off-axis speckles is equal to the on-axis speckle.

Subjective and Objective Speckles. The previous discussion holds for the free propagation of the diffused field depicted in Fig.5-2. This case is referred to as one generating *objective* speckles, in the sense that speckles are only dependent on the diffusing object or target.
Another case of interest is when an optical element, for example a lens as in Fig.5-3, is interposed between the conjugated image and source planes. This is the case of *subjective* speckles, so called because then the speckle depends on the observer, too.

We may analyze the optical conjugation of the diffuser (with diameter D) to an image plane by a lens (with diameter D_L and focal F_L) as drawn in Fig.5-3. For ease of notation, let us take $z \gg F_L$ so that the distance p from the lens is $p=F_L$.
Along its diameter, the lens accommodates a number of speckles $N = D_L/s_t$ given by the ratio of lens diameter to speckle transversal size. Explicitly, here we consider N as a ratio of linear size, not of surface.

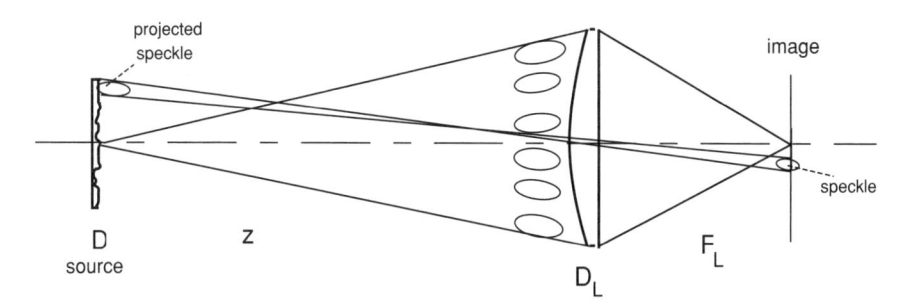

Fig.5-3 A lens looking at the source collects N speckles and, because of mode invariance, forms N (smaller) speckles inside the image. As the image is conjugated to the source, N virtual speckles appear projected back on the source. This is the *subjective* speckle pattern.

Because speckles point to the diffuser center, they are each oriented at a different angle and thus come to a different position at the focal (or image) plane, filling there the image of the diffuser. The image has a diameter $D(F_L/z)$ and contains $N=D_L/s_t$ speckles. Then, the diameter of speckles (transversal size $s_{t(fp)}$) found in the image at the focal plane is:

$$s_{t(fp)} = D(F_L/z)(s_t/D_L) = \lambda F_L/D_L \tag{5.2}$$

This speckle dimension is the same that we get for objective speckle with a diffuser of diameter D_L at distance F_L.

Now, from the focal plane we can see a virtual image of the source, equal to the one in the focal plane multiplied by z/F_L. On the diffuser, the apparent dimension (transversal size $s_{t(pr)}$) we see projected is given by:

$$s_{t(pr)} = (\lambda F_L/D_L)(z/F_L) = \lambda z/D_L \qquad (5.3)$$

Eq.5.3 explains why this speckle regime is called subjective: the diameter of speckles seen on the source is dependent on the lens diameter D_L. Moreover, the speckle size is that generated by the lens aperture D_L, not the target aperture. In addition, the longitudinal size of virtual speckle appearing on the diffuser is $s_{l(pr)} = \lambda(2z/D_L)^2$.

The effect of subjective speckle can be visualized easily, looking at a laser spot projected on a wall through a pinhole and varying the diameter of it. For a pinhole, you may use your own index finger, closing it fully folded leaving a small aperture in the middle of the skin fold that you can finely adjust.

5.1.2 Statistical Analysis

We can analyze the statistical properties of the speckle-pattern field starting from the Fresnel approximation to diffraction equation (see also Appendix A5).
Let the diffuser be illuminated by a uniform coherent beam in the plane ξ,η. The complex amplitude of the source can be written as $E_1(\xi,\eta)=E_1 \exp i\phi(\xi,\eta)$ rect $(\xi,\eta,D/2)$, where:
A_1 is the constant field amplitude of the illuminating beam,
$\phi(\xi,\eta)= k \, \Delta z(\xi,\eta)$ is the random phase function, associated with Δz,
$\Delta z(\xi,\eta)$ is the surface roughness, and
rect (x) is the step function defining the limits of integration, defined as:
rect (x)=1 for $x=\xi^2+\eta^2 \leq (D/2)^2$, and rect (x)=0 for $x=\xi^2+\eta^2>(D/2)^2$.

The function $\phi(\xi,\eta)$ has a $\delta(\xi,\eta)$ spatial correlation and a variance σ_z^2 such that $2k\sigma_z^2 >> 2\pi$. Each elemental area $d\xi d\eta$ contributes with a spherical wave delayed by kr_{12} to the field collected at point (x,y) in the image.

In the Fresnel approximation, we may assume $r_{12} \approx z+(x^2+y^2)/2z+(\xi^2+\eta^2)/2z-(x\xi+y\eta)/z$ and $r_{12}/z \approx 1$. By summing up all the elemental contributions in $d\xi d\eta$, we have in the image plane x,y at a distance z (Fig.A5-1):

$$E(x,y,z) = (E_1/\lambda z) \, \exp ikz \, \exp ik(x^2+y^2)/2z \int\!\!\int_{-\infty,+\infty} d\xi d\eta \, \text{rect}(\xi,\eta,D/2) \times$$
$$\exp ik(\xi^2+\eta^2)/2z \, \exp -ik(x\xi+y\eta)/z \, \exp i\phi(\xi,\eta) \qquad (5.4)$$

At the right hand side of this equation, after the amplitude $E_1/\lambda z$ the first term is the phase shift of propagation down the distance z, and the second is the field curvature term. Then, we find the integral with the boundary truncation term (rect). The fourth term is again a field curvature term, the fifth is the mixed-coordinate term originating the Fourier transform kernel, and the last term is the diffuser random-phase term.

As a first property we can deduce from Eq.5.4, the field E is a complex quantity with a real E_R and an imaginary E_I part. Both of them are affected by the random part of the phase $\phi(\xi,\eta)$, as it can be seen by developing the exponential under the integration $\exp i\phi(\xi,\eta)=\cos\phi(\xi,\eta)+i\sin\phi(\xi,\eta)$. As $\phi= k\Delta z >> 2\pi$, we have $\langle\cos\phi\rangle=\langle\sin\phi\rangle=0$ and, as a consequence:

5.1 Speckle Properties

$$\langle E_R \rangle = \langle E_I \rangle = 0 \quad (5.5)$$

Thus, the speckle field has zero-average real and imaginary parts at all points in space.

Second, in the Δz profile, the number of scatterers is very large and, therefore, in view of the central-limit theorem [3], Δz obeys a normal (or Gauss) distribution. At the same way, $\phi = k\Delta z$ and its sine and cosine functions are normal distributions, too.

It is then not surprising that, after linear operations as those performed through the integration in Eq.5.4, the real and imaginary parts of the field, $E = E_R + i E_I$ are both normal distributed and uncorrelated. Accordingly, their probability distributions can be written as:

$$p(E_R) = (2\pi\sigma_E^2)^{-1/2} \exp{-E_R^2/2\sigma_E^2}, \quad p(E_I) = (2\pi\sigma_E^2)^{-1/2} \exp{-E_I^2/2\sigma_E^2} \quad (5.6)$$

In Eq.5.6, we have used Eq.5.5 and indicated with σ_E^2 the variance of the field components that shall be the same for both E_R and E_I for symmetry reasons. Writing the σ_E^2 variance as the mean of squares reveals that it is coincident to the mean intensity:

$$\sigma_E^2 = \langle E_R^2 \rangle = \langle E_I^2 \rangle, \quad 2\sigma_E^2 = \langle E_R^2 \rangle + \langle E_I^2 \rangle = \langle E_R^2 + E_I^2 \rangle = \langle I \rangle \quad (5.7)$$

Another quantity of interest is the field amplitude $|E| = \sqrt{(E_R^2 + E_I^2)}$. By transformation of the variables it is easily found that $|E|$ is Rayleigh distributed, or it can be written as:

$$p(|E|) = (2\pi\sigma_E^2)^{-1/2} |E| \exp{-(|E|^2/2\sigma_E^2)} \quad (5.6')$$

About the intensity $I = |E|^2 = E_R^2 + E_I^2$, knowing that both E_R and E_I are normal distributions and using the rules of variables transformation [3], the probability density $p(I)$ easily follows (see below) as:

$$p(I) = \langle I \rangle^{-1} \exp{-I / \langle I \rangle} \quad (5.8)$$

Because of the negative exponential distribution, weak (or dark) speckles are more probable than intense (or bright) ones. For example, 10% of speckles have ≤10% the average intensity, 1% have ≤1% the average intensity, and so on.

This is just *amplitude fading,* a feature hampering measurements that we shall appropriately tackle in all systems working on a diffusing surface rather than a mirror.

Last, the phase of the electric field at a given point P is a random distribution with no information on the initial distance-related phase kz. Computing the phase as $\varphi = \mathrm{atan}\, E_I/E_R$ gives as a result a uniform distribution of phase on $0-2\pi$ as:

$$p(\varphi) = 1/(2\pi) \quad (5.8')$$

Recall on the change of variables. We want to find the probability distribution $p(E)$ of a random variable $E = f(I)$ related to another variable I, whose probability distribution $p(I)$ is given. We equate the differential probability dp around I and E by writing $dp = p(E)dE = p(I)dI$. The desired $p(E)$ is then found as: $p(E) = p(I)dI/dE = p(I)/[df(I)/dI]$. For example, let $|E| = \sqrt{I}$ and $p(I) = I_0^{-1} \exp{-I/I_0}$. As $df(I)/dI = 1/2\sqrt{I} = 1/2|E|$, then $p(|E|) = I_0^{-1} 2|E| \exp{-|E|^2}$, which is the Rayleigh distribution following from the exponential. For joint distributions, $p(I,\theta) = f(A_1, A_2)$, the generalization of the method is

$p(I,\theta)=p(A_1,A_2)|J|$, where $|J|$ is the Jacobian determinant of the variable transformation, from A_1 and A_2 to I and θ [2].

A summary of the statistical properties of the speckle pattern is reported in Fig.5-4.

The quantities described so far are the first-order statistical properties of the speckle pattern, because they are related to the field in a specific point P. Further information is gathered with the bivariate or joint probability function, which relates the statistical properties of the speckle in two points in space, for example P and P+ΔP in Fig.5-4. The starting point for analyzing the second-order statistical properties of the speckle is the joint probability of intensity and phase in two points P_1 and P_2. This is written as [1,3]:

$$p(I_1,I_2,\varphi_1,\varphi_2) = [16\pi^2\sigma_E^4(1-\mu^2)]^{-1} \times$$
$$\times \exp\{[I_1+I_2-2(I_1I_2)^{1/2}\mu\cos(\varphi_1-\varphi_2-\psi)]/2\sigma_E^2(1-\mu^2)\} \quad (5.9)$$

In this equation, $\mu \exp i\psi = \mu_c$ is the *complex coherence factor* for the field at points P_1 and P_2. The complex coherence factor is defined as the ratio of the mutual intensity normalized to the product of rms values of fields at points P_1 and P_2, or:

$$\mu_c = \langle E(P_1)E^*(P_2)\rangle / [\langle |E(P_1)|^2\rangle \langle |E^*(P_2)|^2\rangle]^{1/2} \quad (5.10)$$

where $\langle .. \rangle$ indicates the operation of ensemble average on the speckle set, and * is the complex conjugate. The coherence factor μ_c has a modulus μ that may go from 0 (complete uncorrelation) to 1 (full correlation). It is complex because, simply, $E(P_2)$ may be delayed respect to $E(P_1)$ and then $\psi=ks(P_2-P_1)$, where s(..) is the distance between P_2 and P_1.

The numerator in Eq.5.10 is the *correlation* of the field, which is also called *mutual intensity* by some Authors [1]:

$$C_\mu = \langle E(P_1)E^*(P_2)\rangle \quad (5.11)$$

The correlation has the important physical meaning of a beating signal between the fields at points P_1 and P_2. It is maximum for full correlation, or P_1 approaches P_2, and decreases to zero when fields are uncorrelated, or P_1 moves far away from P_2. The correlation length is defined as the distance $s(P_2-P_1)$ at which the correlation C_μ has decreased to a specified value (usually 0.5 of the maximum) or has reached the first zero (if C_μ oscillates).

Let us now calculate the speckle size starting from Eq.5.4. To simplify notation, we normalize the field E_{st} to $E_1/\lambda z$, and write the integration $\iint_{-\infty,+\infty} d\xi d\eta$ rect(D/2) as $\iint_{rect(\xi\eta,D)}$. With these positions, we have:

$$E_{st}(x,y,z) = \exp ikz + ik(x^2+y^2)/2z \times$$
$$\times \iint_{rect(\xi\eta,D)} \exp -ik(x\xi+y\eta)/z \exp ik(\xi^2+\eta^2)/2z \exp i\phi(\xi,\eta) \quad (5.12)$$

First, we consider the transversal speckle size and write the correlation C_μ for two points with a displacement Δ along x, or:

5.1 Speckle Properties

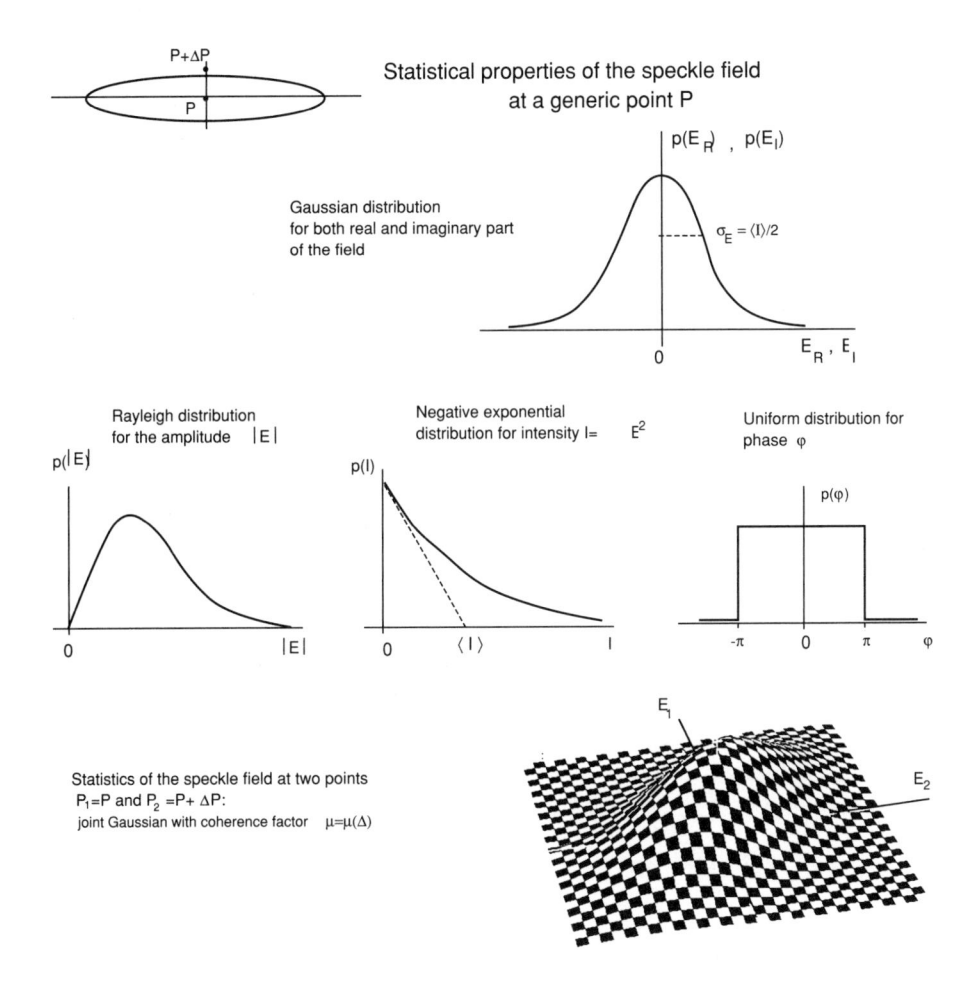

Fig.5-4 Summary of the speckle pattern statistics: field components E_R and E_I, amplitude $|E|$, intensity I and phase φ, and joint distribution E_1, E_2

$$C_\mu = \langle E_{st}(x+\Delta,y,z) E_{st}^*(x,y,z) \rangle \tag{5.11'}$$

Inserting Eq.5.12 in this expression, we get:

$$C_\mu = \langle E_{st}(x+\Delta,y,z) E_{st}^*(x,y,z) \rangle = \exp ik(\Delta^2+2\Delta x)/2z \iiiint \mathrm{rect}(\xi\eta,D)\mathrm{rect}(\xi'\eta',D)$$

$$\exp\text{-}ik[(x+\Delta)\xi+y\eta]/z \; \exp\text{+}ik(x\xi'+y\eta')/z \; \langle \exp i\phi(\xi,\eta) \exp\text{-}i\phi(\xi',\eta') \rangle \tag{5.13}$$

Here, we have used that, by complex conjugation, the terms in x,y,z have opposite phases and cancel out. Also, the average ⟨..⟩, applied to the entire integral at the right-hand side, commutes with the integration and moves into the random-phase terms (the other terms are deterministic).

In view of the uncorrelation from point to point in the source plane, it is:

$$\langle \exp i\phi(\xi,\eta) \exp{-i\phi(\xi',\eta')}\rangle = \delta(\xi-\xi',\eta-\eta') \quad (5.14)$$

By substituting Eq.5.14 into Eq.5.13, and recalling that the δ-function has the property of saturating the integration variable, i.e.:

$$\int\int\int\int_{\text{rect}(\xi\eta,D)\text{rect}(\xi'\eta',D)} F(\xi,\xi',\eta,\eta')\,\delta(\xi'-\xi,\eta'-\eta) = \int\int_{\text{rect}(\xi\eta,D)} F(\xi,\xi,\eta,\eta),$$

we obtain:

$$\langle E_{st}(x+\Delta,y,z)E_{st}^*(x,y,z)\rangle = \exp ik(\Delta^2+2\Delta x)/2z \times$$

$$\times \int\int_{\text{rect}(\xi\eta,D)} \exp{-ik[(x+\Delta)\xi+y\eta]/z}\,\exp{+ik(x\xi+y\eta)/z} =$$

$$= \exp ik(\Delta^2+2\Delta x)/2z \int\int_{\text{rect}(\xi\eta,D)} \exp{-ik\Delta\xi/z} \quad (5.15)$$

In the last line, the integral expression is recognized as the Fourier transform of the rectangle rect($\xi,\eta,D/2$) of diameter D, calculated in the plane ξ,η, and specialized at r=Δ. The multiplying factor is a pure delay, accounting for a phase $\varphi = k\Delta(\Delta+2x)/2z \approx k\Delta(x/z)$. This is the deterministic error we shall expect because of field curvature in an interferometric measurement performed in the speckle regime.

From Appendix A5.3, the integral in Eq.5.15 is readily evaluated as:

$$|\langle E_{st}(x+\Delta,y,z)E_{st}^*(x,y,z)\rangle| = \text{somb}\,\Delta D/\lambda z \quad (5.16)$$

and it gives the real part of the correlation we are interested in.

In Eq.5.16, the somb function (bidimensional generalization of the sinc) is the function generating the Airy's disk (App.A5.3). The somb x drops to half the initial value at x=±0.71, and thus, the speckle size at 50% correlation is $s_t = 2\Delta = 2\times0.71/(D/\lambda z) = 1.42\,\lambda z/D$. If the aperture were square instead of circular, the step function defining the limits of integration would have been rect(x,D/2) rect(y,D/2), and all the previous arguments would still hold, with the final result changed to sinc $\Delta D/\lambda z$. The sinc function drops to half the initial value for x=0.60, whence in this case $s_t = 2\Delta = 2\times0.60/(D/\lambda z) = 1.20\,\lambda z/D$. These slightly different values show that the factor multiplying the $\lambda z/D$ dependence is close to unity.

In conclusion, we simply take as reference value for the transversal size:

$$s_t = \lambda z/D \quad (5.17)$$

Second, we can repeat the calculation for the longitudinal size. Let us consider a displacement Δ along z and write the correlation as $\langle E_{st}(x,y,z+\Delta)E_{st}^*(x,y,z)\rangle$.

5.1 Speckle Properties

Writing Eq.5.12 for this case yields:

$\langle E_{st}(x,y,z+\Delta)E^*_{st}(x,y,z)\rangle = \exp ik\Delta \ \exp\text{-}ik\Delta(x^2+y^2)/2z^2 \times$

$\times \int\int\int\int_{\text{rect}(\xi\eta,D) \ \text{rect}(\xi'\eta',D)} \exp ik(\xi^2+\eta^2)/2(z+\Delta) \times \exp\text{-}ik(\xi'^2+\eta'^2)/2z^2 \times$

$\times \exp\text{-}ik(x\xi+y\eta)/(z+\Delta) \ \exp\text{+}ik(x\xi'+y\eta')/z \times \langle \exp i\phi(\xi,\eta)\exp\text{-}i\phi(\xi',\eta')\rangle =$

$= \exp ik\Delta\text{-}ik\Delta(x^2+y^2)/2z^2 \int\int_{\text{rect}(\xi\eta,D)} \exp\text{-}ik\Delta(x\xi+y\eta)/z^2 \text{-}ik\Delta(\xi^2+\eta^2)/2z^2$ (5.18)

Also in this case the first term multiplying the integral is a pure phase term φ. This phase φ includes the expected (and correct) delay $k\Delta = k[(z+D)-z]$ as well as the curvature (deterministic) error $k\Delta(x^2+y^2)/2z^2$, which is small with respect to $k\Delta$ but will show up when we perform an interferometric measurement in the speckle regime.

The term given by the integral is real and, using Eq.A5.5, is given by:

$|\langle E_{st}(x,y,z+\Delta)E_{st}^*(x,y,z)\rangle| = \text{somb} (\Delta/\lambda z^2)[D^2/4+D\sqrt{(x^2+y^2)}]$

$\approx \text{somb} (\Delta D^2/4\lambda z^2) \quad \text{for } x^2+y^2 \ll D^2$ (5.19)

As for the transversal dimension, we have a somb function for the correlation of the field. The somb x drops to half the initial value at x=±0.71, and thus, the speckle size at 50% correlation is: $s_t = 2\Delta = 2\times 0.71/(D^2/4\lambda z^2) = 1.42 \ 4\lambda z^2/D^2$. If the aperture were square, the result would change to sinc $\Delta D^2/2\lambda z^2$, whence $s_t = 2\Delta = 2\times 0.60/(D^2/4\lambda z^2) = 1.20 \ 4\lambda z^2/D^2$ in this case. In conclusion, we can take as reference value for the longitudinal size:

$$s_l = \lambda \ (2z/D)^2 \qquad (5.20)$$

5.1.3 Speckle Size from Acceptance

In previous section, the transversal and longitudinal sizes of the speckle were calculated on the base of correlation width of the field. Correlation is tightly related to coherence, and we may interpret s_l and s_t as the coherence lengths (along longitudinal and transversal directions) and the speckle grain as a coherence region, or a mode of the field.

This interpretation is indeed correct, and in this section we show that speckle size can be traced to the size of the spatial mode found at a distance z from a diffuser.

A theorem on the invariants of radiometry (Ref.[5], App.A2.3) states that the acceptance a of an aperture receiving power under a solid angle Ω and on an area A is equal to N times λ^2, or:

$$a = A \ \Omega = N \lambda^2, \qquad (5.21)$$

where N has the meaning of number of modes, or of degree of freedom associated with the area-solid angle aperture.

Let us apply this theorem to the speckle propagation, with reference to Fig.5-5. We let $N=1$ for a single mode, and note that radiation is received at distance z under the solid angle $\Omega = \pi(D/2z)^2$. The receiving area associated with a (transversal) spatial mode of width s_t is $A = \pi(s_t/2)^2$. Inserting in Eq.5.21, we get $\lambda^2 = \pi(s_t/2)^2\pi(D/2z)^2$ whence $s_t = (4/\pi)\lambda z/D$. Second, the bundle of rays in the solid angle Ω keeps limited transversally to an extent $\leq s_t$ (Fig.5-5) for a longitudinal width s_t/θ, where $\theta = D/2z$ is the angular aperture of the bundle. Combining these expressions gives $s_l = (2/\pi)\lambda(2z/D)^2$ as the longitudinal size. A comparison with Eq.5.1 shows that these values are the same already found, except for a minor multiplying factor of the order of unity.

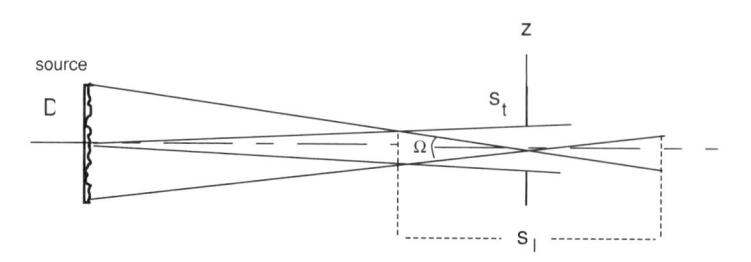

Fig.5-5 Geometry for the calculation of speckle size as a spatial mode

5.1.4 Joint Distributions of Speckle Statistics

The joint statistics of the speckle pattern describes the properties of the field in two points $P_1 = P$ and $P_2 = P + \Delta P$ in space (Fig.5-4). When ΔP is comparable with the speckle size, the joint statistics gives relative-amplitude and phase-difference information important to displacement and vibration measurements made with interferometers on diffuser like target surfaces.

The starting point is the joint probability density of intensities and phases $p(I_1,I_2,\varphi_1,\varphi_2)$ given by Eq.5.9. This probability is a direct consequence of the Fresnel diffraction integral and of the diffuser surface randomness.

We can get the joint probability density of intensities $p(I_1,I_2)$ by a double integration of Eq.5.9 on φ_1 and φ_2 from 0 to 2π. In this way, we obtain [1,3,6]:

$$p(I_1,I_2) = [4\sigma^4(1-\mu^2)]^{-1} \exp\{-[(I_1+I_2)/2\sigma^2(1-\mu^2)]\} \mathfrak{J}_0[\mu(I_1I_2)^{1/2}/\sigma^2(1-\mu^2)] \quad (5.22)$$

In this equation, $\mathfrak{J}_0$ is the modified Bessel's function of first kind, zero order, μ is the modulus of the coherence factor (Eq.5.10), and σ^2 is the variance of the (real and imaginary) field components (Eq.5.7). About this variance, we may recall that it is also $2\sigma^2 = \langle I_1 \rangle = \langle I_2 \rangle$. Similarly, the joint probability density of phases $p(\varphi_1,\varphi_2)$ is obtained by a double integration of Eq.5.9 on I_1 and I_2 from 0 to ∞, and the result is [3,6]:

5.1 Speckle Properties

$$p(\varphi_1,\varphi_2) = [(1-\mu^2)/4\pi^2](1-\beta^2)^{-3/2} \, [\sqrt{(1-\beta^2)}+\beta \arcsin\beta+(\pi/2)\beta] \quad (5.22')$$

Here, $\beta=\mu\cos(\varphi_1-\varphi_2-\psi)$, and ψ is the argument of the complex coherence factor $\mu_c=\mu\exp i\psi$. The probability $p(\varphi_1,\varphi_2)$ is brought to the probability density of the phase difference $\varphi_1-\varphi_2$ by changing the variables φ_1,φ_2 in their sum and difference and integrating on the sum. This gives as a result $p(\varphi_1-\varphi_2)=2\pi \, p(\varphi_1,\varphi_2)$.

The joint probability $p(\varphi_1-\varphi_2)$ is plotted in Fig.5-6 as a function of $\varphi =\varphi_1-\varphi_2-\psi$ and with the coherence factor μ as a parameter. In the diagram, we start at $\mu=0$ with the uniform distribution $p=1/2\pi$. As μ increases, the probability density becomes more and more peaked around $\varphi =0$, up to the limit case $\mu=1$ (not shown in Fig.5-6) when it becomes a Dirac delta, $p(\varphi_1-\varphi_2)= \delta(\varphi_1-\varphi_2)$.

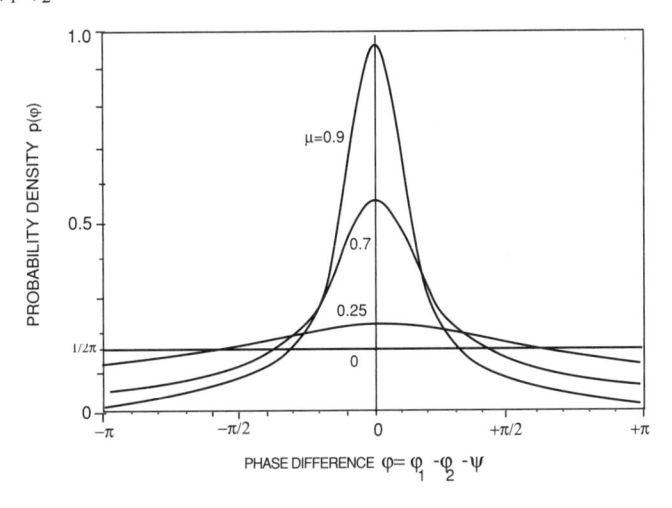

Fig.5-6 Probability density $p(\varphi_1-\varphi_2)$ of the phase difference plotted for some values of μ, the modulus of the coherence factor.

From the probability density, we can calculate the variance σ_φ^2 of the phase difference $\varphi_1-\varphi_2$, that is, the error in a phase measurement performed in the speckle regime between two points with a coherence factor μ.
As it is $\langle\varphi_1-\varphi_2\rangle=0$, the variance is given by $\sigma_\varphi^2 =\int_{0-2\pi} \varphi^2 p(\varphi)d\varphi$, where we have let $\varphi=\varphi_1-\varphi_2$. Collecting the terms and using Eq.5.22', one is left with the evaluation of an integral that appears intractable at first sight, but is finally solved with the result [6]:

$$\sigma_\varphi^2 = \pi^2/3 - \pi \arcsin \mu + \arcsin^2 \mu - (1/2) \sum_{n=1,\infty} \mu^{2n}/n^2 \quad (5.23)$$

Let us now deal with conditional probabilities. They are useful to consider when we can add any knowledge of one of the variables $(I_1,I_2,\varphi_1,\varphi_2)$, or combination of them. Taking account of this knowledge, we may restrict the set of speckle realization to the subset of conditioned values and be able to obtain a smaller variance or measurement uncertainty.

For example, let us assume that the intensity I_1 in point P_1 is known, and we want to compute the probability of intensity at point P_2.

The conditional probability $p(I_2|I_1)$ of having I_2, once I_1 is known, is given by Bayes' theorem [2] as the ratio of the joint probability to the probability density of the conditioning variable:

$$p(I_2|I_1) = p(I_2,I_1)/p(I_1) \qquad (5.24)$$

Combining Eqs.5.24 and 5.22 we obtain [6]:

$$p(I|I_1) = [2\sigma^4(1-\mu^2)]^{-1} \exp[-(I_2+\mu^2 I_1)/2\sigma^2(1-\mu^2)] \mathfrak{I}_0[\mu(I_1 I_2)^{1/2}/\sigma^2(1-\mu^2)] \qquad (5.25)$$

We can now calculate the first and second moments of this distribution, that is:

$$\langle I_2 \rangle|_{I1} = \int_{0-\infty} I_2^2 \, p(I_2|I_1) dI_2, \quad \text{and} \quad \sigma_{I2}^2|_{I1} = \int_{0-\infty} I_2^2 \, p(I_2|I_1) dI_2, \qquad (5.26)$$

and obtain the conditioned mean and variance of intensity, that is:

$$\langle I_2 \rangle|_{I1} = 2\sigma^2(1-\mu^2) + \mu^2 I_1, \quad \text{and} \quad \sigma_{I2}^2|_{I1} = [2\sigma^2(1-\mu^2)]^2 + 4\sigma^2\mu^2(1-\mu^2)I_1 \qquad (5.27)$$

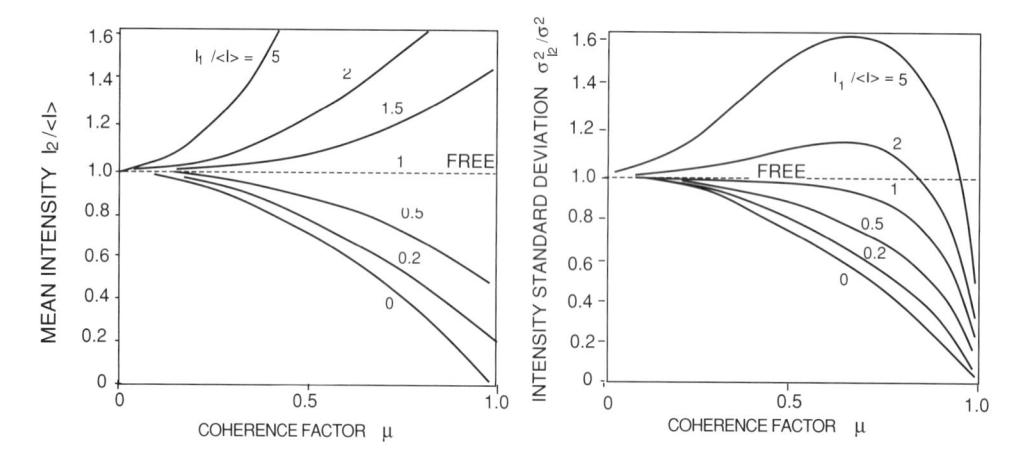

Fig.5-7 Mean value (left) and variance (right) of intensity I_2 conditioned to I_1, as a function of μ, and for some values of the intensity I_1 as a parameter. The 'free' values of mean and variance of the unconditioned distribution are also indicated.

In Fig.5-7 we report the diagrams of mean and variance of intensity I_1 conditioned to intensity I_2 versus the modulus of the coherence factor. The mean value is standardized to $\langle I_1 \rangle = \langle I_2 \rangle = 2\sigma^2$ and the variance to σ^2, the 'free' value. From Fig.5-7 we can see that when the

5.1 Speckle Properties

conditioning intensity I_1 is larger (smaller) than the mean, also I_2 is, on the average.
The variance for large conditioning intensity I_1 is larger than the 'free' value, but not so much larger. For example, $I_1=5\langle I \rangle$ gives only $\sigma_{I2}^2 \approx 1.6\sigma^2$ (at $\mu \approx 0.6$), or a relative standard deviation $\sigma_{I2}/I_1 \approx 0.25$. Of course, if we move to the high-correlation region $\mu \approx 1$, this relative standard deviation is even smaller.
In summary, a bright speckle has comparatively less intensity noise.

Another conditioned probability is that of intensity I_2, when we know both intensity I_1 and phase φ_1 in the other point. This probability can be computed starting from Eq.5.9, and the result is given in [6]. Using the definitions (Eq.5.26), mean and variance are found as:

$$\langle I_2 \rangle |_{I1,\varphi 1}=\sigma^2(1-\mu^2)[3-D(\delta)+2\delta^2], \quad \sigma_{I2}^2|_{I1,\varphi 1}=\sigma^4(1-\mu^2)[6-(1-2\delta^2)D(\delta)-D^2(\delta)+8\delta^2] \quad (5.28)$$

In Eq.5.28, erf is the standard error function [2] and we have let $D(\delta)=[1+\delta\sqrt{\pi}\,(1+\mathrm{erf}\,\delta)\exp\delta^2]^{-1}$ and $\delta=\mu\cos(\varphi_1-\varphi_2-\psi)\sqrt{[I_1/\sigma^2(1-\mu^2)]}$.
Mean and variance of the intensity are plotted in Fig.5-8 for some values of the conditioning parameter $I_1\cos^2\varphi$, the projection of intensity I_1 on I_2, being $\varphi=\varphi_2-\varphi_1$ as usual. The abscissa variable is the modulus of the coherence factor extended to negative values by the multiplication to the sign of cos φ to account for anti-correlation when $\varphi \approx \pi$.
As we can see from Fig.5-8, similar to intensity, also the phase becomes more regular in correspondence to bright speckles.

Last, we want to compute the conditional probability density of phase φ_2, with the intensities I_1 and I_2 and the phase φ_1 as conditioning variables. Before proceeding, let us remark that, if phase φ_1 is known, then the phase difference $\varphi=\varphi_2-\varphi_1$ is the only statistical variable of significance.

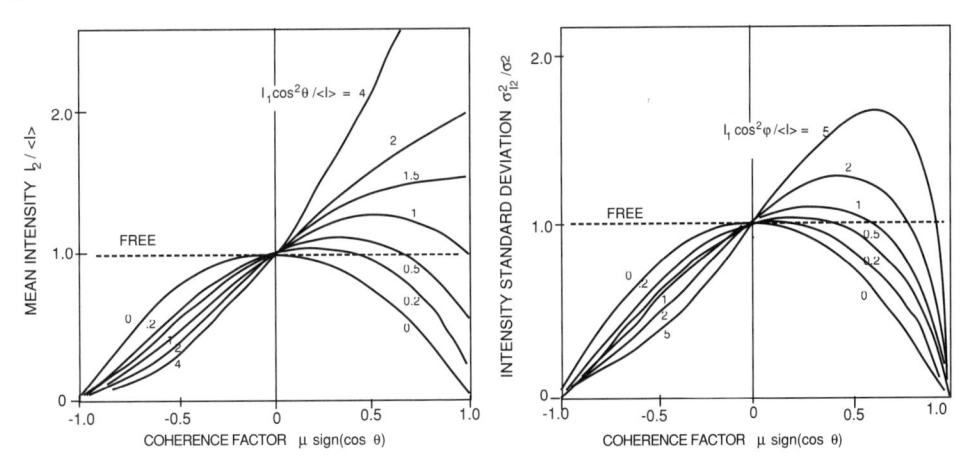

Fig.5-8 Mean value (left) and variance (right) of intensity I_2 conditioned to I_1 and to the phase difference φ, as a function of the coherence factor and for some values of the projected intensity $I_1\cos^2\varphi$ as a parameter.

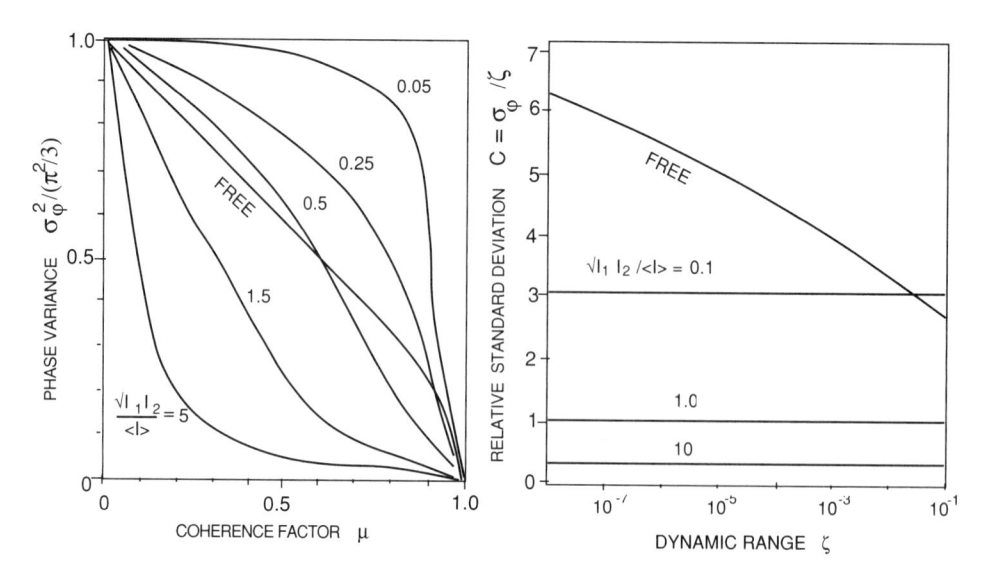

Fig.5-9 Variance of the phase difference σ_φ^2 normalized to $\pi^2/3$, as a function of the coherence factor μ (left), and of the dynamic range factor $\mu=1-\zeta^2$ (right), with the relative intensity $\sqrt{(I_1I_2)}/<I>$ as a parameter. The 'free' variance given by Eq.5.23 is also plotted for comparison. The right hand diagram is an expansion near $\mu\approx1$, the region of high coherence, and is plotted for $\mu=1-\zeta^2$.

Incidentally, the phase difference is just the quantity we are interested in for a measurement of optical path length. Thus, we consider the conditional probability $p(\varphi|I_2,I_1)$. As before, by integration of Eq.5.9, we find [6]:

$$p(\varphi|I_2,I_1)= (4\pi^2)^{-1} \exp[(\mu\cos\varphi)(I_1I_2)^{1/2}/\sigma^2(1-\mu^2)] \{\Im_0[\mu(I_1I_2)^{1/2}/\sigma^2(1-\mu^2)]\}^{-1} \quad (5.29)$$

Because of the same symmetry of Eq.5.22', the mean value of phase $\langle\varphi\rangle|_{I2,I1}$ is still given by the 'free' value of Eq.5.23, for which the diagram is plotted in the left-hand part of Fig.5-9. The variance $\sigma_\varphi^2|_{I2,I1}$ follows from the definition (Eq.5.26) and is found as:

$$\sigma_\varphi^2|_{I2,I1} = \pi^2/3 + [4/\Im_0(z)] \sum_{n=1,\infty}(-1)^n\Im_n(z)/n^2 \quad (5.30)$$

In Eq.5.30, we have let $z=\mu(I_1I_2)^{1/2}/\sigma^2(1-\mu^2)$, and $\Im_n$ is the n-th order modified Bessel function of the first kind (usually denoted with I_n).

Fig.5-9 (left) shows the probability density of phase conditioned to intensity against the coherence factor μ. Also plotted is the 'free' value of phase variance, given by Eq.5.23, and relative to the full set of speckle realization. We can see from the diagram that the free value decreases from $\pi^2/3$ to zero as μ varies from 0 to 1. The conditioned variance is always

5.1 Speckle Properties

smaller than the free variance for $\sqrt{\langle I_1 I_2\rangle}/\langle I\rangle > 1$, that is, where the speckle is locally more intense than the average. The decrease of variance — and of measurement error — is particularly sizeable near $\mu \approx 1$ where the speckle field keeps well correlated.

Equations 5.23 and 5.30 become indeterminate forms when $\mu \to 1$, and cannot be used directly in the region of high coherence. Then, we let $\mu = 1 - \zeta^2$ and obtain the following asymptotic behavior for small ζ:

$$\sigma_\varphi^2 \big|_{free} = \zeta^2 (3 - 2 \ln \zeta^2) \qquad (5.23')$$

$$\sigma_\varphi^2 \big|_{I2,I1} = \zeta^2 [\sqrt{\langle I_1 I_2\rangle}/\langle I\rangle]^{-1} \qquad (5.30')$$

The main dependence of both free and conditioned phase variances is on ζ^2. The multiplying factor steadily increases for $\zeta \to 0$ for the former and is given by the inverse of the relative speckle intensity $\sqrt{\langle I_1 I_2\rangle}/\langle I\rangle$ for the latter. The ratio of rms phase error σ_φ to dynamic range factor:

$$C = \sigma_\varphi \big|_{I2,I1} / \zeta \qquad (5.31)$$

is plotted versus ζ in Fig.5-9 (right).

5.1.5 Speckle Phase Errors

Let us now evaluate, from the results of speckle pattern statistics found in previous section, the noise-equivalent-displacement (NED) associated with the interferometric measurement. Let the phase measured on a displacement $z_2 - z_1$ be $\varphi = k(z_2 - z_1)$, where $k = 2\pi/\lambda$, and the error superposed be σ_φ. Then, we may recall from Sections 4.4.1 and 4.4.7 that the NED can be written as:

$$NED = \sigma_\varphi / k = C \zeta / k, \qquad (5.32)$$

where the second equation follow from the definition of Eq.5.31.

The factor ζ is found from the coherence factor as $\mu = 1 - \zeta^2$. For the two cases considered in Section 5.1.2, of longitudinal and transversal displacements from an initial point P to a final point P+Δ, the coherence factor has the expression (see Eqs.5.16 and 5.19):

$$|\langle E_{st}(P) E_{st}^*(P+\Delta)\rangle| = somb\, X, \qquad (5.33)$$

In the two cases, it was $X = \Delta D / \lambda z$ and $X = \Delta D^2 / \lambda z^2$. Recalling Eqs.5.17 and 5.20, we may write the factor as $X = \Delta / \lambda s_t$ and $X = \Delta / s_l$, s_t and s_l being the transversal and longitudinal size of the speckle. In a generalized form, we can write:

$$X = \Delta/s_{spckl}, \qquad (5.33')$$

and s_{spckl} is the current size for the experiment at hand.

Now, we may develop at the third order of X the somb function and obtain μ as:

$$\mu = \text{somb } X = 2J_1(\pi X)/\pi X = 2\,[\pi X/2 - (\pi X)^3/16]/\pi X = 1 - (\pi X)^2/8 \qquad (5.34)$$

By comparing with $\mu = 1 - \zeta^2$ we get:

$$\zeta = \pi X/2\sqrt{2} \qquad (5.34')$$

Finally, we go back and insert Eqs.5.33', 5.34' into Eq.5.32. The equivalent displacement error due to the speckle turns out to be given by the expressive form:

$$\text{NED} = C\,\zeta/k = (C/4\sqrt{2})\,\lambda\,\Delta/s_{spckl} \qquad (5.35)$$

Eq.5.35 is interpreted as follows. The speckle error, expressed in terms of wavelength (NED/λ), is given at first instance by the ratio of displacement Δ to speckle size s_{spckl}, either longitudinal or transversal.

The statement is refined by considering the multiplying factor $C/4\sqrt{2}$. This factor is not too far from unity and depends on the modality of measurement (that is, a free speckle statistics versus a conditional statistics), as indicated in Fig.5-9 (right).

5.1.6 Speckle Errors Due to Target Movement

The treatment considered so far holds for a displacement of the observation point in the speckle field. That is, it describes the error made by moving the observer from an initial point P to a final point P+Δ in the speckle field (Fig.5-2), while the diffusing target is still.

A different and more frequent case is that of a still observer and a moving target. The target movement may be either transversal or longitudinal, as shown in Fig.5-10. For both movements, we assume that the illumination beam does not move (this is the actual reference frame). Then, the illumination on the target is unaltered in a longitudinal movement, whereas the target sample illuminated by the beam changes in a transversal movement.

We can calculate the correlation function $C_\mu = \langle E(P)E^*(P+\Delta)\rangle$ along the steps that lead to Eqs.5.16 and 5.20. In particular, for a transversal displacement Δ of the target, we shall change ξ to ξ+Δ in all terms of Eq.5.12, whereas, for a longitudinal displacement Δ, we shall change z into z+Δ.

By repeating the calculations carried out in Sect.5.1.2, we obtain in either case the same result that reads $\langle E(P)E^*(P+\Delta)\rangle = \text{somb}(\Delta/s_{spckl})$, in which s_{spckl} is the longitudinal (or transversal) size of the field projected by the target.

Thus, it is like the target drags along its speckle field when it moves with respect to the laser beam of illumination (Fig.5-9).Therefore, Eqs.5.22 to 5.35 apply also to interferometric measurement on a moving target and give the error caused by the speckle regime.

5.1 Speckle Properties

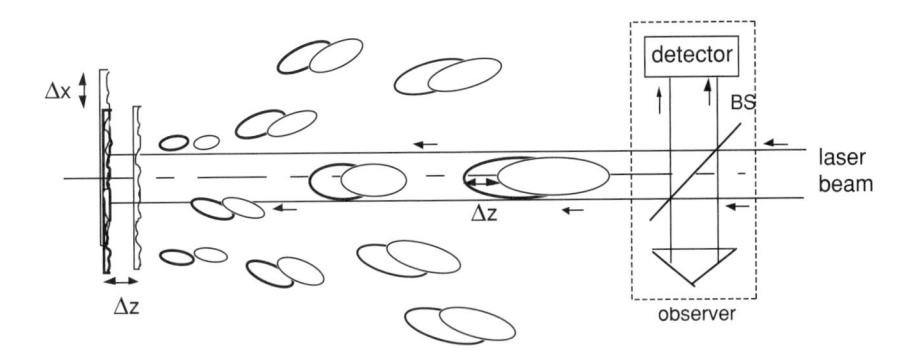

Fig.5-10 A moving target drags the speckle pattern field projected on the measuring system (for longitudinal displacement as shown here, the field is dragged along Δz). The correlation lengths for transversal Δx and Δz displacements are coincident to the speckle sizes s_t and s_l.

5.1.7 Speckle Errors Due to Beam Movement

Another case of concern is the movement of the illuminating beam on the target, when all other parts are still. The movement is not intentional, usually, but may be imparted to the beam by wandering and turbulence effects (App.A3.2) especially when operation is on a sizeable path length (say L>50 m).

We can treat this case by calculating the correlation function $C_\mu = \langle E(P)E^*(P+\Delta)\rangle$ when the diffuser coordinate ξ is changed to $\xi+\Delta$. Again by the calculations carried out in Sect.5.1.2, we obtain in this case the result $\langle E(P)E^*(P+\Delta)\rangle = \text{somb}(\Delta/s_t)$, where s_t is the transversal speckle size of the field projected by the target. This result is intuitive and confirms that moving the target is equivalent to moving the illuminating spot on it.

5.1.8 Speckle Errors with a Focusing Lens

We may use an objective lens to focus the illuminating beam onto the target, as illustrated in Fig.4-16 and generally used in laser vibrometer (Sect.4.6).
To treat this case, we shall just recall the results of Sect.5.1.1 about subjective speckle. At the focal plane of the objective lens, the transversal and longitudinal size of the speckles are given by (Eqs.5.2 and 5.3):

$$s_{t\,(fp)} = \lambda\, F_L/D_L, \qquad s_{l\,(fp)} = \lambda(2F_L/D_L)^2 \tag{5.36}$$

The parameters governing the speckle size are the lens focal length F_L and diameter D_L.
At the object plane, because of the optical conjugation provided by the objective lens, the speckles are seen stamped on the target with apparent dimensions:

$$s_{t\,(trg)} = \lambda\, z/D_L, \qquad s_{l\,(trg)} = \lambda(2z/D_L)^2 \qquad (5.37)$$

Eq.5.37 tells us that, when a lens is used, the aperture governing the speckle dimension is the lens diameter, not the target size.

5.1.9 Phase and Speckle Errors Due to Detector Size

In all interferometers, after the beam recombination we end up with a photodetector that provides the signal carrying the path length information. Let us now consider the effect of finite detector size on (i) the speckle error, a random contribution, and (ii) the field curvature error, a deterministic phase error.

Let us first consider the speckle statistics versus detector diameter D_{det}. In the image plane of the focussing lens (Fig.5-3), if the detector is smaller than the speckle size $D_{det} \leq s_{t(fp)} = \lambda F_L/D_L$, we can assume that the speckle distribution is point-wise sampled. Then, the results of preceding sections directly apply.

A small detector delivers a small signal, however. We might increase the detector size so that it collects $N_{sp} = (D_{det}/s_{t(fp)})^2$ speckles. In this case, what about the statistics? Integration of the received field on several speckles can be treated as the sum of independent samples of the same statistical ensemble. We can consider two cases:
- incoherent addition of speckles intensity when we perform a (normal) direct detection,
- coherent detection of the speckle field when we add a local oscillator or reference beam on the detector (see also Sect.8.1 of Ref.[5]) with the aim of performing a phase measurement.

In the case of incoherent addition, the sum of N_{sp} speckles gives N_{sp} times the intensity I of the single speckle, as an average. The probability of the sum of N terms $p_N(I)$ is the convolution of the single p(I), repeated N times, or p(I)* p(I)*... *p(I). For the exponential distribution $p(I) = <I>^{-1} \exp -I/<I>$, we get [2,4] the result $p_N(I) = (I/<I>)^N (N!)^{-1} \exp -I/<I>$. Thus, small amplitude speckles are no more the most probable, and for large N we go to the well-known result of the central-limit theorem [2]. This is a Gaussian (or normal) distribution of intensity with a relative standard deviation equal to $1/\sqrt{N}$.

In the case of coherent detection, we have to deal with the sum of N independent field components E, each with a normal distribution p(E) (Eq.5.6) having zero mean value and $\sigma^2 = <I>$ variance. The sum of such contributions is again the convolution of the single p(E), repeated N times, or $p_N(E) = p(E)* p(E)*... *p(E)$. The resulting distribution $p_N(E)$ is still normal, with a mean value and a variance both equal to N times the mean and variance of the single p(E). Thus, the mean value of the sum remains zero, while the variance is increased to $N\sigma^2$.

Considering the readout configuration of interferometers (Sect.4.5), the external configuration can even use a large area photodetector, and the previous arguments are fully applicable. On the contrary, the self-mixing configuration has an acceptance determined by the laser cavity aperture, receiving the retro-reflected wave.

This acceptance is λ^2 or, in other words, the equivalent area of the detector is that of the single mode. Therefore, to use a self-mixing configuration in the speckle regime, we shall match the spatial mode of the laser to the speckle transversal size.

Second, let us consider the deterministic phase error given by field curvature. This is caused by the term $\exp ik(x^2+y^2)/2z$ that, in Eq.5.4, multiplies the correct phase term $\exp ikz$ due to propagation. As the detector coordinates x,y have a radial extension from $r=\sqrt{(x^2+y^2)}=0$ to $D_{det}/2$, we integrate to field $\exp ikr^2/2z$ on $2\pi r\, dr$ and average on the area $\pi D_{det}^2/4$. Thus, we obtain the field collected by the detector as:

$$E_{det} = 16 \exp i[k(D_{det}/4)^2/z] \, \text{sinc}\, (D_{det}/\lambda z) \qquad (5.38)$$

The phase is contained in by the exponential term of Eq.5.38 as $\varphi = k(D_{det}/4)^2/z$. The sinc function is close to unity in the reasonable assumption of a small detector, $D_{det} \ll \lambda z$, and can be dropped.

The phase φ term adds to the correct phase kz of the propagation (term $\exp ikz$ in Eq.5.4), and thus we have a deterministic (or systematic) error. We may translate this error in an Equivalent Displacement Error (EDE) letting EDE=φ/k, and we get from Eq.5.38 EDE $=(D_{det}/4)^2/z$. Last, if our measurement goes from distance z_1 to distance z_2, the error is EDE=$(z_2-z_1)(D_{det}/4)^2/z_2 z_1$. With a reasonably small detector, for example D_{det}= 1mm, the error is small on a wide range, say, z_1=100mm to z_2=500mm. In this case, we get: EDE= 400(1/4)^2/100.500mm =0.5µm. However, if it were D_{det}=4mm, z_1=50mm, z_2=100mm, the error would be serious, EDE=10 µm.

5.2 SPECKLE IN SINGLE-POINT INTERFEROMETERS

With regard to operation of interferometers on diffusing surfaces, we have already outlined several topics in Chapter 4, that consider the extension of the basic interferometer (Sect.4.2.3.4), speckle–related errors (Sect.4.4.7), and vibrometers (Sect.4.6). Here, we treat the effect of speckle statistics in *vibration* and *displacement* measurements made with interferometers. By vibration we mean the small amplitude range (less than ≈100 µm), and with displacement we mean the range of large amplitudes (>0.1 mm).
Our aim is to keep the speckle effects under control, starting from the easy task of detecting small-amplitude displacements, and then trying to extend operation on large displacements as required in machine-tool control.

5.2.1 Speckle Regime in Vibration Measurements

The first consequence of the speckle statistics is *amplitude fading*. In fact, the intensity of the speckle pattern field is distributed as a negative exponential (Eq.5.8). Then, in a vibrometer aimed to a target, it will be not so unlikely that we fall on a weak or dark speckle, with the intensity I substantially smaller than the average value $\langle I \rangle$.

Because of the negative exponential, the probability to get $I<0.1\langle I\rangle$ is p=0.1, that of $I<0.01\langle I\rangle$ is p=0.01 and so on. This is amplitude fading because in a small, but not negligible, fraction η of cases, the signal returning from the target will be smaller than $\eta\langle I\rangle$. No matter how sensitive or low threshold the signal-processing circuits are, we cannot avoid falling below the minimum signal condition.

There are several ways to get around the amplitude fading. First we may incorporate in our instrument a circuit sorting the intensity level and warning the operator when the speckle intensity is too small. The operator will then move a little bit the laser beam projected on the target to another spot. Thus, we fall on another realization of the speckle statistics, and this can be a normal or even luckily an intense speckle, with $I>\langle I\rangle$ and a good conditional statistics (see Eq.5.30').

The second approach uses an Automatic Gain Control (AGC) to restore the signal amplitude so that it is adequate for circuit levels. Of course, if M is the dynamic range improvement of the AGC, the threshold of signal loss is moved to a lower value, $\eta\langle I\rangle/M$, but not eliminated.

A third strategy is exploiting sensor duplication, a technique usefully employed in radio engineering and known as *diversity*. We use two receivers, slightly apart in space, to sample different realization of the speckle statistics. With this, the probability of fading on both signals, below $\eta\langle I\rangle$, is η^2, a smaller value than a single channel, yet not zero.

From the point of view of applications, the first strategy is usually quite acceptable for a vibration-detecting instrument. The second and the third can also conveniently be incorporated and reduce the probability of fading down to very acceptable levels.

A fourth possibility, if allowed by the application, is that we can use a superdiffuser as the target surface. A superdiffuser is a piece of back-reflecting tape or varnish of the type used for enhancing the visibility of traffic sign at night. Also known as Scotchlite™ in a commercial product, the superdiffuser provides a gain of a factor 20 to 50 (typically) in the back-reflected signal reaching the interferometer. However, application of the tape or of a varnish drop on the target is a quasi-invasive operation that cannot be accepted in all circumstances. A final option mostly used in interferometers and described later (Sect.5.2.2) is that of dynamical tracking of a bright speckle, one that at the same time cures fading and phase error.

Another concern is speckle *phase error*. This error affects the accuracy of the measurement performed on a diffusing surface. The error is usually small, but, as we can go down to measure very minute amplitudes of vibration (Sect.4.6), it may become important as well. As already seen in Sect.5.1.5, the error translates itself in an NED given, in units of wavelength, by the ratio of vibration amplitude Δ to speckle longitudinal size $s_{l(trg)}$ (Eq.5.35). In a vibrometer, the speckle size can be made much larger than the dynamic range Δ_{max} (i.e., the maximum amplitude of vibration we want to detect). With $s_{l(trg)}>>\Delta_{max}$, the phase error is made negligible.

As a practical example, assuming the reasonable values $\lambda=1\mu m$, $D_L=30mm$, and $z=300mm$, the speckle longitudinal size is $s_{l(trg)}=\lambda(2z/D_L)^2=400\,\mu m$. The phase error $\lambda\Delta_{max}/s_{l(trg)}$ is 2.5 nm for a swing of 1μm, is 0.25 μm for a swing of 100 μm, and so on. Moreover, if we choose a bright speckle, we could be able reduce the phase error further, taking advantage of the factor C<1 in Eq.5.35.

5.2.2 Speckle Regime in Displacement Measurements

In displacement measurements, fading and phase error problems are more serious because we want to span a large dynamic range and cannot make $s_{l(trg)} >> \Delta_{max}$.
This is the case of application to tool-machine numerical control, calling for displacement from tens of cm to meters. These equipment may employ an optical rule (Ref.[5], Sect.9.3.4) as the least expensive solution, or a multiaxis interferometer equipped with corner cubes (Fig.4-13) for increased performance. The use of a plain diffuser has the advantage of eliminating the problem of cleaning the lens and optical parts.

If we are to measure a large displacement, it is likely that we are compelled to use a digital readout, by counting $\lambda/2$- or $\lambda/4$-transitions, to develop the several-digit figure. Then, our measurement is incremental and cannot tolerate a count loss. Nor are we allowed to read just the spot moving it away from a dark speckle like in vibrometer. Count loss due to signal fading is the main concern because, when it happens, we are forced to reset the instrument to its mechanical zero. Thus, watching phase errors to be reasonably small, we focus on fading.

Fading shall be counteracted by a combination of the techniques discussed in Sect.5.2.1, including speckle size tailoring, ACG, and detector diversity. A super-diffusing target will be used first if allowed by the application.

As a first issue, we have to choose the *beam size*. A small beam size gives a large speckle that keeps the phase error small, but the signal collected is small and is affected by quantum noise. A large NA objective lens helps collect a substantial return signal so that the quantum noise is low, but increases the speckle error.

Example of evaluation. We may start requiring a large speckle size. With the external configuration, we shall collimate the laser beam on the full dynamic range Δ_{max} to be covered. Using Eq.2.4, we get a spot size $w = \sqrt{(\lambda \Delta_{max})}$. Then, the longitudinal speckle size is $s_l = \lambda(2\Delta_{max}/w)^2$ and, inserting the previous expression, turns out as $s_l = 4\Delta_{max}$. In this way, we are inside a single speckle on the full dynamic range. The signal collected in this last speckle is very weak, however. If P_L is the power leaving the laser, and we use half of it in the reference path, $P_L/2\pi$ is the intensity at the target, and $(P/2\pi)\pi(w/\Delta_{max})^2$ is the power collected by the receiver. The attenuation respect to the power leaving the laser is accordingly $(w/\Delta_{max})^2 = \lambda/\Delta_{max}$. Using $\lambda = 1\mu m$ and $\Delta_{max} = 1m$, we shall be prepared to a –60 dB loss.

On the other side, if we take a sizeable diameter of the objective lens, let's say $D_L = 30mm$, in place of the small w (typ.$=\sqrt{\lambda \Delta_{max}}=1mm$), we limit the loss to -30 dB, but are faced with a small speckle dimension, $s_l = \lambda(2\Delta_{max}/D_L)^2 = 4mm$ only. In a $1\text{-}m = \Delta_{max}$ dynamic range, we would find 250 speckle passages (each with a $\approx \lambda$ error each).
Another point to consider is that a practical system should be able to tolerate residual walk-off movement of the target or of the illuminating beam (the corresponding errors are given in Sections 5.1.6-7). To illustrate a reasonable compromise, we take a dynamic range $\Delta_{max} = 1m$, and let $n=10$ speckle passages in it, so that the error is just $10\lambda \approx 10\mu m$ (or 10^{-5}) and the average speckle size is $s_l = 100mm$. At $\lambda = 1\mu m$, this means that we can use a lens diameter $D_L = 6\text{-}mm$ in $s_l = \lambda(2\Delta_{max}/D_L)^2$. The chosen speckle size gives room for a reasonable lateral walk off of the target on the $\Delta_{max} = 1m$ dynamic range.

As an efficient method to cure fading, let us now consider *dynamical tracking* of the speckle, a technique that has been recently introduced [7]. The basic idea consists in moving the beam projected on the target to maximize the signal amplitude or, to keep it locked to a bright speckle while the target eventually undergoes its movement. Locking to a bright speckle has two advantages: the probability of fading is greatly reduced, as it will be shown later with experimental results, and the phase statistics improves because we work consistently at a high $I/\langle I \rangle$ ratio (Fig.5-9).

The hardware to perform speckle tracking with a self-mixing interferometer configuration is shown in Fig.5-11. We use a pair of microactuators, in the form of two small bars mounted inside the objective lens holder along the X and Y direction. Specifically, the actuators we used are made of lead-zircon-titanate $Pb(Zr,Ti)O_3$ (PZT) ceramics. By actuating the PZT elements, we generate a (small) lateral displacement Δ of the lens (along X and/or Y), and the incoming beam is deflected by an angle Δ/f, with f being the focal length of the objective lens. On the target located at a distance z, the beam deflection is $z\Delta/f$. In the practical implementation, it suffices to deflect the beam of just $\approx 5\text{-}20\mu m$ on the target to follow the local maximum intensity.

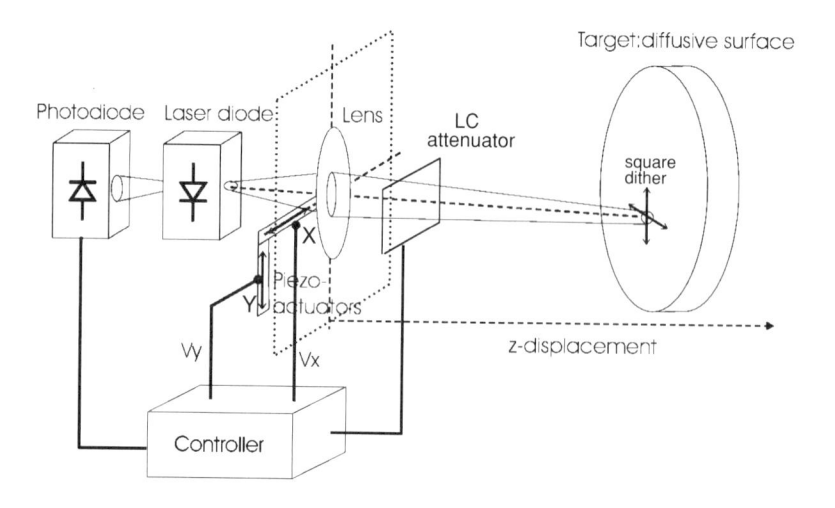

Fig.5-11 Arrangement to track the speckle maximum. Two small bars of PZT ceramic are mounted in the lens fixture and move the lens along the X and Y axes, changing the spot position on the target just of a few micrometers (adapted from Ref.[7], by courtesy of IEEE).

To detect the direction of signal increase and actuate the deflection accordingly, we use the well-known method of small signal sensing and phase detection. For example, to track the maximum along X, we feed the X-axis PZT ceramic with a small-amplitude square wave at audio frequency. The photodiode on the laser rear facet detects the signal amplitude, a square

5.2 Speckle in Single-Point Interferometers

wave that will be in-phase or antiphase (shifted by π) with respect to the PZT drive. When we find it in-phase, we will increase the actuator dc drive, whereas, in antiphase, we will decrease it. When we get the maximum, the photodiode signal is at twice the drive frequency. Thus, demodulation of the photodiode signal with the PZT drive square wave, followed by a low-pass filtering (Fig.5-12) provides the required tracking of the maximum, but along X for the moment.

To derive the two X and Y error signals from a single photodiode output, we take advantage that phase and quadrature signals are orthogonal. Indeed, we use the same square wave, with 0° and 90°-phase shift, as the drive of the X and Y actuators. When the photodiode signal is demodulated by the corresponding X and Y waveforms, the contribution from the other channel is not seen because it is dephased by 90° (or, orthogonal).

This concept is called *dither of phase tracking* and can be implemented as shown in Fig.5-12. With the dither, the optical axis of the laser beam projected out the objective lens is moved along a square path in the X-Y plane. On the target, the side of the square is adjusted to be a few μm, i.e., much less than spot size, yet enough to track the speckle maximum.

In the schematic shown in Fig.5-12, a circuit rectifies the self-mixing signal out from the photodiode and then multiplies it with the two square waves driving the X and Y PZT-actuators. After a low-pass filter, the signals are sent to the PZT actuators. As we get a beam movement in the direction of increasing signal for both axes, a final state of equilibrium is reached when both axes are on a local maximum, or a bright speckle.

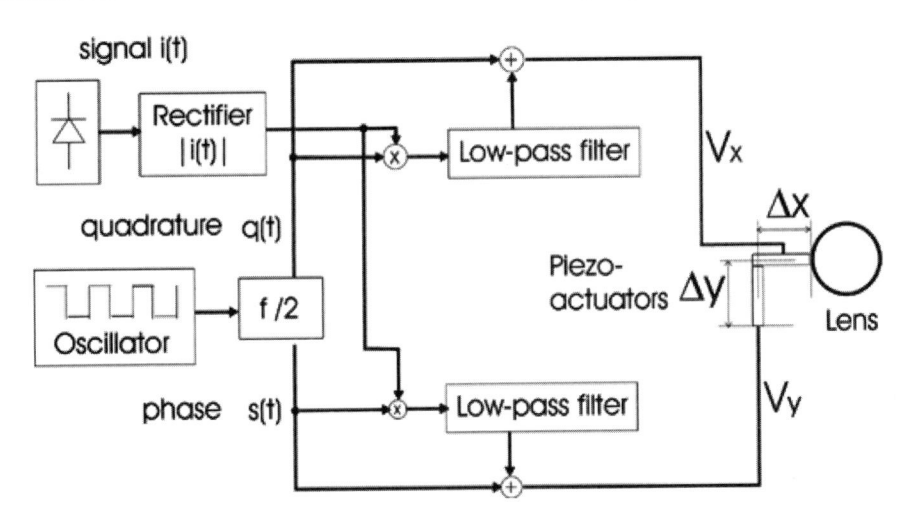

Fig.5-12 Block scheme of the speckle-tracking circuit. The signal from the photodiode is rectified peak-to-peak and demodulated with respect to the dither frequency, in phase and quadrature. The results are the X and Y error signals that, after a low-pass filtering, are sent to the piezo X and Y actuators to track the maximum amplitude or stay locked on the bright speckle (from Ref.[7], by courtesy of IEEE).

The feedback loop is limited by the response time of the PZT actuators in the range of τ_{PZT} =0.1-0.3 ms for small ($\approx$2×2mm) bars. This response time is adequate to follow the target movement up to a speed s_l/τ_{PZT}. For a speckle size of s_l= 100 mm, the target speed we can track without errors is s_l/τ_{PZT} =0.3-1 m/s.

To illustrate the result obtained by the tracking method, we start considering the statistical distribution of the field amplitude |E|. As given by Eq.5.6', the ideal diffuser has a Rayleigh-distributed probability density p(|E|) of field amplitude, and this is one of the curves plotted in Fig.5-13 (left, thin line). Also plotted in Fig.5-13 (left, bars) is the experimental result measured on a white paper target (specifically, plain letter paper). To improve the match, a small nonideality of the real diffuser is accounted for. This is done by assuming the target re-diffuses 98% of incoming radiation and reflects (without altering the phase) the remaining 2%. The calculated distribution of field amplitude for such a real diffuser is also shown in Fig.5-13 (left, thick line).

On the abscissa, the field |E| is standardized to the value that makes C=1 in the self-mixing intereferometer configuration. From Eq.4.35, we have C=(E/zE$_0$)n$_1$L$\sqrt{(1+\alpha_{en}^2)}$, where E$_0$ is the field leaving the laser. Thus, the field for C=1 is E=zE$_0$/[n$_1$L$\sqrt{(1+\alpha_{en}^2)}$].

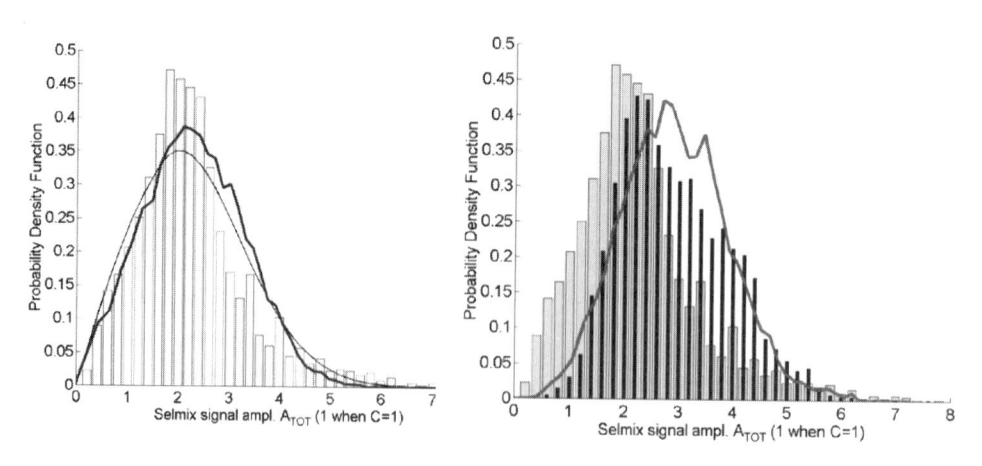

Fig. 5-13 Probability density function p(|E|) of the field amplitude |E|, with the amplitude normalized to 1 for a value C=1 of the self-mixing parameter.
Left: Thin line is the Rayleigh distribution for an ideal diffuser, and thick line is the result of a numerical simulation for a real surface, with 2% reflection and 98% diffusion, and fully developed speckle statistics. Bars are experimental data for plain white paper (z=50cm, w$_{las}$=2 mm).
Right: With the bright-speckle tracking on, the experimental distribution moves to significantly larger |E| (dark bars) as predicted by simulations, with respect to the circuit-off distribution (gray bars). Adapted from Ref.[7], by courtesy of IEEE.

5.2 Speckle in Single-Point Interferometers 207

When the circuit tracking the bright speckle maximum is switched on, the p(|E|) distribution becomes that of the right-hand diagram of Fig.5-13. The improvement of tracking is evident looking at the small-amplitude part of the diagrams. Occurrence of small amplitudes is greatly reduced.

In particular, with the speckle tracking circuit, we can immediately work with the self-mixing configuration of the interferometer, one demanding a given minimum signal amplitude (the condition C≥1) to stay in the up/down switching regime of counts (Sect.4.5.2.5).

The data of Fig.5-13 are relative to a working distance z=50-cm and a laser spot w_{las}=2-mm on the diffusing target. In this condition, the probability of C<1 (loss of counts) is ≈10% in normal conditions, but drops to 0.5% with the tracking-circuit on.

Thus, we can operate on a substantial span of distance without amplitude fading, even with the self-mixing interferometer, and can expect that the phase error is kept low.

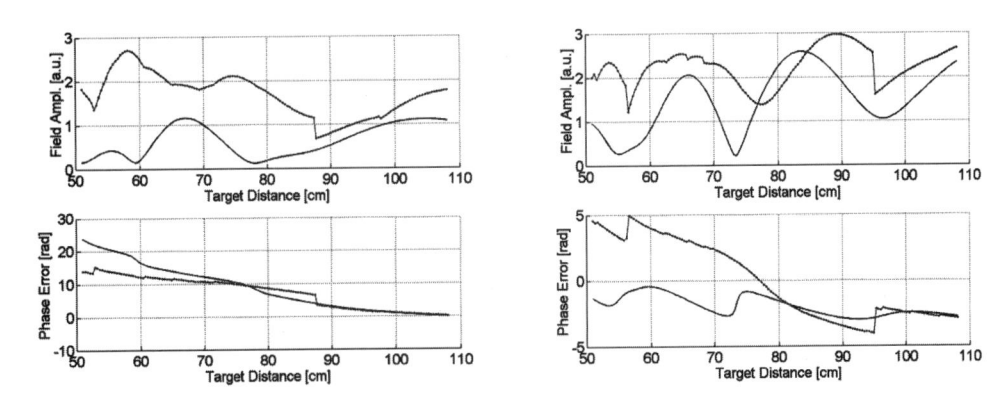

Fig.5-14 Examples of simulations of amplitude (top curves) and phase (bottom curves) of the returning field. The diffusing target moves from 108 to 51-cm. In the amplitude diagrams, top curves are with the tracking circuit on, and the bottom with the circuit off. In phase diagrams, the curves that vary less are with the tracking circuit on. Abrupt jumps (near z=87-cm on the left and 95 cm on the right) are correct switches decided by the tracking system, which skips from one speckle becoming too weak to the next adjacent speckle, being brighter. From Ref.[7], by courtesy of IEEE.

In Fig 5-14, we plot the result of a simulation of speckle-regime amplitude and phase of the returning signal in a generic interferometer. In the calculation, the target moves of about 50 cm. The phase error includes both speckle-related error and field curvature error. Note the jumps at z=87 cm on the left and 95 cm on the right diagrams. These are correct results, and are due to the dither that discovers a better speckle adjacent to the current one, and jump on it. As the simulation goes from 108 to 51 cm, the jump is actually going upward.

Another result of the simulation is the total phase error. In a 1000-sample set of computed displacement (each similar to the outcome shown in Fig.5-14), the unconditioned ensemble average phase error is 9.7-rad, and the standard deviation is 3.9 rad. In the same situation, when the bright-speckle tracking is on, we get an average error of 6.16 rad and a standard deviation of 3.6 rad. Most important, amplitude fading is not a problem any more.

Translating the previous phase error to a displacement, we are left with a $\approx\lambda$ error on a displacement from 50 to 100 cm. This means a $\approx 10^{-6}$ relative accuracy, which is a value that can be accepted or is even very good in several applications.

The self-mixing interferometer has been calibrated against an optical ruler, while in operation on a diffuser with the speckle tracking provision.

Fig.5.15 shows an example of the signal waveform versus displacement, acquired while the target is moved from 70 to 80 cm at a 1 mm/sec by a motor-driven mechanism. Without speckle control, the signal amplitude has a strong fading at ≈ 76 cm, whereas fading is eliminated when the bright speckle control is on. In the same run, the jump at ≈ 73 cm is simply due to the tracking system that decides to change the speckle being tracked and to move to a brighter adjacent speckle.

An interesting remark is that, because of fading, without bright-speckle tracking, the interferometer incurs in a count-loss error near z=76 cm, whereas the error is eliminated when the bright-speckle tracking is active (Fig.5-15 bottom).

In Fig.5-15, the signal with speckle control looks noisier because of the dither of the spot position. Of course, this does not imply a loss of accuracy of the interferometer.

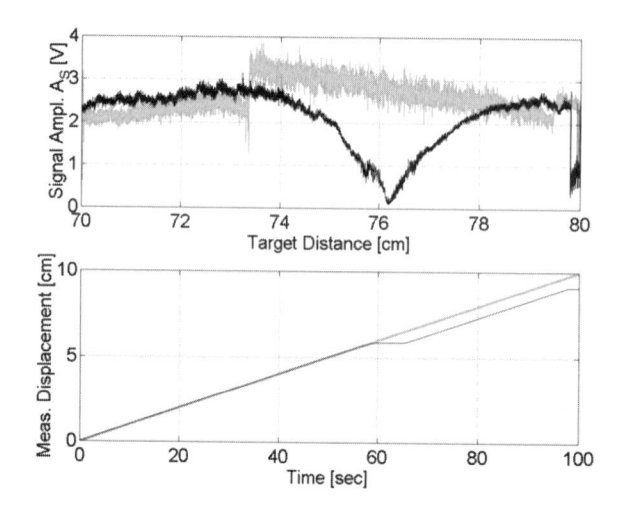

Fig.5-15 Top: comparison between the signal amplitude with (thin trace) and without (thick trace) bright-speckle tracking. Bottom: the measured displacement measured by the interferometer displays an error near z=76 cm because of the speckle fading, whereas the error is removed with the bright-speckle tracking system. The target was moved from 70 to 80 cm at a 1-cm/sec speed. (From Ref.[7], by courtesy of IEEE).

5.2 Speckle in Single-Point Interferometers 209

In the measurements, care was taken to avoid transversal movement of the target while actuating it by the z-axis motor driver. The transversal movement changes the speckle sample introducing an extra fluctuation that increases the chance of fading in normal (nontracked) operation, whereas the effect is readily tolerated when the bright-speckle tracker is working, as shown in Fig.5-16, left.

Another experimental result is reported in Fig.5-16, right. The figure shows the general increase of signal field amplitude that we can obtain with the speckle tracking with respect to a normal interferometer. The experimental results in Fig.5-17 are a sample, representative of the average behavior of the many statistical outcomes we have observed.

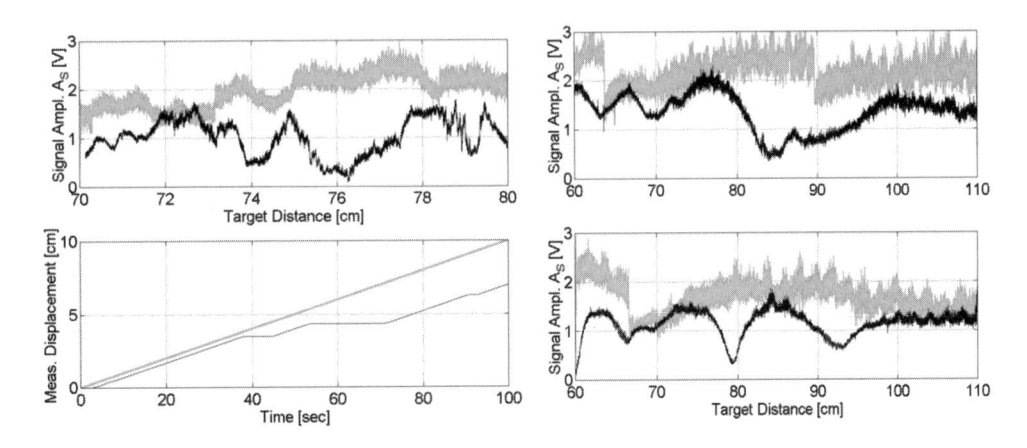

Fig.5-16 Left: Transversal movement of the target worsens the fading problem in a normal interferometer, but can be tolerated when the bright-speckle tracker is added (top); loss of counts and error without tracker versus no error with the bright-speckle tracker. Right: Typical experimental samples of signal amplitude with and without the bright-speckle tracker. (From Ref.[7], by courtesy of IEEE).

A final experimental result worth reporting is about the improvement we can obtain in interferometric measurements by the use of a superdiffuser varnish or tape.
When a superdiffuser varnish or tape is used, the field amplitude level increases by the superdiffuser gain, typically G=20 to 40. Although this is good for small signals, if we encounter a comparatively large signal, the gain may bring the amplitude so high that a preamplifier or another circuit enters saturation. To avoid this, we add an automatic gain control (AGC) function. The function may be conveniently implemented by a Liquid Crystal (LC) cell, inserted in the outgoing laser-beam path as indicated in Fig.5-11. The LC is actuated by a control circuit that looks at the signal amplitude and keeps it constant.

In Fig.5-17, we report the probability density function of the field amplitude |E| for a target made of a Scotchlite™ tape. Comparing it with the probability density function of a normal diffuser, Fig.5-13, we can appreciate the remarkable decrease of small-value amplitudes in the superdiffuser statistical distribution.

The probability of having a weak speckle with C<1 is in the range 10% for a normal diffuser without speckle tracking and 0.5% for a normal diffuser with speckle tracking.

The same probability can go down to 2.2% and 0.15% for a superdiffuser without and with speckle tracking, respectively [7].

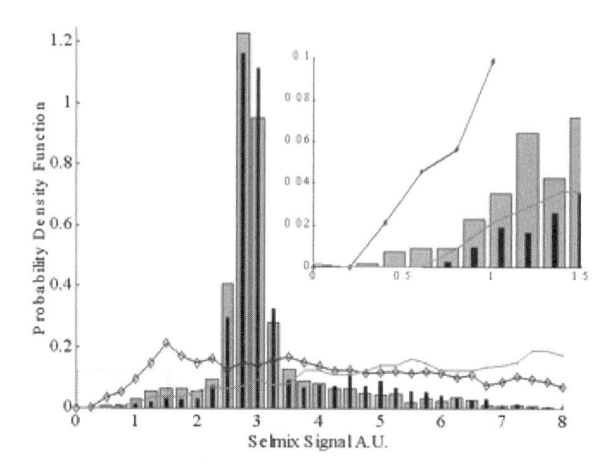

Fig.5-17 Probability density function of the field amplitude from a superdiffusing target, that is, Scotchlite^{TM} tape with a superdiffuser gain G=20. Amplitude on the abscissa is normalized to the C=1 condition as in Fig.5-13. Line with aces is for the normal speckle statistics, and dotted line is for speckle with tracking on. Gray bars are for AGC on and speckle tracking off. Dark bars are for both AGC and speckle tracking on. (From Ref.[7], by courtesy of IEEE).

5.2.3 The Problem of Speckle Phase Error Correction

After amplitude fading has been cured, we are left with the problem of the speckle phase error we have considered in Sect.5.1.5. After taking advantage of the specific methods discussed in Sect.5.2.2 to reduce the speckle effects, we may wonder if the residual statistical error may be corrected by any other general method.

A reason to hope this correction possible lies in the availability of an amplitude signal |E|, in addition to the phase φ we look at in the interferometer. If a relation exist connecting |E| and φ, then we may use the information in |E| to correct the error in φ.

In fact, consider an interferometric measurement of displacement made on the field E received in a point P (Fig.5-10). The paradigm of the measurement is that we measure the field phase term $\varphi(P)$ and get the displacement as a phase difference, $k(z_2 - z_1) = \varphi(P_2) - \varphi(P_1)$.

5.2 Speckle in Single-Point Interferometers

In general, the field given by Eq.5.12 can be written as $E_{st}/|E| = \exp ikz \exp i\varphi_{sp}$, where kz is the propagation phase on distance z, and φ_{sp} is the speckle error of remaining terms. Usually, by beating with the local oscillator of the reference beam, we detect the real E_R and imaginary component E_I of the field, $E = E_R + iE_I$, and compute amplitude and phase as:

$$|E| = \sqrt{(E_R^2 + E_I^2)} \qquad \varphi = \operatorname{atan}(E_I/E_R) = kz + \varphi_{sp} \qquad (5.39)$$

Now, to be able correcting the phase error, there should be a connection between φ_{sp} and $|E|$.

In optics, such a connection is known as Kramer-Kronig relation [8]. We find it when we calculate the phase shift $\phi(\nu)$ associated with the absorption/gain line $g(\nu)$ of a medium (see App. A.1.1), or when we connect real and imaginary parts of the index of refraction $n(\lambda) = n_R(\lambda) + i\, n_I(\lambda)$ (this effect is connected to the line-width enhancement factor α_{en} in Eq.4.46b). In electronics, the Hilbert transform [9] relates the real and imaginary parts of a transfer function $F(\omega)$, ratio of output to input signals of a linear network in the frequency domain. It also relates the real and imaginary parts of a driving-point impedance, ratio of voltage to current.

Specifically, if the network is physically realizable (that is, obeys the causality of effects) and has zero-excess delay (with respect to the physical minimum), the real and the imaginary parts of $F(\omega)$ are Hilbert transforms of each other.

Mathematically, the condition for the real and imaginary parts of a function $F(z)$ of a complex variable z be Hilbert-transforms is that the function $F(z)$ is analytic [9,10] or, written $z = u + iv$ and $F(z) = U + jV$, that the Cauchy-Riemann equations holds $\partial U/\partial u = \partial V/\partial v$, $\partial U/\partial v = -\partial V/\partial u$.

The Hilbert relation can be extended to amplitude and phase of the frequency response. For a zero-excess-delay network or system, the log of amplitude (or attenuation) and the phase are a Hilbert-transform pair. This statement easily follows from the previous, because by writing the attenuation as $\ln E = \ln[|E| \exp i\varphi] = \ln|E| + i\varphi$, we see that the real and imaginary parts of $\ln E$ are just the log-amplitude and the phase.

Moreover, for a system with an excess phase $\exp i\psi$ with respect to the zero-excess-delay response, the Hilbert-transform of the log-amplitude is given by the minimum-phase term plus the excess phase ψ.

The Hilbert-transform $F(z)$ of a function $f(z)$ of the complex variable z is defined [9,10] as:

$$F(z) = H[f(z)] = \pi^{-1} \int_C f(\zeta)/(z-\zeta)\, d\zeta \qquad (5.40)$$

In this equation, the variables z and ζ are homogenous. The integral is a line integral around the origin of the complex plane, or, it is extended on $\zeta = -\infty, +\infty$ as a Cauchy principal value [10]. The inverse Hilbert transformation is:

$$f(z) = H^{-1}[F(z)] = -\pi^{-1} \int_C F(\zeta)/(z-\zeta)\, d\zeta \qquad (5.41)$$

Interpreting Eqs.5.40 and 5.41 as convolution integral, we can also write $F(z) = (1/\pi z) * f(z)$ and $f(z) = -(1/\pi z) * F(z)$.

Noting that the Fourier transform of $1/\pi t$ is $i(\operatorname{sign} \omega)$, we can use Fourier numerical routines to compute the Hilbert transform. We first compute the Fourier transform of f(z), Φ(ω), and then change the sign of Φ(ω) for ω<0 and compute the inverse Fourier transform of the result, obtaining the Hilbert transform F(z).

Examples of transforms. As an illustration, we list here a few Hilbert-transform (HT in the following) pairs. For additional cases, the reader may consult Ref.[10]. Sinusoidal functions cos z has the HT of sin z, and sin z has the HT of cos z. Thus, the complex exponential exp iz has real and imaginary parts that are HTs of each other. The impulse function (or Dirac delta) δ(t) has the HT $(\pi t)^{-1}$. A rectangular pulse centered at t=0 and of width 2T, that is, rect(t,T)=1 for –T/2<t<+T/2, =0 for |t|>T/2, has the HT π^{-1}ln |(t-T)/(t+T)|. A gaussian pulse exp-$t^2/2\sigma^2$ has the HT of Ei(t,T), where Ei is a special function (exponential-integral) [10]. The Lorentzian $1/[1+\omega^2T^2]^{1/2}$ has the HT of $\omega T/[1+\omega^2T^2]^{1/2}$, and these are the real and imaginary parts of a parallel-RC impedance.

Now we can go back to the field E(z) given by Eq.5.12. Under the integration sign, the diffuser function φ(η,ξ) is real and well-behaved, and all other operations are linear and analytic. The terms out of the integral are analytic, too, so we may conclude that the field propagated at point P(z) is analytic with respect to the variable z.

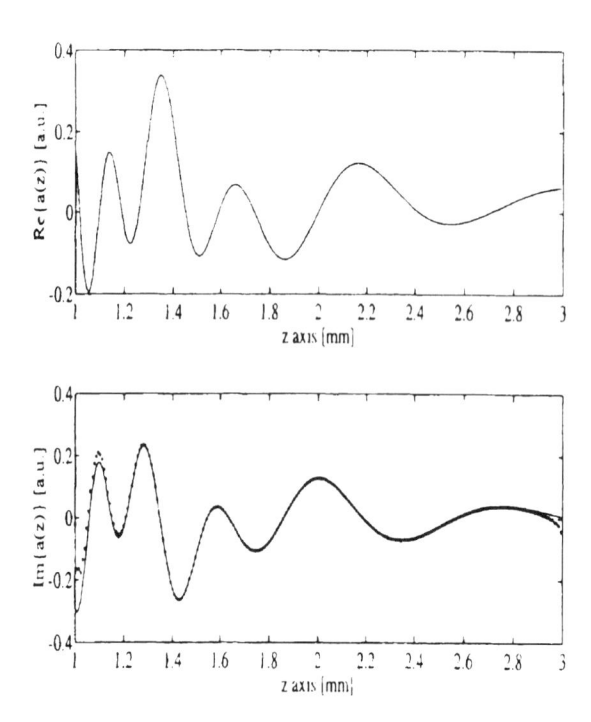

Fig.5-18 Real (top) and imaginary (bottom) part of the field E°(z) at distance z, for a speckle field at λ=1µm generated by a square diffuser, 100-µm by side. Lines are results of simulation, and points in the bottom diagram are the results of a Hilbert transform of the real part. (from [11], by courtesy of IEEE LEOS).

5.2 Speckle in Single-Point Interferometers

The same statement holds if we drop out the pure delay term exp ikz in Eq.5.12 and consider the remaining field E°(z), which is given by E(z)= exp ikz E°(z) and represents the reduced field carrying the speckle-induced phase error. It is easy see that E°(z) is analytic, too, with respect to the variable z. At this point the reader may wonder how z, the distance, is also a complex variable. It can be both at the same time: real when it's distance and complex when it's argument of the complex function, or is the *analytic continuation* of distance z.

In conclusion, the real and the imaginary parts of total and reduced field, E and E°, are Hilbert-transform pairs.

To test this statement, we have carried out numerical simulation of the reduced field E°(z). In the simulation, a square target, 100-μm by side, diffuses outward the illumination received by a coherent source (at λ=1μm). The phase on each elemental area, a square of 1-μm by side, has been randomized on $-\pi,+\pi$. The real and imaginary parts of the field are calculated at several distances z according to Eq.5.12. An example of the results, representative of the actual statistics we have built up on a much larger sample size, is reported in Fig.5-18.

As we can see from this figure, the agreement between HT-transform of the real part and the true imaginary part is very good.

The connection of real and imaginary parts does not help correcting the phase error, however. In fact, using the true E_R and either $HT(E_R)$ or the true E_I to calculate the phase φ=atan (E_I/E_R)=kz+φ_{sp}, we end up with the same result, affected by the speckle error φ_{sp}.

On the contrary, if we measure the intensity I= $E_R^2+E_I^2$ and calculate the HT of ln I, we have φ=HT(ln I). If this expression is applied to the total field, we have φ=kz+φ_{sp}, and the measurement is affected by the speckle-error φ_{sp}. If we apply the expression to the reduced field, we get φ=HT(ln I) =φ_{sp}, and we get the error alone. The strategy for speckle error correction is then subtracting HT(ln I) from the true phase measured by the interferometer.

With the same values given previously, a numerical simulation of amplitude and phase of

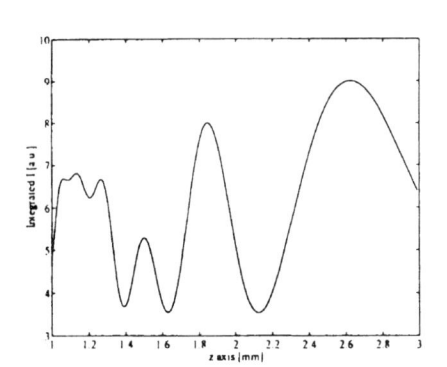

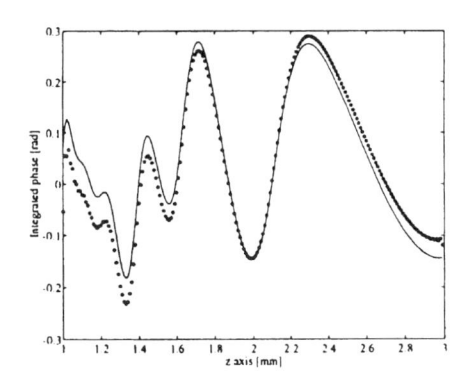

Fig.5-19 Intensity (left) and phase (right) of the speckle-pattern field from a diffuser as in Fig.5-18. The lines represent the results of a numerical simulation, and the dots in the right-hand diagram are the result of the Hilbert transform of log-intensity data of the left diagram (from [11], by courtesy of IEEE LEOS).

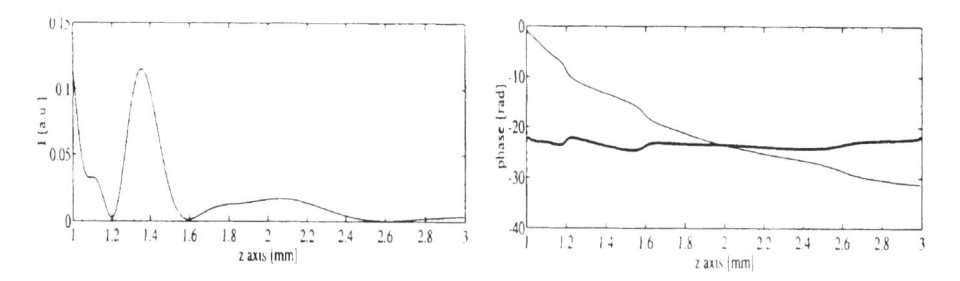

Fig.5-20   If the intensity versus z has a point where it falls to zero (left), then its log has a singularity and the Hilbert transform of log intensity does not give the correct phase any more, as shown at left by the thick line HT(ln I) compared to the thin line, the correct phase (from [11], by courtesy of IEEE LEOS).

the field from an ideal diffuser as a function of distance has been carried out. A sample of good results is reported in Fig.5-19 as a case representative of the statistical trend we may expect when the intensity never goes through a zero on the distance excursion considered. In Fig.5-19, the phase error φ_{sp} swings from –0.2 to +0.4 radians as distance is increased from 1 to 3 mm, and the HT(ln I) closely tracks φ_{sp}. Thus, if we subtract HT(ln I) from the measured phase, we cancel out a large fraction of the error, and a residue of $\approx \pm 0.01$ rad is left.

A serious limit of the HT method, at least until now, is that when intensity is zero, ln I becomes infinity, and the singularity of the complex function E°(z) makes HT(ln I) largely deviate from φ_{sp}. This is shown in Fig.5-20, where the unlucky case of two deep zeros in $E^2(z)$ is reported for illustration.

The HT correction has also other questionable features that we cannot consider here. Additional work is needed to make the HT method truly useful in interferometers, but we think the principle is intriguing and worth reporting.

5.3 ELECTRONIC SPECKLE PATTERN INTERFEROMETRY

In the previous section, we considered the speckle as a source of error in single-channel interferometeric measurement. This is not the only situation, however, because we can also regard each speckle as a separate interferometric channel. Indeed, as coherence is maintained inside a speckle, each speckle is a spatial region within which a phase measurement can be carried out or, it is an individual pixel of an interferometric image. The only point to care about is that, from speckle to speckle, coherence is lost, so the measurement range cannot go outside the speckle. Also, as the phases in each speckle or pixel are uncorrelated, the interferometric image will be 'speckled' with a point like pattern as that of a normal diffuser shown in Fig.5-1.

Therefore, we can actually build an instrument measuring interferometric displacements or vibrations pixel by pixel on an image, all simultaneously, and this instrument is called ESPI,

5.3 Electronic Speckle Pattern Interferometry

acronym for Electronic Speckle Pattern Interferometry.

The development of the ESPI concept can be traced back to the 1970s, when it was recognized as the electronic version of holography [12-15]. As such, ESPI is capable of supplying the same information without the burden of an anti-vibration table and of operation in darkness as required by the exposure of photographic plate, typical of holography.

The basic setup of ESPI is shown in Fig.5-21. It comprises a laser emitting a power up to 20...50 mW in a single spatial mode. Traditionally, HeNe and Ar lasers have been used, with the HeNe being the best for low power.

In recent years, two alternatives have become practicable: (i) quaternary (GaAlAsP) lasers emitting in the red (λ=620-670 nm) up to 100 mW and (ii), diode-pumped, frequency-doubled Nd-YAG lasers emitting in the green (λ=530 nm) up to several W.

Both sources require a beam quality control, e.g., through a pinhole spatial filter, to yield an emission close to the single mode.

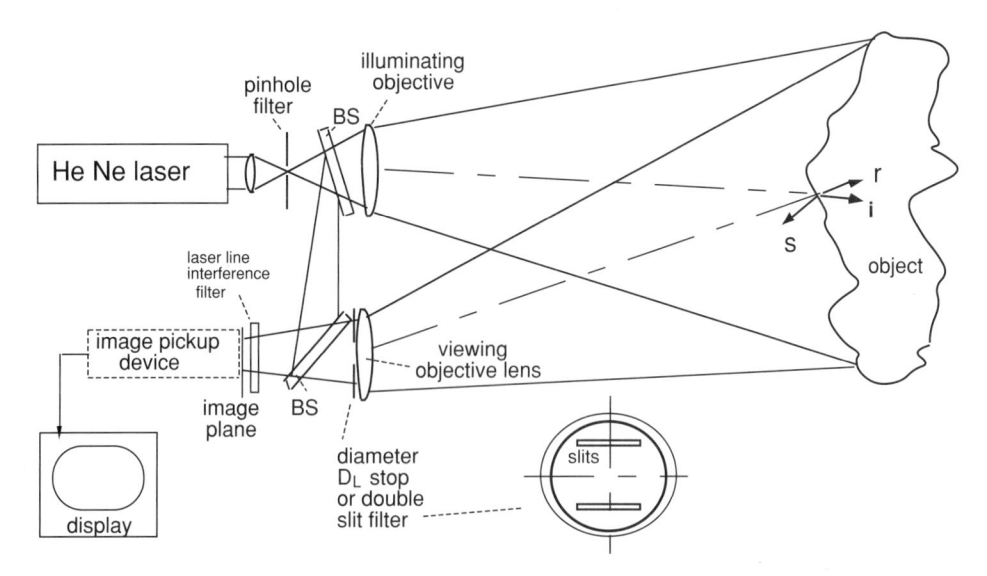

Fig.5-21 Basic setup of ESPI for pickup of displacement and vibration. Light from a He-Ne or other laser with good beam quality is shed on the object. By means of two beam splitters, part of the beam is superposed to the image formed by the viewing objective lens. Spatial filtering of the speckle is provided at the viewing objective lens by a circular stop or a double-slit stop.

The beam is projected on the object through a telescopic two-lens arrangement that eventually incorporates the spatial filter (Fig.5-21). A fraction of the beam is picked out by a beamsplitter and is diverted to the viewing objective lens. Here, the beam is superposed by a second beamsplitter to the image of the object. Thus, like in an interferometer, we recognize a measurement and a reference path. The measurement path is that going to a pixel on the object and back to the viewing lens, and the reference path is that of light coming from the beam splitters to the viewing lens.

Of course, a condition shall be met: that the coherence length of the source is larger than the imbalance of reference and measurement paths. Eventually, we will adjust the balance with an extra path inserted in the reference path.

The power requirement may range from a few mW for small-area objects (up to perhaps 10×10 cm^2) to hundreds of mW and more for large-area testing. The required power depends, of course, also on the sensitivity of the detector (CCD vs vidicon camera) and on the integration time allowed by the application.

In front of the photodetector, in Fig.5-21, we find an interference filter that allows laser light through and blocks out ambient light. In this way, the ESPI instrument can operate without darkening, in the normal laboratory environment, differently from a holographic system.

Let us now consider the image formed on the display. Individual pixels that can be distinguished in the image are the transversal subjective speckles (Sect.5.1). The size of them, on the image, as given by (Eq.5.37), is $s_{t\,(trg)} = \lambda z / D_L$. By adjusting the observing lens diameter D_L, we can bring the speckle size to an optimal trade-off between fine (good resolution) and well visible (better contrast).

We may also recall the results of the speckle analysis (Sect.5.1.6-8), by which the on-axis movement of the target drags along the speckle field. Then, the dragged speckle will interfere correctly (without phase error) with the reference field up to a pixel displacement given by the longitudinal size of the subjective speckle, given by (Eq.5.37) $s_{l\,(trg)} = \lambda (z/D_L)^2$.

Thus, the longitudinal size $s_{l\,(trg)}$ is the allowed dynamic range in which we may perform the interferometer measurement of displacement, and the transversal size $s_{t\,(trg)}$ is the size of our pixel-level interferometer channel. The speckle image is then equivalent to a multiplicity of interferometers, that work in parallel (but with uncorrelated initial phase) on the displacement s(x,y) of the target, from pixel to pixel in the x,y plane.

In each pixel of the image, the photocurrent generated by the detector is the usual interferometric signal (see for example Eq.4.3) made up of a constant term multiplying the phase dependent term 1+cosφ. For the in-line geometry considered so far, it was φ=2ks, but now we may easily consider a generalization. Looking at the geometry of illumination $\underline{i}$ and viewing $\underline{r}$ unit vectors (Fig.5-21), it is easy to write the phase generated for a displacement $\underline{s}$ as:

$$\varphi = k\,(\underline{i} + \underline{r}) \cdot \underline{s} \qquad (5.42)$$

At all pixel locations, the phase difference $\varphi_M - \varphi_R$ of measurement and reference channels follows the speckle statistics and is randomly distributed on $-\pi..+\pi$ (Sect.5.1). Accordingly, the image of a still object viewed by the system of Fig.5-21 has the usual speckle appearance that carries no information.

In a particular pixel, we may initially find either a bright, dark, or gray speckle. Now, let the object vibrate, at a speckle position, with an amplitude Δs such that φ in Eq.5.42 is larger than 2π. Then, we can see the pixel undergo a full cycle of brightness variation, e.g., from dark to bright and back, in analogy to a normal fringe of an interferometer subjected to a full 2π-phase cycle.

5.3 Electronic Speckle Pattern Interferometry 217

By an averaging of the detected signal followed by a high-pass filtering, we can cancel out the speckle constant field and obtain a map of vibrating modes.

Indeed, if the object under test is subjected to a vibration excitation, antinodal points have a large displacement and, in their pixels, the speckles appear smeared out by averaging. In nodal points, the vibration is small, and its speckle is nearly unchanged. We then obtain [14] the images displayed in Fig.5-22. This mode of operation of the ESPI instrument is called *time-averaging mode*.

Fig.5-22 Typical ESPI images taken in the time-averaging mode. The target is a loudspeaker covered with a plain cardboard sheet and is driven into vibration at increasing frequency (in the audio range). At nodes the speckle is unchanged whereas at anti-nodes it is washed out. A few patterns of the many coming out from 200 to 2000 Hz are shown. Upper left picture is the speckle image without excitation (adapted from Ref.[14]).

With the setup shown in Fig.5-21, the time-averaging mode has an interferometric sensitivity to target movement, but the dynamic range is limited to $s_{l\,(trg)}$.

If we want to increase the dynamic range at the expense of sensitivity, we may simply remove the reference beam in the setup of Fig.5-21. By doing so, we again obtain a signal averaging at an anti-nodal pixel location, but now smearing is due to the loss of correlation, which we incur when the displacement is larger than $s_{l\,(trg)}$. The longitudinal size $s_{l\,(trg)}$ can be made in the range of, say, 50 to 200μm by an appropriate choice of lens diameter D_L and observation distance z. This mode is called time-averaging *without reference*, and using it we can observe large-amplitude vibration patterns even in direct sight by eye.

Returning to Fig.5-21, let us comment on the effect of the aperture stop of the objective lens.

The stop affects the speckle size through the diameter D_L because it is a low-pass filter in the frequency (or Fourier) domain of the object. We can use other filtering functions to improve the speckle-image quality. For example, a double-slit stop [12,15] (Fig.5-21) allows increasing the signal amplitude while enhancing the high spatial frequency content of the speckle image picked up by an image photodetector scanned in a raster fashion.

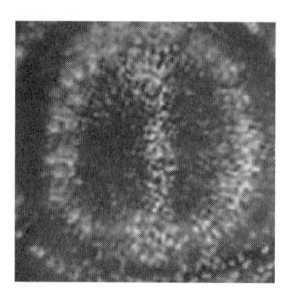

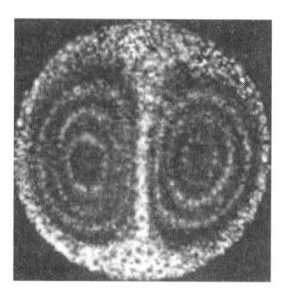

Fig.5-23 ESPI image of a loudspeaker taken with a circular stop and a video high-pass filter (left) and the same with the double-slit stop, high pass, and peak-to-peak rectifier (right).

When we combine a structured stop to filter the optical image with an electronic processing of the photodetected signal, the quality of the image is dramatically improved as shown in Fig.5-23. Here, the video output from the photodetector is high-pass filtered and peak-to-peak rectified. This processing yields a more detailed fringe pattern than with the simple high pass and averaging of Fig.5-22.

Now let us consider the pickup of interference images associated with a displacement distribution s(x,y). This is achieved by the *frame-subtraction mode* of operation of ESPI. As previously noted, the image of a still object is a speckle image that carries no information. Each speckle is an individual interferometer channel, but all channels are uncorrelated because the initial phase $\varphi(x,y)$ is randomly distributed from pixel to pixel.

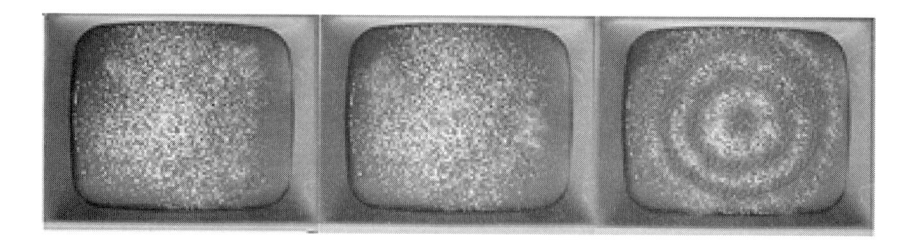

Fig.5-24 ESPI image of an aluminum plate before (left) and after (center) the application of a deformation in the center of it. The image at right is the difference of the two and unveils the displacement fringes.

5.3   Electronic Speckle Pattern Interferometry 219

Now, if the target is subjected to a displacement s(x,y), we get another speckle image, again with an intensity distribution uncorrelated from pixel to pixel. However, and this is the important point, the pixel undergoes a brightness *variation* according to the interferometer phase φ=k (i+r)·s. The variation is singled out by subtracting the initial image from the final image, pixel by pixel.

The result is exemplified in Fig.5-24 for an aluminum plate, observed at rest and after a deformation push at the center. Both initial and final images show no detail, but their difference unveils displacement fringes, each marking a φ(x,y)=2π increment.

The apparatus for the frame-subtraction mode of operation of ESPI is shown in Fig.5-25 [14]. Besides the already described spatial filter in the objective lens stop and the electrical signal processing (high pass and peak-to-peak rectifier), we find a video image recorder. In the recorder, we store the initial image of the object, corresponding to the quiescent state.

The recorded image is read continuously, in synchronism with the photodetector that picks up the image of the object. Then, we may subtract pixel by pixel the initial intensity to the live image intensity. This generates an image containing the fringes associated to the displacement s as, φ=k (i+r)·s.

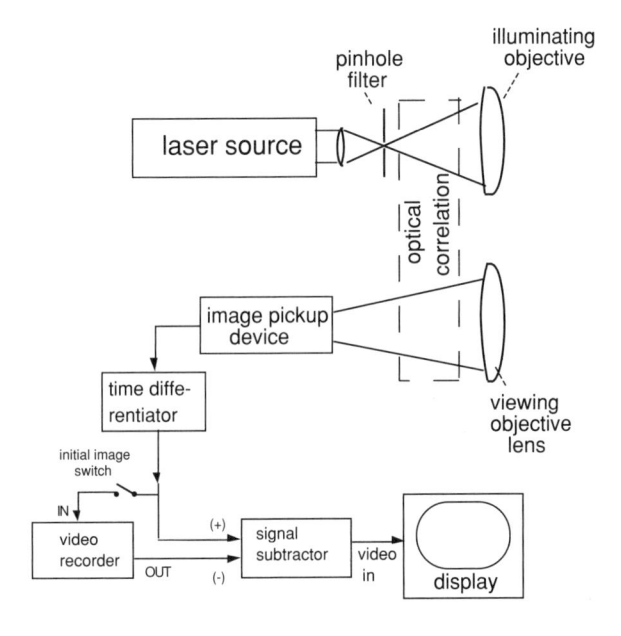

Fig.5-25   Operation of ESPI in frame subtraction mode. The high-frequency spatial content is enhanced with the appropriate lens stop and by time differentiation of the video signal. The initial speckle image is stored in the recorder so that we can subtract it from the live image presented on the display. As a generalization of the reference beam, an optical correlation block can be used.

The optical correlation block indicated in Fig.5-25 can take several forms. In general, these forms are intended to develop fringes more complicated than the simple displacement measurement [14-20]. For example, instead of measuring the total displacement Z=Z(x,y) of a pixel in the image, we may want to get the relative displacement $\Delta Z=Z(x,y)- \langle Z(x,y)\rangle$, where $\langle Z(x,y)\rangle$ is the average displacement of the target. This mode of operation removes the fringes due to rigid-body translation, a contribution that carries little information usually. We can obtain a speckle map of ΔZ by using, as a reference for superposition to the target image, the scattered light from a ground glass placed in front of the objective lens (Fig.5-26).

Second, in front of the objective lens we may place a beamsplitter with a slight tilt so that two images of the target are formed, laterally shifted by a quantity Δx (Fig.5-26). Now, let us assume that Δx is less than the transversal (projected) speckle size. Then, we are correlating the fields received by pair of pixels distant of Δx. The result is a measurement of the $\Delta Z/\Delta x$ strain component, in which each fringe indicates an increment $\approx \lambda/2$ of the out-of-plane displacement ΔZ relative to points distant of Δx.

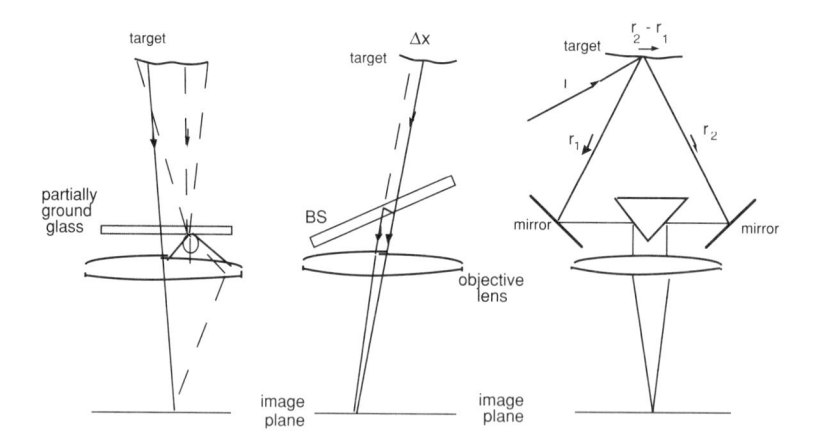

Fig.5-26 Optical correlation may take several forms. Left: A partially ground glass provides the reference for the average position of the target. Center: A tilted beamsplitter in front of the objective lens superposes contributions of image pixels shifted of Δx for strain measurement. Right: A mirror pair allows superpose images with different fields of view to get a shearing interferometer.

Another mode of operation of the ESPI is that of *shearing* measurement. In this interferometer, we want to measure the in-plane strain component $S_x=\Delta X/\Delta x$ following the application of a stress distribution. Here, the displacement X is that of a material point belonging to the target, and x is the laboratory coordinate in which the displacement is viewed. The ratio of their differentials is known as one component of the strain.

One of the several correlation methods developing shear fringes [14,15] is the double mirror [12] shown in Fig.5-26. The target is illuminated from a fixed direction $\underline{i}$ and viewed from two directions $\underline{r}_1$ and $\underline{r}_2$. In this way, the phase-sensitive term governing the speckle is

5.3 Electronic Speckle Pattern Interferometry

$\varphi=(\underline{r}_1+\underline{i})\cdot\underline{s} -(\underline{r}_2+\underline{i})\cdot\underline{s} = (\underline{r}_1-\underline{r}_2)\cdot\underline{s}$, and it only depends on the difference of observation vectors $\underline{r}_1-\underline{r}_2$. As it can be seen in Fig.5-26, the difference vector $\underline{r}_1-\underline{r}_2$ is perpendicular to the lens optical axis or, it is in the plane of the target surface.

From the setup of Fig.5-26, one component of the in-plane strain, let us say $S_x=\Delta X/\Delta x$, is obtained. Of course, by rotating the setup by 90°, we can also get the other in-plane component of the strain, $S_y=\Delta Y/\Delta y$. We do not report here examples of measurement of strain components, which are rather obvious. The interested reader may consult Refs.[12-19].

Another mode of operation of ESPI instruments is the *pulsed mode*. In it, we use a pulsed-laser source with adequate coherence to take snapshots of the target. Speckle fringes are formed by *frame subtraction* of the initial image off the live one. At each instant, we take a snapshot, and a speckle image is acquired. Thus, we can examine objects subjected to periodic stresses, like rotating machinery or turbine blades, and get information on the dynamical deformation undergone along a full cycle of the applied stress. This is valuable information, because it may reveal a pattern of strain evolution forewarning failure.

In processing the ESPI images, we face the problem of phase ambiguity like in any interferometer. Indeed, speckle fringes developed after a displacement s(x,y) obey the equation cos $\varphi(x,y)$= cos k[($\underline{i}+\underline{r}$)·$\underline{s}$(x,y)]. As we know, a single trigonometric function of the type cosine leaves an ambiguity in the reconstruction of displacement from the phase $\varphi(x,y)$.

From the practical point of view, the ambiguity can be removed if we assume that the displacement distribution $\underline{s}$(x,y) shall have the least peak-to-peak swing. Then, we may associate a large-width fringe with a local stationary point of $\underline{s}$(x,y), that is, a minimum or a maximum. If we also know which point or points in the image are still, we can start from there to draw the paths of steepest variation of $\underline{s}$(x,y). In principle, this is the operation, called *phase unwrapping* by which we are able to reconstruct the $\underline{s}$(x,y) distribution [15-20].

Another clean approach to remove ambiguity is to make available a pair of orthogonal signals of the phase shift of the type sine and cosine. This can be easily obtained by applying a *phase dither* φ_{dith} to the reference channel [19,21].

To implement the dither, we add a PZT ceramic phase modulator in the reference beam path of Fig.5-21, so that the video signal at the photodetector output becomes cos[$\varphi(x,y)+\varphi_{dith}$]. If the ceramic is driven by a square wave voltage swinging between 0 and $V_{\pi/2}$, then the phase shift φ_{dith} swings between 0 and $\pi/2$.

By synchronizing the square wave to half the frame frequency of the image pickup, we obtain that, in one frame, the signal is cos k($\underline{i}+\underline{r}$)·$\underline{s}$(x,y) and, in the next frame, is cos [k($\underline{i}+\underline{r}$)·$\underline{s}$(x,y)+$\pi/2$] =-sin k($\underline{i}+\underline{r}$)·$\underline{s}$(x,y). Even and odd frames are stored separately and, after averaging, the reconstruction algorithm can be applied to obtain s(x,y).

REFERENCES

[1] J.W. Goodman, "*Statistical Properties of Laser Speckle Patterns*", in Laser Speckle and Related Phenomena, edited by J.C.Dainty, Springer Verlag: Berlin, 2nd ed., 1981, pp.9-74.

[2] A. Papoulis, "*Probability, Random Variables and Stochastic Processes*", 3rd edition, McGraw Hill: New York, 1991.

[3] D. Middleton, "*Introduction to Statistical Communication Theory*", McGraw Hill: New York, 1960; reprinted by IEEE Press: New York, 1996.
[4] J.D. Gaskill, "*Linear Systems, Fourier Transforms and Optics*", J.Wiley: New York, 1978.
[5] S. Donati, "*Photodetectors*", Prentice Hall: Upper Saddle River, 2000.
[6] S. Donati and G. Martini, "*Speckle-Pattern Intensity and Phase Second-Order Conditional Statistics*", Journal of the Opt. Soc. of America, vol. 69 (1979), pp.1690-1694.
[7] M. Norgia, S. Donati, and D. D'Alessandro, "*Interferometric Measurements of Displacement on a Diffusing Target by a Speckle Tracking Technique*", IEEE Journal Quant. Electr., vol.QE-37 (2001), pp.800-806. See also "*A Displacement-Measuring Instrument Utilizing Self-Mixing Interferometry*", IEEE Trans. Meas. Instrum., vol.IM-52 (2003), pp.1765-1770.
[8] A. Yariv, "*Quantum Electronics*", 3rd ed., J.Wiley and Sons: New York, 1989, App. 4.
[9] H. Blinchikoff and A. Zverev, "*Filtering in the Time and Frequency Domain*", J.Wiley and Sons: New York, 1976.
[10] S.L. Hahn, "*Hilbert Transforms in Signal Processing*", Artech House, Inc.: Norwood, MA, 1996.
[11] M. Sorel, G. Martini, and S. Donati, "*Correlation between Intensity and Phase in Speckle Pattern Interferometry*", in Proc. ODIMAP II, 2nd Workshop on Distance Measurement and Applications, edited by S. Donati, Pavia (Italy), 20-23 May 1999, p.132-137.
[12] J.N. Butters, R. James, and C. Wykes, "*Electronic Speckle Pattern Interferometry*", in Speckle Metrology, edited by R.K. Erf, Academic Press: New York, 1978, pp.111-158.
[13] R. Jones and C. Wykes, "*Holographic and Speckle Interferometry*", 2nd ed., Cambridge University Press: Oxford, 1989.
[14] S. Donati, "*A Speckle Pattern Instrument for Real Time Visualization of Vibration and Displacements*", Alta Frequenza, vol.44 (1975), pp.384-386.
[15] O.J. Lokberg and G.A. Slettemoen, "*Basic Electronic Speckle Pattern Interferometry*", in Applied Optics and Optical Engineering, vol.X, edited by R.R. Shannon and J.C. Wyant, Academic Press: New York, 1987, pp.455-503.
[16] K.J. Gasvik, "*Optical Metrology*", 2nd ed., J.Wiley: Chichester, 1995.
[17] R.S. Sirohi, (editor), "*Speckle Metrology*", Marcel Dekker: New York, 1993.
[18] P.K. Rastogi, "*Digital Speckle Pattern Interferometry and Related Techniques*", J.Wiley: New York, 2000.
[19] J. Burke, T. Bothe, H. Helmers, C. Kunze, R.S. Sirohi, and V. Wilkens, "*Spatial Phase Shifting in ESPI: Influence of Second-Oorder Speckle Statistics on Fringe Quality*", in Fringe '97, Automatic Processing of Fringe Patterns, edited by W.J. Jüptner and W. Osten, Wiley Europe: London, 1997, pp. 111-119.
[20] D. Paoletti and G. Schirripa Spagnolo, "*Interferometric Methods for Artwork Diagnostics*", Progress in Optics, vol.35 (1996), pp.197-255.
[21] G. Martini, M. Facchini, and D. Parisi, "*Automatic Phase-Stepping in Fiber-Optic ESPI by Closed-Loop Gain Switching*", IEEE Trans. Meas. Instr., vol.IM-49 (2000), pp.823-828.

CHAPTER **6**

Laser Doppler Velocimetry

The Laser Doppler Velocimeter, (LDV) was invented in 1964 by Yeh and Cummins [1] and soon became another bright chapter of optoelectronic instrumentation. The first commercial velocimeter appeared in the market soon after the laser interferometer, that is, about the year 1970. Since then, velocimeters sold in several thousand unit-per-year [2] volume and have been recognized as an unparalleled tool for noncontact measurements of flow velocity in fluids [3]. The fluid can be either a liquid or a gas, and the velocity is desired for test, diagnostics and design applications to hydraulics, fluidics and wind tunnel. Because of the application, velocimeters are also called Laser Doppler Anemometers (LDA) and sometimes Particle Flow Velocimeters (PFV).

Conceptually, a velocimeter is nothing else but another kind of interferometer. Instead of looking at the usual out-of-plane component of displacement, given by the normal signal $2\underline{k} \cdot \underline{s}$, the velocimeter is designed to sense the in-plane component, perpendicular to the line of sight.

To obtain the in-plane component, two beams are used to illuminate the fluid. The target is made up of the small particles (μm-size) left back as residual particulate or intentionally dispersed in the medium. Though there is no fundamental reason to conceive the velocimeter as a Doppler-based instrument rather than a two-beam interferometer, the terminology has gained acceptance internationally, and we will therefore use it.

6.1 PRINCIPLE OF OPERATION

The basic schematic of a velocimeter is shown in Fig.6-1. The preferred sources have been traditionally the He-Ne laser or the Ar-ion laser, like in other instruments requiring a CW emission with a good spatial quality.

The beam leaving the laser is filtered spatially and then collimated by a telescope to the desired size. Then it comes to a beamsplitter where it is divided in two parts. With a folding mirror, the two parts of the beam are brought parallel for off-axis incidence on a focusing lens. In this way, two beams are brought to cross each other on the axis of the lens at the focal distance F from it.

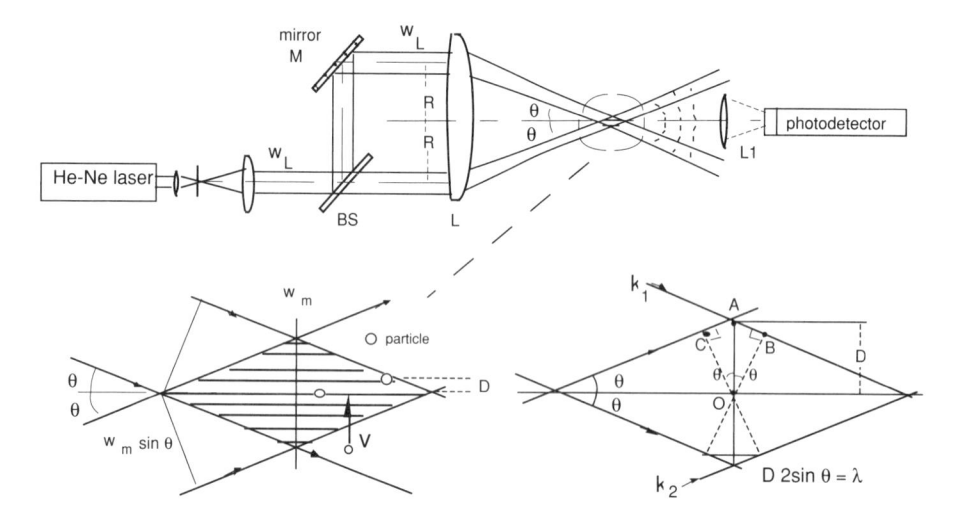

Fig.6-1 In an LDV, the laser beam is first collimated and divided into two parallel beams by the beamsplitter BS and folding mirror M. The beams are sent to the focussng lens L, with an off-axis shift R. The lens brings the beams to cross the axis at the focal distance F, under an angle $\tan\theta = R/F$. Horizontal fringes are formed at the beam crossover. The fringe spacing D is found setting the path length difference AB+AC equal to λ. A particle crossing the fringes at a speed v has brightness oscillating with a period $T=D/v$. The field scattered by the particle is converted by the photodetector in an electrical signal at a frequency v/D.

6.1 Principle of Operation

If we now expand the region of beam superposition (Fig.6-1, bottom) we can see that stationary fringes are formed. In fact, the loci of equal phase difference between the two wave fronts are horizontal planes, in the reference frame of Fig.6-1.

By drawing the wave front of the two beams, it is easy to derive the fringe spacing D. Let us go from point O on a fringe to point A in the next fringe, along the interfering beams. Considering first the beam k_1, we move along the wave front OB with no phase shift and then along BA perpendicular to the wave front with a path length delay D sinθ.
For beam k_2, we get with no phase shift along OC, and a path length delay -Dsinθ along CA.

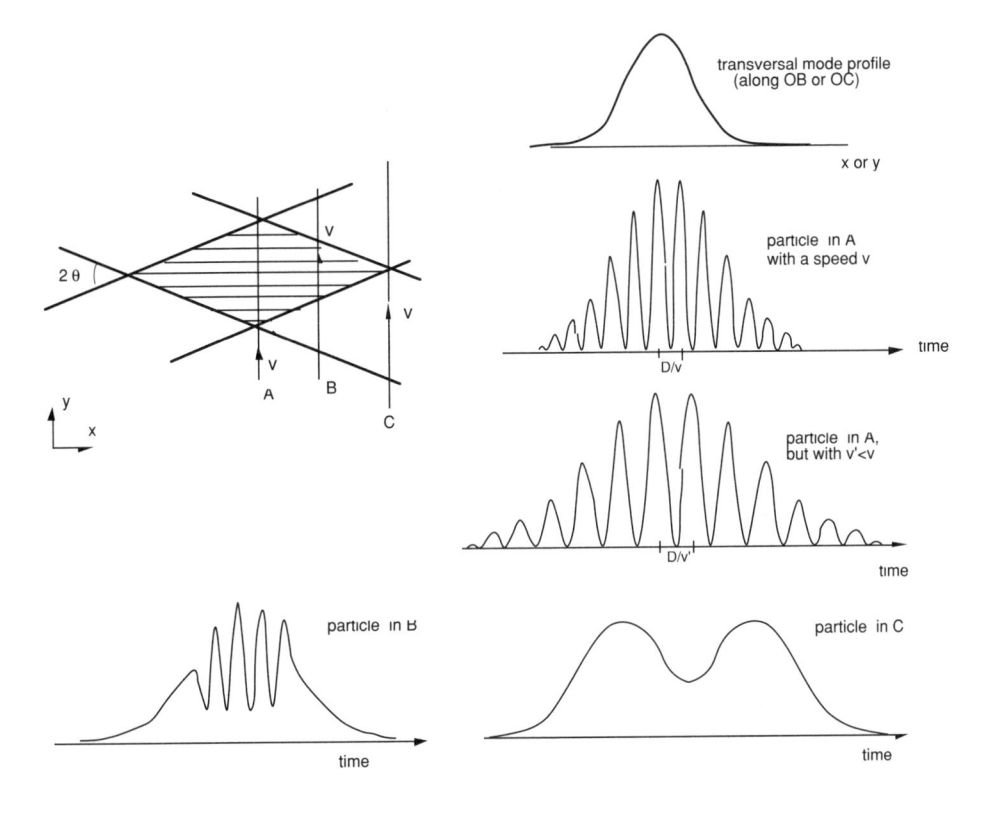

Fig.6-2 Particles crossing the fringes in the superposition region (top left) with a velocity v (the y-component) develop an oscillating signal at frequency f_D=v/D, where D is the fringe spacing. The envelope of the oscillation replicates the Gauss distribution of the beam. Particle A, crossing the superposition region at the midpoint, develops the largest number of periods, and is the ideal response. Particle B gives fewer periods and its signal has less contrast. Particle C, outside the superposition region, only samples the beam distributions and its waveform carries no information.

The total path length difference of interfering beams is then 2 Dsinθ. By equating this quantity to λ for adjacent fringes, we get the fringe spacing as:

$$D = \lambda / (2\sin\theta) \qquad (6.1)$$

If a particle moving along the y-axis crosses the fringes with a velocity v, then it is illuminated by alternate bright and dark fringes, with a period T=D/v (Fig.6-1). The particle scatters radiation outward (see Appendix A3), and we can collect scattered power by a lens and photodetector combination (Fig.6-1), which provides an electrical signal for subsequent processing. The electrical signal contains information on velocity in form of the frequency of the oscillations developed by fringe crossing in the superposition region. The frequency is given by:

$$f_D = v/D = 2\sin\theta\, v/\lambda \qquad (6.2)$$

The envelope of the signal is a Gaussian distribution, a replica of the beam spatial profile of interfering beams.

As we can see in Fig.6-2, not all the particle positions are equally good to develop the Doppler signal, however. Particle A at the middle of the superposition region will cross the largest number of fringes and develop a clean oscillation with a Gaussian envelope. Particle B is halfway between the middle and the border of the superposition region and crosses half as much fringes, and in addition passes through a portion of the each beam alone. Thus, particle B develops fewer cycles, has a long Gaussian tail, and its oscillation has nonunity modulation depth because of beam intensity unbalance. Particle C is just outside the border of beam superposition and crosses no fringes. Thus, particle C develops a sample of the beam profile intensity instead of the desired Doppler signal. In the processing of signals, particle A has the best and most desired waveform, B is marginally useful, and C is to be discarded.

Returning to Eq.6.2, we can see that, in the laser Doppler velocimeter, the fluid velocity measurement is brought to a frequency measurement on the electrical signal, with a scale factor given by $R = f_D/v = 2\sin\theta/\lambda$ [Hz/(m/s)].

This circumstance is very favorable, because it is well known that the frequency measurement is the best we can perform in electronics. It may cover 9 to 12 decades, from mHz or μHz to GHz and above. With a single counter circuit, we may easily perform a frequency measurement on 8 decades, for example from 1 Hz to 100 MHz. In addition, a frequency meter can be easily traced to standards of frequency, and the result is that our measurement has also a *precision*, not only accuracy, with typical values of $\Delta f_D/f_D \approx 10^{-6}$ and better.

Returning to the frequency measurement, let us consider the typical values of the scale factor *R* and of the velocity range we can attain. Using λ=0.633 μm and θ= 0.1 (5 deg)...0.5 (30 deg), we get *R* = 0.3...1.6 MHz/(m/s).

Because the angle dependence is not that strong, let us take *R* = 1 MHz/(m/s) as a reference value. This figure means that we can go from very small velocity, ≈μm/s, in the low range

6.1 Principle of Operation

of frequency (Hz's) to very high velocity, up to ≈100m/s, in the high range of frequency (100MHz's). The wide span of frequency becomes translated in the wide span of velocity we can measure with the laser Doppler velocimeter.

6.1.1 The Velocimeter as an Interferometer

We can now show that the following descriptions of the velocimeter signal are equivalent: (i) the fringe crossing; (ii) the Doppler effect; and (iii) the interferometric phase shift.
About fringe crossing, the result of the analysis is Eq.6.2 we have already considered.
About the Doppler effect, it is well known that the frequency deviation Δf_D experienced when the source moves with respect to the observer at a speed v is $\Delta f_D/f_D = v/c$, or $\Delta f_D = v/\lambda$. Around the direction of illumination (Fig.6-1), the particle is seen to approach by one beam ($\underline{k}_1$), and remove by the other beam ($\underline{k}_2$). Then, considering that the obliquity factor is sinθ, we have $\Delta f_D = v \sin\theta/\lambda$ and $\Delta f_D = -v \sin\theta/\lambda$ for the two beams.
In conclusion, the total frequency difference is 2v sinθ /λ, the result of Eq.6.2.

Third, we can use the interferometric phase shift to derive the equivalence to Eq.6.2. When a field E_0 impinges on the particle from direction $\underline{k}_1$ and we look it from direction $\underline{k}_0$, the field is given by:

$$E_0 \exp(\underline{k}_1 - \underline{k}_0) \cdot \underline{s}$$

This expression is recognized as the generalization of the usual term $E_0 \exp 2ks$, to the case of different directions of illumination and observation, where $\underline{s}$ is the displacement vector, and the dot "·" means scalar product.
We have two fields of illumination in the LDV velocimeter, $\underline{k}_1$ and $\underline{k}_2$. Therefore, the total field observed from direction $\underline{k}_0$ is:

$$E = E_0 \exp(\underline{k}_1 - \underline{k}_0) \cdot \underline{s} + E_0 \exp(\underline{k}_2 - \underline{k}_0) \cdot \underline{s}$$

The signal found at the detector is proportional to the square modulus of the field, $I \propto |E|^2$. Developing the modulus, we easily get:

$$I \propto E_0^2 \, 2 + E_0^2 \, 2 \cos[(\underline{k}_1 - \underline{k}_0) \cdot \underline{s} - (\underline{k}_2 - \underline{k}_0) \cdot \underline{s}] = 2E_0^2 [1 + \cos(\underline{k}_1 - \underline{k}_2) \cdot \underline{s}]$$

Except for a constant term, the signal depends on the scalar product of the displacement $\underline{s}$ and the vector difference $\underline{k}_1 - \underline{k}_2$ of the illuminating beam. Note that the result is independent from the observation vector $\underline{k}_0$.
With the aid of Fig.6-3, it is easy to see that the difference $\underline{k}_1 - \underline{k}_2$ is parallel to the y-axis (the vertical axis in Fig.6-2), and its modulus is given by (see Fig.6-3):

$$|\underline{k}_1 - \underline{k}_2| = 2k \sin\theta.$$

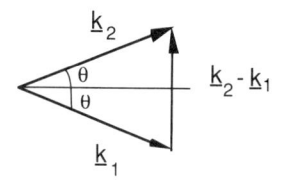

Fig.6-3 The vector difference of the illuminating wave vectors $\underline{k}_1$ and $\underline{k}_2$ is the argument of the phase shift $\phi = (\underline{k}_1 - \underline{k}_2) \cdot \underline{s}$ observed for a particle under a displacement $\underline{s}$.

For a particle making an in-plane displacement $\underline{s}$ parallel to the y-axis, the phase of the cosine function is then $\phi = (\underline{k}_1 - \underline{k}_2) \cdot \underline{s} = 2ks \sin \theta$. By time differentiating the phase shift ϕ, we obtain angular frequency (or pulsation) $\omega = 2\pi f$. Using the expression of ϕ, we can obtain frequency as:

$$f = (2\pi)^{-1}(d\phi/dt) = (2\pi)^{-1} 2k(ds/dt) \sin\theta = 2 \sin\theta \, v/\lambda$$

The result coincides with Eq.6.1 once more, and this demonstrates the equivalence of the Doppler effect and interferometric phase shift descriptions.

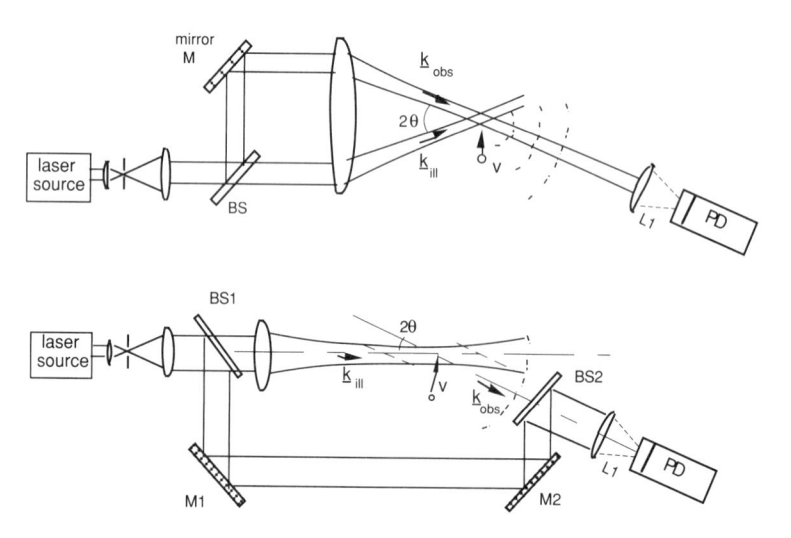

Fig.6-4 Top: An LDV configuration named 'referenced' because the photodetector is positioned to collect one of the illuminating beams and thus acts as the local oscillator for the homodyne detection of the fringe signal. Bottom: Another referenced configuration uses a beam deviated from the laser to the detector. Though fringes are not formed, the vector difference $\underline{k}_{ill} - \underline{k}_{obs}$ determines the axis of sensitivity to the speed component.

Two other configurations related to interferometric arguments are reported in Fig.6-4. The first one (top in Fig.6-4) is called *referenced detection* and is obtained by moving the detector so that it intercepts the portion of the upper beam passing beyond the fringes. This beam acts as the local oscillator beam of a coherent homodyne detection [4], with the result that sensitivity is much increased compared to the placement of Fig.6-1, that operates in the direct detection regime.

The configuration is sensitive to the optical phase shift ($\underline{k}_{ill}$-$\underline{k}_{obs}$)·$\underline{s}$, that is, to the velocity component along the vertical axis in the drawing of Fig.6-4.

This circumstance is no surprise because two interfering beams are in the superposition region that generates the usual horizontal fringes, like in the setup of Fig.6-1. The only basic difference with Fig.6-1 is the presence of the reference beam.

The second configuration in Fig.6-4 (bottom) is truly different from the previous ones, conceptually. The reference beam is again present, but illumination is provided only by one beam, and no fringes are formed. This configuration, one proposed and experimented in the early times of LDV [3], is itself sensitive to the optical phase shift ($\underline{k}_{ill}$-$\underline{k}_{obs}$)·$\underline{s}$.

A last comment is on the difference between velocimetry and vibrometry (Sect.4.6). As already pointed out, both are actually interferometers, and it makes no fundamental difference if we look at the frequency signal rather than the phase signal. The true difference is the direction along which the component of $\underline{s}$ or $\underline{v}$ is measured: parallel or perpendicular to the line of sight aimed to the measurement volume. Another less basic difference is that the LDV looks at a velocity of a fluid moving in front of it, whereas the vibrometer looks at small periodic displacements.

6.2 PERFORMANCE PARAMETERS

Let us now discuss the parameters of operation of the velocimeter and the layout options affecting performance. We deal here with issues such as accuracy of the velocity measurement, sampling volume, and effect of alignment errors. In next sections, we consider the electronic processing and the optical setup configurations.

6.2.1 Scale Factor Relative Error

The scale factor of the Doppler velocity measurement is $R = f/v = 2\sin\theta/\lambda$. The relative error of R, or $\Delta R/R$, is the sum of the relative errors $\Delta\theta/\theta$ and $\Delta\lambda/\lambda$ (absolute values). Usually, the wavelength error is not a problem because λ is known to a very good accuracy, at least $\Delta\lambda/\lambda \approx 10^{-4}$ or better (see also App.A1).

The angle of beam crossing can be expressed as $\theta = \operatorname{atan} R_{os}/F$ where R_{os} is the axial offset of the beams and F is the focal length of the objective lens in Fig.6-1.

F is usually known with a typical 10^{-4} accuracy, whereas R depends on the position of beamsplitter and folder mirror. R_{os} can be measured to 10^{-4} accuracy as well, but the long-term stability of mechanical mounts may limit the repeatability to probably a few 10^{-3}.

In conclusion, from the measured frequency, we may expect to go back to velocity $v=f_D/R$ with an accuracy $\Delta R/R \approx 10^{-3}$.

6.2.2 Accuracy of the Doppler Frequency

Several sources limit the accuracy of the measured Doppler frequency f_D. Let us now consider the ultimate accuracy, the one set by the information content in the waveform being measured.

As we have seen previously, several types of waveforms are generated by particles crossing the fringe region in Fig.6-2. The cleanest waveform is that generated by particle A at the fringe middle, and this waveform is likely the best for information content. Thus, we now take particle A as a reference for our considerations.

In crossing the beams, particle A is shone by a Gaussian distribution with spot size w_m (Fig.6-1). Therefore, the waveform s(t) of the signal scattered out to the detector is a sinusoidal function multiplied by a Gaussian envelope, or:

$$s(t) = \sin 2\pi f_D t \; \exp -t^2/2\sigma_t^2,$$

where $\sigma_t = w_m/v$ is the standard deviation of the Gaussian (or $1/e^2$ half-width), and v is the speed of the particle.

By calculating the Fourier transform of s(t) [5], we find that the frequency spectrum of the Doppler signal is again a Gaussian distribution, centered at the Doppler frequency f_D, or:

$$s(f) = \exp -[(f-f_D)^2 (2\pi \sigma_t)^2/2]$$

In this expression, the standard deviation σ_f of frequency f_D is $\sigma_f = 1/(2\pi \sigma_t)$. Substituting $\sigma_t = w_m/v$ in this expression gives $\sigma_f = [2\pi w_m/(f_D D)]^{-1}$, and by rearranging we get:

$$\sigma_f/f_D = 1/(2\pi N_f) \qquad (6.3)$$

Here, we have let $N_f = w_m/D$ for the number of fringes contained in w_m.

In conclusion, we have shown that the relative accuracy σ_f/f_D of the Doppler measurement is proportional to the inverse of the number N_f of fringes crossed by the particle.

Of course, because it is $v=f/R$, the same result holds also for the relative accuracy of the speed measurement, $\sigma_v/v = 1/(2\pi N_f)$.

In point A, the height w_m of the fringe region (Fig.6-1) and the number N_f of fringes are maxima, and therefore the relative error σ_f/f_D is a minimum. This confirms the initial assumption that point A is the best for accuracy.

Now, let us consider those particles crossing the fringes off the optimal point A. Even if we are able to get rid of waveform artifacts (Fig.6-2), the number of fringes N' progressively decreases from the maximum N_f to zero at the crossing border (point C in Fig.6-2).

6.2 Performance Parameters

By repeating the above calculation, we find that the relative accuracy is still given by Eq.6.3, but with the actual number N' of fringes (or periods of the Doppler waveform) in place of N_f. Thus, as the particle crossing gets farther away from the middle of the superposition region, accuracy gets worse and worse.

Then, it will be appropriate to make a selection of 'good' particles and discard 'bad' ones.

As described later in next section, this selection will be actually performed by means of an *a-posteriori* electronic validation, which is equivalent to the limit of the lateral offset of particles allowed in the measurement. If the selection restricts the particles to those crossing at least N/2 fringes, for example, the relative accuracy is intermediate between $(2\pi N_f)^{-1}$ and $(\pi N_f)^{-1}$.

The maximum number of fringes in the superposition region (particle A in Fig.6-2) is given by the ratio of radial offset R_{os} to the spot size w_L of the beams to be recombined:

$$N_f = (2/\pi) R_{os}/w_L \tag{6.4}$$

This result is derived on next section in the case of a focusing lens, but has a general validity.

Last, let us consider what happens if the measurement is repeated on N_M particles. If we take the average value of N_M individual measurements, the relative accuracy of the measurement improves as $1/\sqrt{N_M}$.

Combining the $1/\sqrt{N_M}$ factor and the $\sigma_f/f_D \propto 1/N_f$ dependence, it is easy to see that the laser Doppler velocimeter can attain very low values of the relative error (in velocity), for example 10^{-3} or less. The caution is here that Eq.6.3 is an ultimate limit for accuracy that is approached only after properly curing other systematic errors, for example, the unequal spacing of fringes (see Sect.6.2.4).

6.2.3 Size of the Sensing Region

With the aid of Fig.6-5, we can easily calculate the size of the sensing region determined by the beam superposition, where fringes are formed. This is the measurement region, also called the sampling region or volume of the LDV. A feature specific of the laser Doppler velocimeter is the small size (mm or less) of the sampling region. Together with the noncontact operation, this is a definite advantage of LDV compared with other instruments. Let w_L be the beam size produced by the beam expander and reaching the entrance of the focusing lens. In the focal plane, the beam is focused to a spot size w_f (Fig.6-5). We can find w_f by applying the invariance of the acceptance $a = a\omega$ [4], which is the product of the area and of the solid angle through which the bundle of rays is passing. For a single-mode spatial distribution, acceptance is equal to λ^2 [4].

Because $a = \pi w_f^2$ and $\omega = \pi(w_L/F)^2$, we may apply the previous statement and write for the spot size:

$$a\omega = \pi w_f^2 \pi(w_L/F)^2 = \lambda^2$$

Solving this expression for w_f, we obtain:

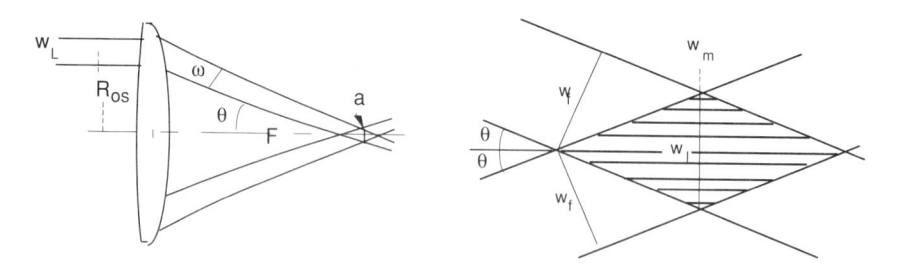

Fig.6-5 The sensing region is defined by the longitudinal and transversal sizes, w_m, and w_l. These quantities depend on beam size w_L and offset radius R_{os}.

$$w_f = \lambda F/\pi w_L \tag{6.5}$$

Now, we can take into account the obliquity factor (Fig.6-5, right) connecting the spot size to the transversal size w_m (a factor $1/\cos\theta$) and to the longitudinal size w_l (a factor $1/\sin\theta$). Doing so, we can write the following results:

$$w_m = (\pi \cos\theta)^{-1} \lambda F/w_L, \text{ and } w_l = (\pi \sin\theta)^{-1} \lambda F/w_L. \tag{6.6}$$

In these expressions, the angle θ is determined by the offset radius R_{os} (Fig.6-5, left), and is given by $R_{os}/F = \tan\theta$.

Recalling the fringe spacing $D=\lambda/2\sin\theta$ from Eq.6.1, we can work out the number of fringe as:

$$N_F = w_m/D = (\pi \cos\theta)^{-1} \lambda F/w_L [\lambda/2\sin\theta]^{-1} = (2/\pi) \tan\theta \, F/w_L = (2/\pi) R_{os}/w_L$$

The last term in the expression coincides with the result of Eq.6.4.

To illustrate the typical values that can be found in a velocimeter, we may consider the case $R_{os}=50$mm and $F = 200$mm, calling for a lens with a diameter $D_{lens} = 2R_{os} = 100$mm (and with an F-number [4] F/ =0.5).

The angle of superposition is found from $R_{os}/F =\tan\theta = 0.25$, as $\theta= 14°$.

With a laser spot size $w_L= 1$mm on the focusing lens, we obtain a number of fringes $N_F =(2/\pi)50=31$.

The accuracy of the single velocity measurement is therefore $\sigma_v/v =\sigma_f/f_D=(2\pi N_f)^{-1}=0.5\%$.

At the He-Ne wavelength $\lambda=0.633\mu$m, the fringe spacing is evaluated as $D= 0.633\mu$m$/2\sin14°= 1.30\mu$m, and the scale factor is $R= 2\sin\theta/\lambda= 0.766$ Hz/(μm/s) or, if preferred, 0.766 kHz/(mm/s) or MHz/(m/s). The focused spot size is $w_f = (0.633\mu$m$) 200/\pi\cdot1= 40\mu$m. The sizes of the sensing region are: $w_m = w_f /0.97= 41\mu$m, and $w_l = w_f /0.24 = 166\mu$m.

6.2 Performance Parameters

As we can see, these central design values provide a good intrinsic accuracy and a small sampling volume. The sampling volume is not so far away from the focusing lens, however.

If we need to sense the fluid farther away, we may use a larger focal length, for example F=500mm. With this new value, the angle of superposition is tan θ= 0.1, or θ= 5.7°. With the same laser beam size w_L, the number of fringes is unchanged, as well as the accuracy σ_v/v. Fringe spacing and scale factor are increased to D=3.16µm, and R= 0.316 Hz/(µm/s). The focused spot size changes to w_f= (0.633µm) 500/π 1= 100µm, and the sizes of the sensing region become w_m = w_f /0.995= 101µm, and w_l = w_f /0.1= 1000µm. Should this large value of the longitudinal size w_l be inconvenient, we can always trim it by the a-posteriori validation of fringe (see Sect.6.3.1).

6.2.4 Alignment and Positioning Errors

Errors in alignment of the beams sent to the focusing objective have the consequence of distorting the fringes in the superposition region.
As illustrated in Fig.6-6, an angular error of the parallelism of the beams results in an unequal spacing of fringes, whereas an unbalance of the lateral offset radius R tilts the fringes with respect to the optical axis.

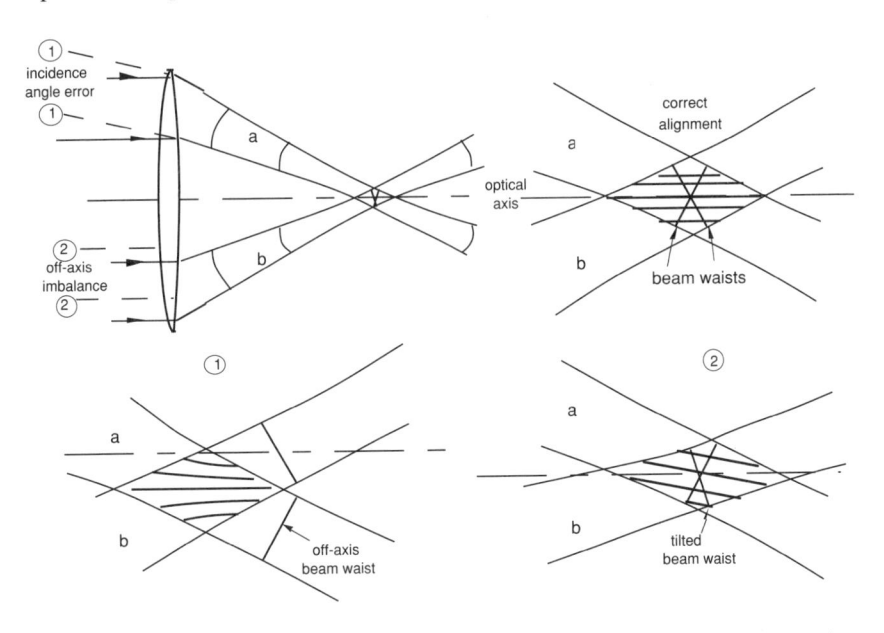

Fig.6-6 At the focusing objective, errors of the incidence angle or of the lateral offset of the two beams result in fringes with unequal spacing (bottom left) or with a tilt with respect to the optical axis (bottom right).

Of course, there are several other possible types of alignment or positioning error. All of them result in a specific distortion of the fringe pattern. Therefore, accuracy of the velocity measurement is reduced. To minimize the effect of alignment or positioning errors, we should employ an optical configuration that is easy to trim and mechanically stable in time. From this point of view, the configuration of Fig.6-1 is not the best because mirrors are affected by an angular error of their position, and each of them can go out of alignment.

A preferred configuration is shown in Fig.6-7 (left). Here, a cube beam splitter BS divides the incoming beam, and folding prisms are used in place of mirrors to derive beams b1 and b2. A prism, used as a mirror by the internal reflection at the 45° surface, is much sturdier and mechanically stable in time than a mirror on a flat. In addition, using the other two surfaces as a reference for the mechanical mount, the parallelism of P1, P2, and P3 is much easier to achieve. If desired, we could improve the configuration further with the use of Dove prisms as shown in Fig.6-7 (right).

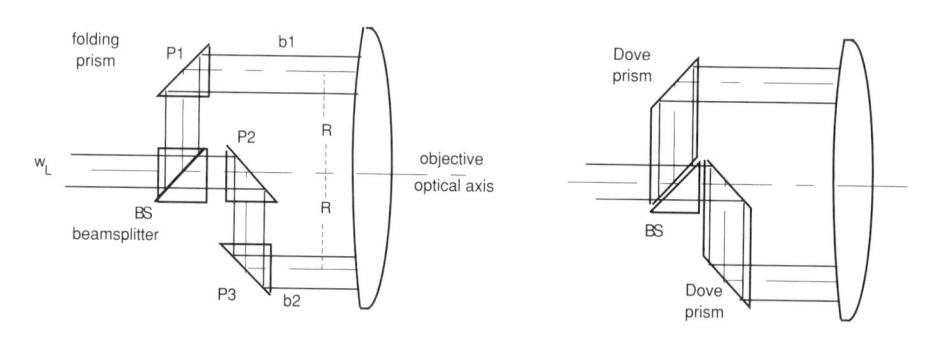

Fig.6-7   Using a beamsplitter prism BS and the folding prisms P1-3 as mirrors (left), we obtain a splitting configuration much more stable and easily trimmed as compared with that of Fig.6-1. Even better, we can employ Dove prisms to minimize parts count and improve stability further (right).

A last interesting feature of the prism configuration is optical path-length balance. Looking at Fig.6-1, we can see that path lengths to the superposition region are different, because the lower beam reaches the lens straight, whereas the upper beam has the extra vertical path to cover. The difference is not so large, being of the order of 2R (typ. ≈100mm). However, if the laser source is multimode in frequency, that is, emits several longitudinal modes, we may face a reduction of the coherence factor.
In this case, the fringe visibility is not a maximum and we get a sort of pedestal in the Doppler signal, as illustrated in Fig.6-8.

Using the prism configuration to generate the beam superposition, we enter parallel to the lens optical axis so that path lengths are balanced (Fig.6-7).
In addition to the configurations of Figs.6-1 and 6-7, several other variants have been proposed [3] and used in commercial LVDs.

6.2 Performance Parameters

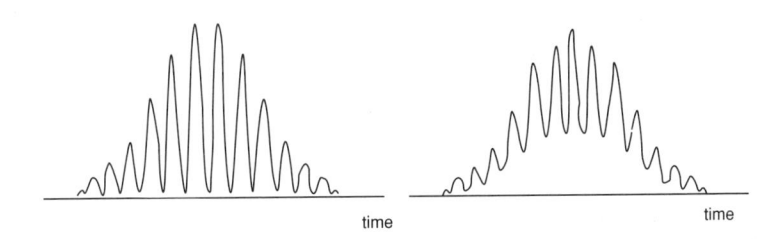

Fig.6-8 The Doppler waveform has 100% modulation (or V=1 fringe visibility) if the two superposed beams have a path-length difference less much than the coherence length of the source (left). When coherence is not unity, or beams have unequal intensity, waveforms exhibit a pedestal (right).

6.2.5 Placement of the Photodetector

The photodetector collects the signal scattered out by the particle crossing the fringes and converts it to an electrical signal for processing. Because scattered light radiates in any direction, the photodetector could be in principle positioned everywhere around the beam crossing region. However, because we want to collect a large signal and the fluid flows in front of the objective lens, we have in practice just two options for the photodetector placement. As shown in Fig.6-9, we may place it beyond the fluid to look at the forward-scattered radiation or behind the objective lens to look at the backward-scattered radiation. In both cases, we will use an objective lens L' to improve light collection and to define a certain solid angle of collection Ω.

In the two cases, at equal distance from the scattering volume and equal aperture of the objective lens, the signal collected is different because it is proportional to the value of the scattering function $f(\theta)$ (App.A3.1) in the forward ($\theta \approx 0$) and in the backward ($\theta \approx \pi$) direction, respectively.

The scattering function changes considerably when we go from the Rayleigh to the Mie regime. In the small-particle ($r<<\lambda$) Rayleigh regime, the scattering function is nearly isotropic, and then the forward-scattering and backward-scattering photodetectors are equivalent. However, we do not prefer working with very small particles, because the amplitude of the scattered signal, or the scattering cross section Q_{ext} (App.A3.1), is very small. Indeed, Q_{ext} is proportional to $(r/\lambda)^4$ in the Rayleigh regime.

If the particle radius is comparable to wavelength or larger, we are in the Mie regime, in which the cross-section factor is comparatively large ($Q_{ext} \approx 2$) and the scattered signal is much increased. In the Mie regime, however, the scattering function $f(\theta)$ is strongly peaked forward.

The ratio $p(\pi)/p(0)$, proportional to the relative photodetected signals in the backward and forward directions, is in general strongly dependent on the particle characteristics, but usually lies in the range 10^{-3} to 10^{-2}.

Clearly, in this case, we will prefer collecting the signal in the forward direction, if the placement of the detector beyond the fluid is accessible. If it is not, we shall work in the backward direction and will use a detector of increased sensitivity to compensate for the reduced signal amplitude.

In practice, both photomultipliers and photodiodes (pin and avalanche) are used according to the required sensitivity [4].

The amplitude of the signal supplied by the photodetector can be calculated as follows. Let P_L be the laser power, split in the two beams and recombined in the scattering volume. A particle crossing the scattering volume is illuminated with a power density P_L/w_1^2 and scatters out a total power $P_{ts} = (P_L/w_1^2)Q_{ext}\pi r^2$ (App.A3.1).

Denoting with $f(\theta_{meas})$ the scattering function evaluated at the measurement angle θ_{meas} (=0 or π), the signal collected in the solid angle Ω is written as $4\pi P_{ts} f(\theta_{meas}) \Omega$. Last, recalling that the photogenerated current is σ times the detected power, σ being the spectral sensitivity [4], we get:

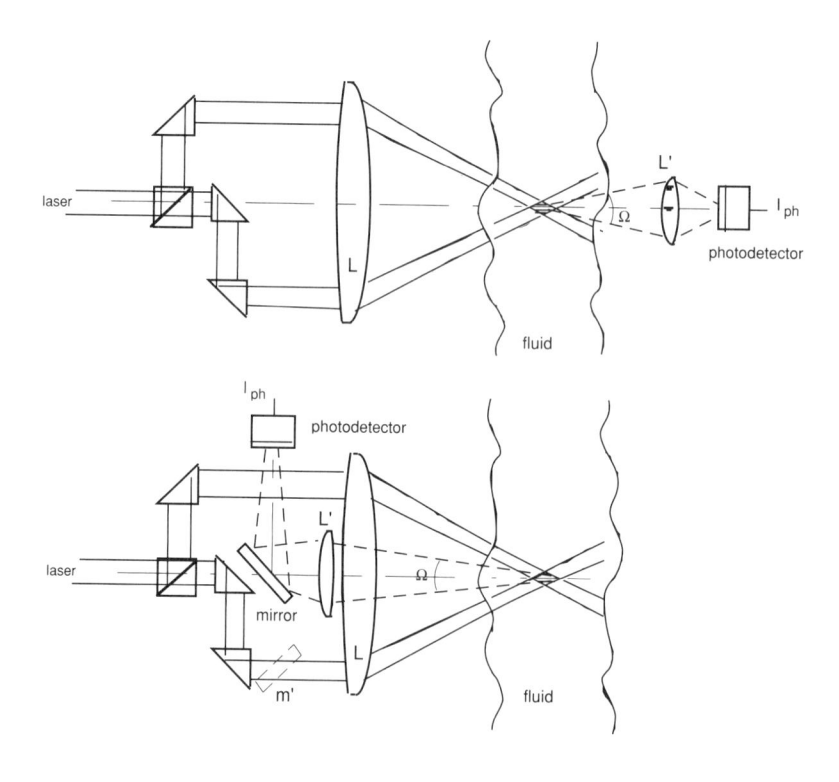

Fig.6-9 Light scattered by the particle in the measurement region can be collected by a photodetector looking at the forward scattering (top) or at the backward scattering (bottom).

6.2 Performance Parameters

$$I_{ph} = \sigma \, (P_L/w_1^2) \, Q_{ext}\pi r^2 \, 4\pi \, f(\theta_{meas}) \, \Omega \qquad (6.7)$$

In most schemes considered so far, we have assumed using direct detection to avoid unessential complication.

Coherent detection can be employed as well, at the expense of some extra complexity in the optical setup, and it generally improves the sensitivity of photodiodes without internal gain. Coherent detection requires that a fraction of the outgoing laser beam be superposed to the scattered return on the photodetector. (see also Ch.7 of Ref.[4]).

An example of this arrangement has been already provided in Fig.6-4.

In addition, most direct detection schemes of LDV can be easily converted to coherent ones. For example, with reference to Fig.6-9, we can convert to coherent detection by these three steps: (i) removing lens L' so that the returning beam comes out collimated from L, (ii) adding beamsplitter m' so that part of the reference beam is superposed on the photodetector, and (iii) changing the folding mirror in a beamsplitter.

6.2.6 Direction Discrimination

As in any interferometer, in a single-channel LDV we cannot distinguish the direction of the velocity, or tell v from –v. Actually, few cases exist in which direction discrimination is necessary, for example, vortex study.

In these cases, we may use the configuration of Fig.6-10 where a Bragg cell is inserted on one of the two recombining beams. The Bragg cell is a frequency shifter based on the acousto-optical effect. If the frequency of the input beam is f, from the output of the Bragg cell we get a frequency f+Δf. The typical shift in the range Δf =20-100 MHz.

Let us now consider the superposition of the two beams (Fig.6-10). We may repeat the reasoning of Sect.6.1.1, modified to take account of the different frequencies of the two beams.

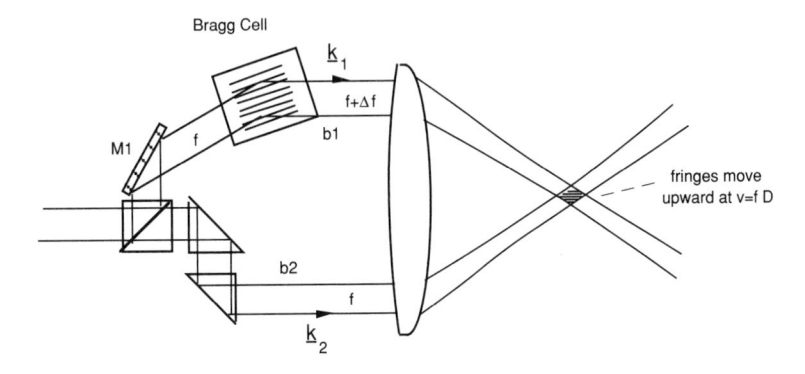

Fig.6-10 By adding a frequency shifter on the path of one of the beams of the velocimeter, we generate a moving fringe pattern in the measurement region and are able to discriminate the direction of velocity v.

The total field is given by:

$$E = E_0 \exp i [2\pi ft+(\underline{k}_1-\underline{k}_{obs})\cdot\underline{s}]+ E_0 \exp i [2\pi(f+\Delta f)t+(\underline{k}_2-\underline{k}_{obs})\cdot\underline{s}]$$

The signal at the detector, $I \propto |E|^2$ is given by:

$$I \propto E_0^2 \cdot 2 + E_0^2 \cdot 2 \cos [2\pi\Delta ft + (\underline{k}_1-\underline{k}_0)\cdot\underline{s} - (\underline{k}_2-\underline{k}_0)\cdot\underline{s}] = 2E_0^2 [1+ \cos 2\pi\Delta ft+(\underline{k}_1-\underline{k}_2)\cdot\underline{s}]$$

By time differentiation of the phase in the cosine term, we obtain frequency as:

$$f = (2\pi)^{-1}(d\phi/dt) = (2\pi)^{-1}(d/dt)[2\pi\Delta ft +(\underline{k}_1-\underline{k}_2)\cdot\underline{s}] = \Delta f + 2 \sin\theta \, v/\lambda$$

The result tells us that the velocity dependent frequency, $2\sin\theta \, v/\lambda$, is superposed to the frequency shift Δf. Thus, as long as $2\sin\theta \, v/\lambda < \Delta f$, we can determine the sign of v.
We can interpret the frequency bias introduced by the Bragg cell as an upward movement of the fringe pattern. Indeed, the shift Δf corresponds to a fringe velocity $v_{fr}= \lambda\Delta f/2 \sin\theta$.

6.2.7 Particle Seeding

About particles generating the scattered signal, we can either have them already present in the fluid as unintentional contaminant or can seed them purposely in the fluid under measurement. The seeding operation is preferable if practicable, because we can choose the diameter best suited for generating a clean and strong signal.
Usually, seed particles are Latex spheres of calibrated diameter, typically a few microns. These spheres are available from polymer companies in a range of sizes, from submicron to tens of micrometer in diameter.
The diameter is chosen large enough to give a large scatter signal in the Mie regime, but small enough compared to the fringe period to avoid averaging of the illumination spatial modulation.
The concentration of the spheres is readily calculated by assuming no more than one particle at the time in the scattering volume, for the reasons seen in next section. In view of the random spatial distribution the particles, we may take 0.1 as the design value of the average number. Then, we may write the concentration (in number per unit volume) as C_N =$0.1/V_{meas}$= $0.1 \tan\theta /w_1^3$. The corresponding relative volume concentration (or particle dilution) is C_V =$0.4 \, r^3\tan\theta/w_1^3$. Typical design values of C_V turn out to be $\approx 10^{-5}$ to 10^{-4}, that is, very small.
These small values explain why we can easily find the fluid already disseminated by a lot of scattering particle. When concentration is much larger than the desired one particle per scattering volume, we have two possibilities. One is to clean the fluid by filtering and then artificially seeding it, the other is increasing the scattering volume and using signal processing techniques tolerant to multiparticle superposition.

6.3 Electronic Processing of the Doppler Signal

There are two options for the electronic processing of the Doppler signal obtained from the photodetector: time-domain and frequency-domain processing.

We prefer processing the waveform in the time domain when there is, on the average, much less than one particle at the time in the measurement volume, as in the case represented in Fig.6-11, top. Then, the output of the detector is a sequence of Gaussian-envelope oscillations, well distinct and separated in time. Each oscillation indeed carries a lot of information.

As we have found in Sect.6.2.2, a waveform with N_f periods allows us to measure frequency or velocity with a relative error $\Delta f_D/f_D = \Delta v/v = 1/2\pi N_f$. There are also 'poor' waveforms with low N_f due to particles crossing near the edge of the fringe region, but these can be easily discarded by electronic processing in the time domain.

When the particle concentration increases, first we have the superposition of waveforms (Fig.6-11, middle), but the Gaussian envelopes can again be recognized. Then, we could still use time-domain processing, and wait for those waveforms that occasionally come isolated.

At a further increase of particle concentration (Fig.6-11, bottom), we cannot resolve individual waveforms any more and must revert to a frequency-domain method. With this method, we will look at the average frequency contained in the random-like superposition of Gaussian-envelope oscillations.

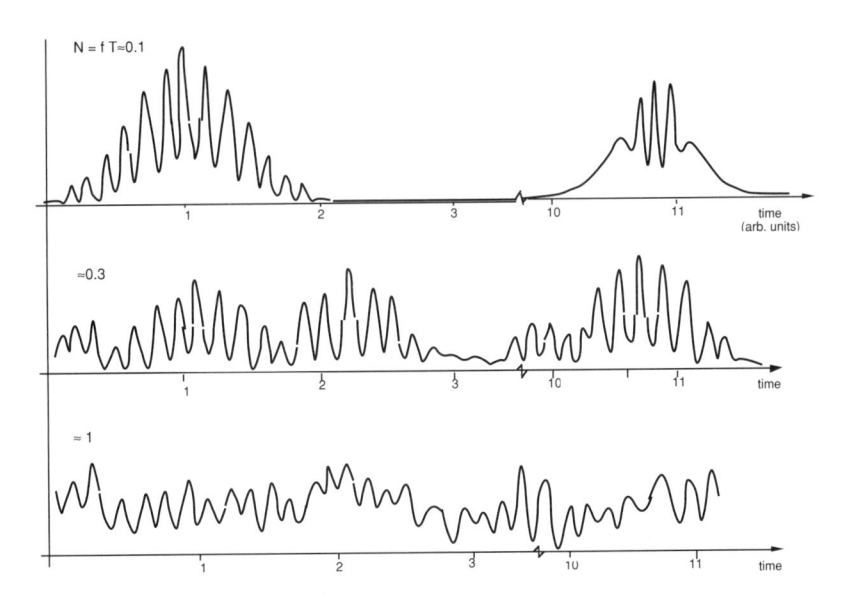

Fig.6-11 According to the average number $N=fT$ of particles in the measurement volume, the waveform is best processed in the time domain (top) or in the frequency domain (bottom)

6.3.1 Time-Domain Processing

We assume that waveforms are distinct in time and do not overlap so that the periods contained in the waveform are well correlated to the frequency $f_D = v/D$.

Basically, the approach used to perform the velocity measurement follows the one outlined in Fig.6-12.

First, we perform a low-noise preamplification of the photodetector output signal and make available a clean Doppler waveform $S(t)$. Then we time-differentiate $S(t)$ so that eventual pedestal or slow drifts are canceled out. In the waveform $S'(t)$ (Fig.6-13), the periods of oscillation are centered around the zero level and last $1/f_D$. Next, we pass the signal through an amplitude discriminator with a threshold S_0 close to zero and obtain a square-wave signal S_{SW} that contains a number of periods $N_f \approx w_L/D$ equal to the number of crossed fringes.

Frequency f_D is determined by two counters, enabled by the presence of the Doppler signal $S(t)$. To form the enable command, we make the envelope of $S(t)$, obtaining S_{env} (Fig.6-13) and pass it through a discriminator with threshold S_0 (Fig.6-12). The result is S_{enabl}, a signal lasting a time T_C, which enables the gates to open the counters.

During time T_C, a main counter counts the number of periods N_f, and, simultaneously, an auxiliary counter counts the N_C pulses of a clock running at a suitable frequency f_C (Fig.6-12). The counts of the signal are $N_f = f_D T_C$ and those of the clock are $N_C = f_C T_C$. By taking the ratio of the counts, we get a quantity $V = N_f/N_C = f_D/f_C$ proportional to $f_D = v/D$.

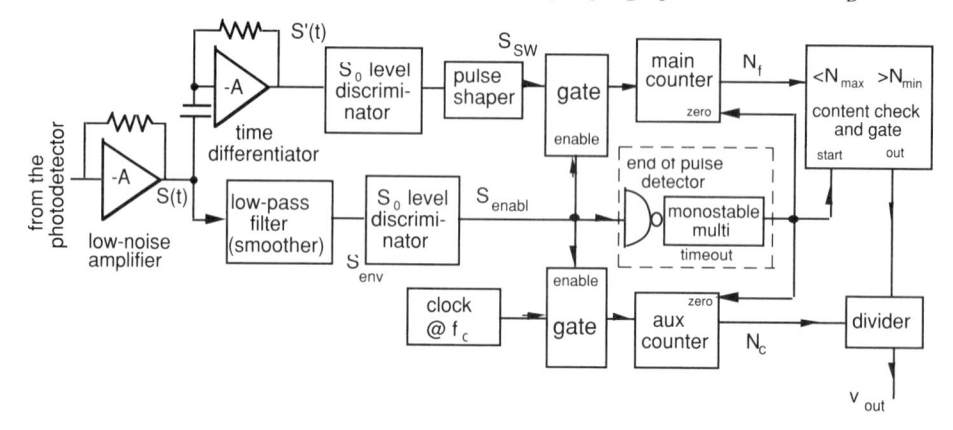

Fig. 6-12 In the time-domain processing, the Doppler signal is first time-differentiated and then squared by a discriminator with a threshold near to zero. Another discriminator works on the signal envelope to determine the presence of the signal and to enable the counters. One counter is for the Doppler signal periods, and the other is for the clock. When the signal is ended, an inspection circuit compares the counter content to the reference minimum. If content is high enough, the result is validated and passed on to calculate velocity, whereas if it is low, it is discarded.

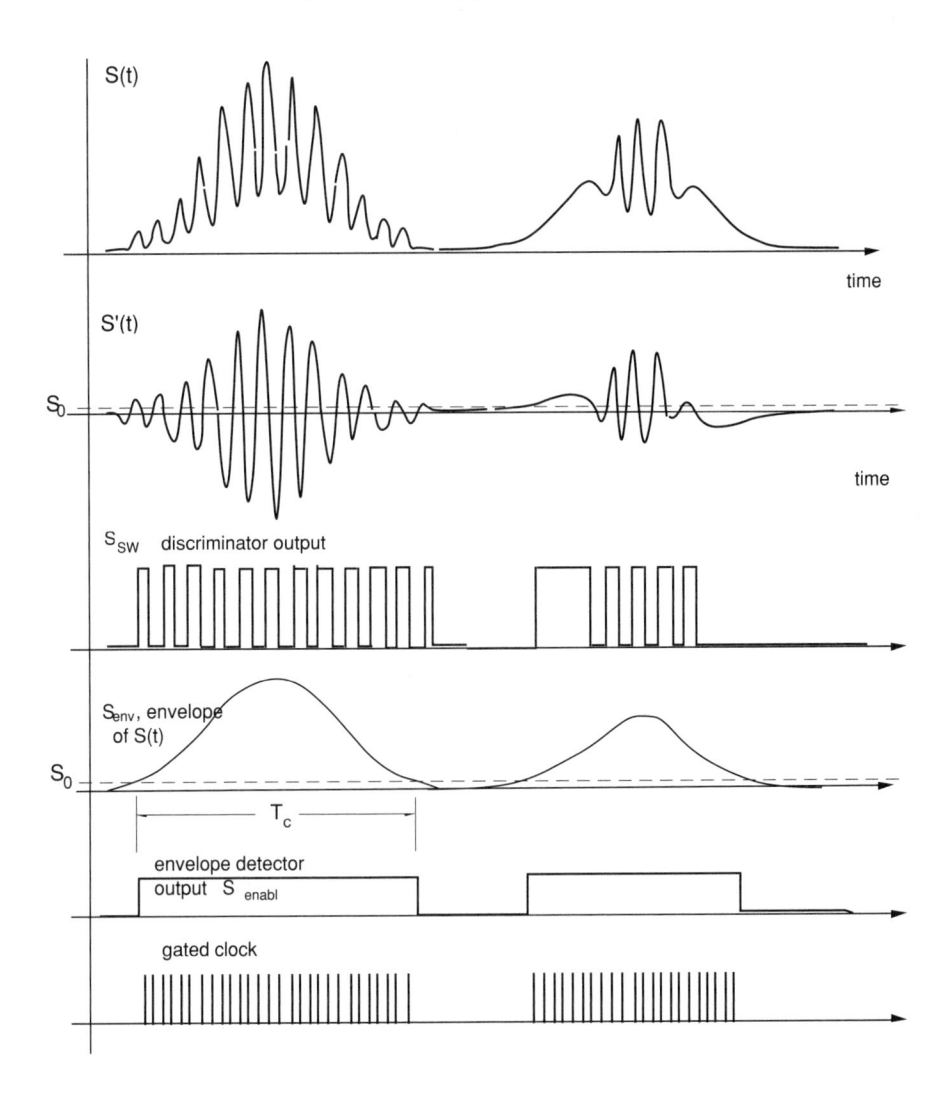

6.3 Electronic Processing of the Doppler Signal

Fig.6-13 Waveforms in the time-domain processing. First diagram is the Doppler signal S(t) at the photodetector output. Second: Computing the time derivative S'(t), we are able to cancel out any pedestal and drift and get the Doppler frequency in form of zero-crossings. Third: After discrimination with a threshold S_0, we get N_f pulses in the time duration T_c. Signal end is obtained by an envelope detector, and yields a gate square wave lasting a time T_c. When signal is over, N_f is validated by an inspection circuit to stay within the expected limits of good signals. Clock pulses (N_c) are counted during T_c. By computing the ratio N_f / N_c, velocity is obtained.

We can also adjust the clock frequency so that the ratio v/f_cD is a decimal in the velocity v.

At the end of the signal, after S_{enabl} has returned to zero, an end-of-pulse sequencer gates the division operation (Fig.6-12) and resets the counters for the next Doppler signal S(t). When a 'poor' Doppler signal is received, as exemplified by the second waveform in Fig.6-13, the number of counts N_f is low, and the content-check circuit of Fig.6-12 rejects the result at the end of computing period. This operation is carried out by digital comparators and registers of minimum/maximum counts (Fig.6-12). If $N_f > N_{min}$ and $N_f < N_{max}$, the waveform is validated, and the content is passed on to the subsequent processing.

The minimum threshold N_{min} is usually set at half the maximum number of fringes N_f. This is a good compromise between accuracy of the single measurement and waste of useful data. With the threshold placed at $N_f/2$, all particles crossing the fringe region farther than half the longitudinal width w_1 (Fig.6-5) are discarded. Therefore, the actual measurement volume is determined by a longitudinal size $w_1/2$. About the upper threshold N_{max}, this is set to a little bit more than N_f (for example $1.2N_f$) to avoid taking two successive particles for one.

After the validation, the data processing may proceed in two ways, according to the type of measurement we are carrying out. If we deal with a single point measurement, for best accuracy, we will integrate the results of N successive measurements and exploit the $1/\sqrt{N}$ dependence.

If we are measuring a velocity profile, we will use one or a few measurements and then move to the next point of the scan, either along a line or in a raster pattern. A line scan is readily accomplished by translation of the focusing lens, which moves the measurement volume axially. A raster scan or a more complicated pattern will require a suitable moving mirror arrangement.

6.3.2 Frequency-Domain Processing

When particles crowd up the measuring volume as in shown in Fig.6-11 bottom, time-domain processing can no longer be used. The signal oscillations carrying the Doppler frequency are damped out in amplitude and have a frequency jitter. In this case, given an average number $N \gg 1$ of particles in the measuring volume, the signal exhibits amplitude and phase fluctuations resembling those of the speckle pattern.

Indeed, the total signal is the summation of the contributions of N particles, and each particle yields a waveform of the type shown in Fig.6-11, top. Because the time of occurrence of each particle is randomly distributed, we have a summation of N uncorrelated contributions. The result is a field that replicates, for high N, the speckle pattern statistics, with zero mean value of the phase and quadrature components. Consequently, the intensity distribution is a negative exponential with a decay constant equal to the mean value.

It is then clear that, should we attempt a time-domain processing, the number of counts N_f per period T_c will be affected by a large error.

More appropriately, we shall look directly at the frequency content of the signal S(t), as obtained at the output of the preamplifier stage following the photodetector.

6.3 Electronic Processing of the Doppler Signal

There are two approaches for frequency-domain processing: autocorrelation, and phase-locked loop.

In the first approach, we compute the autocorrelation function $c(\tau) = \int_{0-\infty} S(t)S(t+\tau) \, dt$. The calculation can be performed by converting the signal S(t) to digital, with the aid of an Analog to Digital Converter (ADC), and then acquiring the data and computing $c(\tau)$ numerically in a computer unit.

The ADC does not need a very high resolution (usually, 8 bits will be adequate) because the round-off error is usually small respect to that of waveform noise. Rather, the ADC and the acquisition interface need to be fast if we are to cover the high-velocity range allowed by the LDV instrument (see Sect.6-2). For example, if we get 100 kHz of conversion and acquisition rate, we are then bound to a maximum velocity $v \approx 1\mu m \cdot 100 kHz = 0.1$ m/s.

From the autocorrelation function $c(\tau)$, we may already obtain the velocity by looking at the first zero of $c(\tau)$. If $c(\tau)=0$ for $\tau=\tau_z$, then the main frequency contained in S(t) is $f=1/\tau_z$. Alternatively, we may compute the Fourier transform of $c(\tau)$, for example by means of a FFT algorithm that will be easily available if we are using a personal computer for acquiring S(t) and computing $c(\tau)$. Because of the well-known Wiener-Khintchine theorem [5,6], the transform of $c(\tau)$ is the power spectrum of the signal S(t). Thus, the power spectrum is a well-defined line at the Doppler frequency f_D if the fluid has a uniform velocity or is a distribution p(f) of spectral content if the fluid has a distribution of velocity.

When velocity is very high and the signal S(t) has a frequency content too high for the ADC and interface, we can resort to a second approach based on the PLL technique. As shown in Fig.6-14, the signal S(t) at the preamplifier output is compared to the oscillation of a Voltage Controlled Oscillator (VCO). The phase error is used to correct the frequency of the VCO until it is dynamically locked to the frequency of the Doppler oscillation contained in S(t). The gain and filter sections following the phase discriminator act as the feedback loop of the frequency-tracking operation. With them, we can filter signal S(t), which may vary in frequency, yet keep a narrow band around the average frequency, which results in a very effective cleaning-up of the signal.

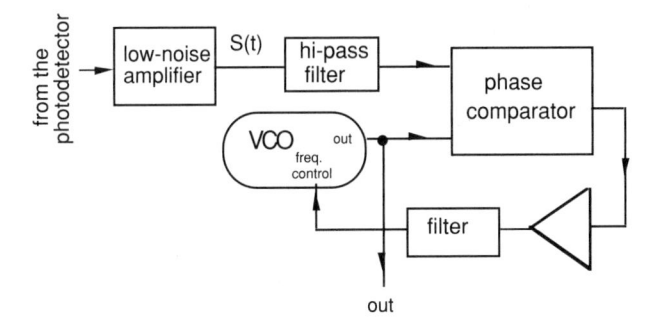

Fig.6-14 The Doppler signal of an LDV can be filtered for f_D extraction by a simple PPL (phase locked loop) arrangement.

244   Laser Doppler Velocimetry   Chapter 6

6.4 OPTICAL CONFIGURATIONS

Lot of configurations have been experimented with through the years and reported in the literature. Several of them have gained acceptance for specific applications and are incorporated in commercially available products.

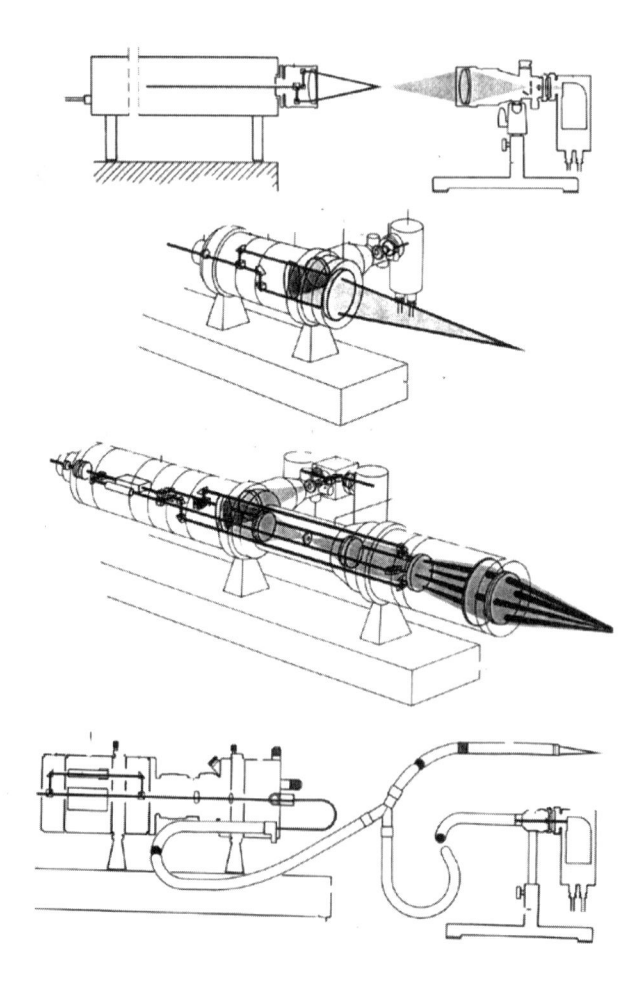

Fig.6-15 LDV configurations incorporated in products. From top to bottom: (i) forward-looking configuration ending on a photomultiplier; (ii) backward configuration; (iii) two-components LDV using a two-lines Ar-ion laser; (iv) fiber optics configuration that connects the laser and the receiver sections to the measuring volume (by courtesy of Dantec Dynamics, Copenhagen).

6.4 Optical Configurations

As we can see in Fig.6-15, we have a forward-looking LDV (first line in Fig.6-15) that uses a He-Ne laser as the source and an S-20 photomultiplier (see Ref.[4], Ch.4) as the detector. This is the simplest and probably the cheapest configuration of LDV. The same arrangement, but designed to collect the back-scattered signal, is shown in the second line of Fig.6-15.

As already pointed out, the forward-scatter configuration collects a larger optical signal compared to the backward-scatter one, but of course can be used only if we are allowed to access the region under measurement from two opposite sides. If we cannot do that, then we shall use the backward-scatter configuration.

The third line illustrates a dual-component (v_x and v_y) measurement of the velocity field, developed with a dual-λ Ar-ion laser emitting at 488 and 514 nm. The beams at the two wavelengths are separated by an interference filter, and then are directed to two beamsplitter-and-mirror combinations arranged along the x- and y-axis. Thus, at the focusing lens, we find two pairs of beams arranged in perpendicular planes. The measurement volume is thus illuminated with a double pattern of fringes, in the planes xz and yz, sampling the components v_x and v_y of velocity.

The same λ-selective operation is carried out in the backward path. Light collected by the objective lens comes to an interference filter oriented at 45 degrees, so that the beam at one wavelength is reflected and the other is transmitted. The detector section is duplicated to sort the velocity signals separately out along the x-axis and the y-axis.

The last line depicts a fiber optics version of the LDV. Here, source and detector are optically delivered to the superposition volume by means of a bundle of optical fibers.

These fibers are usually multimode to provide a large acceptance surface for the returning optical field. This maximizes the collected signal, but may impair the spatial coherence of the beams to be superposed in the measurement region.

A refinement is to use a single-mode fiber for the downlead path, so that the single-mode spatial distribution is preserved, and surround the single-mode fiber with an annulus of multimode fibers, so that collection of scattered field is maximized.

All the instruments described in Fig.6-15 are based on the external configuration of the interferometer (Sect.4.5). However, we can also envision using the other two configurations described in Sect.4.5 as well.

The internal configuration is outside this discussion because of the high attenuation suffered in the back-scattering path, preventing oscillation of the laser.

The self-mixing configuration can be used, and it has been reported in the literature [7,8] as an example of applying the injection concept since the early experiments. Some of the papers reporting a self-mixing velocimeter were actually on what we have called a vibrometer (Sect.4.6). That is, some of them described an instrument providing the v_z component of the target motion (the z-axis being the line-of-sight) and operating on a surface rather than on particles dispersed in a fluid.

To have a true velocimeter (or LDV) measuring the out-of-plane (v_x or v_y) velocity component, we will start with a laser with a good self-mixing effect and add an appropriate optical section. This section will be composed of the usual beamsplitter, mirror, and objective lens arrangement shown in the basic schematic of Fig.6-1. In this configuration, the field returning from the scattering volume is collected by the objective lens and fed into the laser cavity.

Here, the injection generates the usual AM- and FM- induced modulation. If we limit ourselves to detect the easy amplitude modulation component, it will suffice to place a photodetector on the rear mirror of the laser to get the desired LDV signal.

Of course, the self-mixing LDV can only operate in the backward-scatter configuration. Compared to the external interferometer configuration, there is the extra loss of mirror transmittance, but the optical part count is the minimum possible.

REFERENCES

[1] Y. Yeh and H.Z. Cummins, *"Localized Fluid Flow Measurements with a He-Ne Laser"*, Appl. Phys. Lett. vol.4 (1964), pp.176-178.

[2] S. Donati, *"Electro-Optics in the year Y2K"*, web paper at http://ele.unipv.it/~donati.

[3] L.E. Drain, *"The Laser Doppler Technique"*, J. Wiley and Sons: Chichester, 1980.

[4]. S. Donati, *"Photodetectors"*, Prentice Hall: Upper Saddle River, 2000.

[5] J.D. Gaskill, *"Linear Systems, Fourier Transforms and Optics"*, J.Wiley and Sons: New York, 1978, Chapter 7.

[6] M.J. Buckingham, *"Noise in Electronic Devices and Systems"*, Ellis Horwood: Chichester, 1983.

[7] M.J. Rudd, *"A Laser Doppler Velocimeter Employing the Laser as a Mixer-Oscillator"*, Journal of Physics E, vol.1 (1968), pp.723-726.

[8] P.J. de Groot, G. Gallatin, and S.H. Macomber, *"Ranging and Velocimetry Signal in a Backscattering Modulated Laser Diode"*, Applied Optics, vol.27 (1988), pp.4475-4480.

CHAPTER 7

Gyroscopes

The gyroscope is a sensor intended for the measurement of the angular rotation of a mobile vehicle with respect to the inertial frame of the fixed stars.

The rotation is measured with respect to the inertial reference frame, not the laboratory or a generic relative frame, and this gives the gyroscope a special place among sensors. Since the early times of electronic aids to navigation, the gyroscope has been widely used in a range of applications, including inertial navigation, dead reckoning, attitude heading, horizon sensor, etc., which all require the inertial-sensing property.

Before the advent of the laser, mechanical gyroscopes (MG) were employed with success. An MG is based on the measurement of the Coriolis force acting on the axis of a fast-rotating mass. This sensor worked nicely and reached a technical maturity after decades of improvements. It has been used as the only choice for inertial sensing up to about the 1970s and is now abandoned because it is technically obsolete.

The MGs had the following drawbacks: (i) a long (≈ 10 s) switch-on time to the regime speed of rotation, (ii) the dependence of performance and lifetime on fine rotational balancing of the mass, and (iii) a bulky and power-consuming structure.

7.1 OVERVIEW

When the first He-Ne laser was demonstrated in 1961, the idea of reading the Sagnac phase shift in a closed-cavity path was readily recognized as a very promising approach for a new type of gyroscope, the *laser or electro-optical gyroscope*.

Indeed, already in 1962 Macek and Davis [1] were able to demonstrate the detection of the earth rotation by the Sagnac phase shift induced in a square cavity, 1-m by side, made up of He-Ne laser tubes (Fig.1-7). They used what we today call an internal-readout configuration of interferometry (Sect.4.5).

From this encouraging result, a rush started to develop in a short time a compact, well-engineered electro-optical gyroscope with nearly quantum-limited sensitivity for the use in avionics. Contrary to expectation, the road to such a device was paved with obstacles, not only technological but also conceptual.

Indeed, the first experiments unveiled the problem of locking of counterpropagating modes and the consequent washout of signal in the region most important to application, low-speed rotations.

After fifteen years of efforts by the international scientific community, sized in thousands man-year work and hundreds of papers published in scientific journals, the locking problem was finally circumvented [2]. Other technological improvements were also crucial in achieving the low-scatter mirrors and low thermal-expansion cavities required for a good sensitivity performance.

These progresses marked the coming of age of the modern *Ring-Laser-Gyroscope* (RLG), a sensor that since 1975 to 1980 has entered the mass-production stage and has been incorporated in the Inertial Navigation Unit (INU) of any newly fabricated aircraft (Fig.1-8).

The RLG has then rapidly become an undisputed success of electro-optical instrumentation. This is true in terms of scientific sophistication incorporated in a product and also in terms of market — a rich niche totaling about US$1 billion in sales per year since its inception.

In a fitting comparison [3], if the MG is the like a mechanical clock, the RLG is the like an electronic clock, with the advantages and drawbacks that each of them may have.

While the RLG was maturing to a well-understood, producible and high-performance sensor as we know it today, suddenly another approach to the electro-optical gyroscope came out, the FOG (fiber optics gyroscope). Proposed in 1976 by Vali and Shorthill [4], the FOG took advantage of the single-mode fiber, which had just become available at that time, to develop a Sagnac fiber-interferometer based on the external-readout configuration as opposed to the internal-readout of the RLG (Sect.4.5).

Initially, the hint was to exploit a long-fiber cavity to substantially increase the Sagnac signal and hence the sensitivity. As another point of interest, the FOG presented a modular structure promising ease of fabrication, scalability, and low cost, as compared to the RLG.

Like the RLG, the early FOG also started with performance far from those of interest to applications. And again, a decade of research efforts has been necessary to solve basic as well as technological problems [5-8] before the FOG at last has become an established device, though not yet approaching the ultimate sensitivity limit.

7.1 Overview

Progress has been in understanding and curing the small effects disturbing the balance of propagation path lengths in the fiber loop, with the first breakthrough coming after the reciprocity concept was clarified by Ulrich and Johnson [5].

Shortly afterward, specialty fibers with polarization control were developed expressly for use in the FOG. At the same time, the semiconductor source was tailored to a special new device, the super-luminescent LED (or SLED), which matches the low-coherence and high radiance requisites of the FOG fiber interferometer [6-8].

Last, much work was devoted to improve the signal processing so that the FOG could attain a very wide dynamic range without sacrifice of linearity and minimum detectable signal. The matching of measurement process to the optical configuration has been investigated in depth by Lefevre, and concepts of closed-loop and serrodyne [7] operation were developed as the result. Thus, in the late 1980s, the FOG had finally gained technical acceptance, and several companies started production in quantity and deployment in the field.

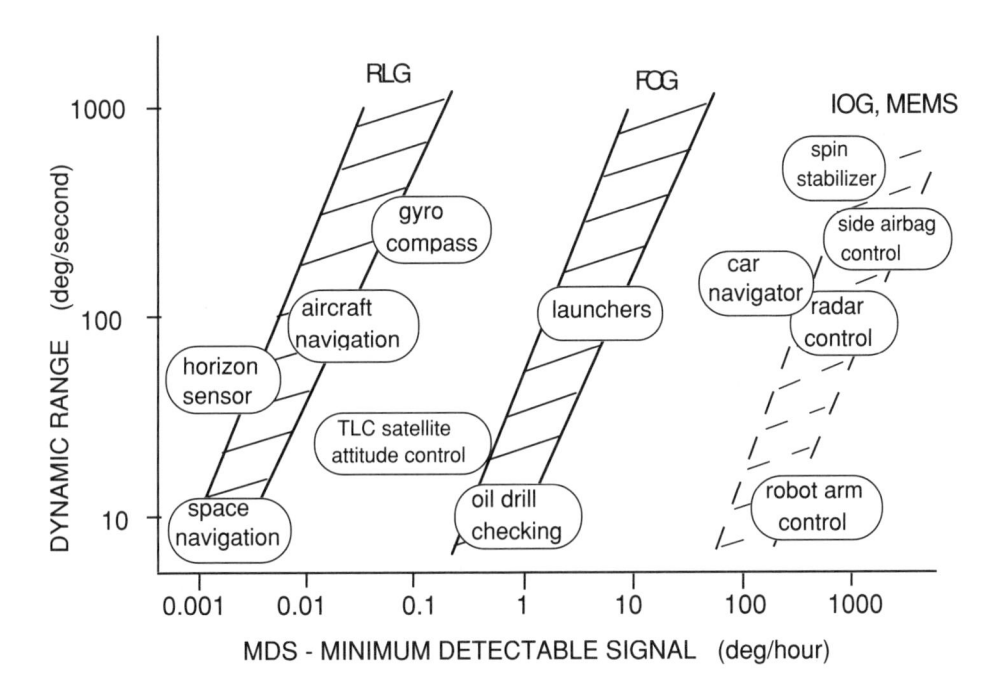

Fig.7-1 Applications of gyroscopes in a nutshell. The main requirements of several applications are indicated in diagram. The scales are: dynamic range of rotation speed (in degree per second), and of minimum detectable signal of rotation speed (in degree per hour). The basic performances are shown for various classes of gyroscopes: RLG (ring laser gyro), FOG (fiber optics gyro) and the emerging IOG (integrated-optics gyro) and MEMS (micro-electro-mechanical-system) technologies.

Many parameters need to be taken into account when we assess a gyro or want to compare different gyros. However, to gain a first-hand insight we may consider just sensitivity, or better minimum detectable signal of rotation (MDS), and dynamic range (maximum signal measured without a serious impairment of linearity), as in the diagram of Fig.7-1.

From Fig.7-1, we can see that the RLG outperforms the FOG by two orders of magnitude, whereas both attain about the same dynamic range. The ratio of dynamic range to MDS is $\approx 10^6$ for the FOG and $\approx 10^8$ for the RLG, which are very respectable figures that qualify the electro-optical gyroscope as a sensor with an unusually wide dynamic range of measurement. The RLG and the FOG are complementary rather than competitive approaches because the former attains a better accuracy while the latter is lightweight, has a larger projected lifetime, has better environmental characteristics, and is potentially cheaper (Table 7-1).

A survey of gyroscope families is presented in Fig.7-2. We will discuss the details of the families in the next sections.

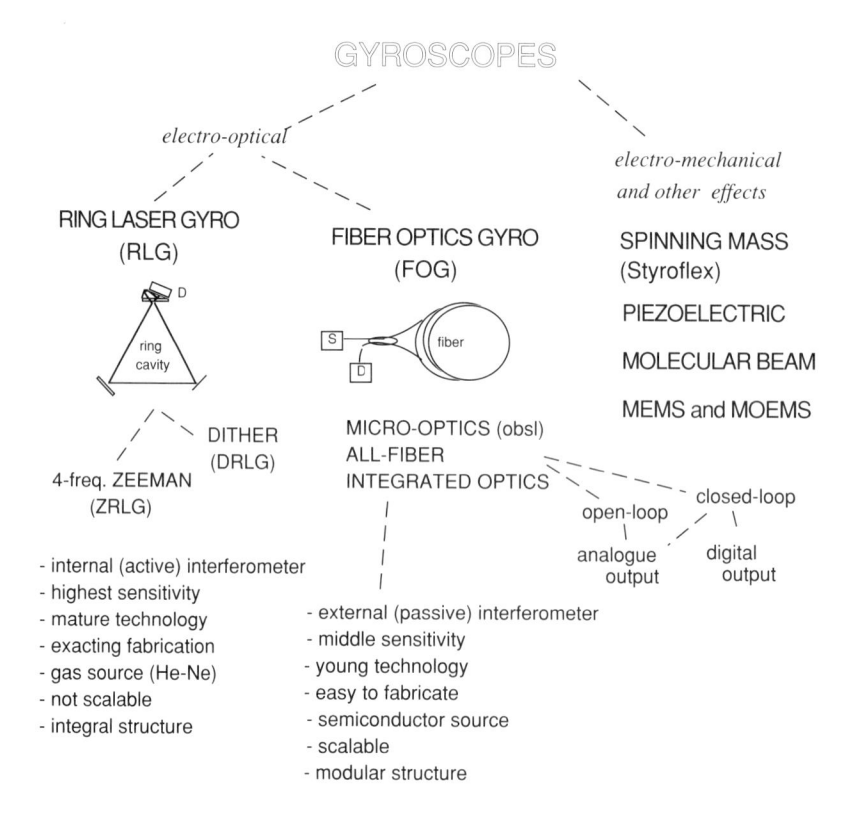

Fig.7-2 Classification of the gyroscope families and highlights of their main technical features. The inertial-grade gyros are the RLG and the FOG, while the spinning mass MG is not used in new designs.

7.1 Overview

In the 1990s, the performances of the standard RLG and FOG gyroscopes continued to improve slowly, thanks to a steady effort of improvement of the fabrication technology. About the FOG, another effort has been devoted to shorten the fiber and to increase responsivity, and the result was the Resonator Fiber Optics Gyro (R-FOG), a device promising to surpass the traditional Sagnac-ring FOG. In the area of active RLG, attempts have dealt with the miniaturization of the laser cavity, using a semiconductor (typically GaAlAs) active waveguide, but no conclusive result has been yet obtained for devices aimed at the inertial grade performance.

In addition, new approaches have been studied for application to the less-demanding robotics and automotive areas. These approaches are aimed at a strong reduction of weight and size, even at a reduced sensitivity (Fig.7-1), which is a specification of applications.
After early attempts to miniaturize the FOG, with the so-called minimum-configuration or 3×3-coupler FOG, the approaches most pursued recently have been the Integrated Optics Gyroscope (IOG), and the MEMS gyro.

The IOG aims to integrate on a suitable substrate, either active like a semiconductor chip or passive like Silica on Silicon (SOS), the functions of either the internal interferometer (RLG) or the external interferometer (FOG). The research efforts are still under way, but IOGs appear promising in the field of robotics and automotive sensors.

The MEMS gyro is more mature, and is a revival of the old mechanical gyro (MG), but fabricated in the modern technology of silicon micromachining, one that allows squeezing the device size down to 0.1 mm.

Table 7-1 Comparison of inertial-grade gyroscopes, RLG and FOG

	RLG gyro	FOG gyro
Sensitivity, or MDS (°/h)	0.001 – 0.01	0.1 – 1
Dynamic range (°/s)	100-1000	100-1000
Scale factor accuracy	10^{-5}	10^{-3}
Typical size (cm)	15 (dia) × 6	8 (dia) × 4
Useful lifetime	typ. of He-Ne laser	typ. of laser diode
Immunity to EM disturbance	high	very high
Immunity to microphonics	high	middle
Optical structure	monolithic	modular
Scalability	poor	good
Manufacturing req'ment	very tight	middle
Clean room fabrication	yes	no
Fab plant cost	very high	middle

MEMS gyros have been already incorporated in commercial car navigators, although performances are marginally adequate, and a sensitivity improvement would be highly desirable. This circumstance stimulates continuing the research in the field.

7.2 THE SAGNAC EFFECT

All electro-optical gyroscopes are based on the Sagnac phase shift. Despite the different interferometer configuration used and the different signal obtained, the sensing mechanism is always the same.
The Sagnac effect is a nonreciprocal effect that takes place when light propagates on a curved path, as in Fig.7-3, and the plane π is subjected to a rotation with angular velocity Ω. Then, a small variation ϕ_S of the optical phase shift is observed, with respect to the normal path length measured at rest. The phase shift ϕ_S is proportional to Ω, sign included. Thus, the phase shift is nonreciprocal, and propagation in the clockwise and counterclockwise directions results in a variation $+\phi_S$ and $-\phi_S$, respectively.

We may evaluate the Sagnac phase shift ϕ_S by reference to Fig.7-3. Classically, let us assume that radiation undergoes the Doppler effect and write $\Delta\lambda/\lambda = v/c$. Changing to the wave number k and adding the vector dependence, we have:

$$\Delta \underline{k}/k = \underline{v}/c \qquad (7.1)$$

where $\underline{v} = \underline{\Omega} \times \underline{r}$ is the tangential velocity. The corresponding optical phase shift for the elemental path $d\underline{s}$ is:

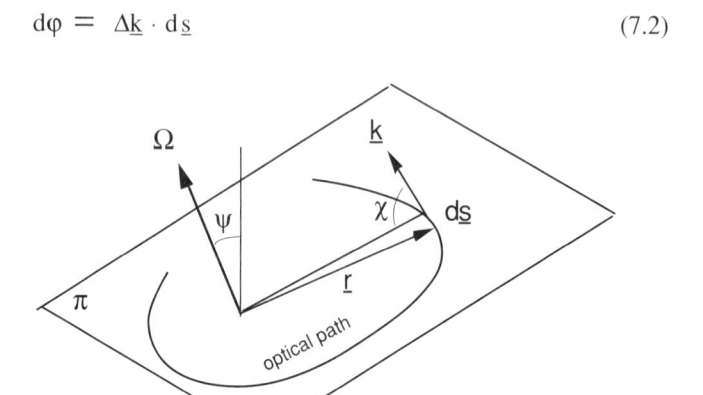

$$d\varphi = \Delta \underline{k} \cdot d\underline{s} \qquad (7.2)$$

Fig.7-3 An optical path schematizing the propagation in a Sagnac-based gyroscope. The path (and $\underline{k}$) is contained in the plane π, and the rotation rate Ω is inclined of ψ respect to the perpendicular to π.

7.2 The Sagnac Effect

Inserting Eq.7.1 in Eq.7.2 yields:

$$d\varphi = k/c \; \underline{v} \cdot d\underline{s} = k/c \; (\underline{\Omega} \times \underline{r}) \cdot d\underline{s}$$

$$= \Omega \; (k/c) \; r \sin \chi \; \cos \psi \; ds \qquad (7.3)$$

The last expression is obtained by developing the vector and scalar products involving angles ψ and χ.

From Eq.7.3, we see that the Sagnac phase shift is sensitive to the rotation component perpendicular to the plane of propagation, or $\Omega_\psi = \Omega \cos \psi$. Keeping this in mind, we omit in the following the factor $\cos \psi$ to simplify notation.

Let us now considering the factor $r \sin \chi \, ds$ in Eq.7.3. From Fig.7-3, we can see that this factor is twice the area of the elemental triangle of base ds and height r. By integrating on a closed loop, we get $\int r \sin \chi \, ds = 2A$, where A is the area enclosed by the optical path. Therefore, we obtain for the Sagnac phase shift:

$$\phi_S = \Omega \; (k/c) \; 2A \qquad (7.4)$$

Noting that the total phase shift Φ_S between the two counterpropagating waves is $2\phi_S$, and substituting $k = 2\pi/\lambda$, we finally get the total phase signal as:

$$\Phi_S = 8\pi \; A\Omega \; /\lambda c \qquad (7.5)$$

In this expression, A is the physical area of the RLG gyro, and the linked area in the FOG, also written as NA to put in evidence the number of turns N of the sensing coil of area A.

7.2.1 The Sagnac Effect and Relativity

Before proceeding, it is appropriate to comment about the method used to derive Eq.7.5. From the derivation, it would be reasonable to assume that wavelength λ and speed of light c in Eq.7.5 are those for the propagation medium. Thus, if n is the index of refraction of the medium, we should use values n times smaller than λ and c in vacuum.

Actually, Eq.7.5 holds, but with λ and c for the vacuum at all times.
We may then wonder what is incorrect in the derivation based on the Doppler effect. The point is that a rotating system is an accelerated reference, for which a rigorous treatment requires the theory of general relativity. Thus, the simple classical vision based on the Doppler effect is not applicable, strictly speaking. It is just an approximation, which is not so bad after all, if it predicts correctly the $8\pi \; A\Omega \; /\lambda c$ dependence, although with λ and c to be interpreted.

The treatment of the Sagnac effect with the general relativity is rather complex, and we refer the interested reader to specific publications [9,10].

However, there exist a simple derivation, based on a conceptual experiment devised by Schultz DuBois [11], which is rigorous from the standpoint of relativity.

The derivation of Schultz DuBois starts with a lossless toroidal cavity (Fig.7-4) with totally reflecting walls and vacuum inside. The cavity schematizes the sensor, whereas an observer integral with the cavity represents the detector.

Now, suppose that light is somehow inserted in the cavity so that counterpropagating waves form a standing wave pattern with a spatial period $\lambda/2$.

At rest, the standing wave pattern is still, and the observer sees no movement of it. If the ring rotates at an angular Ω, the standing wave pattern is again still, because the waves do not exchange energy with the cavity. As the observer rotates integral with the cavity, he sees the standing wave pattern moving. At pattern speed ΩR, the observer sees $N = \Omega R/(\lambda/2)$ periods per second passing under him. He ascribes the movement to a difference of frequencies arisen in the two counterpropagating waves. The frequency difference, just the one observed in an RLG, is:

$$\Delta f_S = 2 \Omega R / \lambda \qquad (7.6)$$

To show the equivalence of this result and Eq.7.5, let us recall that the periodicity of the longitudinal modes (App.A1) of a ring cavity is $\Delta f = c/p = c/2\pi R$, with p being the perimeter. If the two counterpropagating modes are phase shifted by Φ_S, then the corresponding frequency difference Δf_S is related to Φ_S by the proportionality:

$$\Delta f_S / \Phi_S = \Delta f / 2\pi \qquad (7.7)$$

Inserting $\Delta f = 2\pi R/c$ in Eq.7.7 and using Eq.7.6, we obtain:

$$\Phi_S = 2\pi \Delta f_S / \Delta f = 4\pi \ (\Omega R / \lambda)(2\pi R/c) = 8\pi \ \Omega A / \lambda c \qquad (7.8)$$

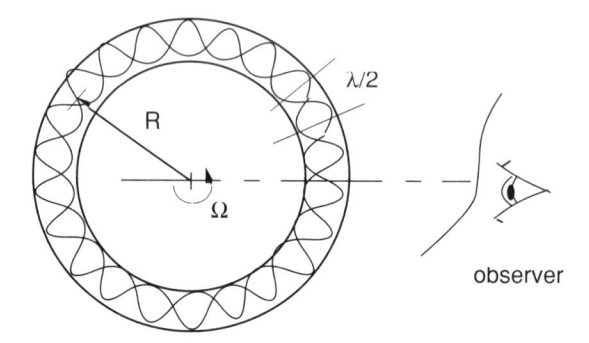

Fig.7-4 In a rigorous derivation of the Sagnac effect, we consider the standing wave formed in a toroidal cavity. An observer rotating at speed Ω sees the pattern running under him at a speed ΩR. Thus, a number of periods $\Omega R/(\lambda/2)$ per second are observed, which is equivalent to a frequency difference.

where we have used the expression $A = \pi R^2$.

In conclusion, Eqs.7.5 and 7.7 are equivalent. We may regard the gyro as a device supplying a phase shift $\Phi_S = R_\Phi \Omega$ or a frequency difference $\Delta f_S = R_F \Omega$, where R_Φ and R_F are the frequency and phase responsivity of the conversion from rotation to optical quantity.

In closing, let us consider that the cavity of the Schultz DuBois ideal ring is filled with a material of index of refraction n.

Now, we have no reason to refuse changing λ to λ/n and c to c/n in Eq.7.5. This increases the responsivity to $n^2 R_\Phi$ (or $n^2 R_F$). This looks correct, if the standing wave pattern is still, because wavelength in the material becomes shorter and we see more periods passing across.

Actually, relativity warns us that, in this case, light is partially dragged along by the material. This is the Fizeau effect [9-12], another effect known since the early times of relativity. The Fizeau effect develops a phase shift similar to the Sagnac and has a responsivity equal to $(n^2-1) R_\Phi$ [12].

Because the drag subtracts fringe crossings, we shall subtract the Fizeau phase shift $(n^2-1) R_\Phi$ from the Sagnac phase shift $n^2 R_\Phi$ to get the total phase shift resulting in the case of propagation through a material. Nicely, the result is $n^2 R_\Phi - (n^2-1) R_\Phi = R_\Phi$, or, it is independent from the index of refraction of the material, as of course it is well verified experimentally.

Other conceptual cases of deviation of the responsivity from the value given by Eqs.7.5 and 7.7 are those of (i) medium moving and detector still, and (ii) medium still and detector moving. From the previous arguments, the responsivity is found $n^2 R_\Phi$ in case (i) and $(n^2-1) R_\Phi$ in case (ii) [12].

In conclusion, we need relativity and not just the classical Doppler effect to treat the gyroscope correctly. When the cavity is filled with a medium, the phase shift Φ_S is still given by Eq.7.5, with λ and c in the vacuum. This is a consequence of the combination of Sagnac and Fizeau effects.

The phase shift Φ_S also changes if the cavity or the observer (the photodetector) is still. This case is of little practical interest, however, because the gyroscope will invariably be used as a strap-down sensor, with all parts integral with the vehicle whose inertial rotation is being measured.

7.2.2 Sagnac Phase Signal and Phase Noise

Let us now discuss the typical values of the Sagnac signal we are going to measure with the gyroscope. The signal is a phase shift in the FOG and a frequency difference in the RLG. Because Φ_S and Δf_S are related through Eq.7.7, we will consider only the phase shift in the following.

Plotting Φ_S against Ω for several values of the total area NA, we obtain the diagram of Fig.7-5. Dotted lines in the diagram are representative of typical RLG and FOG.

As we can see in the diagram, values of the phase shift are very small indeed. A big rotation speed, say $\Omega = 1$ rad/s, already gives a small Φ_S, $\approx 10^{-4}$ rad for the RLG and ≈ 0.1 rad for the FOG. If we go to the earth rotation rate, $\Omega = 15$ deg/h, the corresponding phase shift becomes $\approx 10^{-8}$ rad for the RLG and $\approx 10^{-5}$ rad for the FOG.

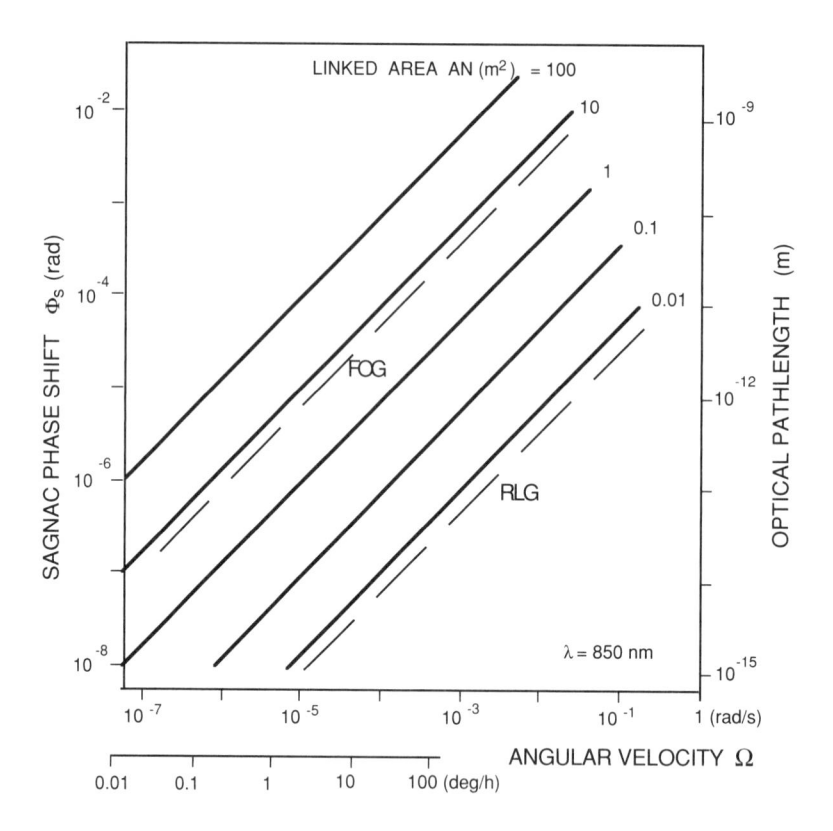

Fig.7-5 The Sagnac phase shift Φ_S developed in a gyroscope with linked area NA, as a function of angular velocity Ω. Dashed lines are for typical devices, a FOG using a 200-m length fiber coil, and RLG using a cavity 10-cm by side. On the right-hand scale, the optical path change corresponding to Φ_S is indicated (adapted from [13] with permission)

Important to note, the measurement of these small phase shifts shall include the dc term, because the measurand Ω may well have a very low frequency content that prevents filtering. Thus, the measurement is much more difficult than that of a very small ac phase. We shall therefore devise very robust techniques of phase measurement, and be extremely careful in avoiding even very small bias, either optical or electronic, that can enter in the setup and corrupt the minute phase shift Φ_S.

A point deserving a comment, measuring the previously quoted phase shift corresponds to detecting optical path length changes of the order of pico- and femtometer (or, 10^{-12} m and 10^{-15} m), as we can see in Fig.7-5.

This is a remarkable record, not only for interferometric measurements, but also in the area of instrumentation science.

7.2 The Sagnac Effect

Indeed, to obtain a resolution of $\approx 10^{-15}$m in a RLG with a 0.3-m perimeter requires that we are able resolving 1 part out of $3 \cdot 10^{15}$. This performance is not so far from the sensitivity of the LIGO gravitational antenna (Fig.1-11) and of the recently announced spatial 6-m gyro experiment [14].

Returning to the Sagnac signal, we shall check that the phase noise permits the measurement we want to perform.

The phase noise in interferometric measurement has been already treated in Sect.4.4.1 (see Eqs.4.13 to 4.16 and Fig.4-18). Recalling the result of Eqs.4.15 and 16, written in terms of the noise equivalent phase ϕ_n, and with R=V=1, we have for the phase noise:

$$\phi_n = k \cdot NED = (S/N)^{-1} = (2 \kappa h\nu B/\eta P_0)^{1/2} \qquad (7.9)$$

Here, B is the measurement bandwidth, η is the quantum efficiency of the photodetector, and P_0 is the power received by the photodetector.

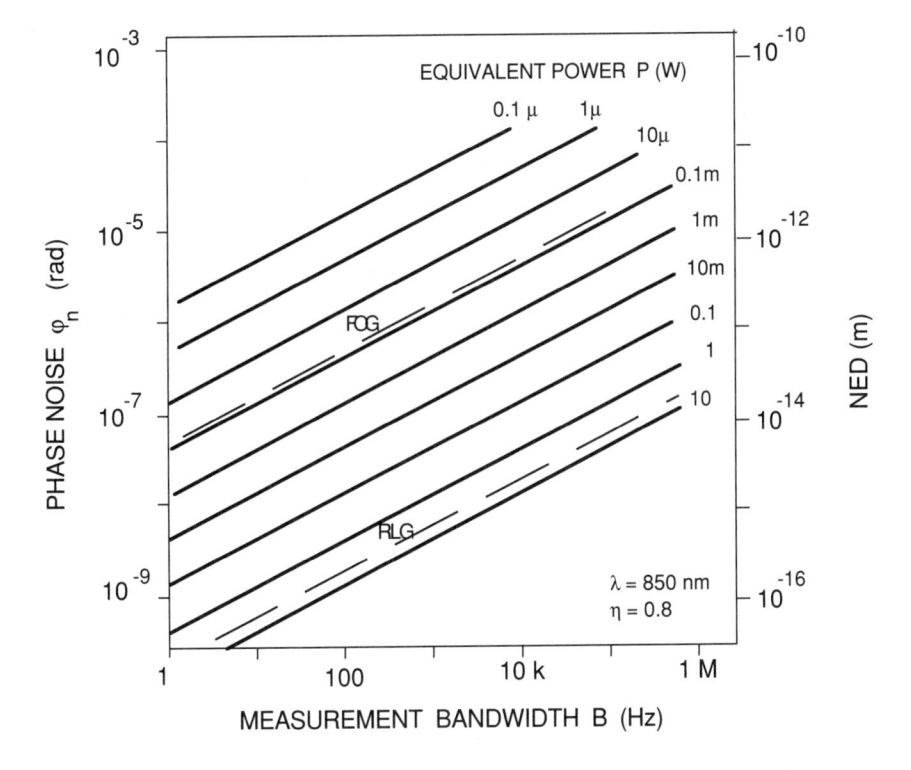

Fig.7-6 Phase noise ϕ_n at the quantum limit plotted as a function of the measurement bandwidth B and for several values of the equivalent power $\eta P_0/\kappa$. Dotted lines indicate the typical vales for RLG and FOG.

The factor κ has been added to be able to write the same expression for the FOG and the RLG. For the FOG, κ=1 from Eq.4.16.
For the RLG, by analyzing the laser line width and translating the result in a phase noise [13], we obtain:

$$\kappa = (1-r)^2/(1+\alpha_{en}^2) \qquad (7.10)$$

where r is the (power) reflectivity of the output mirror, the other two mirrors being assumed ideally reflecting, and α_{en} is the linewidth enhancement factor (see Sect.4.5.2.4).

The phase equivalent noise ϕ_n given by Eq.7.9 is plotted in Fig.7-6 versus bandwidth B and with the equivalent power P=ηP_0/κ as a parameter. Typical equivalent powers of FOG and RLG are indicated by dotted lines in Fig.7-6. In the FOG, the equivalent power is appreciably less than the optical power available from the source (≈mW) because of the attenuation of the Sagnac interferometer and of the components used in it. In the RLG, the small $(1-r)^2$ factor brings up the equivalent power of 3-4 decades, with respect to the ≈mW power level that the source would be able to emit.

Thus, though the RLG starts from a smaller responsivity due to the smaller linked area respect to the FOG, it readily recovers the disadvantage through the much larger equivalent power and ultimately surpasses the FOG in sensitivity by 2-3 decades (Fig.7-6). In addition, the RLG approaches the quantum limit more easily than the FOG because the mirror cavity has much less residual non-idealities than a long fiber made of solid glass.

Comparing Figs.7-5 and 7-6, we can see that there is room to achieve the desired Minimum Detectable Signal (MDS) performance indicated in Sect.7.1 if we are not too far from the quantum limit.

The MDS is so called because it is made up of two terms: the phase noise ϕ_n and the zero-bias error ϕ_{ZB}. The zero-bias error is defined as the phase equivalent of the gyro output when Ω=0. Thus, we have:

$$\text{MDS} = \phi_n + \phi_{ZB} \qquad (7.11)$$

The reason for considering ϕ_{ZB} is because the gyroscope shall work down to the zero-frequency, or dc term, included. Then, the zero-bias error ϕ_{ZB} is an integral part of the minimum detectable signal.
Usually, we will trim the zero-bias error to null with a calibration operation before using the gyroscope. After calibration, ϕ_{ZB} drifts away from null because of residual fluctuations that cannot be controlled, like: dependence from temperature and other parameters, aging effects, switch-on/off, 1/f-noise components, etc.

In practice, while the phase-noise contribution scales with the bandwidth as $\sqrt{B}$ (see Eq.7.9), or with the integration time as $1/\sqrt{T}$ (because B≈1/T), the zero-bias error is about independent from time or slowly increases with T.
Thus, on long integration times, the ultimate performance of the gyroscope is likely be limited by ϕ_{ZB} rather than ϕ_n.

7.3 BASIC CONFIGURATIONS OF GYROSCOPES

The basic schemes of RLG and FOG gyroscopes are shown in Fig.7-7. In both schemes, we find a propagation path, a triangle cavity in the RLG and a coiled fiber in the FOG. In the path, two counterpropagating waves CW (clockwise) and CCW (counter-clockwise) cumulate a phase difference Φ_S because of the Sagnac effect. At the output, with the prism and mirror combination in the RLG and the beamsplitter BS in the FOG, the CW and CCW waves are superposed onto a photodetector. The signal output from the photodetector carries the Sagnac phase Φ_S.

There are significant differences however, in the way the two configurations treat the phase signal, as already discussed in Section 4.5. The RLG is an active, internal configuration interferometer providing a frequency signal in response to a path length (or phase Φ_S) difference, whereas the FOG is an external configuration interferometer providing a phase modulated, baseband signal in response to Φ_S.

In the RLG, the readout of the cavity length is performed through the frequency difference of two counterpropagating modes of the triangular mirror cavity.
When the gyroscope is at rest ($\Omega=0$), because of resonant condition met by the laser oscillation, we find an exact number of wavelengths in the perimeter p of the cavity, and the CW and CCW frequencies are equal.
When the gyroscope rotates, the Sagnac effect changes the perimeter length seen in the CW and CCW direction, by $\Delta p=+\phi_S/k$ and $\Delta p=-\phi_S/k$, with $k=2\pi/\lambda$ being the wave number.
Accordingly, one frequency (f_{CCW}) increases, and the other (f_{CW}) decreases.
By differentiation of the resonance condition $kp=(2\pi f/c)p=$const., we get $f\Delta p + p\Delta f = 0$ for the frequency difference $\Delta f_S = |f_{CCW}-f_{CW}|$. By inserting $\Delta p=\phi_S/k$ in $\Delta p/p=\Delta f_S/f$, we obtain $\Delta f_S = (c/p)\phi_S/2\pi$, which is equivalent to Eq.7.7.

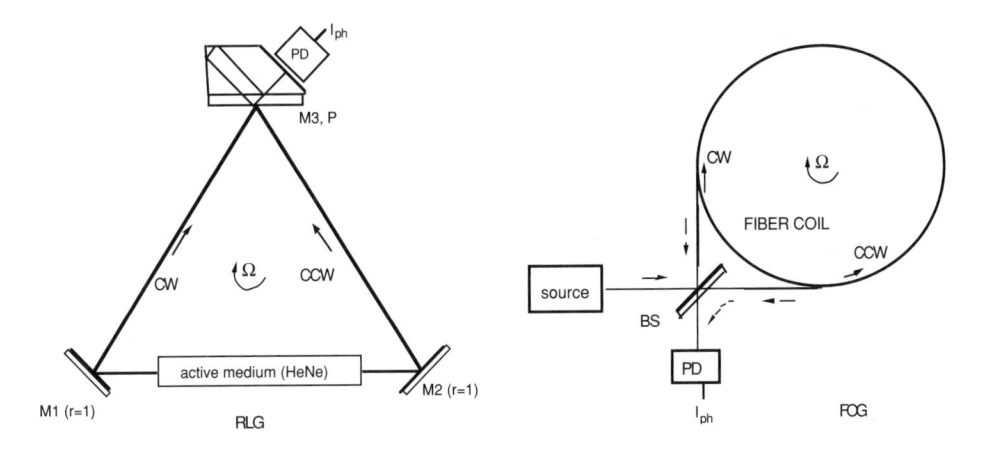

Fig.7-7 Basic configurations of the RLG (left) and of the FOG (right)

Using $\phi_S = 4\pi A\Omega/\lambda c$, this expression can also be written $\Delta f_S = 2(A/p)\Omega/\lambda = 2R\Omega/\lambda$, where $R=A/p$ is apothem of the triangle.

If we now recombine the CW and CCW waves at one of the cavity mirrors (M3 in Fig.7-7) by the prism arrangement, we get a photodetected signal given by:

$$I = I_0 [1 + \cos 2\pi \Delta f_S t] =$$
$$= I_0 [1 + \cos 2\pi \phi_S(c/p)t] = I_0 [1 + \cos 2\pi(2R\Omega/\lambda)t] \qquad (7.12)$$

Thus, the RLG supplies the Sagnac signal in form of a sinusoid whose frequency is proportional to the angular velocity of rotation Ω. This is an advantageous feature because frequency is measured easily and accurately by standard electronic circuits on a very wide dynamic range. In addition, the frequency signal $\Delta f_S = R_F \Omega$ has a scale factor $R_F = 2R/\lambda$ that only depends on size R and wavelength λ and can be known with a very good precision and without the need of a calibration.

The scale factor is large, too. For example, a gyroscope with R=50 mm λ=1µm, has $R_F = 10^5$ Hz/(rad/s), and a rotation of 1 deg/h (or 4.7 µrad/s) is translated into a 0.47-Hz signal. By considering a measurement time interval of, say, 1000 s, the minimum observable frequency is 10^{-3} Hz, which corresponds to a minimum measurable rotation speed Ω_{min}=0.002 deg/h. On the opposite end, the maximum measurable frequency is that corresponding to the frequency spacing Δf_{Smax}= c/p of the longitudinal modes. Using p= 30 cm, we get Δf_{Smax}= 1GHz or also $\Omega_{max}=10^4$ rad/s, a very large theoretical limit.

A final advantage is that frequency is integrated without any low-frequency cutoff simply by counting the periods contained in the photodetected signal. This allows us to obtain numerically the quantity $\int_{0-t} \Omega \, dt = \Psi$, that is, the angle of (inertial) rotation Ψ of the gyroscope. The minimum angle of rotation that can be resolved with an RLG operating in the digital regime of signal processing is the angle Ψ_{min} corresponding to 1 count, or $\int_{0-t} \Delta f_S \, dt = 1$. Using the expression $R_F \Omega$ for Δf_S, we get for the minimum resolved angle:

$$\Psi_{min} = 1/R_F = \lambda/2R \qquad (7.13)$$

Note that Ψ_{min} is the same as the diffraction-angle from the gyro aperture 2R, with a typical value of 10^{-5}= 2 arc-sec with the numbers of the example. This value is of interest for attitude heading and gyrocompass applications in avionics [15].

Up to this point, we have avoided considering the pitfalls of the basic scheme of the RLG. Now, it is the right place to discuss them.

First, the signal supplied by the photodetector (Eq.7.12) is a sine or cosine function of a frequency proportional to Ω. If Ω changes to -Ω, the Sagnac phase ϕ_S of each CW and CCW wave changes sign, and formally the frequency Δf_S does, too. As we cannot distinguish between positive and negative frequency, the sign of Ω is lost.

We may think of solving the sign problem by adding somehow a frequency-difference bias Δf_{bias} in the mode oscillations, to have $\Delta f_{bias} \pm \Delta f_S$ for $\pm \Omega$. This approach does not work, however, because the stability required to Δf_{bias} is far beyond practical feasibility.

7.3 Basic Configurations of Gyroscopes

Another serious problem is that of signal washout. The signal Δf_S is observed to disappear at small angular rotations Ω, and the reason for that is the frequency locking of the counterpropagating modes.

The locking phenomenon is inherent to the internal configuration of interferometry, as already pointed out in Sect.4.5.1, and is due to the even very minute coupling of the two oscillating modes. Any phenomenon producing an exchange of energy between the modes contributes to coupling. The most important sources of coupling are the scattering at the mirror surfaces and in the medium filling the cavity and the cross-saturation of gain provided by the active medium.

Let us briefly describe the locking regimes by considering two modes oscillating with an unperturbed frequency separation Δf_0. When coupling is negligible, we observe an unperturbed sinusoidal waveform of beating at frequency Δf_0. This is the unperturbed regime. Then, increasing the coupling, we first observe frequency attraction of the two modes, whose separation becomes $\Delta f < \Delta f_0$, and the waveform of beating is a distorted sinusoid. This is the injection modulation region of Fig.4-21. By increasing the coupling further, attraction and distortion effects become very strong, and we finally observe the two modes collapse in one and the beating waveform disappear. This is the locking range or dead band in Fig.4-21.

An analysis of the locking regime can be carried out with the Lamb (or the Lang and Kobayashi) equations, written for the two modes and modified to include the coupling term (see Ref.[16], Sect.8.4). Solving these equations for the phase φ_1 and φ_2 of the modes, and with the notation used in Sect.4.5.3.4, the result is the following, known as Adler's locking equation:

$$(d/dt) (\varphi_1 - \varphi_2) = \Delta f_0 [1 + K \sin (\varphi_1 - \varphi_2 + \zeta)] \qquad (7.14)$$

Here, K is the coupling factor. For K<<1, the oscillation frequency is Δf_0 and the beating waveform is sinusoidal. As K approaches unity, the beating waveform is more and more distorted [16]. At K≥1, we get locking because the solution of Eq.7.14 is $(d/dt)(\varphi_1-\varphi_2)=0$, or the phases of the two modes are locked at $(\varphi_1-\varphi_2)$=const. For modes of equal intensity, the expression of the coupling factor is [16]:

$$K = a_c (c/p)/\Delta f_0 \qquad (7.15)$$

where a_c is the (field) coupling coefficient between the modes. Eq.7.15 reveals why the locking problem is so hard. No matter how small a_c is, if the modes are closer and closer in frequency (that is, Δf_0 is small), we will ultimately get a large K.

Of course, to cure the locking problem, we will start trying to decrease coupling as much as possible. The most serious source of coupling is scattering at the mirrors, caused by minute residual irregularity of the surface. Scattering is described by the mirror diffusion coefficient δ, the fraction of power diffused according to Lambert cosine law [17] instead of being reflected. In a normal multilayer interference mirror, we may have $\delta \approx 0.3$-1%, but, with a special manufacture for the gyro application, we may achieve $\delta \approx 10^{-5}$. The coupling coefficient is easily related to δ using the acceptance invariance [17]. This yields $a_c^2 = \delta \omega / \pi$, where ω is the solid angle associated with the oscillating mode, $\omega = \pi (\lambda / \pi w_0)^2$.

Collecting the terms, we have:

$$a_c = (\lambda/\pi w_0)\sqrt{\delta} \qquad (7.16)$$

Using the condition K<1 in Eq.7.15, we get the minimum frequency separation that prevents locking as $\Delta f_0 < a_c(c/p) = (c/p)(\lambda/\pi w_0)\sqrt{\delta}$.

We can appreciate by a numerical example the technological effort in reducing δ, one of the key issues leading to the success of the RLG. Taking $\lambda=1\mu m$ and $w_0= 1$ mm and a normal mirror with $\delta=0.01$, we have the typical value $a_c = 3\cdot 10^{-5}$. The corresponding delocking frequency is $\Delta f_0 < 3\cdot 10^{-5} \cdot 1$ GHz = 3 kHz, which corresponds to a dead-band rotation speed of $\Omega_{db} = \Delta f_0/R_F = 3$ kHz/10^5 = 0.03 rad/s (or, ≈ 1.8 deg/s, a very large value indeed).

When a good low-scatter dielectric mirror is used, for example with $\delta=10^{-5}$, we improve the previous figures by a factor $\sqrt{(0.01/10^{-5})} = 30$, going down to a more reasonable value of dead-band rotation, $\Omega_{db} = 1.8/30$ deg/s ≈ 200 deg/h.

As an illustration, Fig.7-8 shows the typical response curve of the basic scheme of the RLG, that is, one without solution for locking and sign detection.

Last, we can circumvent locking by separating somehow the modes in frequency by more than Δf_0. The practical solutions following this hint, which also add the sign of Ω, will be discussed in Sect.7.4.

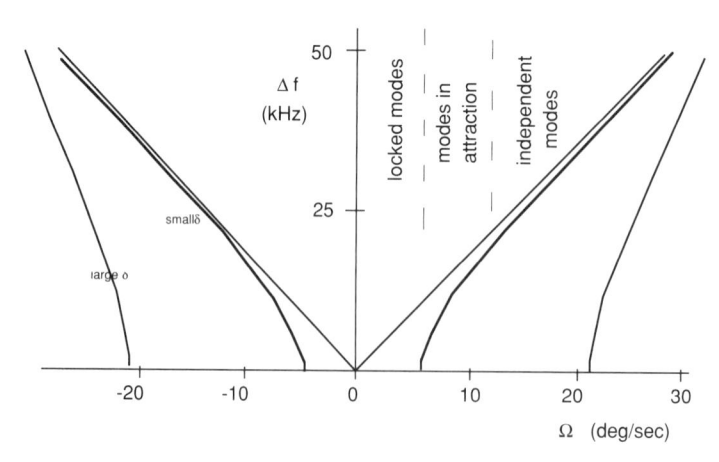

Fig.7-8 The response of the basic scheme of RLG. There is locking dead-band around $\Omega \approx 0$, and no sign for Ω is supplied

Now, let us turn to consider the basic scheme of the FOG (Fig.7-7). The FOG is developed around an external configuration called a Sagnac interferometer (see App.A2), using a relatively long single-mode fiber coil as the propagation medium.

Light from a semiconductor laser source is launched in the fiber pigtails of the coil with the aid of a beamsplitter (BS in Fig.7-7). Two modes, CW and CCW, travel down the fiber in oppo-

7.3 Basic Configurations of Gyroscopes

site directions (full arrows) and, after completing the coil path length, they re-emerge at the beamsplitter (dashed arrows). After recombination of the modes on the photodetector PD, an output signal I_{ph} is generated.

The total phase of each two modes is the sum of the Sagnac phases, $+\phi_S$ and $-\phi_S$, and of the phase shift introduced by the beamsplitter ϕ_{BS}. Indeed, a beamsplitter delays the transmitted beam by $\phi_{BS}=\pi/2$ with respect to the reflected beam [18].

Considering the routes traveled by the CW and CCW modes (Fig.7-9), we can see that the CW is reflected twice by the beamsplitter and suffers no extra delay, while the CCW crosses the beamsplitter twice and suffers an extra delay π.

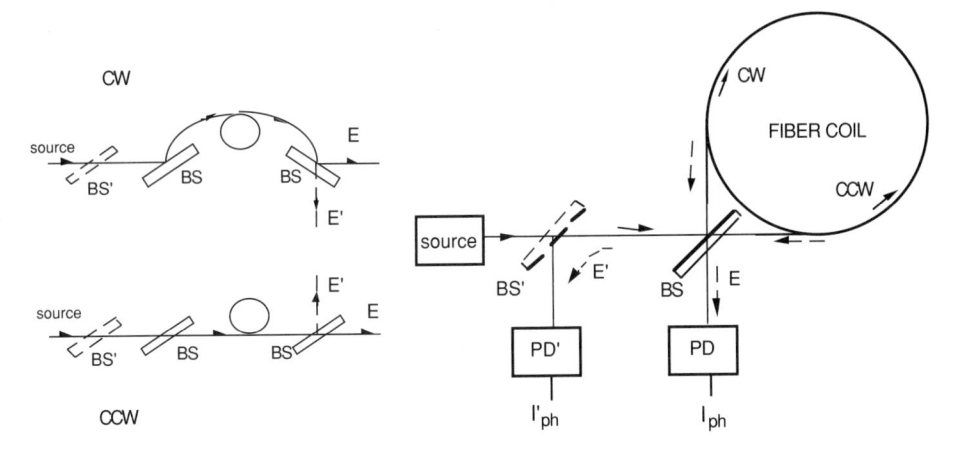

Fig.7-9 In the FOG, using the output E for the recombination of the counter-propagated beams, the BS is passed twice in reflection by the CW beam and twice in transmission by the CCW. To ensure reciprocity, it is better to use output E' in which both the CW and CCW beam make one reflection and one transmission at BS.

Thus, we may write the modes recombined on the photodetector as:

$$E_{CW} = E_0 \exp i(+\phi_S), \quad E_{CCW}=E_0 \exp i(-\phi_S+2\phi_{BS}) \quad (7.17)$$

Accordingly, the detected current is given by:

$$I_{ph} = |E_{CW}+E_{CCW}|^2 = E_0^2[1+\cos(2\phi_S+\pi)] = E_0^2[1-\cos 2\phi_S] \quad (7.18)$$

As we can see from the last term in Eq.7.18, the FOG basic scheme supplies a phase signal $2\phi_S=R_\phi \Omega$ proportional to the rotation speed Ω, with a cosine dependence (Fig.7-10).
The scale factor is given by Eq.7.8 as $R_\phi = 8\pi A/\lambda c$. With the typical values of $\lambda=1\mu m$, and a 300-m long fiber wound on a 8-cm diameter coil, we have $R_\phi \approx 0.5$ rad/(rad/s). This value means that, to read a ≈ 1deg/h rotation speed, we get a phase variation $2\phi_S \approx 1\mu$rad and a very small signal, $\Delta I_{ph}/I_{ph} \approx 10^{-6}$, in the point of maximum sensitivity of the cosine.

The point of maximum sensitivity cannot be reached so easily. In the cosine function, it is at a phase $=\pm\pi/2$, not 0, and we should be able to add a phase bias Φ_{bias} exactly equal to $\pi/2$ to reach the condition. However, the requirement of a bias $\Phi_{bias}=\pi/2$ with a accuracy of the order of the μrad is far beyond the practical feasibility.

In other words, with the cosine-type signal, we face again the problem of a vanishing sensitivity for small $2\phi_S$ and are unable to detect the sign of $2\phi_S$, as seen in Fig.7-10.

Another drawback of the FOG basic scheme (Fig.7-7) is the very noisy signal found when the output is obtained out from the same beamsplitter used to launch in the coil. The excess noise is due to fluctuations of the phase shift ϕ_{BS} introduced by the beamsplitter.

Indeed, a real beamsplitter gives a phase shift $\phi_{BS}=\pi/2+\varepsilon$, where $\varepsilon=1-(t+r)$ is the loss, given by the complement to one of transmission plus reflection [18]. As ϕ_{BS} directly impacts the signal (see Eq.7.17), even very minute fluctuations of ε (down to a few part per million) are of importance.

To avoid spoiling the ultimate quantum noise performance of the sensor, we shall use a reciprocal configuration for the propagation of CW and CCW waves, as first pointed out by Ulrich and Johnson [5]. To do so, we use the recombined wave E' going toward the source (Fig.7-9), deviated by BS' onto photodetector PD'. Now both CW and CCW waves pass the beamsplitter BS once in transmission and once in reflection, that is, the entire path is seen symmetrically by the counterpropagating waves. The output I'_{ph} is given by:

$$I'_{ph} = E_0^2 [1+ \cos 2\phi_S] \qquad (7.18')$$

The waveform of I'_{ph} is the complement to one of I_{ph} given by Eq.7.18, as shown in Fig.7-10. Unfortunately, the new signal does not help to circumvent the vanishing sensitivity and sign problems, yet it exhibits a much improved noise performance because of reciprocity.

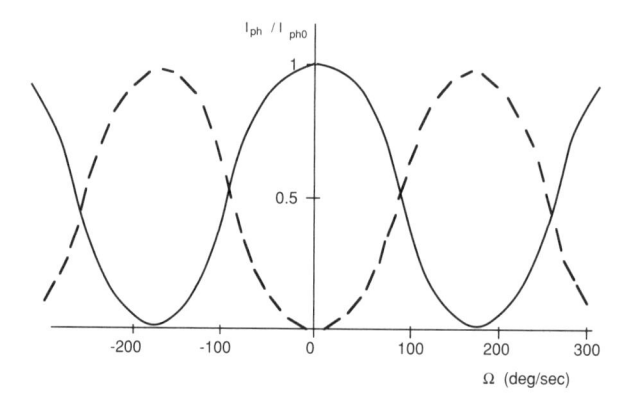

Fig.7-10   The response of the basic FOG scheme is a cosine function of the Sagnac phase, a functional dependence that is nonlinear for small Ω and that can not detect the sign of Ω. Moving from output E to E' only changes the response in its complement to one (dotted line).

In the basic scheme of the FOG, there are several other issues to be improved before a practical configuration is developed. These issues will be discussed in Sect.7.5.

7.4 DEVELOPMENT OF THE RLG

In the previous section, we discussed the working principle of external and internal interferometer configuration for the readout of the Sagnac effect and the problems inherent to both configurations. In this section, we present the solutions devised to develop the RLG concept in a commercial product and outline some technological hints that have been recognized as fundamental to achieve inertial-grade performance.

All practical configurations of the RLG employ a He-Ne ring laser, built around either a triangular or square cavity. The basic elements are shown in Fig.7-11.

The cavity is machined out from a single block of material, like Cervit or Zerodur, which are vitro-ceramic composites with very low thermal expansion ($\approx$ ppm/°C typically).

The block is drilled to obtain the capillary bores for the He Ne mixture, and the edge surfaces are precisely lapped flat to accommodate the three (or four) mirrors of the cavity. The angular placement of the edge surfaces is made to coincide within $\approx$ arc-second to the nominal value required by the cavity. In this way, by leaning the mirrors on the block edges they are already aligned and no further adjustment is necessary.

Mirrors are not required to be cemented to the edge surface. By pressing them against the block, air is squeezed out from the interface (a joint known as anaerobic), and the mirror is kept firmly in place. Just a little stripe of vacuum-grade glue (like Tor-Seal) is used to protect the rim of the mirror and block interface.

The gas is the standard 10:1 He:Ne isotopic mixture using ^{20}Ne as the active atom. Natural Ne is not adequate because it is composed of two isotopes ^{20}Ne and ^{22}Ne with atomic lines slightly offset in frequency. Thus, natural Ne has an atomic line with a secondary peak disturbing frequency stabilization. Using the isotope, a smooth atomic line is obtained, allowing a clean frequency stabilization and multioscillator operation.

The pressure of the gas mixture is the usual 1-mbar value of linear He-He lasers. The discharge characteristic and the supply circuits are those described in Appendix A1.1.2. Typically, about 20 kV will be needed to switch the discharge on, and $V_{ak} \approx$ 1500 V will be the working voltage for an optimal discharge current $I_a \approx$ 5-7 mA. The cathode and the anode of the discharge tube are made by the usual Al-cylinder (a cold cathode for low thermal disturbance) and a W-wire, common to the standard He-Ne laser technology. Two anodes and two cathodes are provided in the block (Fig.7-11), arranged for opposite current flow to cancel the so-called Langmuir flow.

The Langmuir flow is due to the ions flowing from anode to cathode and dragging light because of the Fizeau effect. The Langmuir flow contributes with a $\Omega_{BE} \approx 10°$/h zero-bias error per leg of discharge. Using two opposite discharges, the error is canceled out, provided the currents in the two legs are balanced within a $\approx \mu A$ difference (typical).

Mirrors of the RLG are multilayer interference mirrors manufactured for very low scattering.

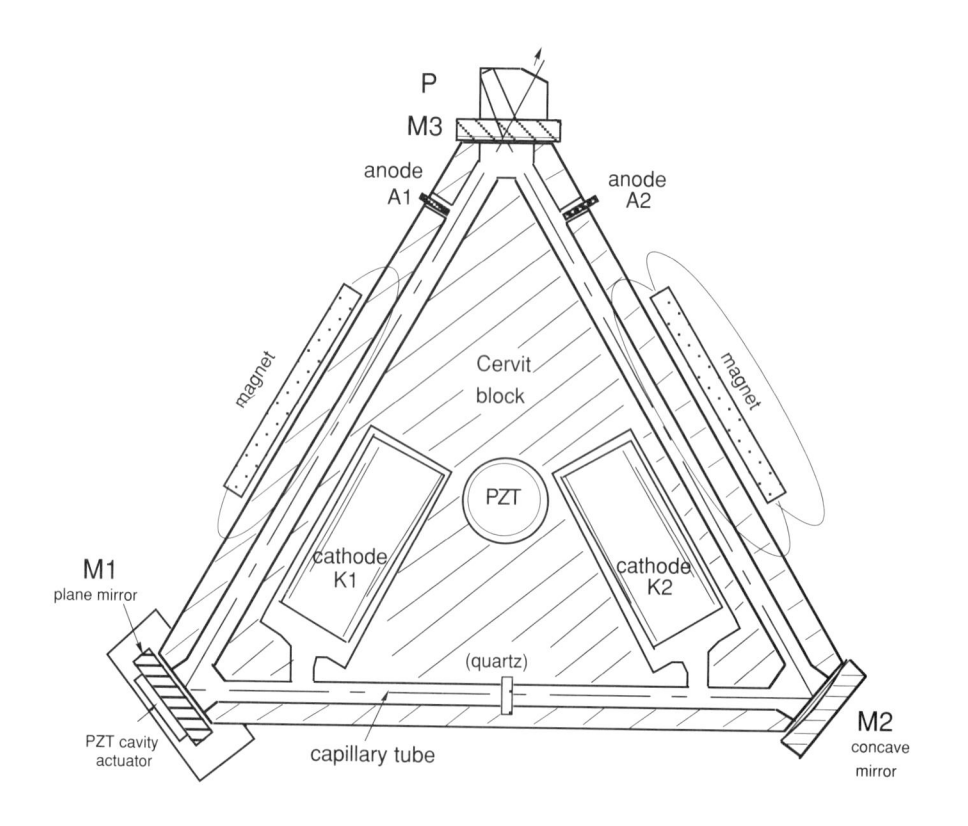

Fig.7-11 Construction elements of a ring laser gyroscope. A Cervit block hosts the capillary bores of the triangle (or eventually square) cavity. Mirrors (two plane and one concave, long radius) are cemented to the block edge surfaces, lapped flat and with the correct angles for cavity alignment. One mirror (M1) is actuated by a piezo PZT disk, and another (M3) is partially transmitting (≈0.5%) for the beam output. Prism P is cemented to M3 and serves for the recombination of CW and CCW mode. Flow discharges (A1-K1 and A2-K2) are arranged opposite to cancel out the Langmuir-flow error. Magnets are used in the ZRLG and a vibrating PZT in the DRLG.

Their curvature radii are chosen to ensure that the cavity stability condition is met and that the mirrors are easy to fabricate. Like with linear He-Ne units, the preferred choice is one long-radius concave mirror and plane mirrors for the remaining elements. The radius of the concave mirror is chosen a few times larger than the perimeter of the cavity, usually 1-3 m.

Of the two plane mirrors, one is maximum reflectivity (in practice 99.999%) and will be the mirror (M1 in Fig7-11) for cavity length control, the other will be the output mirror (M3). Transmission of M3 is chosen in the range $t=1-r=0.3-1\%$ to get an output power of ≈0.2-1 mW out of the cavity. It is not advisable to increase the output power beyond the minimum required

7.4 Development of the RLG

for detection, because output power represents a loss for the cavity and affects ϕ_n (see Eqs.7.9 and 7.10). The concave mirror (M2) is a maximum reflectivity too.

The frequency stabilization scheme may be one of those discussed in Appendix A1.2.
In the Zeeman RLG, it is obvious to choose the Zeeman-splitting as the mechanism of marking the atomic line, whereas, in the dithered RLG, it is better to use the Lamb's dip or the polarization-mode arrangement.

A practical RLG also contains additional items like (i) permanent magnets applying a longitudinal magnetic field of 300 to 500 Gauss, to split the atomic line in two; (ii) a quartz flat, to add a circular birefringence in the cavity; (iii) a PZT (lead-tin-zirconate) ceramic disk, to exert a rotational vibration to the device; (iv) magnetic shields (not shown in Fig.7-11); and (v) one ore more temperature sensors (thermistors) to sense the cavity temperature and correct the zero bias. Items (i) and (ii) are specific of the ZRLG, whereas item (iii) is specific of the DRLG.

A final comment on the cavity shape is worth reporting. The two major companies, which are active worldwide in the field of RLG for avionics and sharing a very wealthy market, use one square and the other triangle cavity (Fig.7-12).
Both shapes have advantages and disadvantages. The triangle cavity saves one mirror respect to the square, and works with a 30-deg incidence angle instead of 45-deg, so scatter is less and the lock-in range is smaller. The square cavity has a better surface to perimeter ratio A/p, and thus a larger responsivity (Eq.7.13) at equal size. Also, a square cavity is aligned more easily than a triangle, but it uses four mirrors instead of three.

In good substance, all these differences are minor, but they are good enough to claim the superiority of one choice over the other. So, patents on both cavity shapes have been filed and accepted, and were apparently expected to waive one company from the other's charge of infringement. Actually, a lawsuit filed by one big manufacturer against the other big manufacturer, known in the field as the '*gyro dispute*' has lasted ten years in the United States, and was recently concluded with a compensation of about $500 million [19].

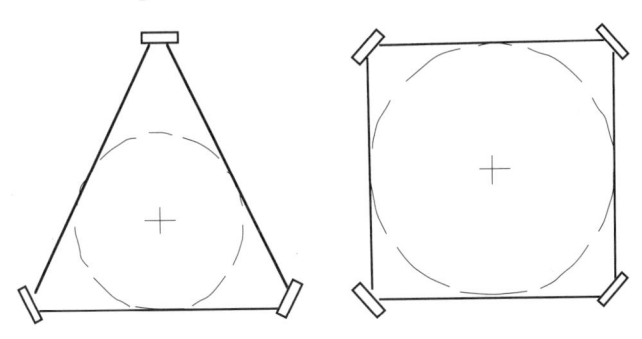

Fig.7-12 Both triangle and square cavities are used in the RLG by different manufacturers. There are minor differences in the two. The triangle saves a mirror and has less scattering, but the square has a better responsivity at equal size and alignment is easier.

7.4.1 The Dithered Laser Gyro

In the Dithered RLG (DRLG), we circumvent the problem of the lock-in deadband by applying an external small angular rotation to the device. The rotation is supplied by a piezoceramic actuator mounted in the middle of the Cervit block (Fig.7-11). The piezo may be a single piece of ceramic, vibrating in the shear tangential mode (Fig.7-13 top) or a bimorph composite with the same function. The vibration consists in an angular movement of the top surface, integral with the gyro, with respect to the bottom surface, and integral with the reference body of the vehicle to which the gyro is strapped down.

The applied angular velocity Ω_d is such that the device exits from the deadband at each period of vibration (Fig.7-13 bottom), whence the name of dither given to the technique.

The required angular amplitude Φ_0 of vibration supplied by the piezoceramic need not be so large to unlock the gyro.

Indeed, if we use a reasonable frequency of excitation ω_0 (say ≈10 Hz), and thus have the instantaneous rotation $\Phi = \Phi_0 \sin \omega_0 t$, the corresponding angular velocity is $\Omega_d = d\Phi/dt = \omega_0 \Phi_0 \cos\omega_0 t$. Even with a small $\Phi_0 = 1$ mrad, and $\omega_0 = 60$ rad/s, we get an amplitude of rotation speed $\Omega_{d0} = \omega_0 \Phi_0 = 0.06$ rad/s, which is largely in excess of the normal locking range (Sect.7.3).

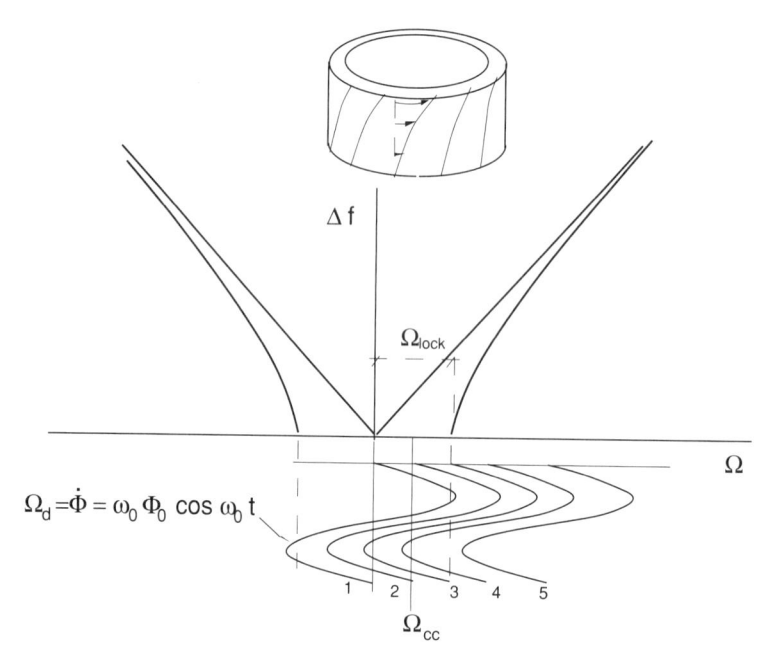

Fig.7-13 A ceramic tube vibrating in the tangential shear mode (top) is used to dither the RLG out of the locking deadband (bottom). Even small vibration amplitude Φ_0 (≈ mrad) is enough to unlock the gyro at audio frequency ω_0.

7.4 Development of the RLG

The reason for a high frequency of dither is that ω_0 should be larger than the highest frequency contained in the signal. Otherwise, beating may occur, and the dithered signal may disappear. However, as the power required to actuate the gyro is $(1/2)I_{RLG}\omega_0^2$, where I_{RLG} is the momentum of inertia of the gyro, we shall limit ω_0 to the smallest value necessary.

Processing of the output signal of the DRLG follows the principle illustrated in Fig.7-14. We take the measurement of mode beating Δf for each semiperiod of the dither waveform $\Omega_d = \Omega_{d0}\cos\omega_0 t$, let us call these signals C_1 (for $\Omega_d > 0$) and C_2 (for $\Omega_d < 0$).

First, we determine the sign of Ω by looking at which semi-period has the largest signal. When $C_1 > C_2$, Ω has the same direction of the dither rotation Ω_d (let's say $\Omega > 0$), whereas for $C_1 < C_2$, Ω has the opposite direction of Ω_d (and $\Omega < 0$).

Second, let us consider the dependence of C_1 and C_2 from the baseline angular velocity Ω_{cc}, the signal now representing the relatively slow rotation velocity to be measured.

We can see that, as Ω_{cc} increases, C_1 increases and C_2 decreases. For a small Ω_{cc} internal to the lock-in range (cases 1 and 2 in Fig.7-14), a good approximation to Ω_{cc} is the signal difference $C_1 - C_2$. When Ω_{cc} increases further and is external to the lock-in range (cases 3 and 4 in Fig.7-14), it finally converges to the average of C_1 and C_2, and we take their sum $C_1 + C_2$ as the approximation to Ω_{cc}.

The break point between difference and sum is best placed at the amplitude marked 3 in Fig. 7-14, that is, where about half the dither peak comes out of the locking region.

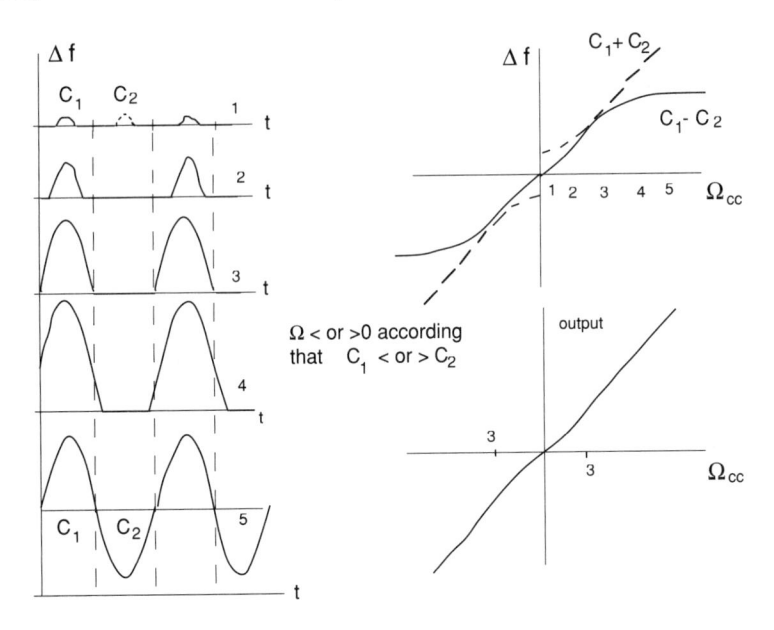

Fig.7-14 In the semiperiods of the dither drive $\cos\omega_0 t$, the outputs C_1 and C_2 vary with the dc baseline Ω_{cc} as shown at left. By computing sum and difference of these signals, we get a good approximation for Ω_{cc} (right).

This condition ($\Omega_{d0} \approx \Omega_{lock}$) corresponds also to the minimum nonlinearity error of the transfer characteristics (Fig.7-14 right).

To correct the residual nonlinearity error, we may store the dependence of Ω_{cc} from the sum and difference signals in a look-up table and use the measured values of C_1-C_2 and C_1+C_2 to compute Δf and then Ω_{cc}.

The dither approach may appear disagreeable to the researcher because we start with an inherently static structure (the electro-optical gyroscope) and end up forced to apply a mechanical movement to have it work. The objection is founded, we shall admit. However, in engineering, we should never care about the beauty of our solutions. The best solution is definitely the one that works better, and this should be the concept of beauty for the engineer. We should not prefer a conceptually appealing approach to one working better, because this could simply result in a loss for our company.

In actual products, manufacturers use the dither concept to realize the RLG with preference respect to the more appealing ZRLG, because the DRLG is cheaper, uses fewer components, and is easier to fabricate.

7.4.2 The Ring Zeeman Laser Gyro

The Zeeman RLG (ZRLG) circumvents the lock-in problem in a very elegant way. It makes use of the combined effect of a double splitting of the oscillating modes to produce in the cavity four oscillations well separated in frequency [20,21]. The four oscillations are circularly polarized with a suitable order and are obtained by introducing appropriate circular birefringence [22] in the cavity.

Recalling circular birefringence. Circular birefringence is also called rotatory power and is described by rotation $\Psi=\rho L$ of the polarization plane of a linearly polarized wave propagated through a length L, where ρ is the rotatory power of the material. A material with $\rho>0$ rotates the polarization plane of a positive angle, or counterclockwise, and is called levogyrate, whereas for $\rho<0$ the material is dextrogyrate and rotates a linear polarization in the clockwise direction.

We can attribute a rotating sense to the material because the rotatory power is independent from the sense along which we look through the material. Thus, if we take a flat of quartz as in Fig.7-11, we get the same rotation $\Psi=\rho L$ for both the left-bound and the right-bound wave passing through it.

This circumstance is also described by saying that the rotatory power of crystals is *reciprocal*. Usually, effects are reciprocal and, for example, no effect violating reciprocity is known in transmission optics or for linear birefringence. The only example of nonreciprocity in optics is the rotatory power associated to the Faraday or Zeeman effect.

The Faraday circular birefringence is again described by angle $\Psi=\rho L$, but now we have $\rho=V\underline{H}\cdot\underline{\xi}$, where $\underline{\xi}$ is the unity vector associated with the propagation path through the material, and $\underline{H}$ is the vector of magnetic field intensity. Now, the material is levogyrate or dextrogyrate according to the sense we cross it (the sign of $\underline{H}\cdot\underline{\xi}$), or we deal with a *nonreciprocal* circular birefringence.

Another equivalent description of rotatory power is when considering circular polarization states.

7.4 Development of the RLG

Along a path with rotatory power, the LC and RC (left- and right-circular) states become phase shifted of $+\Psi/2$ and $-\Psi/2$, respectively. Thus, we have a circular birefringence with the two LC and RC states.

Now, this is the crucial point for the ZRLG working principle. When we have nonreciprocal circular birefringence, the path length is different for the two modes (in the same state) propagating in the $+\xi$ and $-\xi$ directions. Thus, we have the chance of distinguishing between the CW and CCW waves and will use the nonreciprocal birefringence effect to separate the modes in frequency, finally defeating the mode locking.

As a remark, the Faraday effect is strictly about the circular birefringence case, whereas the Zeeman effect is the splitting of the atomic line in two lines, one for the LC and one for the RC. It can be shown [23] that the tails of the lines split by the Zeeman effect are responsible, off resonance, for just the extra path length described by the Faraday birefringence.

Now, going to the schematic of Fig.7-11, the non-reciprocal circular birefringence (or Zeeman effect) is provided by the means of the permanent magnets. These are designed to apply a modest field ($\approx$200 to 500 Gauss) to a portion ($\approx$5 to 10 cm) of the capillary tube.

The reciprocal circular birefringence is provided by the means of the quartz flat. Normally, a $\approx$1-mm thick flat is used, for a zero-order delay of about $\approx$45 deg, the exact value being not so critical.

Because of the Zeeman effect, the atomic line is split in two lines (Fig. 7-15b) separated by $\approx$200 MHz, each sustaining a determined polarization state. As discussed in Appendix A1-2, when the magnetic field is applied longitudinally, the lines sustain circular polarization states, for example the LC of the lower-frequency line, and the RC of the upper-frequency. Thus, two modes with LC and RC polarization should be able to oscillate.

If we assume that the cavity line is somehow kept centered at the middle of the unperturbed atomic line, we should expect that oscillation is located right there in frequency, at v_0 for both modes.

Actually, because of the pulling effect (App.A1.1), each polarization mode is slightly offset toward the center of the respective atomic line (Fig.7-15b). The frequency pulling Δv_a depends on the cavity line width and on the detuning from the atomic line center [24]. With no detuning and with a typical Zeeman splitting of $\approx$200 MHz, we may have frequency pulling in the range $\Delta v_a \approx$20-50 MHz.

In these conditions, the ring cavity sustains two oscillations in opposite propagation directions, and each oscillation is degenerate in polarization, or it is double.

Indeed, in the CCW direction (Fig.7-15b), the oscillation with the LC polarization is at $v<v_0$ and that with the RC polarization is at $v>v_0$. The oscillation propagating in the CW direction has the LC mode at $v>v_0$ and the RC mode at $v<v_0$, contrary to the CCW.

This is no surprise, however, because a RC polarization propagating in the CCW direction is the same as a LC polarization propagating in the opposite CW direction, whatever is the convention used to define the handiness of a circular polarization. Because of this, the CCW-RC and CW-LC share the same population of atoms split by the Zeeman effect and oscillate at the same frequency (Fig.7-15b).

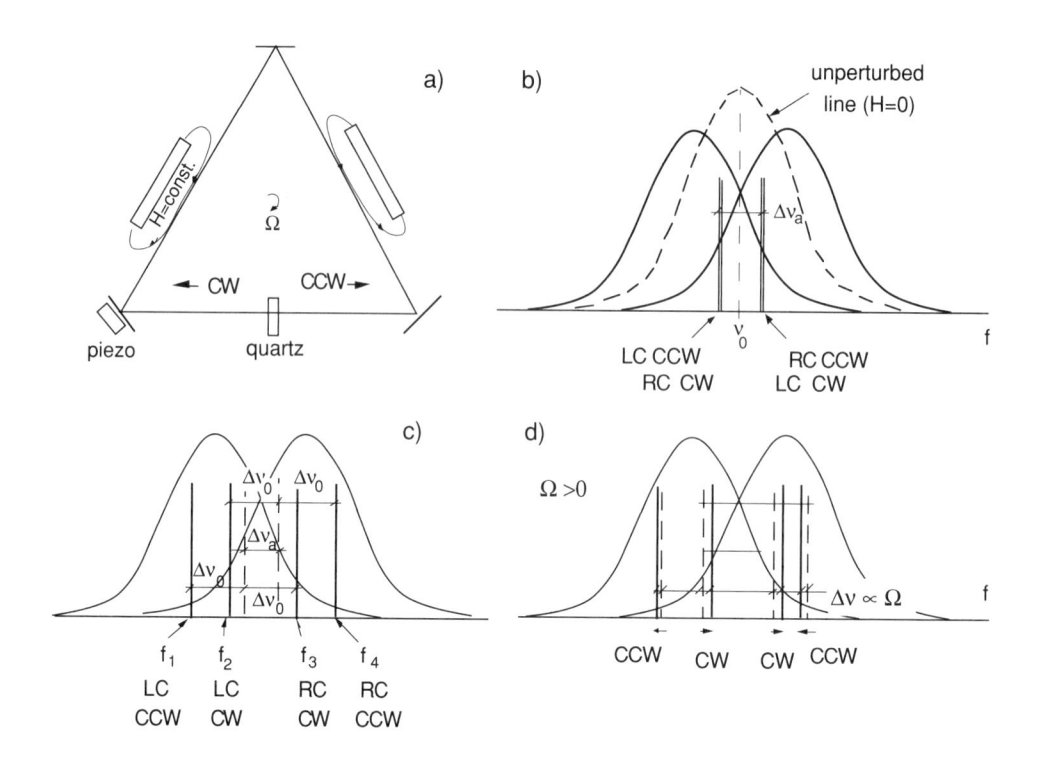

Fig.7-15 In the ZRLG, we get four oscillating modes by introducing (a) a reciprocal circular birefringence (by the quartz flat) and a nonreciprocal circular birefringence (by the magnetic field). (b) The atomic line (dotted line) is split in two by the Zeeman effect produced by the axial magnetic field, and there is one line for LC (left circular) and one for the RC (right circular) polarization (dotted lines). Because of frequency-pulling effect, the oscillations are moved apart of a frequency Δv_a, toward their atomic line center. Introducing the quartz flat (c), the LC and RC waves are split further by Δv_0. When the gyro rotates (d), frequency of the CW waves increases, and the frequency of the CCW decreases. Thus, f_1-f_2 and f_3-f_4 carry a frequency difference Δv proportional to Ω and allow recovering its sign.

Now, consider the effect of adding the quartz plate in the cavity, and with it a circular birefringence Δl, of a fraction of wavelength. Because of the birefringence, the ring cavity perimeter is seen to increase for one mode (say, the LC) and to decrease for the other mode (the RC). The quartz birefringence is reciprocal, so both the CW and CCW modes undergo the same Δl.

As consequence of the added Δl, to maintain the zero round-trip phase-shift (Barkhausen condition, App.A1.1.1), the oscillation frequency shall now change a little, by -Δv_0 for the LC mode and +Δv_0 for the RC mode (Fig.7-15c).

7.4 Development of the RLG

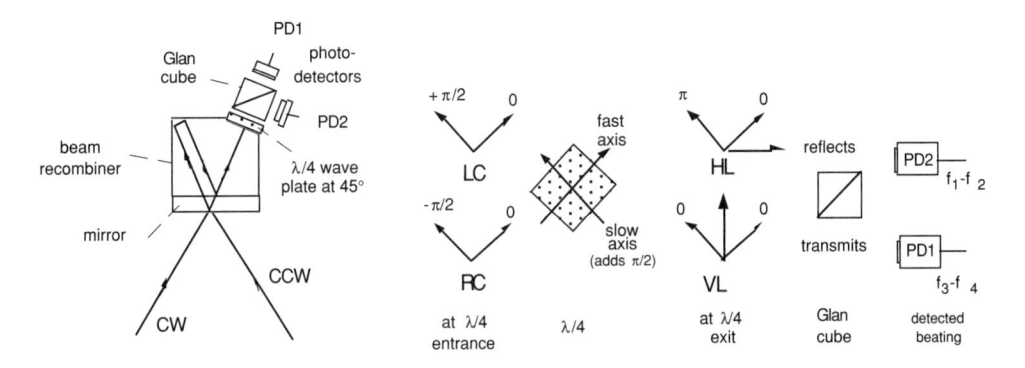

Fig.7-16 At the output mirror, total reflection at the prism corner is used to recombine the CW and CCW beams, each of which contains two LC and two RC polarizations. A quarter-wave plate oriented at 45 degree converts the polarization states to linear (HL and VL), and a Glan-cube polarizing beamsplitter transmits one linear component to PD1 and reflects the other to PD2. At photodetectors, optical signals at different frequency produce a beating, and electrical signals at the frequency difference f_2-f_1 and f_4-f_3 are generated.

The quantity Δv_0 is easily evaluated by recalling that $\Delta v=c/p$ is the frequency variation corresponding to $\Delta l=\lambda/2$, whence in the proportion $\Delta v_0=(2\Delta l/\lambda)(c/p)$.
We choose the quartz birefringence Δl so that $\Delta v_0 \approx 50$-100 MHz, a value less than, but comparable to, the He-Ne atomic line width $\Delta v_{at} \approx 1500$ MHz.

Now, we have four oscillating modes in the cavity, as shown in Fig.7-15c. If splitting is symmetrical, the oscillation frequencies are so arranged that the differences of LC and RC modes are the same, or $|f_1$-$f_2| = |f_3$-$f_4|$ at rest ($\Omega=0$).
In addition, the frequency separation of the four modes is now large enough to prevent frequency attraction or locking between them. This assume of course, that we have accurately cured all coupling effects, going down to a residual locking range of about 1MHz or less.

Worth noting, the scheme lends itself to a method for frequency stabilization. In fact, when the mode pattern is symmetrical around the original atomic line v_0 center, the amplitudes of oscillation E_1, E_2 at frequency f_1, f_2 are equal to amplitudes E_3, E_4 at frequency f_3, f_4. If a cavity drift causes a decrease in frequency of the mode pattern, then E_1, E_2 become larger than E_3, E_4. Thus, the difference $(E_1+E_2)-(E_3+E_4)$ is a good error signal to feed the piezo actuator of the cavity length control (mirror M1 in Fig.7-11).

We may now consider the dynamical mode oscillation when the gyroscope is put into rotation. Let $\Omega>0$ in the reference parallel to magnetic field $\underline{H}$. Then, the CCW waves decrease their frequency of an amount $\Delta v=R_F\Omega/2$ while the CW waves increase theirs of $\Delta v=R_F\Omega/2$, as indicated in Fig.7-15d.

The strategy to compute Ω now comes out easy. We will make the beating of the two CS and CD modes on two photodetectors PD1 and PD2, so to obtain the two signals:

$$I_{PD1} = I_{01}(1+\cos 2\pi F_1 t), \quad I_{PD2} = I_{02}(1+\cos 2\pi F_2 t)$$

where $\quad F_1 = |f_1-f_2|+\Delta\nu, \quad F_2 = |f_3-f_4|-\Delta\nu \qquad (7.19)$

By counting the periods of signal I_{PD1} and I_{PD2}, we can process the signal in digital format. First, we compute the sign of Ω as >0 if $F_1>F_2$, and <0 if $F_1<F_2$. Then, the difference $F_1-F_2 = |f_1-f_2|-|f_3-f_4|+2\Delta\nu$ is the desired Sagnac inertial signal, and we get twice the responsivity of the normal ring (Sect.7.1), or $R_F=4R/\lambda$.

We may easily integrate the angular velocity signal and obtain the angle of rotation Ψ (see also Eq.7.13) by counting the zero crossings of I_{PD1} and I_{PD2} so that we integrate the frequency difference $F_1-F_2=2\Delta\nu$.

From the previous considerations, it can now be appreciated why we wanted to match the mode frequency difference at rest.

Indeed, if it is $|f_1-f_2|=|f_3-f_4|$, and we get $F_1-F_2= 2\Delta\nu$ exactly, whereas if we incur a residual small error $|f_1-f_2|-|f_3-f_4|=\Delta f_{offset}$ in the oscillation frequency pattern, the signal $F_1-F_2= 2\Delta f_S+\Delta f_{offset}$ is affected by an offset Δf_{offset}.

Last, let us describe the beam recombination arrangement (Fig.7-16). The CW and CCW waves, each carrying a LC and RC polarization mode, are brought together at the output mirror like in the basic scheme. A double-reflection 90-deg corner is used to fold one beam (the CCW in Fig.7-16) and to superpose it to the other beam (the CW).

In leaving the prism, the four-mode beam crosses a quarter-wave plate. The plate is oriented (its fast and slow axes) at 45-deg with respect to the incidence plane of the Glan cube that follows. The quarter-wave plate is a birefringence element that introduces a $\lambda/4$ (or 90-deg) phase shift in one linear polarization respect to the other. The result is that circular polarization states are transformed in linear polarization states. For example, the RC becomes a HL (horizontal linear) and the LC becomes a VL (vertical linear), see Fig.7-16.

At the output of the quarter-wave plate, the Glan-cube beamsplitter cube divides the HL and VL components. These components are made available at the transmitted and reflected cube outputs, respectively. Then, two photodetectors convert the HL and VL optical fields into electrical signals.

Note that HL and VL components are correctly made up of the RC (CW plus CCW) and of the LC (CW plus CCW) modes, that is, those that we need to supply the F_1 and F_2 signals.

7.4.3 Performances of RLGs

Both the DRLG and the ZRLG are truly inertial-grade sensors, and provide performances that are still unparalleled by other approaches.

Let us briefly discuss performances and the way they are achieved, in this section.

Minimum Detectable Signal (MDS) In Sect.7.2.1, we have considered it in terms of the phase shift to be measured. More frequently, we may be interested to the minimum detectable angular velocity or $MD\Omega$ (deg/h). Using Eq.7.11, and dividing the phase MDS by the responsivity $R_\Phi=\Omega/\Phi$, we can write:

7.4 Development of the RLG

$$MD\Omega = \sigma_{\Omega n} + \Omega_{ZB} \quad (7.20)$$

The first term $\sigma_{\Omega n}$ is the random contribution coming from noise. At the quantum limit, using Eq.7.9, we can write $\sigma_{\Omega n}$ in the form:

$$\sigma_{\Omega n} = \phi_n/R_\Phi = (2\kappa\, h\nu/\eta P_0)^{1/2}/R_\Phi \sqrt{B} = S_\Omega \sqrt{B} \quad (7.20a)$$

where S_Ω has the meaning of spectral density of angular velocity fluctuation and is measured in (deg/h)/√Hz or also, more frequently, in (deg/√h). Here, let us note that 1 (deg/h)/√Hz = 1 deg/(√h 60√s √Hz)= (1/60) deg/√h, whence the spectral density S_Ω is a number 60 times smaller in the deg/√h unit.

When the random term prevails in Eq.7.20, the minimum detectable Ω increases with bandwidth as indicated by Eq.7.20a. This holds at the quantum limit, strictly, but we may assume the dependence is $S_\Omega \sqrt{B}$ also for other sources of excess noise.

If we average or integrate Ω on a period T of time, then the measurement bandwidth is $B \propto 1/T$, and accordingly we get:

$$\sigma_{\Omega n} = S_\Omega / \sqrt{T} \quad (7.20b)$$

As Eq.7.20b tells us, the minimum detectable Ω decreases at increasing averaging time, and of course this statement holds until the second term Ω_{ZB} in Eq.7.20 is finally reached, or some nonwhite noise components of $\sigma_{\Omega n}$ come into play.

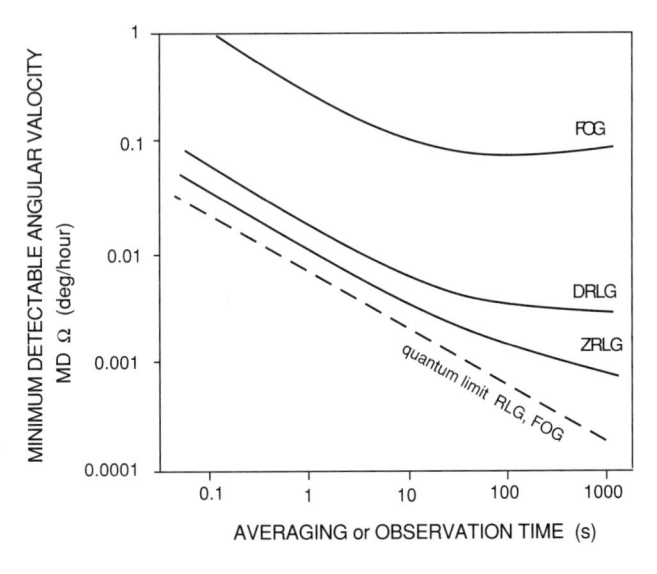

Fig.7-17 Minimum detectable angular velocity MDΩ as a function of integration time T. Results are typical performance for ZRLG, DRLG, and FOG with the data in Figs.7-5 and 7-6. Dotted line is the quantum limit (about the same for both).

In Fig.7-17, we plot data representative of the MDΩ performance achieved in gyroscopes. As we can see there, the two RLG versions are not so far away from the quantum limit, with the ZRLG performing a little bit better than the DRLG, whereas the FOG is a factor ≈30 off. The maximum useful integration time differs as well. It is up to several hundreds s in RLGs and just ≈10 s in the FOG.

The spectral density of angular velocity fluctuations can be read from the diagram of Fig.7-17 using the Nyquist value B=1/2T for the bandwidth. At T=0.5-s, we get the typical values $S_\Omega \approx 0.03$ deg/h√Hz= 0.0005 deg/√h for the RLGs, and $S_\Omega \approx 0.6$ deg/h√Hz ≈0.01 deg/√h for the FOG.

By increasing the integration time, we can attain a few 10^{-3}-deg/h in the RLG, and perhaps 0.1-deg/h (or not much better) in the FOG.

Zero bias and null provision. The above figures of MDΩ are achieved if we are able to keep the second term in Eq.7.20, that is the zero bias error Ω_{ZB}, consistently smaller than the random noise term.

A simple null operation of the Δf signal by a dc correction does not suffice, usually. We need a more sophisticated algorithm to be able reduce the initial zero bias error Ω_{ZB}, perhaps in the range 1-10 deg/h, to the minute ≈10^{-3}-deg/h level. The calibration operation is carried out in a test chamber well shielded from external, ambient-induced, mechanical disturbances. The gyro is oriented perpendicular to the earth rotation axis (Ω=0) and subjected to a well-defined set of physical quantities, or disturbances, M_k (k=1...n). For each M_k, we determine the deviation of the zero bias Ω_{ZB} from the initial value Ω_{ZB0}. In this way, we get the coefficients C_{ki} of the Ω_{ZB} expansion at the first-order:

$$\Omega_{ZB} = \Omega_{ZB0} + \Sigma_{k=1..n} \; C_{ki} (M_k - M_{k0}) \qquad (7.21)$$

Here, Ω_{ZB0} is the zero offset in the standard conditions ($M_k=M_{k0}$). For each disturbance M_k, the measurement is repeated on a set of values (i=1...m) to determine if the linear regression (Eq.7.21) is adequate, or we need a higher order expansion, easily following as an extension of Eq.7.21.

The physical quantities in Eq.7.21 are measured by separate sensors (see also Sect.7.4). Typically, they are: (i) temperature of the cavity and temperature of the magnets (ZRLG only); (ii) currents in the discharge legs; (iii) magnetic field to the bore (ZRLG) or dither amplitude (DRLG); (iv) frequency detuning from the atomic line center. The measurement of coefficients C_{ki} is time-consuming, because each element requires a time of the order of the maximum useful integration time (Fig.7-17).

Following the calibration, the correction of the zero bias error Ω_{ZB} through Eq.7.21 is performed by means of the on-board computer, in which the matrix C_{ki} is written. The correction is effective, and it typically brings down the initial value of Ω_{ZB} by a factor of ≈10^3 or better.

With aging however, the matrix C_{ki} changes, and the correction improvement may worsen to ≈10^2 in a period of two to three years. Then, the sensor shall be recalibrated, and this operation requires disassembling the sensor to and delivery back for the factory.

7.5 Development of the Fiber Optics Gyro

Dynamic range and linearity. The dynamic range we need for inertial applications is 6 to 8 decades, spanning from the ≈0.01-deg/h of the best resolution to the maximum ≈100 deg/s required for handling fast jerks that the vehicle may experience. The scale factor should be calibrated to better than 10^{-4} as a specification for achieving a ≈0.1-km position accuracy after a ≈10^3-km trip. For the same reason, the linearity error should be less than ≈10ppm on the entire dynamic range (1ppm = part-per-million, or 10^{-6}).

Both DRLG and ZRLG have a wide range of frequency available for the signal Δf swing, about 100 MHz under the He-Ne atomic line of 1500 MHz width. However, a small linearity error arises, because of the frequency pulling effect. In addition, a large Δf swing may disturb the frequency stabilization, and hence result in a frequency error. Thus, in the end we get a ≈10^{-3} linearity error as the starting performance. This figure is improved and brought to the 10 ppm level again by calibration. Pendulum-based servos are employed to calibrate the measurement. Like for the zero correction, the linearity correction is also carried out by providing a matrix of coefficients to the on-board computer.

With all the provisions discussed above, the RLG becomes the hearth of INU (inertial navigation units) and AHRS (attitude heading reference systems) extensively used in avionics [15]. Typical performance of an INU is ≈0.5-nm (nautical mile) error in dead reckoning after 1 hour of navigation, during which an aircraft flies approximately 500 nm. The INU is typically a box of 15" by 8" by 8" and includes three gyroscopes and the computer to handle all the signals, inertial and housekeeping, necessary to compute the actual position of the vehicle on the earth. Cost of an INU is in the US$1.2 to 1.8 million range.

7.5 DEVELOPMENT OF THE FIBER OPTICS GYRO

The first problem encountered when developing the FOG from its basic configuration (Fig.7-7) has been that of the cosine-type dependence of the signal (Eq.7.18), which is unsuitable for measuring small Ω [4,7].

Several approaches to circumvent the cosine problem have been proposed. Despite the variety of solutions, all are based on moving the signal Φ_S from the baseband (f=0) to a carrier frequency f_m. In this way, looking at resulting signal $\cos(2\pi f_m t + \Phi_S)$ and comparing it to a reference $\cos 2\pi f_m t$, the Sagnac phase can be recovered.

Among the proposed solutions, we find those based on using two-frequency modes [8], a frequency shift, or a frequency modulation [25]. Over the years, however, the solution finally recognized as the most effective and easily interfaced to the basic scheme has been the one based on the *phase modulation* [6,7,26].

7.5.1 The Open-Loop Fiber Optic Gyro

The argument leading to phase modulation is understood looking at the cosine signal plotted in Fig.7-18. Usually, we are near $\Phi_S \approx 0$ and the photodetected current I_{ph} is about zero.

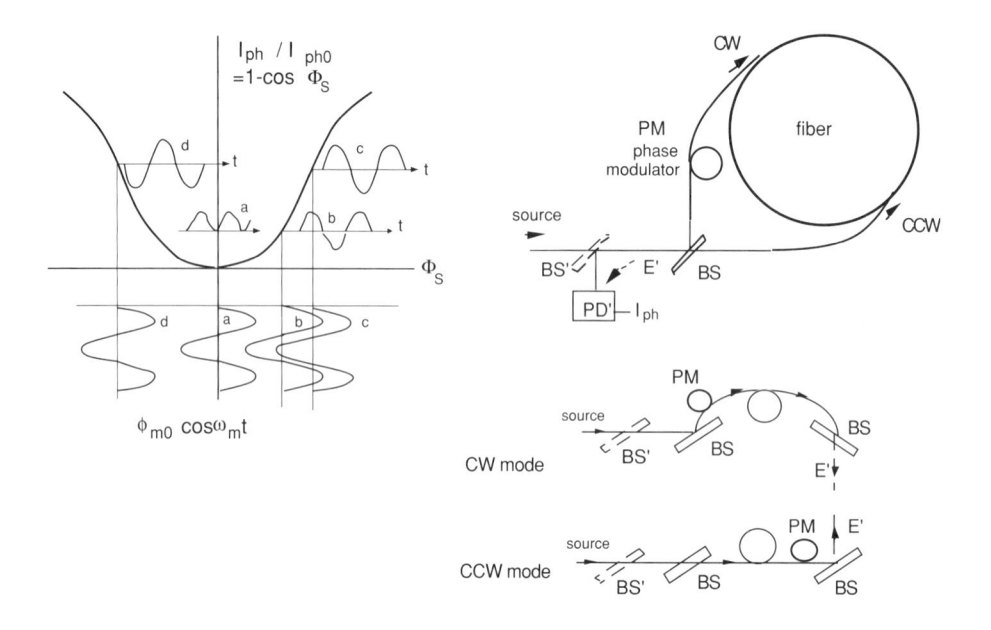

Fig.7-18 The cosine-type signal carrying the Sagnac phase Φ_S can be brought to linear with the aid of a phase modulator, adding a small modulated phase $\phi_m \cos\omega_m t$ to Φ_S (left). Then, the waveform of the photodetected current I_{ph} becomes modulated and carries the sign and the amplitude of Φ_S, as can be seen from the examples (a,b,c,d at left). The phase modulator shall be placed asymmetrical with respect to the coil (right, top) so that the modes pass through it with a delay and are able to collect a phase difference (right, bottom).

If we are able to superpose a phase modulation to the Sagnac-induced phase Φ_S, let us say $\phi_m = \phi_{m0} \cos\omega_m t$, the photodetected current I_{ph} becomes modulated at frequency $f_m = \omega_m/2\pi$. As we can see from Fig.7-18, for $\Phi_S > 0$, the waveform of I_{ph} is in phase with the modulation ϕ_m, whereas for $\Phi_S < 0$, it is in antiphase. Thus, the sign of Φ_S is recovered.

Moreover, the amplitude of the modulated current increases at increasing Φ_S, removing the zero sensitivity near $\Phi_S \approx 0$. In the examples in Fig.7-18, note how the ac component at the modulation frequency f_m increases from *b* to *c*, reverses its phase in *d* and goes to zero in *a*, where a $2f_m$ component is left. Actually, as it will be shown later, the component of I_{ph} at the modulation frequency f_m is proportional to $\sin\Phi_S$ ($\approx \Phi_S$ for small Φ_S) and therefore to the measurand Ω.

A feature of phase modulation measurement scheme is that we end up with the desired $\sin\Phi_S$ dependence starting from ta $\cos\Phi_S$ dependence. This tells us that the phase modulation readout yields the derivative of the initial response dependence.

7.5 Development of the Fiber Optics Gyro

We may also wonder if the working point of the interferometer, $1-\cos\Phi_S \approx 0$ for small Φ_S, is good from the point of view of noise. Indeed, the dc component is zero (or small at first order in Φ_S), but the signal is small at the second order, as $1-\cos\Phi_S \approx \Phi_S^2/2$. From an analysis of S/N ratio [27], one can find that indeed the 'dark fringe' point of the interferometer is the best, and we work in the phase noise limits of Sect.7.2.1.

To implement the phase modulation, we shall insert a suitable element in the FOG propagation path. We need a nonreciprocal element because the Sagnac signal we are dealing with is a nonreciprocal phase. The best and most stable element implementing a nonreciprocal phase shift using a reciprocal phase shift is the asymmetrical phase modulator shown in Fig.7-18.

The phase modulator provides a reciprocal phase shift that we may write as:

$$\Phi_m(t) = \Phi_{m0} \cos \omega_m t \qquad (7.22)$$

The phase modulator is placed at one of the fiber entrance (Fig.7-18), that is, asymmetrically with respect to the coiled fiber middle point. Thus, the CW and CCW beams cross it at different times, collecting different phase shifts. This is equivalent to nonreciprocity.

Let us denote the fiber coil length by L. Then, the propagation phase shift is nkL, where $k=2\pi/\lambda$ and n is the effective index of refraction of the fiber. The propagation time through the coil is $T=nL/c$, and this quantity is the delay that the CW and CCW beams experience in crossing the phase modulator.

After a complete propagation through the coil, the CW and CCW phase is written as:

$$\Phi_{CW} = \Phi_S/2 + \Phi_m(t-T/2) + n_{CW}kL + R^{BS}(t-T/2) + T^{BS}(t+T/2)$$

$$\Phi_{CCW} = -\Phi_S/2 + \Phi_m(t+T/2) + n_{CCW}kL + R^{BS}(t+T/2) + T^{BS}(t-T/2) \qquad (7.23)$$

Here, $\Phi_S/2$ and $-\Phi_S/2$ are the Sagnac phase shifts, $-T/2$ and $+T/2$ are the time delay in crossing the modulator, n_{CW} and n_{CCW} are the effective index of the modes, and R^{BS}, T^{BS} are (field) reflection and transmission of the launch beamsplitter BS.

At the photodetector, the beating of the recombined modes $E_{CCW}+E_{CW}$ yields a signal:

$$I_{ph} = |E_{CW}+E_{CCW}|^2 = E_0^2[1+\cos(\Phi_{CW}-\Phi_{CCW})] \qquad (7.24)$$

Let us now consider the phase difference $\Phi_{CW}-\Phi_{CCW}$. If the modes have the same effective index and the beamsplitter BS does not change appreciably with time, then we may drop the last three terms in the expressions of Φ_{CW} and Φ_{CCW} and get:

$$\Phi_{CW} - \Phi_{CCW} = \Phi_S + \Phi_{m0}[\cos \omega_m(t-T/2) - \cos \omega_m(t+T/2)]$$

$$= \Phi_S + \Phi_{m0} \, 2 \sin \omega_m t \sin \omega_m T/2 \qquad (7.25)$$

Now we let $\Phi_0 = 2\Phi_{m0} \sin \omega_m T/2$ for the effective phase deviation impressed by the modulator. Note that the maximum of phase deviation, $\Phi_{0max} = 2\Phi_{m0}$ is achieved for $\omega_m T/2 = \pi/2$ or also $2f_m T=1$. The most efficient frequency of modulation is then $f_m=1/2T$, the inverse of twice the time of propagation through the coil [for example, with L=200m it is $T=200/3 \cdot 10^8 =0.66\mu s$, $f_m = 0.75$ MHz].

With this position, the photodetected signal follows from Eq.7.24 as:

$$I_{ph}/I_{ph0} = 1+ \cos(\Phi_S+\Phi_0 \sin \omega_m t) \qquad (7.26)$$

Now, we can expand the second-hand term of Eq.7.26 in series of ω_m using Bessel functions [26,28]. In this way we get:

$$I_{ph}/I_{ph0} = 1+ [\ J_0(\Phi_0) + 2 \sum_{n=1,\infty} J_{2k}(\Phi_0) \cos 2k\omega_m t] \cos \Phi_S +$$

$$+ [\ 2 \sum_{n=1,\infty} J_{2k-1}(\Phi_0) \cos(2k-1)\omega_m t] \sin \Phi_S \qquad (7.27)$$

Thus, the photodetected signal contains all the harmonic components of the modulation frequency $\omega_m/2\pi = f_m$. The amplitude of the odd-harmonic components is proportional to $\sin\Phi_S$, the desired dependence linear near $\Phi_S \approx 0$, whereas the even-harmonic components are proportional to the cosine term.

By analyzing the photodetected signal with the aid of a lock-in amplifier, driven by the phase modulator waveform, we can filter out the $\cos \omega_m t$ component. The amplitude of this component yields the $\sin\Phi_S$ signal.

It may be useful to maximize the scale factor $2J_1(\Phi_0)$ of the $\sin\Phi_S$ signal. The maximum is obtained at $\Phi_0=1.8$ and the value is $2J_1=1.16$ [28].

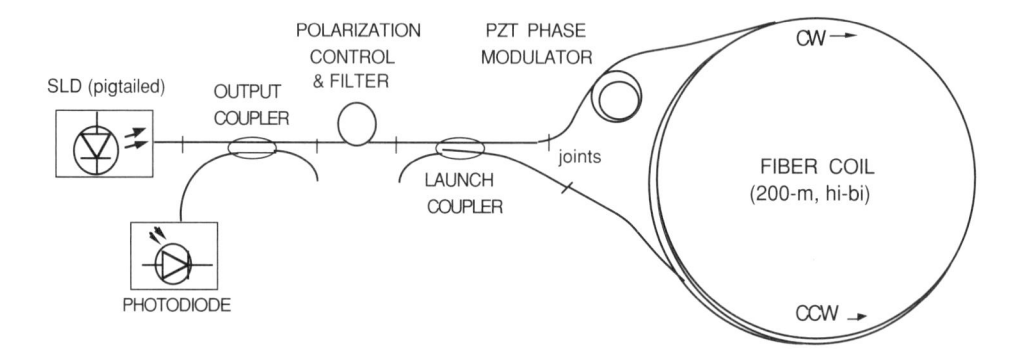

Fig.7-19 Open-loop configuration of the FOG, implemented in all-fiber technology. Beam splitters are realized by fiber fused couplers, and the phase modulator by a coil wound on a PZT piezo ceramic tube. The fiber is a polarization-maintaining fiber with high linear birefringence to minimize mode coupling. The source is a superluminescent light emitting diode (or SLED) with high radiance but low coherence length.

7.5 Development of the Fiber Optics Gyro 281

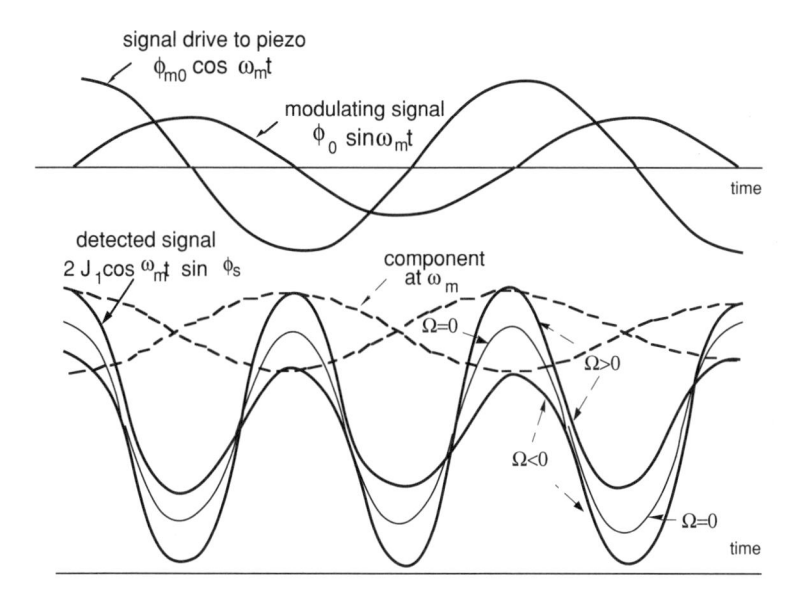

Fig.7-20 Waveforms of piezo drive and phase modulation (top), and of the photodetected signal I_{ph}/I_{ph0} (bottom). At rest (Ω=0), the signal is made up mostly of the second harmonic component, and its peaks have the same height. When the gyro rotates, for Ω>0, odd peaks are higher than the even, and vice versa for Ω<0. The signal at the fundamental frequency (dotted line) is in phase with the drive to the piezo.

As we have let $\Phi_0 = 2\Phi_{m0} \sin\omega_m T/2$, we need a phase amplitude Φ_{m0}=0.9-rad from the phase modulator, once the sine factor has been maximized ($\sin\omega_m T/2=1$).

In Fig.7-19 we report the schematic of the most commonly used FOG configuration, called the *all-fiber FOG*. This schematic is the realization of the conceptual scheme of Fig. 7-18, and has been the first demonstrated as a practical device capable of close-to-inertial performance [6,26]. The name all-fiber is because all the components in the device are made up by pieces of fiber. In particular, we use a coil of fiber wound on the piezo tube to realize the modulator [29], and fused-fiber couplers to realize the beamsplitters. Also shown in Fig.7-19 is a fiber-based device for polarization control and filtering that is intended for a fine selection operation to improve the zero bias and noise.

The waveforms of the drive to the piezo, of the resulting phase modulation and of the photodetected signal, are shown in Fig.7-20. At rest, only the second harmonic of the modulation is contained in the photodetected signal but, when $\Omega\neq0$ the fundamental component shows up, in phase or antiphase with respect to the drive according that Ω<0 or Ω>0.

Fig.7-21 illustrates how the electrical sections of the open-loop FOG are interconnected. An oscillator provides the drive signal $V_{dr}=V_0\cos\omega_m t$ to the piezo modulator, and it also feeds the reference-input of a lock-in amplifier. The measurement input is the photodetected signal, properly amplified by a transimpedance amplifier [30].

The lock-in amplifier sorts out the components in-phase with respect to the reference. We retain that at the fundamental frequency, proportional to $\sin\Phi_S$. The second harmonic component is proportional to $\cos\Phi_S$ and may also be used to introduce a linearity correction to signal $\sin\Phi_S$ when Φ_S is not very small.

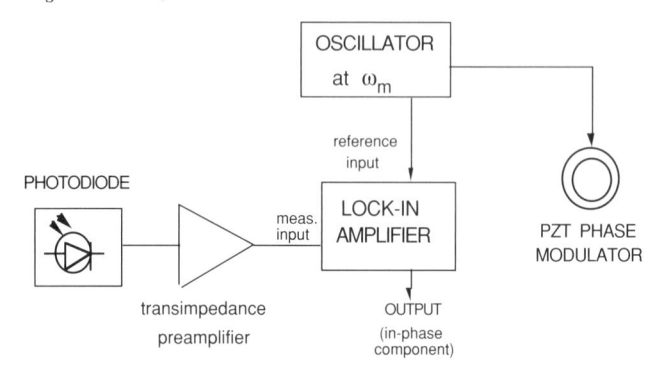

Fig.7-21 A lock-in amplifier is used to extract the output signal $\sin\phi_s$, proportional to the in-phase component at the fundamental frequency.

7.5.2 Requirements on FOG Components

Let us now discuss the elements incorporated in the all-fiber FOG configuration of Fig.7-19, and consider the requirements we need to achieve a good performance.

7.5.2.1 The phase modulator

The *phase modulator* is realized by winding a number (N=5-30 typ.) of fiber turns on a piezoceramic tube. The usual material is PZT (lead zirconate titanate), and typical size is a tube ≈30-mm diameter, ≈2-mm thick and ≈20-mm long [29]. When voltage is applied to the electrodes on tube walls, the tube strains by the piezoelectric effect, and the fiber is stretched to an increased length, say L+ΔL. In addition, the optical path length changes because of the elastooptical effect, which changes the index of refraction by Δn. The two effects combine to give a total phase shift written as $\Phi_{m0}=k\Delta(nL)=k(L\Delta n+n\Delta L)$. Another expression is the form $\Phi_{m0}=\kappa NV$, showing that the phase modulation is proportional to the number of turns N and to the drive voltage V, besides a coefficient κ specific of the material.

With the above size and N=10 turns, the phase shift is ≈2π (or ≈λ in path length) for an applied voltage of a few Volts [29].

The frequency response of the phase modulator is constant up to the first resonant frequency f_R of the mechanical structure. Typically, it is f_R=50-100 kHz (for the above size), but one can reach several tens of MHz with linear (or slab) piezo elements [29], however at the ex-

7.5 Development of the Fiber Optics Gyro

pense of a reduced efficiency κ. For the reason already explained, the piezo phase modulator is placed asymmetric in the propagation route of CW and CCW modes (Fig.7-19) to maximize the modulation impressed to the Φ_{CW}-Φ_{CCW} signal.

7.5.2.2 The fiber

The *fiber* we choose for the propagation coil is the most critical component in a FOG. As already pointed out, we have the exacting requirement that the two counterpropagating waves shall see just the same optical path length or phase delay in opposite directions.

Usually, this statement is a commonplace. For example, in fiber communications, nobody would doubt about the equality of path lengths in opposite directions, even through very long fiber trunks. However, when very minute differences become crucial, say the picometer or the microradiant as needed in a FOG, the matter changes.

In other words, we need ensuring reciprocity of propagation among the two counter-propagating waves, as first pointed out by Ulrich [5]. In the fiber coil, reciprocity means that we shall avoid all sources of diversity in propagation.

Thus, the first choice is that the fiber shall be monomode to avoid any difference of mode mix in the two waves. In fact, different modes experience a different phase delay (Fig.7-22, top).

We can appreciate the consequence of this difference by considering a beam mainly monomode, but with an added small fraction β of a higher order mode.

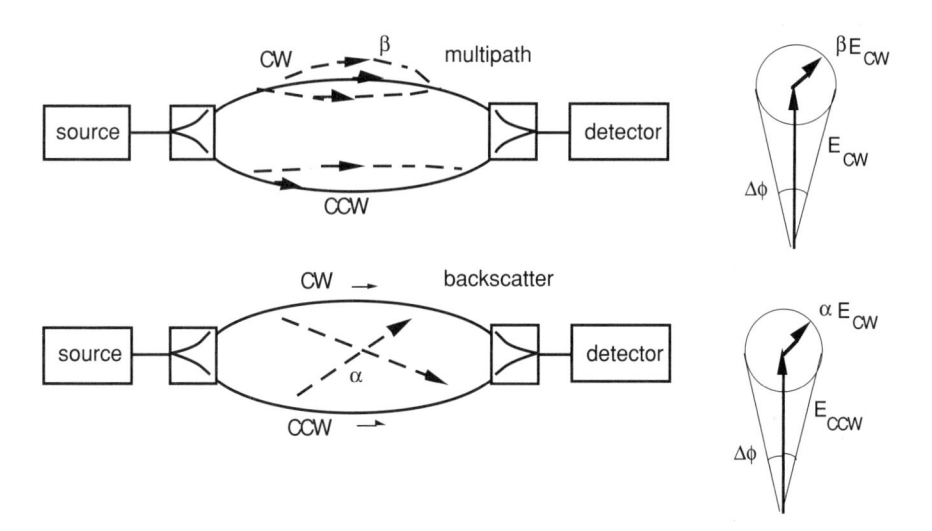

Fig.7-22 Phase errors are generated by multipath effects due to propagation with contributions from multimode or polarization diversity regimes (top). These effects are eliminated by working in single-mode, single polarization. Another phase error comes from very small back-reflections coupling the modes, as due to the Rayleigh scattering of the fiber (bottom).

In this case, at the exit from the coil, the optical field would be the vector sum of a normally delayed contribution, E_{CW}, plus a minute fraction βE_{CW} with a different delay due to the higher order mode.

On the fiber length of a FOG, this optical path length difference is very large ($\approx 1000\lambda$) and the delay spans the whole range 0-2π. We can represent the vector sum as in Fig.7-22, by a circle of uncertainty around the average position of the expected optical field E_{CW}.

The phase error is readily seen to be $\Delta\phi=\beta E_{CW}/E_{CW}=\beta$. As an example of how severe the effect is, let us assume that our beam propagating down the coil has a spurious modal content of a bare 10^{-6} in power. Then, the field ratio is $\beta=\sqrt{10^{-6}}$ and the phase error is $\Delta\phi=1$ mrad, a very large value indeed, in our Sagnac-scale of phase shifts. Fortunately, not all the $\Delta\phi$ error contributes to noise. Should the modes have a very stable propagation constant, $\Delta\phi$ would also be a constant, and all would end with a zero bias error, at most. Instead, because of fluctuations, $\Delta\phi$ has frequency components, and a fraction of it contributes to noise, actually. As default estimate, about 10^{-2} of the total error may end up in noise, and this means that the phase noise is $\varphi_n=10^{-2}$ 1mrad $=10$ μrad. Then, from Fig.7-5 we can read NEΩ≈5deg/h in correspondence to the FOG dotted line.

The state of polarization also affects the propagation constant, and we need a polarization holding fiber in the coil to get rid of phase fluctuations. This type of fiber, also called *polarization maintaining,* or hi-bi (high birefringence) fiber [31], is made by structure similar to a normal fiber, but in the clad stress elements are added. These elements run parallel to the core and their section is double circular (Panda fiber) or double wedge (bow-tie fiber) in shape. Because of the stress imparted to the core, a birefringence is induced by elastooptic effect. Mode propagation has then two different constants along the slow and fast axis of the core.

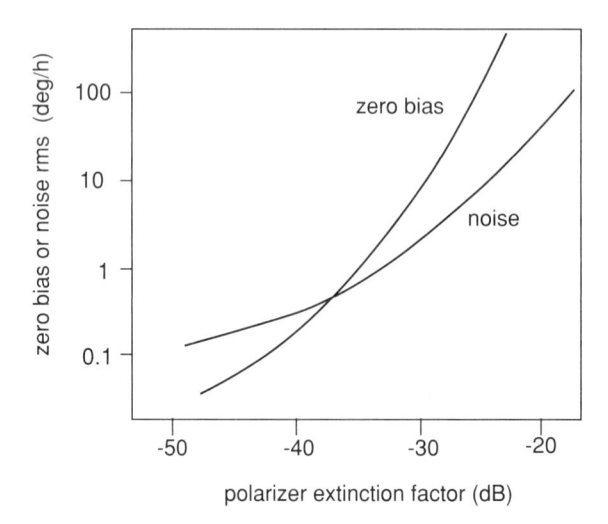

Fig.7-23   Typical experimental dependence, in an all-fiber FOG, of zero bias error and noise from the extinction ratio of the polarizer used as in Fig.7-19.

7.5 Development of the Fiber Optics Gyro

In consequence of this difference, phase matching between the two polarized modes is destroyed. This is also called mode decoupling, and it means that the effect of any distributed source of energy exchange is washed out.

Then, the factor β considered previously becomes much less than in an ordinary fiber (in practice, of a factor $\approx 10^{-3}..10^{-5}$).

From the discussion above, it is clear that, to avoid incurring in a severe noise and zero bias spurious contribution, we shall carefully filter the beam launched into the propagation coil. To this end, we use a filter (Fig.7-19) acting on the spatial distribution as well as on the state of polarization to approach the ideal condition of propagating the cleanest and conversion-resistant mode.

Actually, the fiber itself is a mode filter, when we use it at a normalized frequency V lower than the second-mode cutoff (V_{co}=2.405). Especially for low V (for example 1.3-1.7), we get a good rejection of higher modes eventually generated by geometrical and winding imperfections while keeping low the extra losses due to coil curvature.

7.5.2.3 The polarizer

Using the fiber itself as the spatial filter allows us to place a *polarizer* outside the coil for the filtering function.

The requirement on spurious components not suppressed by the polarizer can follow the arguments developed in the previous section. A practical result is shown in Fig.7-23, and it clearly indicates that the best extinction ratio achieved in practice (≈50dB) is necessary to avoid degradation of the FOG performance. A fiber polarizer can be implemented [32] by winding a coil of a few turns of hi-bi fiber used near cutoff. Because the two polarizations have a different V parameter, the differential attenuation due to curvature can be exploited. On several turns of fiber, we can get a negligible loss (0.2 dB) for one polarization and a large loss (≈50dB) for the other [32].

7.5.2.4 The effect of backscattering

Another source of error associated with the fiber is cross-coupling between the counter-propagating waves (Fig.7-22 bottom). In the fiber coil, this is caused mainly by *Rayleigh scattering*. As in any material, also in the glass of the fiber, we find a minute density fluctuations that act as a scattering center. At wavelengths outside absorption bands, scattering is the main source of loss in the fiber.

The Rayleigh loss is about 3 dB/km at λ=850 nm, the wavelength preferred for the all-fiber FOG operation. This figure amounts to a backward coupling α, between the CW and CCW modes, of about –55dB/m. To bear the full extent of Rayleigh scattering on a 200-m (or +23dB re 1-m) fiber, we should reckon with a (-55+23)=-32-dB or 1/1600 of power transferred from one mode to the other. Again, as calculated above for the multimode case, we would have a phase error of about $\Delta\phi=\alpha E_{CW}/E_{CCW}=1/\sqrt{1600}=25$ mrad, having taken $E_{CW}\approx E_{CCW}$.

Even considering that the effective noise is only a fraction of the Δφ error, the value is so huge that it makes the FOG definitely useless.

The remedy is in avoiding that backscattered contributions can really add up to the field in transit. The assumption by which they add is that superposition is coherent. As the backscattered contribution has run a path different from the path of the field on which it superposes, to have addition, the coherence length of the source shall be larger than this path difference. Now, let us decrease the coherence length to a small value, say of the order of $l_c \ll 1$mm, instead of the large value 100-1000m of frequency stabilized lasers.

Then, only the backscatter arising from locations in the fiber coil that are distant for less than l_c will contribute. The point where the path-length difference is zero is the midpoint of the fiber coil. With a coherence length l_c, only a length l_c of fiber contributes to backscattering. Thus, if we have $l_c = 50\mu$m (=1/20.000-m or –43 dB re 1-m), we get as the Rayleigh scattering a value (-55-43)=-98 dB, now a safe value. Of course, the requirement of slashing the Rayleigh scattering is now transferred from the fiber to the source, which will be a low-coherence one (see below).

Back-reflected contributions are also found in components external to the fiber. All joints (Fig.7-19) between the individual components are potential sources of back-reflection.

Usually, we define a return loss (RL) as the power ratio of back-reflected power to input (forward going) power. Typical RL figures of good joints made by fusion splice are –50 dB for normal joint, -70 dB for angled joints.

Considering that the field coupling factor is half the power factor, the previous figures would yield $\Delta\phi$=3 and 0.3 mrad, respectively. These are again unacceptable values but, fortunately, they are cured the same way the fiber backscatter is, i.e., by lowering the coherence length of the source.

7.5.2.5 The source

About the source, a device developed specially for the FOG is the *super-luminescent LED* or *SLED*. This is a semiconductor device, that has a structure identical to that of a normal Fabry-Perot diode laser (App.A1.3) of comparable power level, but missing the output mirror.

In fact, after cleaving the chip out the wafer, the normal operation that creates the flat mirrors of the cavity, we coat the output facet of the SLED with a single or a multiple anti-reflection layer. Thus, the reflectance of the facet may go down to a residual 10^{-4} from the initial value $(n-1)/(n+1) \approx 0.3$ of the Fresnel interface.

When we drive the diode with an injection current, the laser threshold is never reached. The emitted radiation is nevertheless different from a normal LED. Because the guiding structure provides confinement and the material is pumped to population inversion, the spontaneous emitted radiation is optically amplified. Thus, the radiance (power per unit surface and solid angle) of the emitted light is much larger than that of an LED and is comparable to that of a laser. This feature is very important in the FOG, to be able to launch power in a single mode fiber with a good efficiency.

On the other hand, radiation is amplified spontaneous emission, not a coherent oscillation. The emission line width is about the same of the LED, that is $\Delta\lambda$=25-50-nm (Fig.7-24).

7.5 Development of the Fiber Optics Gyro

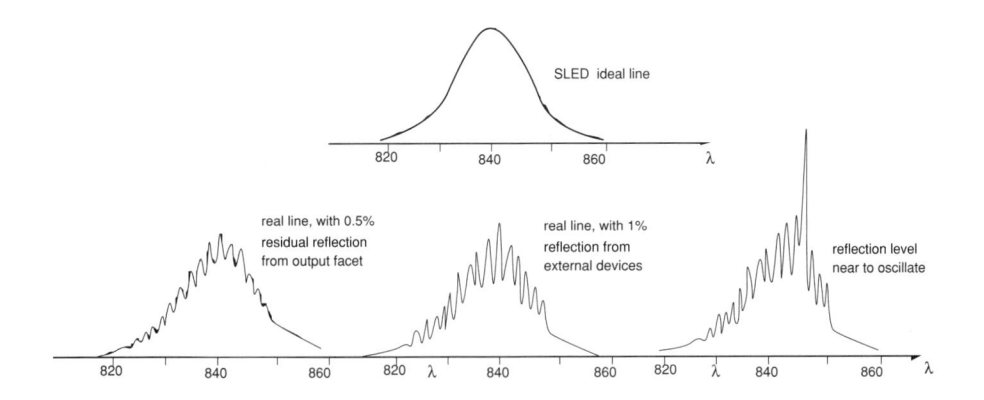

Fig.7-24 The source used in the FOG is the super-luminescent LED (or SLED), because it has a broad spectrum and thus a short coherent length, together with a small emitting area for high efficiency of coupling to fiber. If the output mirror has no residual reflectance, the SLED has the ideal line (top), but at increasing R the spectral distribution becomes distorted (bottom). Distortion causes increase of noise and zero bias error of the FOG.

The corresponding coherence length $l_c = \lambda/\Delta\lambda$ is $\approx$ 20-40 nm, a nicely small value as we wish for the FOG.

A problem with the SLED is back-reflections fed back into the source. Back-reflections originate at components and down the coiled fiber. Those at components will be carefully minimized, specifying an appropriately small return loss. The back-reflection from the coiled fiber is unavoidable, however, and not so small ($\approx$ -32 dB for a 200-m fiber, as in the previous example).

As we can see in Fig.7-24, even small reflections distort the shape of the emission line, eventually resulting in a SLED breaking into oscillation. As the line shape is progressively distorted, coherence length increases back and an excess amplitude noise is found in the emitted power. Both effects contribute to increase noise and zero bias error of the FOG.

To prevent back-reflected light from reaching the SLED chip, the ideal remedy is to insert an optical isolator at the output. Unfortunately, optical isolators are not easily available at the wavelength ($\lambda \approx$ 800 nm) of the GaAlAs source, because they require exotic materials for the Faraday rotator. A way out may be moving to another more favorable wavelength, $\lambda \approx$ 1500-nm for example, where optical isolators based on YIG crystal [33] are readily available. In this case, we may use as a source the output from an Er-doped optical amplifier.

At $\lambda \approx$ 800-nm, another way to reduce back reflection [34] is using a circular polarizer at the SLED output. The circular polarizer is in practice realized by cascading a linear polarizer and a λ/4 wave plate, both made by piece of fibers, in the all-fiber FOG. As circular polarizion changes handiness when it returns back upon reflection or from the Sagnac loop output, all contributions are blocked out when they return at the polarizer and λ/4-plate combination. Only scattering contributions undergoing a polarization change are unaffected, and this is a minor fraction (estimated in 10^{-3}) of the total back reflection.

7.5.2.6 Couplers

Fiber couplers in the all-fiber FOG can be fabricated with the fusion-splice or the lapping technology [35,36]. We may want to fabricate them out of the same hi-bi fiber in the coil, and then lapping technology is better suited because it disturbs the fiber structure less, whereas using a normal fiber, the fusion-splice technology is superior.

To the fiber couplers of the FOG, and especially to the launch coupler, we ask a very stable splitting ratio, as close as possible to 50%, to get an interferometer signal with a nearly 100% visibility factor V (Sect.4.4.1). In addition, we need an exceptionally low excess loss (≈ 0.01 dB) and the highest return loss (better than ≈ 60 dB) because these values impact the noise and zero bias error as discussed in the previous pages.

All these figures are rather demanding in the common practice of the field. However, they can be obtained with a careful control of the fabrication process [36].

7.5.2.7 The receiver

The *photodiode receiver* presents no criticality. The standard amplifier based on the transimpedance circuit [30] is adequate as a front-end of the photodiode and provides quantum-limited performance already at low detected currents (say ≈ 0.1 to 1 µA).

7.5.2.8 Residual errors

When noise and zero bias of an all-fiber FOG are measured and compared to the theoretical quantum limit values (Fig.7-6), a disappointing discrepancy of as much as 1.5 to 2 decades is found. Thus, starting from year 1985, researchers had been studying [7,37,38] the possible effects responsible of the reduced performance in the hope that, following the identification, a breakthrough of performance could be finally achieved.

Three main effects about the fiber worth mentioning are (i) polarization mode dispersion (PMD); (ii) cross phase modulation (XPM), and (iii) thermodynamic phase noise (TPN). All these contributions affect the term n_{CW}-n_{CCW} in Eq.7.23.

The *PMD* deals with the propagation time dispersion δ_P, a quantity that translates directly to phase delay dispersion. The PMD in a normal fiber is typically $\delta_P \approx 0.05$ ps/√km, a figure associated to a birefringence beat-length of ≈ 100m. The corresponding phase dispersion is $\Delta \phi_P = (2\pi f) \delta_P \sqrt{L} = 2\pi \ 400 \ 10^{12} \ 5 \ 10^{-14} \sqrt{0.2} = 55$ rad for a L=200 m fiber length at λ=800 nm (at which f≈400 THz). The main part of $\Delta \phi_P$ is deterministic, and we can estimate that only a fraction, $\approx 10^{-3}$ of it, or 0.05 rad, may contribute to the random fluctuation of phase. In addition, because $\Delta \phi_P$ is read out by two counterpropagating beams, a further reduction (estimated in $\approx 10^{-4}$) is obtained. The final phase noise that can be ascribed to PMD dispersion is ≈0.005 mrad, equivalent to a ≈3 deg/h noise in a typical FOG (see Fig.7-5).

7.5 Development of the Fiber Optics Gyro

The *XPM* is due to the optical Kerr effect, by which the index of refraction weakly depends on the light intensity of the beam in transit through the medium. In a Sagnac interferometer, the $(n_{CW}-n_{CCW})L$ path length difference is proportional to the nonlinear $\chi^{(3)}$ coefficient of silica, and to the difference $P_{CW}-P_{CCW}$ of counter propagating beams powers. For a L=200-m coil, the XPM contributes to an error of $\chi^{(3)}L(P_{CW}-P_{CCW}) \approx 10(deg/h)/mW$.
This requires to carefully balancing the power of beams launched in the opposite directions of the fiber coil. If we are able to keep the unbalance down to $P_{CW}-P_{CCW}<0.1\mu W$, we get a negligible noise, down to $\approx 10^{-3}(deg/h)$.

Thermodynamic phase noise (*TPN*) has been already discussed in Sect.4.4.5. The TPN contribution can introduce a large noise, up to ≈ 10 deg/h [36] in general. However, in the Sagnac interferometer, there is a strong cancellation effect due to the counter-propagation readout, so the effective error or noise can go down to $\approx 10^{-2}$ deg/h.

In conclusion, in the FOG there are several effects, each of which may well justify the excess noise found in experiments. However, after everything is apparently kept under control, and all the discussed effects are seized down to the $\approx 10^{-3}$ deg/h level, a large unexplained noise remains (Fig.7-17).

The excess noise is clearly associated with the fiber coil length L, being proportional to L or $\sqrt{L}$ in practice when an optimal length ($\approx 200m$) is exceeded. The reason for the excess noise, or difference with respect to the quantum noise limit of Fig.7-17 is still unclear.

7.5.3 Technology to implement the FOG

As already pointed out, the term *all-fiber* is used to designate the FOG of Fig.7-19 because of the technology to implement it. All elements are made by pieces of fiber, properly worked or exposed to an external perturbation giving the desired functionality.
There are two more technologies to implement the FOG: the micro-optics and the integrated optics. The three technologies are shared by all types of fiber optic sensors, generally.

The *micro-optics* technology consists in using conventional optical components, interconnected by free space propagation. From one side, this technology offers the widest choice of components to assemble in the device, but the free propagation in between is the bottleneck. We need objective lenses to launch into the fiber, or relay lenses between components. As we attempt to develop a design, we soon realize that launch and mechanical mountings are critical, bulky, and extremely sensitive to external mechanical disturbance. This makes a difficult point because mechanical stability should be guaranteed over the full useful life of the sensor. Therefore, though used in the first developed prototypes of FOG, micro-optics is not a good approach for a device deployed in the field. It is just helpful to carry out experiments in the laboratory.

The *Integrated Optics* (IO) technology aims to put all required functions together, that is, to integrate the individual devices in a single chip. On the substrate of the chip, we may perform functions like (i) passive functions (SoS or silica on silicon), (ii) passive and modulation functions (lithium niobate or LiNbO$_3$), and (iii) active functions (compound semiconductors like GaAs and InP).

Generally, the cost of facilities for an IO technology is high, so we may use it only for substantial volume of production to share the fixed cost on a large number of pieces. Unfortunately, applications with large volumes are scarce, both in the sensor and in the communication segments.

Another point of concern with IO is coupling of the chip to the fiber pigtails we shall make available to the user for accessing the device. This operation, referred to as *pigtailing*, is a critical manufacture step, because we need to align the fiber core and the guide to within ≈0.1 μm and shall ensure that alignment is kept good throughout the useful life of the device. In addition, the fiber joint usually introduces a substantial pigtail loss. This is because of the differences in size, shape, and index of cores even when the residual alignment error is small. With a normal single-mode fiber, the loss may be as low as 1-3 dB per joint when the waveguide is not so different from the fiber (as with SoS and $LiNbO_3$).

On the other hand, the loss may go up to 6 to 10 dB per joint with compound semiconductors, which have a shape (a rectangle with high aspect ratio) and a high index of refraction much different from the fiber.

Examples of IO gyros have been reported in literature [39,40]. The gyro closest to a fully integrated FOG is a 3×3 version (see Sect.7.7) entirely realized on an SoS (silica on silicon) substrate, a rectangle of about 30×50mm (Fig.7-25). The chip includes the coupler and a spiral-shaped propagation guide, totaling a few meters of optical path.

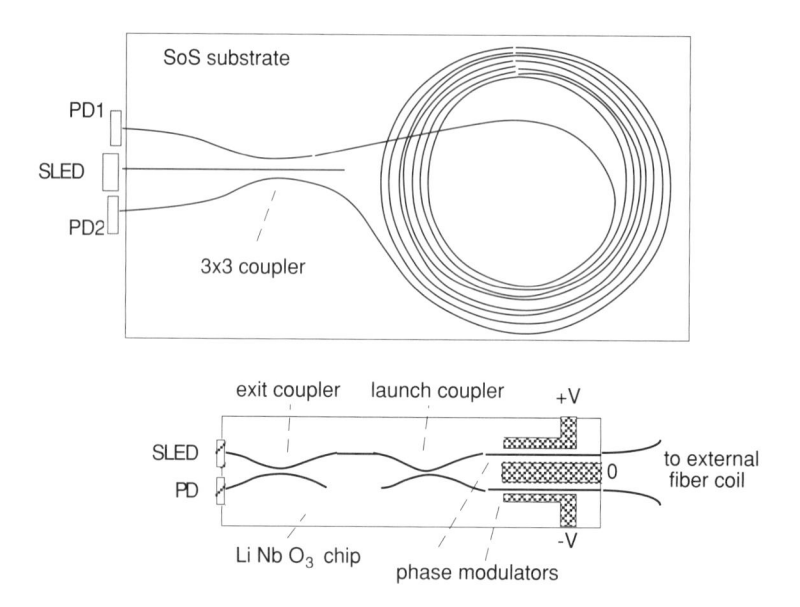

Fig.7-25 Examples of IO technology for FOG: (top) a fully-integrated, minimum-configuration FOG with 3×3 coupler in SoS; (bottom) a lithium niobate chip implementing the splitting and phase modulation functions of a normal open-loop or closed-loop FOG.

The chip is butt-coupled to a SLED and two photodiodes. The device was intended for the low-performance end (Fig.7-1) of the gyro range.

Not a fully IO device, but one incorporating all the components external to the fiber, an electro-optical chip in lithium niobate has been used by several manufacturers of FOG since the mid 1980s. The chip includes the two couplers (Fig.7-19) for source and detector, and two symmetric phase modulators (Fig.7-25). Waveguides are obtained by photolithographic patterning of the chip, followed by a Ti in-diffusion to increase the index of refraction.

The phase modulator is simply realized by a waveguide with a pair of electrode stripes running parallel to it. Applying a voltage to the electrodes, the index of refraction of the waveguide is changed because of the electro-optical effect, and a phase $\Phi = \Delta nL$ is generated. The phase modulators outputs are pigtailed with the usual coil of ≈ 200-m long hi-bi fiber. To get the nonreciprocal phase modulation due to time-asymmetry, the phase modulators are driven with opposite-polarity waveforms.

Called the FOG with integrated chip or *iFOG*, the device has performances equalling those of the best all-fiber FOG. However, performances are now consistently obtained with high ($\approx 95\%$) yield on a production line, not just in the laboratory and after individual trimming. Noteworthy, the iFOG starts from a lower level of received power, typically −10..−15 dB less than in an all-fiber FOG because of the fiber-to-chip joints, but recovers the performance of zero bias and noise in virtue of the cleaner and stable setup.

Another approach based on IO technology aims to implement the RLG (ring-laser-gyro) concept on a semiconductor material like GaAlAs [40]. On a semiconductor chip, the enclosed area A shall be small (≈ 0.1 cm^2) if we aim to get an inexpensive sensor. Therefore, the potential applications of the *IO-RLG* are limited to the low-end sensitivity, like those required in the field of automotive and robotics (see Fig.7-1).

Work is in progress to demonstrate the feasibility of the concept, despite a variety of phenomena observed in semiconductor ring-lasers [41], like locking, self-pulsation and alternate oscillation, that have precluded so far to exploit the Sagnac effect.

7.5.4 The Closed-Loop FOG

The open-loop configuration of Fig.7-19 has a fair performance of linearity and dynamic range. Linearity is good at small ϕ_S, but at increasing ϕ_S is limited by the sin ϕ_S dependence of the signal (Eq.7.27).

Additionally, the maximum range of the Sagnac phase is $\phi_S < \pi/2$ if we are to avoid the sine-function ambiguity.

Applying a postdistortion of the type y=arcsin x to signal x= sin ϕ_S can give a sizeable improvement in linearity, but leaves the dynamic range unchanged.

If we want to improve both linearity and dynamic range, we shall process the pair of signals sin ϕ_S and cos ϕ_S. The two signals are available in the open-loop gyro of Fig.7-19. As we have seen from Eq.7.27, sin ϕ_S and cos ϕ_S are contained in the amplitudes of the odd and even harmonics, respectively.

Thus, we can measure them with the lock-in amplifier as the amplitudes at frequency ω_m (proportional to $\sin \phi_S$) and at frequency $2\omega_m$ (proportional to $\cos \phi_S$).

Signals $\sin \phi_S$ and $\cos \phi_S$ can be processed in a number of ways, similar to those already discussed for interferometric measurements (Chapter 4). Thus, we are able to compute ϕ_S also when it exceeds the $\pi/2$ ambiguity angle, and to reduce the linearity error. Typically, we may reach the 10^{-3} (or 1000 ppm) level of residual error on a range of several 2π.

Yet, this level may not be sufficient for the inertial applications, which call for a 10 to 100 ppm linearity error. In this case, we employ the *closed-loop* configuration.

The basic idea of the closed-loop configuration is that of keeping dynamically zero the total phase shift read by the gyro. We null the gyro by adding a phase shift Φ, equal and opposite of Sagnac phase ϕ_S (Fig.7-26). In this way, the lock-in works around zero at all times, and the linearity error of the $\sin \phi_S$ dependence is eliminated.

The element providing the phase shift Φ shall be a nonreciprocal phase modulator and is discussed below. The condition $\phi_S - \Phi \approx 0$ is ensured by the feedback loop, which is designed to have a large negative gain.

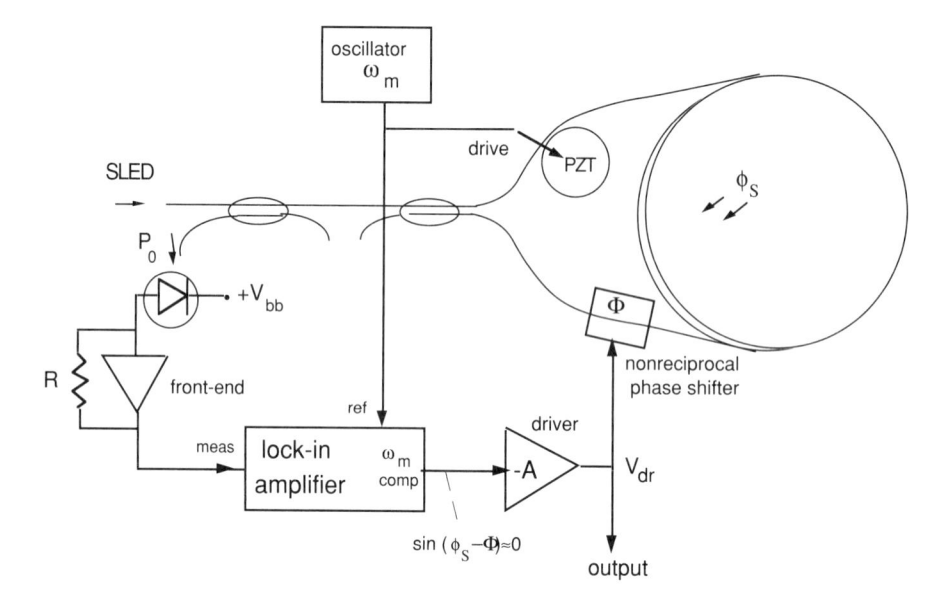

Fig.7-26 In the closed-loop FOG, the output of the lock-in amplifier is fed back to the optical path after amplification and conversion to a nonreciprocal phase shift Φ. The phase shift Φ is obtained by driving a suitable modulator with the amplified voltage V_{dr} of the lock-in output. Because the loop gain is large and negative, the phase difference is kept dynamically zero by the feedback loop, or $\phi_S - \Phi = 0$. The output is then read from the drive voltage V_{dr} as $\phi_S = \Phi = f(V_{dr})$, where f is the transfer characteristic of the phase shifter.

7.5 Development of Fiber Optics Gyro

We can check the negative feedback action by going along the loop in Fig.7-26. Let Φ suffer a negative variation, say it -ε. Then ϕ_S-Φ becomes positive (=+ε), the signal out of the front-end increases, and the lock-in output increases too, and finally we return to the starting point with a phase Φ increased by +Gε. As the counterpart of the electronic follower, we get a feedback loop suppressing any deviations from the nominal ϕ_S-Φ=0 condition, by a factor equal to the loop gain G.

By inspection of Fig.7-26, the loop gain is readily evaluated as:

$$G = 1.16 \, \sigma \, P_0 \, R \, A_{\text{lock-in}} \, A \, \kappa \qquad (7.28)$$

The symbols used in Eq.7.28 are as follows: σ=I/P=spectral sensitivity of the photodiode, P_0 =received power, R =transimpedance of the front-end, $A_{\text{lock-in}}$ =lock-in gain at the first harmonic, $\kappa=\Phi/V_{dr}$ is the modulator efficiency, and $1.16 = 2J_1$ is the modulation factor (Eq.7.27). In the closed loop FOG, the output is voltage V_{dr}, and we go back to the Sagnac phase using the modulator transfer characteristic, $\phi_S = -\kappa V_{dr}$.
In doing so, we have moved from the sin ϕ_S dependence to the new dependence $-\kappa V_{dr}$. Now, we need a well linear transfer-characteristic of voltage to phase, or a constant κ on the whole range of Sagnac phase ϕ_S.

Early attempts [25] employed a double-pass Bragg cell as the nonreciprocal modulator, but the setup was involved, and the linearity level was not satisfactory. The Faraday effect was not much better. Both approaches also suffered from criticality, in avoiding that spurious nonreciprocal phase shifts are inadvertently introduced.

Finally, it was just the asymmetrical phase modulator to be recognized as the best linear element capable of supplying the desired nonreciprocal phase shift. Because a PZT phase modulator is already there for the readout of the Sagnac phase, we may think of using it also for the phase null operation, adding the two signals sent to the PTZ drive.

Alternatively, we may introduce a second PZT located symmetrical with respect to the fiber coil, as in Fig.7-27. Having the two signals on separate PZTs, each with its own requirement of linearity to satisfy, leads in a better linearity over a wider range of applied signals.

The phase-null PZT shall be driven by a periodic sawtooth signal. This waveform is our best effort to approximate a phase ψ=H t steadily increasing with time. As we cannot increase the phase of a modulator indefinitely, we use the sawtooth signal. During the linear ramp period, the signal has the form s(t) = H t (for t=0..τ), and then has a fast flyback to zero, in a time interval $\tau_{fb} \ll \tau$.
If the initial phase of the sawtooth is insignificant to our measurement, we get the equivalent of an indefinitely increasing ramp, provided we gate out the measurement during the flyback period.
Now, when we add the sawtooth phase, the nonreciprocal phase shift $\Phi=\Phi_{CW}-\Phi_{CCW}$ suffered by the two counterpropagating waves can be readily found by repeating the reasoning leading to Eq.7.26. If T=nL/c is the fiber coil transit time, the result is found as:

$$\Phi = H(t+T/2) - H(t-T/2) = H \, T \qquad (7.29)$$

Eq.7.29 tells us that the quantity affecting the generated phase is the sawtooth slope. Therefore, we will use the voltage V_{dr} to drive a voltage-controlled ramp generator (not shown in Fig.7-26), whose output is the actual signal sent to the phase modulator. Thanks to the feedback loop, it is now the slope to be clamped at the correct value to null ϕ_S-Φ.

The above configuration is an *analogue* closed-loop FOG because of the analog processing of signals. The sawtooth arrangement is also referred to as a *serrodyne modulation* [42-44] and in the mid 1980s it was the first demonstration of the FOG potentiality in approaching inertial grade performance.

The analogue processing is objectionable because, if we finally convert the FOG signal to digital for navigation calculations, we shall pass through the nonidealities of an A/D (analog to digital) converter.

Then, we look for an intrinsically digital readout of the Sagnac phase, while retaining the benefits of the closed loop approach. The solution is shown in Fig.7-27 and is called the *digital* closed-loop FOG [7,45]. It employs a multistep sawtooth modulation, also called digital ramp because of its function.

Each step increases the phase by an increment $\Delta\phi$ equal to the least-significant-bit (LSB) we want in the FOG readout. Simply stated, if in the transit time T through the fiber coil the phase modulation changes by M steps or $\Phi=M \Delta\phi$, and the interferometer is balanced (ϕ_S-Φ =0), the Sagnac phase is then read out as ϕ_S= M$\Delta\phi$.

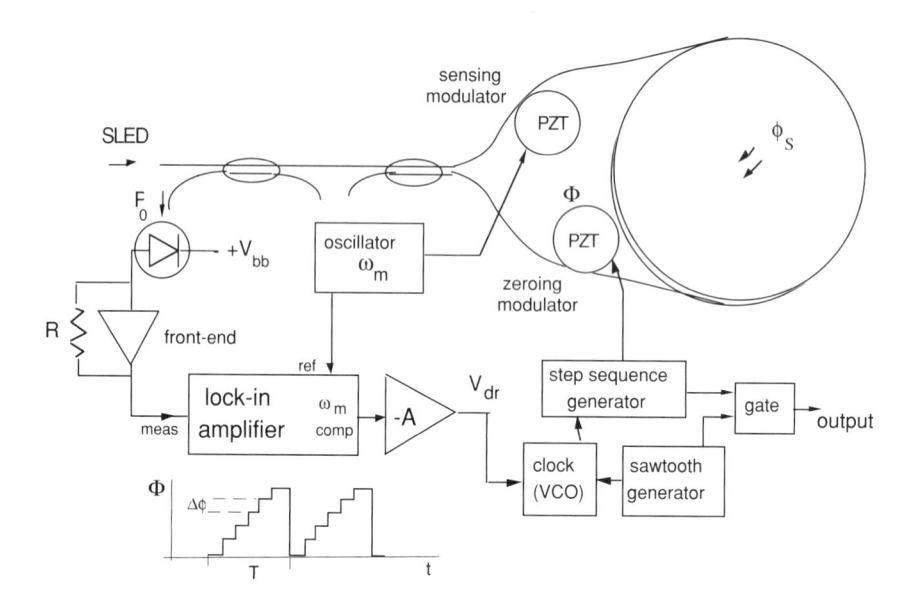

Fig.7-27 In the digital-readout closed-loop FOG, a digital (or multistep) ramp is used to drive the nonreciprocal Sagnac-zeroing modulator. Each phase step corresponds to one LSB of the readout.

7.6 The Resonant FOG and Other Approaches

To drive the PZT modulator, we now need to convert the lock-in output voltage V_{dr} to a digital ramp. In the digital ramp, all the steps have the same amplitude, and their frequency (or inverse of the time spacing) shall increase as the lock-in output V_{dr} increases. These functions are readily implemented by the electronic circuits shown in Fig.7-27. A VCO clock fed by V_{dr} provides the pulse frequency, whereas the sawtooth generator gives the underlying ramp waveform and provides the synchronism to validate the output through a gate.

The digital readout has the additional advantage of allowing an easy increase of resolution (App.A0) by averaging, as we now will illustrate. For a good start, the LSB phase increment is chosen a bare $\Delta\phi=0.1$ mrad, apparently a small value. Yet, the angular velocity corresponding to $\phi_S=0.1$ mrad is not at all large, typically about $\Omega=100$ deg/h (see Fig.7-5). Now, suppose that each measurement takes, say 0.1 ms. We can acquire 10^4 of them in a 1-s measurement time. Summing them all up, the result has a truncation error $\sqrt{10^4}=10^2$ times smaller than the original truncation error of the individual measurement. For this statement to be true, we only need that the random noise superposed to the measurement is larger than 1 LSB. Thus, by averaging we are able resolve $\Omega=1$ deg/h.

As a concluding remark about the digital closed-loop configuration, the frequency pass band required to the phase modulator is not simply $\approx 1/T$ of the open-loop readout, but probably about 20..50/T because of the ramp and multistep features that we need to preserve.
This leads to preferring the integrated-optics approach based on the lithium niobate modulator (Fig.7-25) for implementing the digital closed-loop [7,37,45].

In conclusion, after ten years of maturation, the FOG has evolved to probably its best level of technical appeal as a sensor with the *digital closed-loop* configuration we have discussed in the previous pages. An example of this sensor is shown in Fig.1-9.

The internal configuration counterpart of the FOG, that is, the RLG (Fig.1-8) has won an indisputable position in the field of application of INUs and AHRSs in avionics. On the other hand, the *digital closed-loop* FOG has reached a good acceptance in some space applications (AHRS of low-orbit satellites, and gyro-compass) and other niches (oil drilling, rocket spin stabilization). Yet, the FOG market size has never reached that of the RLG nor has really threatened the RLG on its applications.

7.6 THE RESONANT FOG AND OTHER APPROACHES

The research on the FOG is restless. As mentioned in last section, the FOG is not satisfactorily close to its quantum limit of performance. This circumstance has triggered a number of international R&D laboratories to pursue several alternative approaches, aimed to take to the stage of inertial-grade gyroscopes. A configuration that appears promising is the R-FOG, or *Resonant Fiber-Optic-Gyroscope* [46].
The basic idea is to use the fiber coil not just for propagation, but as an element to build a resonator. The resonator is the ring we obtain when input and output of the launch coupler of the basic FOG are interchanged.

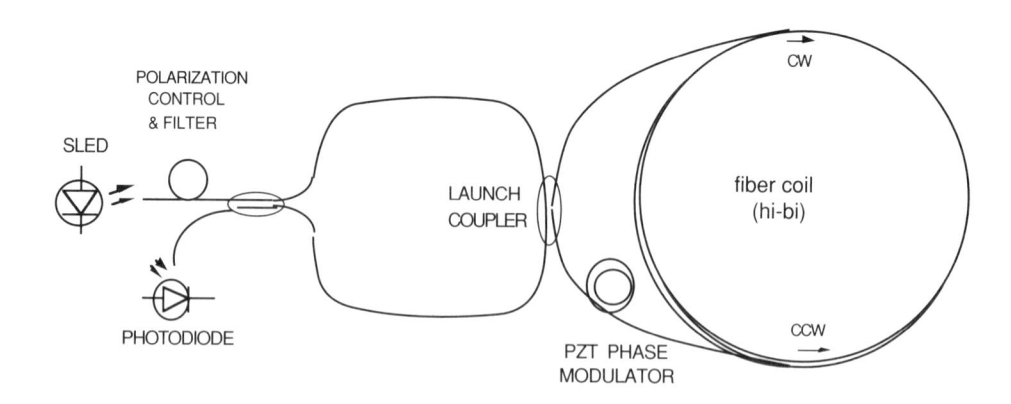

Fig.7-28 The R-FOG uses the fiber coil as a ring resonator. The PZT phase modulator keeps the total path length locked to the resonance peak, and senses the Sagnac phase shift by an added ac modulation.

The result is a long path-length resonator (Fig.7-28), which potentially can reach a high value of finesse F (or cavity Q-factor). As pointed out in Appendix A2.1, the ring resonator has a response curve much sharper than the normal Sagnac interferometer.

If we look at Fig.A2-2, we see that the slope of the relative power versus $2ks=\phi$ curve is steeper than that of the normal sin 2ks response. We have an improvement of slope or responsivity by a factor F, the finesse of the resonator (Table A2-1). We can then achieve from the R-FOG the same responsivity of a normal FOG just using a fiber shorter F times that of a FOG.

As it is (Eq.A2.4) $F=2\sqrt{R}/(1-R)$, where R is the (power) through transmission of the coupler, with the reasonable choice R=0.99, we can easily get F=200. Thus, instead of L=200 m, we can work with a bare L=1 m in a R-FOG with the same signal.

The saving of fiber length and associated cost is not that important, however. Much more important is the strong decrease of nonidealities we get with the short fiber.

Paradoxically, we can get a better performance by a shorter fiber in the ring than from the more responsive full length called from the FOG. In a certain sense, shortening the fiber moves the FOG near to the RLG and its perspective performance closer to the quantum limit.

The readout of the R-FOG is performed by one or more photodetectors sensing the outputs tapped by the launch coupler or the result of their recombination.

The PZT modulator applies a sine phase shift like in the normal FOG. In addition, it also serves to lock the working point at the peak of the ring resonance, by applying a dc bias that dynamically tracks the resonance.

Performance of the R-FOG started in mid 1980s from a relatively unsatisfactory level, but rapidly caught and marginally surpassed the best figures provided by the conventional FOG, at least in laboratory units reported in the open literature.

Another approach to the FOG was explored in the 1990s, following the advent of the EDFA (erbium doped fiber amplifier) technology.

A coil of doped fiber, typically ≈10-30 m long, is optically pumped, using a ≈100 mW laser-diode, to population inversion. The pump laser emits at an absorption band of the active Er ion, and the two most used wavelengths are 980 nm and 1480 nm. The active transition of the Er ion sustains amplification, and eventually oscillation, in the well-known third window of optical fibers, that is in the range λ=1520...1580 nm.

Thus, we may try to fabricate an EDFA-based version of the internal-configuration Sagnac-interferometer, that is, the fiber counterpart of the RLG. Once more, the experiment reveals that the fiber is a medium much dirtier than a He-Ne gas mixture, optically speaking. Thus, instead of a clean narrow-line oscillation, the EDFA-based RLG has a broad line, jittering in wavelength because of longitudinal mode competition and of backscattering down the fiber. In addition, the counterpropagating modes are strongly locked by the much larger backscattering in the fiber.

In view of these problems, the EDFA-based fiber laser is not likely to become the solid-state replacement of the old He-Ne laser for an RLG configuration.

The EDFA can be useful, however, for application in the external-configuration FOG. The pumped fiber is a source of *superluminescent* radiation. This radiation is just the ASE (amplified spontaneous emission) supplied by the fiber [47], and we can use it as the source of the FOG in place of the normal semiconductor SLED. The advantage of such *S-EDFA* is that the fiber is antireflection coated to a much better extent than a semiconductor facet, whence a smoother shape of the line (see Fig.7-24). Second, we can protect our S-EDFA source from back reflections originated by the returned signal simply with the use of an optical isolator that, differently from λ≈850 nm, is a readily available component at λ≈1550 nm. Actually, moving from 850 to 1550-nm almost halves the response of the FOG, but the better control of the source more than compensates for the loss.

An improvement of about an order of magnitude in the NEΩ (down to perhaps some 0.01 deg/h) has been reported for the S-EDFA i-FOG.

7.7 THE 3X3 FOG FOR THE AUTOMOTIVE

Another appealing market for the gyroscope is the automotive segment [48,49]. Some requirements of the automotive are also common to the robotic market (Fig.7-1), and this circumstance adds value to a low-performance application with a potentially large volume.

A feature of the automotive application is that volumes are huge, but price shall be very low, even at a sacrifice of performance. Price comes first, in the automotive segment, and only matching the price you can go ahead and talk of performance. Instead of the US$ ≈5,000 to 10,000 of the best FOG, the automotive FOG calls for a price of just ≈ US$ 10 to 20.

To try matching this severe price specification, we shall think of a very simplified configuration, with the minimum part-count, yet capable of reaching a respectable performance, say ≈300 deg/h ≈0.1 deg/s.

Although we will renounce to a reciprocal configuration because it requires too many components, we shall solve the $\cos\phi_s$ ambiguity of the normal FOG, ensuring that the signal supplied by the sensor is of the form $\sin\phi_s$.

An alternative to the use of the phase modulator is that of employing a *minimum FOG* configuration based on a *3×3 coupler*.

This 3×3 configuration has been proposed [50] since the early times of the FOG as a solution to get the $\sin\phi_s$ dependence, but it was soon abandoned because it is not inherently reciprocal.

In a minimum configuration FOG, we clear the reciprocity requirement and take advantage of the drastic component-count reduction, so the 3×3 FOG becomes very appealing.

About the 3×3 coupler, we rely on the phase shifts introduced between the input and output ports to obtain the sine dependence. To understand how, let us assume that the coupler is made by three fibers with the core sections symmetrically placed (Fig.7-29).

Then, coupling from one fiber to each of the other two fibers shall be the identical, for symmetry reasons. We shall now apply the conservation of electric field vector as well as of the power (square of field amplitude), generalizing the well-known 2×2 coupler results [18].

Entering the coupler with E_0 at port 0 (Fig.7-29), the output vectors E_{1C} and E_{2C} at outputs 1 and 2 shall be parallel because of symmetry, and both shall be orthogonal to the output vector E_D of output 0.

The three vectors then form a right-angle triangle. As we can check, the vector sum gives E_0 as a result, and also the intensity E_0^2 is the sum of the squares E_D^2 and $(E_{1C}+E_{2C})^2$.

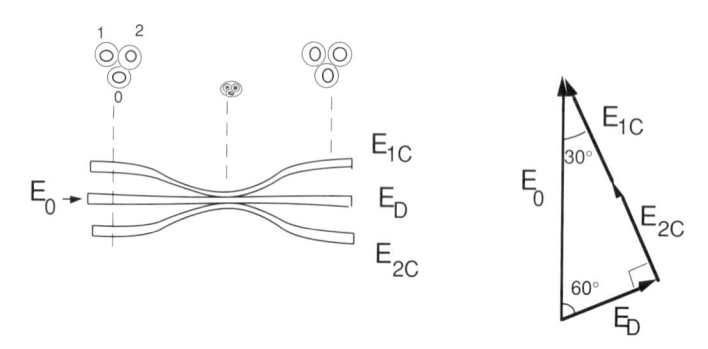

Fig.7-29 The 3×3 coupler splits the input field E_0 in three output ports. The vector at the direct output E_D is perpendicular to the crossed output vectors E_{1C} and E_{2C}, which are parallel to each other. If the splitting ratio is equal for all outputs (=1/√3), then angles at the hypotenuse are 30° and 60°.

We may now assume that the splitting ratio is equal for all ports, that is 1/3 in power and 1/√3 in field amplitude. Then, the triangle of vector sum has one side that is double of the other, and the angles are 30 and 60 deg.

We can write the above results as:

7.7 The 3×3 FOG for the Automotive

$$T_D = E_D/E_0 = (1\sqrt{3}) \exp -i\, 60°$$

$$T_{1C} = E_{1C}/E_0 = T_{2C} = E_{2C}/E_0 = (1\sqrt{3})\, E_0 \exp +i\, 30° \qquad (7.30)$$

Let us now turn to consider the 3×3 FOG configuration of Fig.7-30. As we can see, power from a SLED is launched from one fiber of the coupler, and the propagation coil is connected to the other two fibers of the coupler. When we come back after propagation, the two fields mix again at the coupler, and we collect them on the photodiodes PD1 and PD2.

Worth noting, the configuration uses the minimum possible number of components. The fiber coil is made of normal, telecommunication-grade, single mode fiber. This fiber may cost ≈0.1 $/m and we use just L=100-m of it on a small coil (≈20..30-mm diameter), because increasing L does not improve the NEΩ in a nonreciprocal configuration. An SLED or even a high radiance LED can be used, emitting at a wavelength anywhere in the broad range λ=1200 to 1600 nm of low fiber attenuation αL at the assumed length L.

The coupler is fabricated with the fused-coupler technology [36] out of the same fiber pigtails of the coil and of the SLED. In this way, no joint is necessary, and we save a substantial cost of assembly. Photodiodes are normal, small area ternary (InGaAs) pn or pin PDs, with no special speed requirement. The only expensive component is the pigtailed SLED.

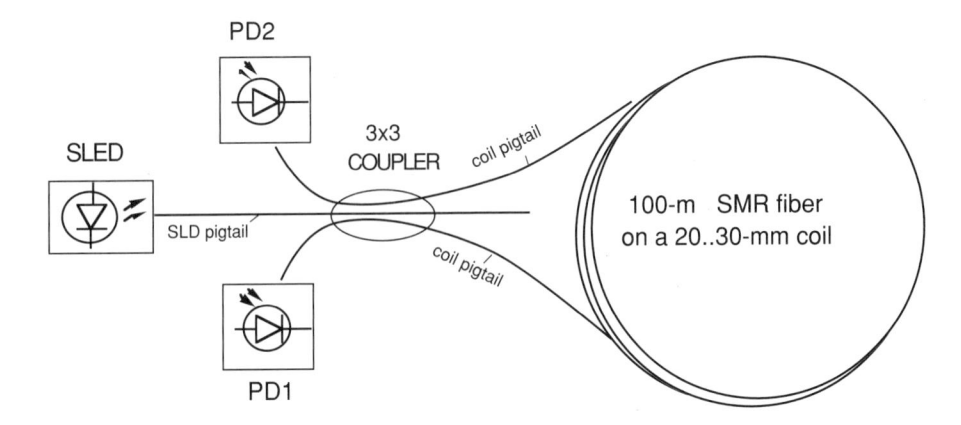

Fig.7-30 The 3×3 FOG is the gyroscope with the minimum part count. It requires just the fiber coil (a normal fiber, shorter than in a normal FOG), a SLED source, two photodiodes, and the 3×3 coupler. To dispense of fiber splices, the coupler is fabricated directly on the pigtails of SLED and coil.

Let us now analyze the fields collected at the photodiodes PD1 and PD2, that is, E_{PD1} and E_{PD2}. They are made up of two contributions, coming from the same or the opposite-side fiber. As we can see from Figs.7-29 through 7-31, we have:

$$E_{PD1} = E_{1CC} + E_{2CD}, \quad E_{PD2} = E_{1CD} + E_{2CC} \qquad (7.31)$$

Recalling Eq.7.30, we can write the terms in Eq.7.31 as:

$$E_{1CC} = E_{1C} \exp{-i\phi_S} \; T_{1C} \quad E_{2CC} = E_{2C} \exp{+i\phi_S} \; T_{2C}$$

$$E_{2CD} = E_{2C} \exp{+i\phi_S} \; T_D \quad E_{1CD} = E_{1C} \exp{-i\phi_S} \; T_D \quad (7.32)$$

By inserting in Eq.7.31, we get:

$$E_{PD1} = E_{1C} \exp{-i\phi_S} \; T_{1C} + E_{2C} \exp{+i\phi_S} \; T_D = (E_0/3) \; [\exp i(+60°-\phi_S) + \exp i(-30°+\phi_S)]$$

$$E_{PD2} = E_{2C} \exp{+i\phi_S} \; T_{2C} + E_{1C} \exp{-i\phi_S} \; T_D = (E_0/3) \; [\exp i(+60°+\phi_S) + \exp i(-30°-\phi_S)] \quad (7.33)$$

The photodetected current follows [16] as the square mean value of the electric field, $I = \langle E^2 \rangle$. Because there are two terms in the fields of Eq.7.33, we shall develop the square. Doing so, we get: $I = \langle E^2 \rangle = \langle E_0^2 (\exp i\xi + \exp i\zeta)^2 \rangle = E_0^2 [1+1+2\text{Re}\{\exp i(\xi-\zeta)\}] = 2E_0^2 [1+\cos(\xi-\zeta)]$. Applying this expression to Eq.7.33 gives:

$$I_{PD1} = 2 \, (E_0/3)^2 \, [1 + \cos(90°-2\phi_S)] = 2 \, (E_0/3)^2 \, (1+ \sin 2\phi_S)$$

$$I_{PD2} = 2 \, (E_0/3)^2 \, [1 + \cos(90°+2\phi_S)] = 2 \, (E_0/3)^2 \, (1- \sin 2\phi_S) \quad (7.34)$$

By computing the difference $I_\Delta = I_{PD1} - I_{PD2}$ of detected currents, we are able to suppress the dc term and obtain a signal proportional to the sine of the Sagnac phase ϕ_S for small ϕ_S:

$$I_\Delta = 4 \, (E_0/3)^2 \sin 2\phi_S \quad (7.34')$$

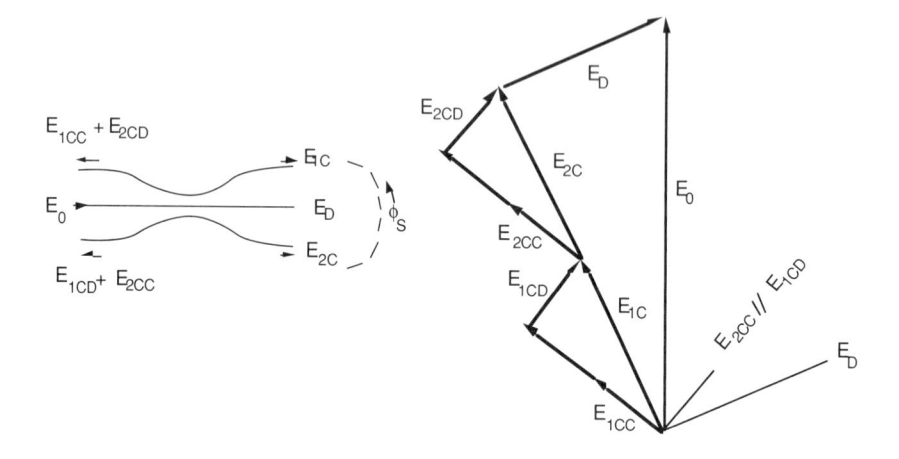

Fig.7-31 Propagation of signals through the 3×3 coupler of a minimum-configuration FOG. Each output port receives two contributions, propagated in opposite directions. Because of the phase shift suffered in the 3×3 coupler, the resulting beating are of the desired form $\pm \sin 2\phi_S$.

7.8 THE MEMS GYRO AND OTHER APPROACHES

A typical 3×3 FOG is built using a 30-mm diameter fiber coil with 100 m of single mode fiber [51]. For such a device the performance is the following: baseline drift ≈0.05 deg/s, NEΩ≈0.01 deg/s (at B=10 Hz), and linearity error <0.5% up to a Ω≈200°/s dynamic range.

7.8 THE MEMS GYRO AND OTHER APPROACHES

The electro-optical gyroscope is the undisputed choice for inertial-grade applications, but in the automotive and robotic segments other technical approaches have been developed, because they are competitive in terms of low cost at reduced performance (Fig.7-1).

In this section, we briefly digress on these approaches to get an insight on the challenge coming from competing technologies.

All the approaches developed so far, based on an effect different from the Sagnac effect, can be traced to the Coriolis' effect as the fundamental mechanism for sensing the inertial rotation speed Ω.

As we recall from physics, Coriolis' effect is about the apparent force F_C developed on a mass m moving with a velocity v, when the laboratory is subjected to an angular velocity Ω:

$$\underline{F}_C = 2m \, \underline{\Omega} \times \underline{v} \qquad (7.35)$$

Here, $\underline{\Omega}$ and $\underline{v}$ are the vectors of angular velocity and spatial velocity, and × indicates the vector product.

In a schematization of the MG, the mass m is suspended with springs to the walls of the sensor box, as illustrated in Fig.7-32.

The constant of the x-axis spring is k_x, so that the displacement x produced in the spring by force F_x is $x = k_x F_x$. Also schematized in Fig.7-32, friction r_x is related to the force needed to keep a speed $x'=dx/dt$, as $F_x = r_x x'$. Last, we know that a force $F_x = m\,x''$ is required to accelerate the mass m at an acceleration $x''=d^2x/dt^2$.

By summing up the contributions, we can write the following balance equation for the external force F_x applied to the mechanical system:

$$m\,x'' + r_x\,x' + k_x^{-1}\,x = F_x \qquad (7.36)$$

For the two other axes, y and z, two more equations similar to Eq.7-36 should be written. Thus, our mechanical system is described by second-order responses for each axis.

To sense the Coriolis' force F_C, the mass is put into vibration by means of a suitable actuator.

Let the actuator move the mass along the x-axis, as shown in Fig.7-32 (left), so that $\underline{v} = v\,\underline{x}$, where $\underline{x}$ is the unit vector of the x-axis.

Then, a rotation $\underline{\Omega}$ parallel to the y-axis develops a force directed along the z-axis, because Eq.7.35 gives in this case $\underline{F}_C/2m = \underline{\Omega} \times \underline{v} = (\Omega v)\underline{y} \times \underline{x} = -(\Omega v)\underline{z}$.

We will measure the displacement z produced by this force through a suitable sensor (Fig.7-32, left) to obtain the angular velocity component Ω_z.

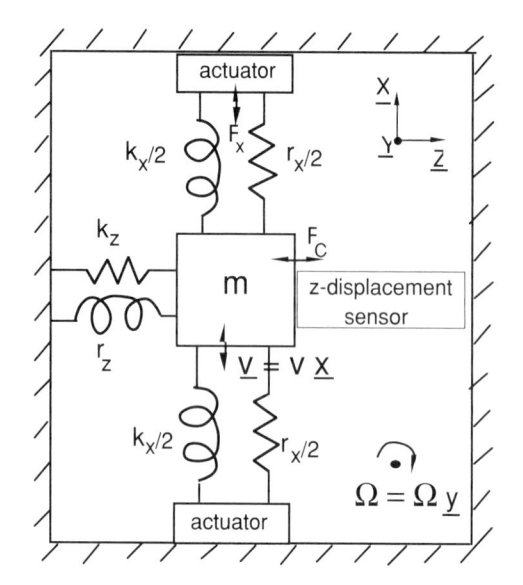

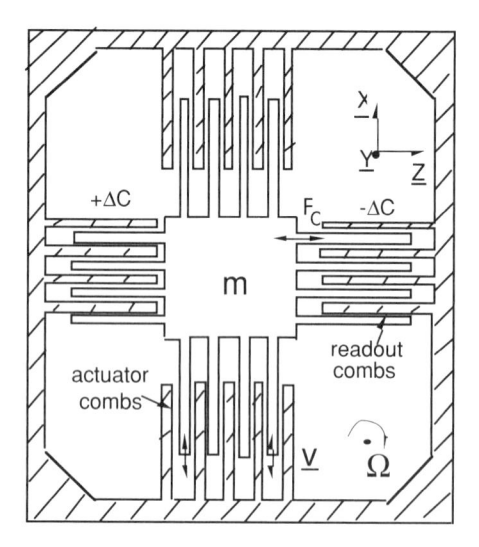

Fig.7-32 Left: A mechanical gyro is based on the Coriolis' force acting on a sensing mass m. The mass is suspended by springs and is put into vibration along the x-axis by a suitable actuator. By rotating the gyro at an angular velocity $\underline{\Omega}$ along the y-axis, a Coriolis force F_C is developed along the z-axis. A suitable sensor measures the z-displacement, and this result is traced back to the value of Ω. Right: In the MEMS gyro, the mass m is suspended above the substrate by a spring hinge (not shown) and has comb expansions along the x- and z-axis. The x-axis combs serve to actuate the mass by electrostatic force. This force attracts the mobile comb inside the fixed comb of the frame. The z-axis combs have a capacitance that varies with opposite sign as the mass moves along z, and this is the electrical readout of the z-displacement.

Alternatively, if $\underline{\Omega}$ were parallel to the z-axis, we would get $\underline{\Omega} \times \underline{v} = (\Omega v)\underline{z} \times \underline{x} = (\Omega v)\underline{y}$, and the signal is developed along the y-axis.

If we are able to sense the vibrations transferred to both the z and y axes, we could in principle make a 2-axis measurement of the angular velocity vector $\underline{\Omega}$ with a single sensor. This was actually the case of the old mechanical gyros, for example, the Styroflex gyro family [15].

On the contrary, in all other gyros and in MEMS, the mechanical axes are not so well decoupled. Equations like $x=\eta_{xy}y$, $x=\eta_{xz}z$ and $z=\eta_{zy}y$ are to be added to Eq.7.36 to describe coupling. The result is a cross-sensitivity at small η's and more complicate chaotic effects at higher η's. To avoid all these errors, we shall make a single-axis measurement and strive to minimize the cross-sensitivity to other axes.

7.8 The MEMS Gyro and Other Approaches

This is done by minimizing the cross-coupling η coefficients by design, and by appropriately increasing the stiffness k of the axes other than the measurement and excitation axes, so that the residual F/k is a small displacement.

The actuator indicated in Fig.7-32 provides a periodic excitation of the proof mass M at a frequency ω_{exc}. This generates, in response to the measurand Ω, a periodic signal of the type $(\Omega v) \cos \omega_{exc} t$ at a carrier frequency ω_{exc}. Of course, an ac signal is better than a dc one because the measurement is moved away from the low frequency range, affected by 1/f and slow drift components. The frequency response of the sensor to the measurand Ω is a fraction 1/Q of ω_{exc}, the carrier frequency for the signal.

The maximum frequency of excitation is determined by the mechanical resonance of the structure, expressed by Eq.7.36. The resonance is given by $\omega_{res}=\sqrt{(k/m)}$, m being the oscillating mass and k the stiffness (or spring constant) of the elastic structure bearing the mass.

Again from Eq.7.36, we find the fractional pass-band of the resonance as $\Delta\omega_{res}/\omega_{res}=1/Q$, where Q is the merit factor given by $Q=\omega_{exc}m/r$, and r is the friction coefficient (r in Fig.7-32).

Along the excitation (x-axis) and the readout (z-axis) in Fig.7-32, the spring coefficients are k_x and k_z. For the best performance, we will require $k_x = k_z$ so that $\omega_{res(x)} = \omega_{res(z)}$.

Working with the excitation at resonance $\omega_{res(x)}$ gives a large velocity signal with a small drive power, reduced by a quantity equal to the Q_z-factor with respect to the power required off-resonance. Working with the readout at resonance $\omega_{res(x)}$ ensures an increase of response amplitude by a factor Q_x.

In general, we may assume that a displacement x is induced by Coriolis' force F_C, as a response to measurand Ω. The displacement is $x = k_z F_C$ in the example considered in Fig. 7-32. Now, because we want a large response $1/k_z$ and a high frequency cutoff $\sqrt{(k_z/m)}$, it is clear that we need a mass m as small as possible. This is what we get in MEMS structures.

7.8.1 MEMS

A MEMS (Micro Electro Mechanical System) is just a micro-system fabricated in a silicon crystal with the technology of semiconductor devices [52].

As it is well known, silicon is the most used electronic material, and we know that devices and integrated circuits are fabricated by a well-established process, involving steps like epitaxy, lithography, etching, oxidation, dielectric and polysilicon deposition, metallization, etc.

When we look to properties other than electrical, silicon is recognized as a very good mechanical material too, with strength comparable to steel. Moreover, silicon has a good thermal conductivity.

We can fabricate a miniaturized mechanical structure on a Si chip by combining the usual processing steps of selective growth and etching to one more specific step, the removal of sacrificial layers. These sacrificial layers can be etched from underneath to leave suspended structures in silicon [52].

As resolution is set by photolithography, the MEMS finest detail may be as small as ≈1μm, and already complex structures may be fabricated with a typical size of only 50 to 100 μm.

Thus, the degree of miniaturization offered by MEMS is very good, and the integration may go up to a complexity only limited by the yield to be attained in the fabrication process.

The resulting technology is called *micromachining* and is currently pursued to develop new components, starting from machines like motors and actuators and including a wide range of devices for communication, displays, bioelectronics, and sensors.

Among sensors, the MEMS gyro can be regarded as a revisitation of the old mechanical gyro (MG), but fabricated on a Si chip with the new technology of micro-machining [53].

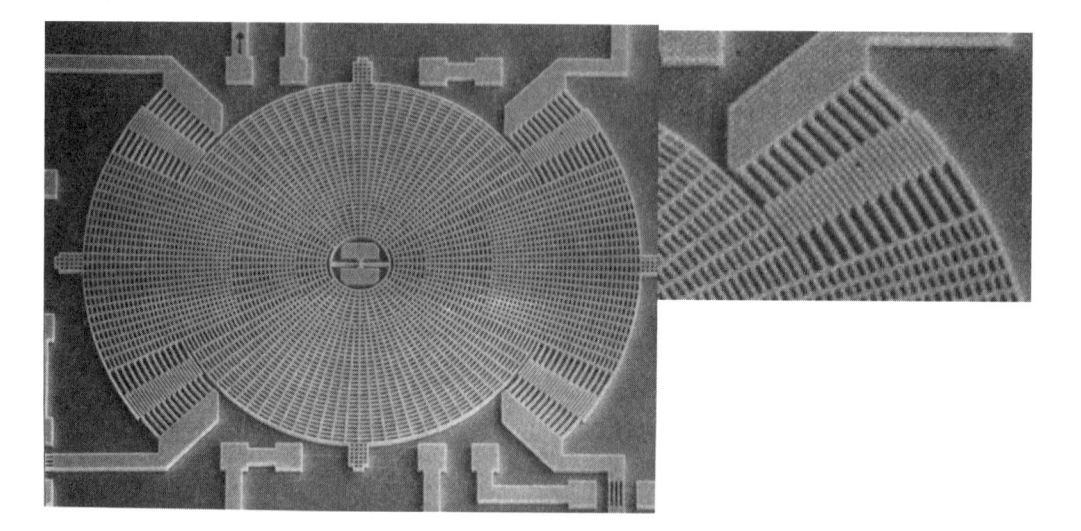

Fig.7-33 A typical MEMS gyro is made by a Si-rotor suspended above the substrate and free to oscillate around the central hinge. Comb fingers (magnified at right) located at the four corners allow electrical excitation of the oscillatory motion. An angular velocity Ω in the plane of the rotor displaces the rotor vertically (out of plane). The finger capacitance is measured to sense this displacement (adapted from [54], by courtesy of Bosch GmbH and IEEE).

The structure of MEMS gyros can take lot of forms. We can use either a linear structure with mass suspended by springs (as in Fig.7-32) [53] or a rotor structure free of oscillating about a central hinge (Fig.7-33) [54]. Common to all structures is the excitation by means of electrostatic actuation and the readout of the displacement by means of capacitance variation.

Electrostatic actuation is very effective because, on the small scale of MEMS size, even a modest voltage V produces an appreciable force $F=\epsilon V/d$ across the small gap d, more than adequate to move the small mass m.

The preferred actuation is a sinusoidal waveform at the resonant frequency of the structure so that the oscillation amplitude of velocity v is high even with modest drive power.

Capacitance between the fixed part and the mobile part subjected to the Coriolis' displacement is increased up to the desired level ($\approx$pF) by appropriately increasing the surface of

7.8 The MEMS Gyro and Other Approaches

the comb structure (Fig.7-33). In this way, a large ΔC/C signal is developed following a transversal displacement Δx. If the gyro is put into vibration parallel to the x-axis (in the reference of Fig.7-32), and Ω is directed along the y-axis, the displacement is along the z-axis and is in ac like the drive along the x-axis.

We then need two small stiffnesses k_x and k_z, to let the structure move easily along the x- and z-axes, whereas the stiffness along the y-axis will be much larger than k_x and k_z.

The capacitance signal ΔC/C is read by arranging the readout comb capacitors in a bridge circuit. If the MEMS is designed so that two combs in it collect opposite signals (+ΔC and -ΔC as in Fig.7-32), we will place these elements at opposite diagonals in the bridge. The bridge is completed by two more dummy elements, similar in shape to the active combs.

By applying dc voltages to the dummy elements, the bridge can be balanced at rest, or the zero-bias is trimmed to zero.

The limit performance of MEMS gyro is approached when we are able to cure a number of disturbances, like cross-axis sensitivity due to mechanical and/or electrical coupling, noise of the read out circuits, etc.

Then, the ultimate limit of NEΩ is set by mechanical-thermal noise because each degree of freedom of the mechanical structure (either rotational of translational) acquires by Boltzmann's theorem an energy (1/2)kT because of thermal agitation.

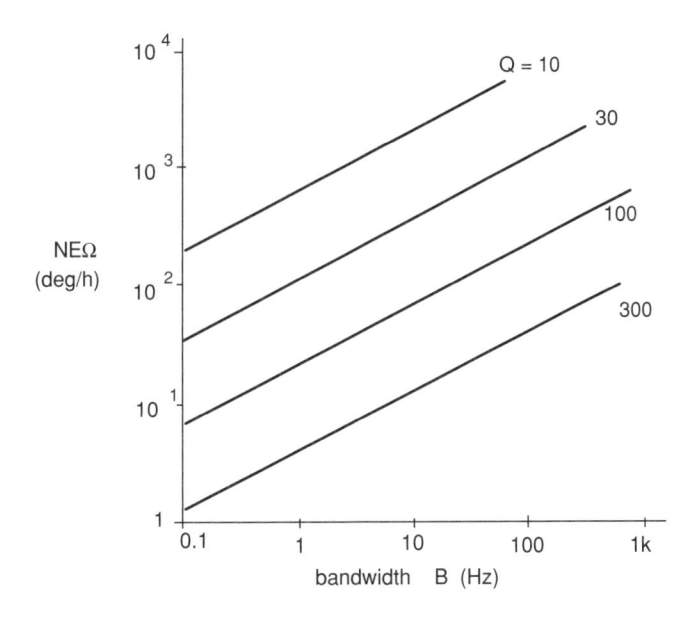

Fig.7-34 Noise equivalent angular velocity NEΩ versus measurement bandwidth B for a MEMS gyro as limited by the mechanical-thermal noise. Data is from Eq.7.36 with m=2μg and ω=10 kHz.

By analyzing the effect of mechanical-thermal noise on a vibrating mass [55], it can be found that the corresponding NEΩ is given by:

$$\text{NE}\Omega = \omega^{3/2} (4kmTB)^{1/2} /2Q^{3/2} F_0 \qquad (7.37)$$

Here, F_0 and ω are the peak force and frequency of excitation, $F= F_0 \cos \omega t$, Q is the quality factor along the sense axis, and m is the vibrating mass.

As we can see from Fig.7-34 where the NEΩ is plotted versus bandwidth [55], for Q=100...300 and a driving frequency $\omega/2\pi$= 5..20 kHz, appropriate to obtain a measurement bandwidth B≈100 Hz, we get a mechanical-thermal noise limit in the range NEΩ= 40-200 deg/h.

This is a satisfactory target for applications to automotive and robotics, but is obviously far away from the inertial-grade specifications, as all other nonoptical approaches are.

In practical MEMS fabricated so far, performance differs from the mechanical-thermal noise limit by a factor ≈5..10.

Perhaps a better result can be obtained with a hybrid MEMS, combining the mechanical structure of a MEM gyro to the optical readout of Coriolis'-induced vibration. The advantage of this approach is the separation of the mechanical structure from the readout, thus a release of design constraints.

As indicated in Fig.7-35, we can read the displacement along the z-axis, following an excitation along the x-axis and a rotation Ω parallel to the y-axis, by means of a self-mixing interferometer [56] placed in front of the mass. A feasibility evaluation of the concept indicates a potential performance of ≈0.6 deg/h√Hz in terms of noise spectral density [56].

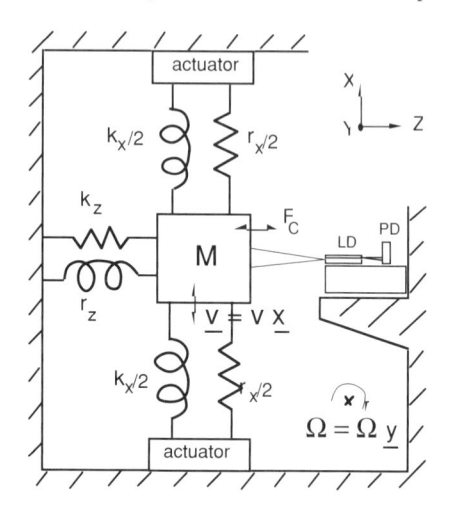

Fig.7-35 A hybrid MEMS with optical readout of the displacement along the z-axis, or MOEMS [56] can alleviate the design requirements and help improve NEΩ.

7.8 The MEMS Gyro and Other Approaches

7.8.2 Piezoelectric Gyro

The piezoelectric gyro (PG) is another example of mechanical gyro (MG). In the PG, both excitation and readout of the mechanical oscillation are obtained through the piezoelectric effect. The piezoelectric effect is the transfer of energy from mechanical to electrical form and vice versa, taking place in piezoelectric materials like quartz, poled ceramics (like PZT), etc. The distinct advantage of PG, compared to other MG approaches, is that we have an easy access to the mechanical quantities of the sensor through electrical counterparts, simply by depositing a metal electrode at the chosen location. The disadvantage is that bulk effects are used in devices reported so far. In consequence, the overall size of the PG is in the range of cm's rather than mm's as for Si-MEMS.

Several different shapes have been proposed for the PG, all referable to the basic MG configuration of Fig.7-32. Perhaps the first has been the tuning fork PG [57], shown in Fig.7-36.

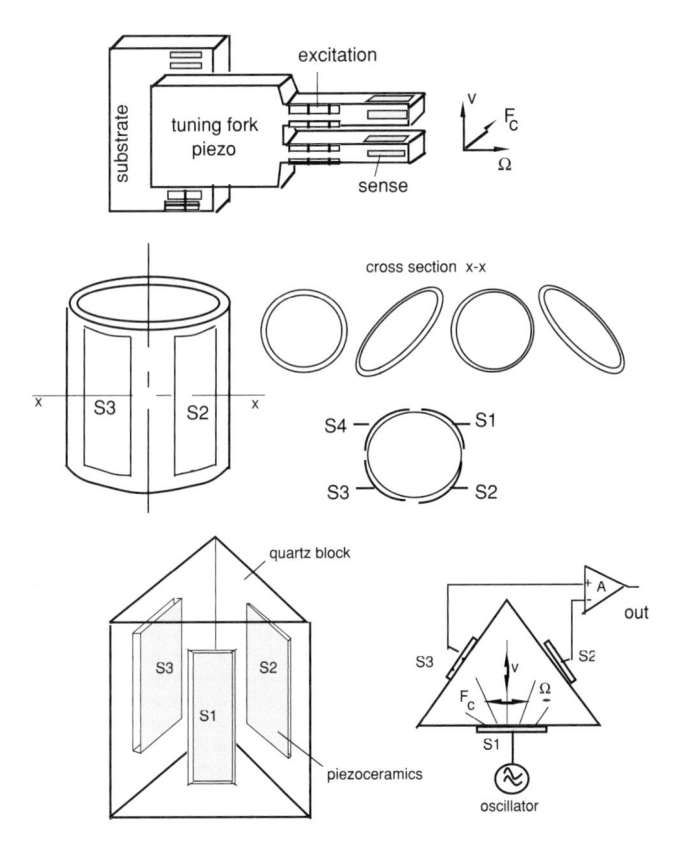

Fig.7-36 Examples of piezoelectric gyros configurations: tuning fork (top), wineglass resonator (center), and triangle resonator (bottom).

This configuration uses an inexpensive quartz crystal, similar to those of electronic watches. Oscillation of the fork is along the axis labeled v, and sensing is along the axis labeled F_c. in Fig.7-36. Then, the device is sensitive to angular velocity Ω in the plane of the fork.

Another popular configuration is that of the wine glass resonator (Fig.7-36, middle) [52]. In this resonator, a tube vibrates with an elliptical-shaped flexure. If the tube is well balanced mechanically, the major axes of the ellipse (represented at 45° and at –45° in Fig. 7-36) may take any inclination. Oscillation is excited by applying ac voltages at the resonant frequency of the tube to the electrodes. When an angular velocity $\pm\Omega$ is applied along the tube axis, the inclination of the ellipse elongation moves clockwise or counterclockwise (according to the sign of Ω). Then, we get the measurement of Ω by comparing the difference of signals S1-S3 and S2-S4 collected at the electrodes.

Another configuration widely employed is that of the resonating quartz triangle. In this device, we excite an acoustical wave by means of the transducer S1 (made of PZT) driven at the resonant frequency of the mechanical structure. Two more PZT transducers sense the amplitude of the incoming acoustical wave. When Ω is applied parallel to the prism axis, one transducer receives more power than the other due to Coriolis' force (see Fig.7-36). By computing the difference of outputs S2 and S3, a signal proportional to Ω is obtained.

About performances of the PGs, data are, of course, compliant with the thermal noise limitation of Fig.7-34. Typical sensitivity of commercial products is 0.2..1 deg/s and the dynamic range is 50 to 200 deg/s.

REFERENCES

[1] W.M. Macek and D.I.M. Davis, *"Rotation Rate Sensing with a Travelling Wave Laser"*, Appl. Phys. Letters, vol.2 (1963) pp.67-69.
[2] V.E. Sanders and R.M. Kiehn, *"Dual Poralized Ring-Laser"*, IEEE Journal of Quantum Electronics, vol.QE-13 (1977), pp.739-745.
[3] A. Mattews, *"The Laser Ring Gyro"*, in: Proceedings Rotation Sensing Conference, Paris 1979, AGARD vol.220, pp.1-21.
[4] V. Vali and W. Shorthill, *"Fiber Ring Interferometer""*, Applied Optics, vol.15, (1976), pp.1099-1100.
[5] R. Ulrich and M. Johnson, *"Fiber Ring Polarization Analysis"*, Optics Letters, vol.4 (1979), pp.152-154.
[6] R.A. Bergh, H.C. Lefevre, and H.J. Shaw, *"All-single-mode Fiber Gyroscope"*, Optics Letters, vol.6 (1981), pp.198-200; see also: Optics Letters, vol.6 (1981), pp. 502-504.
[7] H.C. Lefevre, *"The Fiber Optic Gyroscope"*, Artech House: Boston, 1993, see Ch.11.4.
[8] S. Donati and V. Annovazzi Lodi, *"Fiber Gyroscope with Dual Frequency Laser"*, in: Fiber-Optic Rotation Sensors, edited by S.Ezekiel and H.J.Arditty, Springer Verlag: Berlin 1982, pp.292-297.
[9] E.J. Post, *"The Sagnac Effect"*, Rev. Mod. Phys., vol.39 (1967), pp.475-495.
[10] J. van Bladen,: *"Relativity in Engineering"*, Springer Verlag: Berlin, 1984.
[11] E.O. Schultz DuBois, *"Alternative Interpretation of Fresnel-Fizeau Effect in a Rotating Optical Ring Interferometer"*, IEEE Journal of Quant. Electr., vol.QE-2 (1966), pp.299-306.

7.8 The MEMS Gyro and Other Approaches

[12] V. Vali and W. Shorthill, *"Fresnel-Fizeau Effect in a Rotating Optical Fiber Ring Interferometer"*, Applied Optics, vol.16, (1977), pp.2605-2607.
[13] S. Donati, *"The Electro-Optical Gyroscope: State of the Art and Prospects"*, Alta Frequenza Riv. Elettron. *(in Italian)*, vol.2, (1990), pp.143-154.
[14] G.E. Stedman, *"Ring-Laser Tests of Fundamental Physics and Geophysics"*, Progress in Phys. vol.60 (1997) pp.615–688. See also: *"Ring Laser Gyroscope Receives its Centerpiece"*, Europhotonics (Dec/Jan.2001), p.15.
[15] R.P.G. Collinson, *"Introduction to Avionics"*, Chapman and Hall: London, 1996.
[16] S. Donati, *"Photodetectors"*, Prentice Hall: Upper Saddle River, 2000.
[17] see Ref.[16], Appendix A2.
[18] see Ref.[16], p.261.
[19] *"Gyroscope Battle is Settled"*, Opto&Laser Europe, 92, (February 2002), p.15.
[20] V.E. Sanders and R.M. Kiehn, *"Dual Polarization Ring Laser"*, Journal of Quantum Electronics, vol.QE-13 (1977), pp.739-746.
[21] W.W. Chow et al., *"Multioscillator Ring Laser Gyro"*, Journal of Quantum Electronics, vol. QE-16 (1980) pp.918-935.
[22] M. Born and E. Wolfe, *"Principles of Optic"*, 6th ed., Pergamon Press: Oxford, 1983.
[23] B. Rossi, *"Optics"*, Addison Wesley Publ. Co.: Reading MT, 1957.
[24] S. Donati, *"Laser Interferometry by Induced Modulation of the Cavity Field"*, Journal of Applied Physics, vol.49 (2), 1978, pp.495-497.
[25] S. Ezekiel and H.J. Arditty *"Fiber-Optic Rotation Sensors"*, Springer Verlag: Berlin 1982, pp.292-297.
[26] R.A. Bergh, H.C. Lefevre, and H. J.Shaw *"An Overview of Fiber Optics Gyroscopes"*, IEEE Journal of Lightwave Technology, vol.LT-2 (1991), pp.91-107.
[27] D. D'Alessandro and S. Donati, *"Optimum Bias for Interferometers in the Quantum Noise Regime"*, Alta Freq. Vol.12 (2000), pp.72-75.
[28] A. Angot, *"Complements de Matematiques"*, 6th ed., Masson et Cie: Paris, 1972.
[29] G. Martini, *"Analysis of a Single Mode Fiber Piezoceramic Phase Modulator"*, J. Optical and Quantum Electronics, vol.19 (1987), p.179-190.
[30] see Ref.[16], Sect.5.3.
[31] T. Osaka, K. Okamoto, et al., *"Low-loss Single-Polarization Fibres with Asymmetrical Strain Birefringence"*, Electronics Lett., vol.17 (1981), pp.530-531.
[32] S. Donati, L. Faustini, and G. Martini, *"High-Extinction Coiled-Fiber Polarizers by Careful Control of Interface Reinjection"*, IEEE Photonic Technology. Letters, vol.7, (1995), pp.1174-1177.
[33] G.P. Bendelli and S. Donati, *"Optical Isolators for Communications: Review and Current Trends"*, Europ. Trans. on Telecomm., vol. 3 (1992), pp.63-69.
[34] V. Annovazzi Lodi, S. Donati, S. Merlo, L. Zucchelli, and F. Martinez, *"Protecting a Power-Laser Diode from Retroreflections by Means of a Fiber $\lambda/4$ Retarder"*, IEEE Phot. Techn. Letters, vol.8, (1996), pp.485-488.
[35] V. Annovazzi Lodi and S. Donati, *"Technology of Lapped Optical-Fiber Couplers"*, J. Opt. Commun., vol.11, (1990), pp.107-121.
[36] S. Donati, V. Annovazzi Lodi and F. Picchi, *"Ultra-low Insertion Loss Fused-Couplers Fabricated by a Long Furnace"*, in Proc. LEOS Workshop on Passive Fiber Optics Components, WFOPC'98, edited by S. Donati, Pavia, Sept.18-19, 1998, pp.161-164.

[37] E. Udd (editor), "*Fiber Optic Gyros 10th Anniversary Conference*", SPIE vol.719, (1988).
[38] V. Annovazzi Lodi, S. Donati, and S. Merlo, "*Thermodynamic Phase Noise in Fiber Interferometers*", Journal of Optical and Quant. Electr., vol.28 (1996), pp.43-49.
[39] O. Graydon, "*Integrated Gyro is set to Reduce the Cost of Navigation*", Opto-Laser Europe, Dec.1997, pp.23-25, see also C. Wulf-Mathies, "*Integrated Optics for Fiberoptics Sensor*", Laser und Optoelektronik vol.21 (1989), pp.53-63.
[40] S. Donati, G. Giuliani, and M. Sorel, "*Proposal of a new Approach to the Electro-Optical Gyroscope: the GaAlAs Integrated Ring Laser*", Alta Freq. Riv. Elettron. vol.9 (1997), pp.61-63; see also Alta Freq. Riv. Elettron. vol.10 (1998), pp.45-48.
[41] M. Sorel, P. Laybourn, G. Giuliani, and S. Donati, "*Unidirectional Bistability in Semiconductor Waveguide Ring Lasers*", Applied Physics Letters, vol.80 (2002), pp.3051-3053.
[42] B.Y. Kim and H.J. Shaw, "*Phase–Reading an All-Fiber-Optic-Gyroscope*", Optics Letters, vol.9 (1984), pp.378-380.
[43] A. Edberg and G. Schiffner, "*Closed Loop Fiber-Optic-Gyroscope with a Sawtooth Phase Modulated Feedback*", Optics Letters, vol.10 (1985), pp.300-302.
[44] C.J. Kay, "*Serrodyne Modulator in a Fiber-Optic-Gyroscope*", Proceedings IEE part J Optoelectronics, vol.132 (1985), pp.259-264.
[45] H.C. Lefevre, P. Graindorge, H.J. Arditty, et al., "*Double Closed-Loop Hybrid Fiber Gyroscope with a Digital Ramp*", Proc. OFS'84, Munchen 1984, paper PSD-7.
[46] R. Carroll and J.E. Potter, "*Backscatter and Resonant Fiber-Optic-Gyro Scale-Factor*", IEEE Trans. Lightw. Techn., vol.7 (1989), pp.1895-1900.
[47] J.M. Lopez-Higuera, "*Superfluorescent Fiber Optic Sources*", Ch.10, Handbook of Optical Fibre Sensing Technology, J.Wiley &Sons: Chichester, 2002.
[48] S. Oho, H. Kajioka, and T. Sasayama, "*Optical Fiber Gyroscope for Automotive Navigation*", IEEE Trans. Vehic.Techn., vol.44 (1995), pp.698-704.
[49] N. Barbour and G. Schmidt, "*Inertial Sensor Technology Trend*", IEEE Sensors Journal, vol.1 (2001), pp.332-339.
[50] S.K. Shyeem, "*Fiber Optic Gyroscope with a 3x3 Directional Coupler*", Appl. Phys. Lett., vol.37 (1980), pp.869-870.
[51] V. Annovazzi Lodi, S. Donati, and M. Musio: "*A Fiber Optic Gyroscope for Automotive Navigation*", Proc. Intl Conf. Advances in Microsystems for Automotive Applications, Berlin, 2-3 Dec.1996, pp.123-128.
[52] S.D. Senturia, "*Microsystem Design*", Kluwer: Boston, 2000.
[53] K. Tanaka et al., "*A Micromachined Vibrating Gyroscope*", Sensors and Actuators, Vol.A50 (1995), pp.111-115.
[54] D. Teegarden, G. Lorenz, and R. Neul, "*How to Model and Simulate Microgyroscope Systems*", IEEE Spectrum, July 1998, pp.66-75.
[55] V. Annovazzi Lodi and S. Merlo: "*Mechanical-Thermal Noise in Micromachined Gyro*", Microelectronics Journal, vol.30 (1999), pp.1227-1230.
[56] S. Donati and M. Norgia, "*Hybrid Opto-Mechanical Gyroscope with Injection-Interferometer Readout*", Electronics Letters vol.37 (2001), pp.756-757.
[57] J. Soderkvist, "*Piezoelectric Beams and Vibrating Angular Rate Sensor*", IEEE Trans. Ultras., Ferroel. and Freq. Control, vol.38 (1991), pp.271-280.

CHAPTER **8**

Optical Fiber Sensors

The most important application of optical fibers is undoubtedly in optical communications.
Optical fibers can be used as sensors as well, if we look at how certain parameters of propagation are affected by external perturbations. This is a very interesting niche from both the scientific and technical points of view.

In applications other than sensors, we strive for getting rid of external perturbations and make the fiber as immune as possible. In sensors, we need to make the sensitivity reproducible, and eventually try to increase it. Because we are interested in its measurement, we now call the physical perturbation a *measurand*.

In Optical Fiber Sensors (OFSs), the effect of a measurand may eventually be very minute, yet we are able to devise readout methods capable of responding with very high sensitivity. A very high sensitivity is actually a special feature of some configurations of OFSs, unequalled by any other approach. This is one of the reasons why OFSs have been actively studied and developed internationally by the scientific community since about the beginning of the 1980s, when the early high-quality fibers started to become available.

Perhaps the most successful example of an OFS is the FOG, which was treated in Ch.7. Many other examples of viable OFSs have been reported in the literature, and prototypes of OFSs for measuring quantities such as temperature, pressure, strain, and pollutants have been proposed and developed into commercial products.

The revenues from OFSs have been rather disappointing, however. The sensor market has fierce competition among many different technologies. The market expectation is not just for high-performance sensors, but more commonly for cheap sensors and sensors with a good price-to-performance ratio.

The new OFS technology can offer high performance, but is much more expensive than conventional sensors. Thus, optical fiber sensors have not been accepted for general-purpose applications, the domain of conventional electronic sensors. Only very special circumstances (like harsh environment, high EMI, etc.) may render the OFS acceptable, but this is just a small percentage of the rich and appealing sensor market. Thus, after twenty years of R&D efforts, only a few OFSs are left over as viable and successful.
In this chapter, we will present the fundamentals of OFSs and discuss a few selected examples of application.
More exhaustive references may be found in the copious literature on the subject [1-4].

8.1 INTRODUCTION

The paradigm of an OFS is shown in Fig.8-1. The measurand, through the mechanism of interaction, affects an optical parameter of the fiber.
Specifically, the optical parameter may be the intensity, or the state of polarization (SoP), or the phase of the field propagated through the fiber.

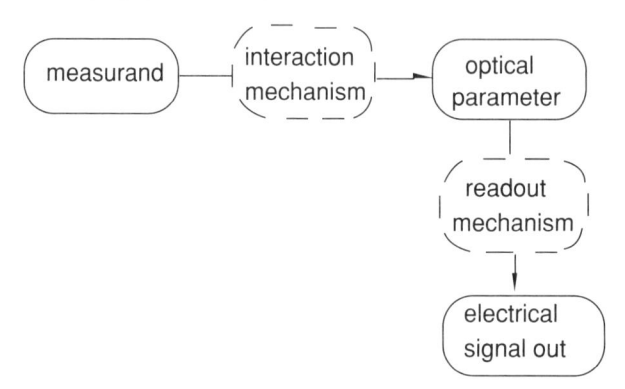

Fig.8-1 The paradigm of an OFS: the measurand, or quantity to be measured, affects an optical parameter, which is translated by a readout configuration into an electrical output.

8.1 Introduction

According to the parameter affected by the measurand, we arrange a readout configuration that transforms the variation of the parameter into an output electrical signal for the user.

8.1.1 OFS Classification

Sometimes the measurand is first converted by a conventional mechanism into an intermediate measurand more conveniently measured by the fiber.

Because of this difference, OFSs are accordingly classified as *intrinsic* or *extrinsic*. Another important classification of an OFS stems from the type of optical parameter involved by the interaction with the measurand.

The most commonly used parameters are (i) intensity (or guided power), (ii) state of polarization (SoP), and (iii) optical phase of the propagated mode.

The corresponding readout (or OFS) is called an *intensity*, *polarimetric* or *interferometric* sensor (Fig.8-2). This classification is not simply nominal, as it implies a type of fiber and of source necessary to implement the readout.

Intensity readout OFS can be implemented with multimode fibers, whereas polarimetric and interferometric OFSs require a monomode fiber to preserve the interaction without the errors originated by the multimode regime.

Indeed, when many modes are allowed to propagate, each tends to collect a slightly different variation, and the modes mix generates an error.

Also, in intensity-based OFSs we do not look at the SoP nor do we need coherence. Therefore, we may use a cheap incoherent, unpolarized source like an LED. In polarimetric and interferometric OFSs, on the other hand, we need to start with a clean input field, with well-defined SoP or coherence length. We shall therefore use a laser or a SLED (super luminescent LED).

READOUT (optical measurand)	fiber type	source type
INTENSITY (power)	multimode (occasionally monomode)	LED
POLARIMETRIC (state of polarization)	monomode	laser
INTERFEROMETRIC (optical phase)	monomode	laser, SLED

Fig.8-2 Classification of the three main readout principles and types of OFSs: intensity, polarimetric, and interferometric. Also indicated are the fibers and sources they use.

8.1.2 Outline of OFS

An outline of the most important OFSs, demonstrated in the laboratory or developed up to the level of units deployed in the field, is presented in Fig.8-3. As we can see, the range of applications is wide, spanning from mechanical to electrical to chemical quantities, and covering virtually all types of measurand.

Among the mechanical quantities, we find force, pressure, stress, strain, acoustical fields, and other derived measurands. The OFSs developed for these quantities are mostly of the intrinsic type.

Some of them are known as the optical strain gage, the fiberoptic vibrometer, and the optical hydrophone. Transducer effects used include the elastooptical effect, evanescent attenuation, and attenuation by proximity.

Optical Fiber Sensors

measurand	OFS type	effect used, Intrinsic vs Extrinsic
mechanical	STRAIN GAGE	elasto-optical I
	VIBROMETER	evanescent coupling I
	HYDROPHONE	coupling attenuation E
thermal	THERMOMETER	birifringence temp coeff E
		index refraction temp coeff I
inertial	GYROSCOPE	Sagnac effect I
	ACCELEROMETER	mass-loaded app force E
electrical	ELECTRIC FIELD	Faraday and Pockels effects I
	MAGNETIC FIELD	electro-, magneto-striction E
chemical	OPTRODES (pH, pCO, pollutants, etc.)	fluorescence E
		chromatic reaction E

Fig.8-3 Outline of the most developed OFSs, with the main effect used and type of interaction.

8.1 Introduction

Temperature OFSs may be intrinsic as well as extrinsic. The temperature coefficient of the fiber is used for interferometric OFSs, whereas extrinsic OFSs have been developed using temperature-dependent birefringence in crystals and blackbody emission.

Electrical OFSs have been developed based either on the Faraday and Pockels effects, both intrinsic to the fiber, or occasionally on the magnetostriction and electrostriction effects, both extrinsic to the fiber.

Inertial sensors include the FOG gyroscope (see Ch.7) which is an intrinsic OFS based on the Sagnac effect, and accelerometers, which are usually realized as extrinsic, mass-loaded strain gage.

Finally, OFSs have been demonstrated for the detection and measurement of chemical species (pollutants, hydrocarbons, etc.) and for chemical quantities like pH, pO_2, etc.
Most chemical OFSs are extrinsic, and they are based on a color-developing reagent conveniently placed on the fiber tip. Because of this, a chemical OFS is also called an *optrode* (optical electrode).

The advantages and disadvantages of OFSs with respect to other conventional sensors derive from the fiber structure.
Because OFSs have a completely passive structure from the electrical point of view, their immunity to electromagnetic interference and chemical species is good.
They can generally withstand a high dose of radiation, in the range of Mrads compared to the 50-krads dose typical of electronic components.
Even their immunity to mechanical disturbance is generally good because of the rugged structure of the fiber.
The overall size and weight of OFSs compare favorably to other sensors and flexibility of design in shape and size is good. Frequently, OFSs are able to operate without physically contacting the measurand, or they are non-invasive.
OFSs can attain sensitivity values unequalled by other types of sensors, especially when using the most sophisticated versions, the interferometric ones.

The list of disadvantages regarding OFSs is not short, unfortunately.
To attain a high-sensitivity performance, OFSs are rather complex and expensive, whereas simple intensity-based OFSs are cheap but not so appealing in performance.

Flexibility and reconfiguration of OFSs are modest.
Ideally, the user's requirement is an OFS composed of a readout electronic box and a probe. The electronic box (or rack-mounted panel) should be capable of performing all three readouts (intensity, polarimetric, interferometeric) with the appropriate fiber probe. In the user's view, there should be many probes available, to perform all the measurements indicated in Fig.8-3 with a single instrument.
Unfortunately, no OFS is able to satisfy this wish list. Frequently, when an OFS is finally completed for a specified measurement objective, if the user changes the specification just a little (in sensitivity, dynamic range, or sensor placement), we need to develop a completely different solution. Thus, while the performance capability offered by OFS technology is quite powerful and should always be considered a potential solution, the range of applications outside the specification is generally quite modest and allows little or no reconfiguration.

8.2 THE OPTICAL STRAIN GAGE: A CASE STUDY

The field of sensors in general, and that of OFSs in particular, is characterized as being open to a variety of disparate contributions. Different from other segments of engineering, the sensor field has many possible solutions to a particular problem, which look at least technically plausible when first scrutinized.
Only after a thorough evaluation of the sensor, the most suitable solution can usually be sorted out. In addition, the best solution is frequently challenged as soon as a minor change of technology occurs.
To illustrate this point, we report in Fig.8-4 an example of OFS development, that is, the optical strain gage.
This sensor has a well-known electronic counterpart, which is the resistive strain gage made of a thin film of conductive material deposited on a substrate, usually a plastic (or polymer) material. When the substrate is subjected to a stress, the film resistance changes. Arranging this resistance in a Wheatstone bridge, we can easily measure the stress or quantities connected to it (like strain, pressure, force, etc.).
Of course, the optical strain gage should offer some special feature warranting its superiority respect to the electrical strain gage if users are to prefer it.
We can start by considering a multimode-based OFS (Fig.8-4), in which two fiber pigtails are aligned in the package and one of them is actuated by a thin membrane to which external force is applied.
We can refer to this device as the reference (=1) sensitivity.
A second OFS easily follows as an extrinsic sensor when we put a grating inside the package and read the relative grating position through the fibers with collimating lenses. If the grating has black and white lines with thickness $\approx 2\mu m$, compared to the $50\mu m$ of the fiber diameter, we get an improvement in sensitivity by a factor ≈ 25.
Further, we may think of sandwiching the fiber between two undulated jigs. If the jig is actuated in compression by the force to be measured, we induce in the fiber a curvature loss proportional to the force. The intensity-based readout of the loss is much more sensitive, perhaps by a factor ≈ 50 with respect to the reference scheme.
Next, we may consider reading the birefringence induced by the jig. Doing so, we get a polarimetric readout of the effect. We can use a Glan cube to separate polarization, detect the two components, and compute the ratio of the two fields to get a quantity proportional to the applied force.
As a last step to increase sensitivity, we can try measuring the optical path length variation produced in the fiber by the applied force that is, we can try converting the strain gage into an interferometric-readout OFS. We need, as illustrated in Fig.8-4, two beamsplitters to build an interferometer around the cell. The sensitivity of the interferometer readout is much larger, say about ≈ 500 times that of the reference configuration.
The sensor works better by removing ambient-related vibration errors. Using two fiber couplers to build the interferometer (Fig.8-4), we are able to improve the stability of the setup and accordingly the working sensitivity.

8.2 The Optical Strain Gage: a Case Study

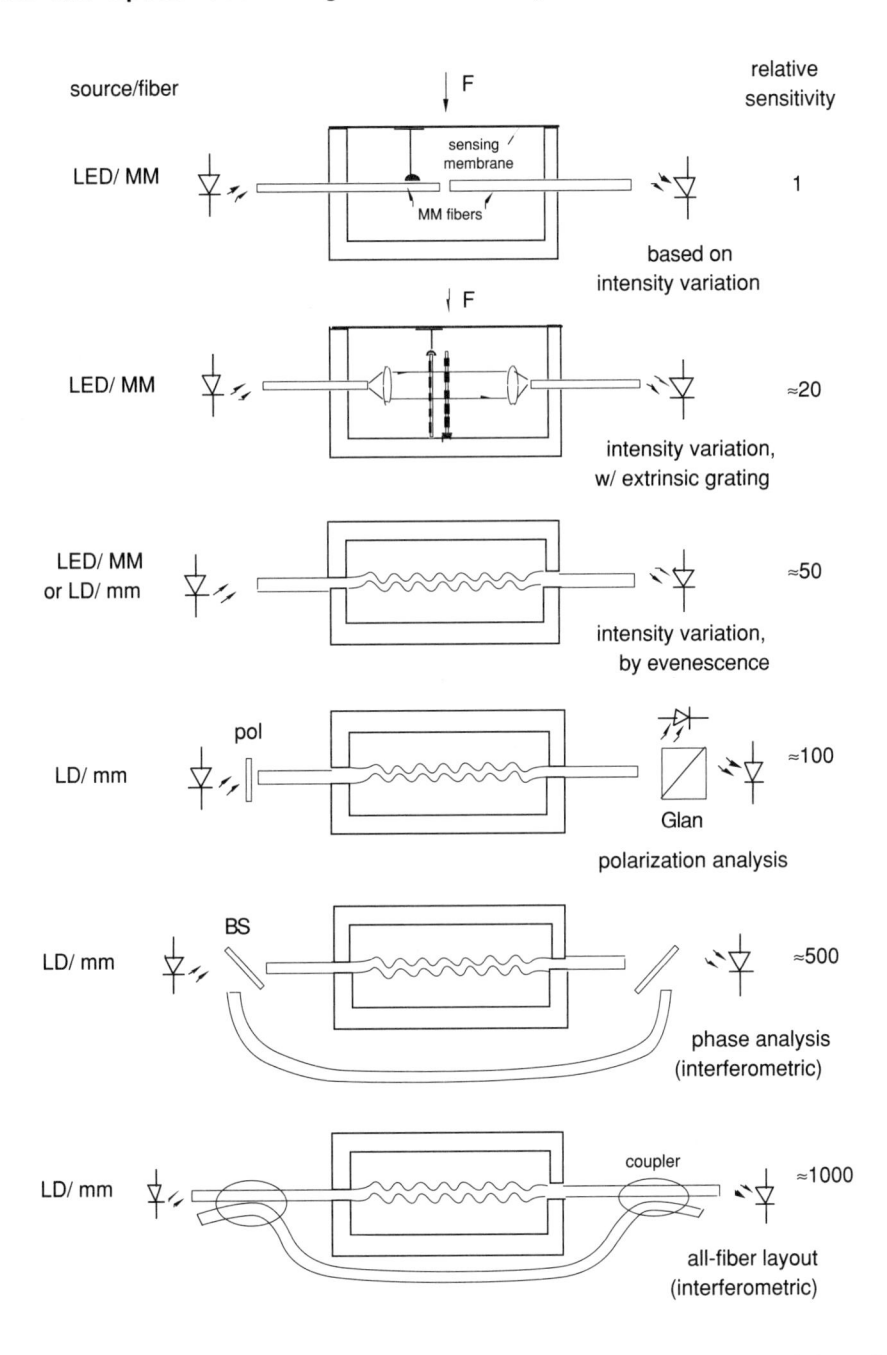

Fig.8-4 Evolution of different types of optical strain gage, through intensity, polarimetric and interferometric readouts.

8.3 READOUT CONFIGURATIONS

Let us now consider sensor readout configurations in more detail. Our interest is assessing the basic performance of each configuration and its limits, as well as showing examples of sensors developed from the concepts.

8.3.1 Intensity Readout

A physical measurand can affect the power propagated down the fiber in a variety of mechanisms. The most commonly used in fiber sensors are illustrated in Fig.8-5.

The measurand M may act on the position of a fiber pigtail and thus affect the alignment loss of a fiber joint (Fig.8-5a). A variant is obtained when the measurand actuates a light stop inserted between two fiber pigtails (Fig.8-5b).

In another mechanism, the measurand is related to the index of refraction n=n(M) of the medium surrounding the joint of two fiber pigtails (Fig.8-5c). A variation of the index of refraction causes the variation of the joint loss and we can read the measurand variation from it.

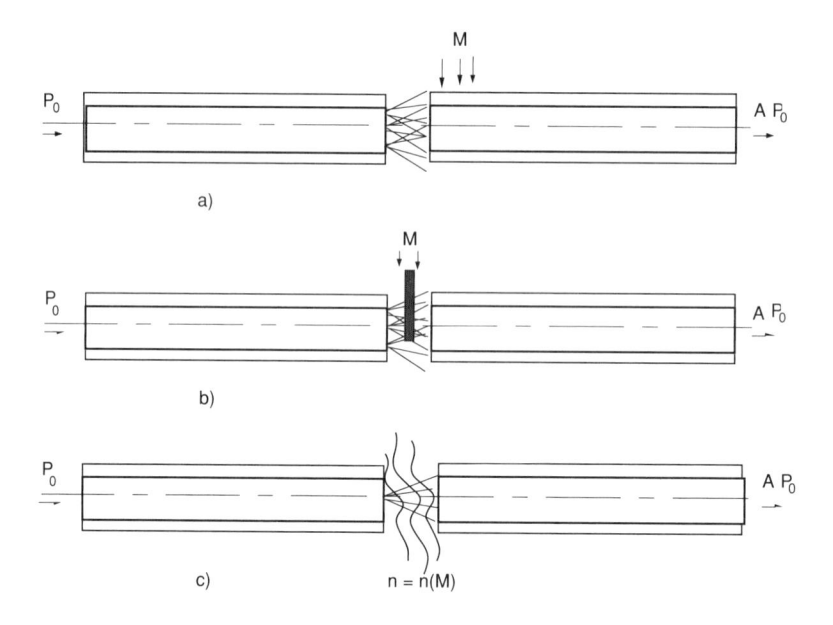

Fig.8-5 Different types of intensity-based sensors may use the measurand to actuate the position of a fiber pigtail in a fiber joint (a), or to exert attenuation by insertion of a light stop (b). A variation of the index of refraction, induced by allowing the measurand in the joint region, can also be read as attenuation (c).

8.3 Readout Configurations

Finally, we can take advantage of curvature-induced loss to exploit intensity readout. This can be done by mounting the fiber in a fixture with a serpentine profile of corrugation, as already considered in the example of the fiber strain-gage (Fig.8-4).

The quantity of response obtained by these transducer mechanisms can be evaluated from the alignment losses plotted in Fig.8-6 [5]. Here, loss is defined as the value of attenuation A in dB, or loss=20 $\log_{10}$ A, where A is the ratio of output to input power (see Fig.8-5).

Three basic errors are considered in Fig.8-6: the longitudinal displacement z, the transversal displacement r, and the tilt angle α of the fiber axes. The fiber is assumed to be multimode with a core diameter 2a, and data are reported with the numerical aperture NA=$[\sqrt{(n_1^2-n_2^2)}]/n_0$ ranging from 0.1 to 0.3. In the loss data, the Fresnel reflection at input/output surfaces is not included. The data of Fig.8-6 also hold true for monomode fibers at first approximation.

The case of n=n(M) can be treated, using the data in Fig.8-6, by considering that the external index of refraction n can affect the numerical aperture NA as n_0=n in this case.

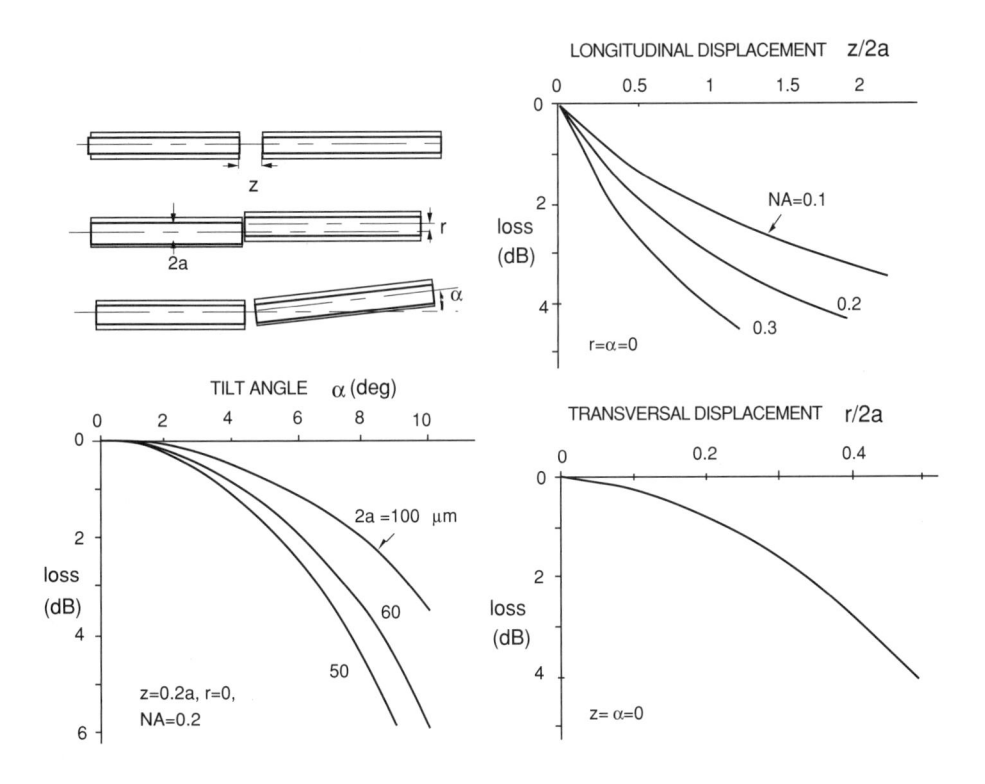

Fig.8-6 Losses due to three misalignments in a joint between two fiber pigtails: longitudinal and transversal displacements and axis tilt. When more than one error is present, the total loss is the sum of individual contributions (at first approximation).

The actuation by a stop can also be treated with the displacement-loss diagrams. If we use a blade-like stop entering by a length w in the fiber gap of width g, the total loss is the sum of a transversal displacement w/2a and a longitudinal displacement g/2a loss.

Fig.8-6 is the starting point for the design of a sensor, because we can determine the displacement leading to a full signal swing as well as the linearity of response.

However, as we start developing a practical sensor design, we soon realize that the conceptual configurations of Fig.8-5 put forth several undesirable features. To illustrate this through an example, let us consider a pressure sensor developed from the blade-like stop concept (Fig.8-7).

This sensor uses two pigtails of multimode fiber (with core diameter 2a=50 or 62.5 µm), aligned by sleeves to face each other precisely, typically with a gap g=20µm. A thin diaphragm membrane is deformed by the pressure under measurement. Integral with the diaphragm, a thin blade is slipped into the gap between the fibers.

From the data of Fig.8-6, we can calculate the attenuation $A=P_{out}/P_{in}$ of optical power passed though the sensor. This quantity is proportional to $I_{ph}=\sigma P_{out}$, the current supplied by the photodiode placed at the exit fiber, σ being the spectral sensitivity of the photodiode [6].

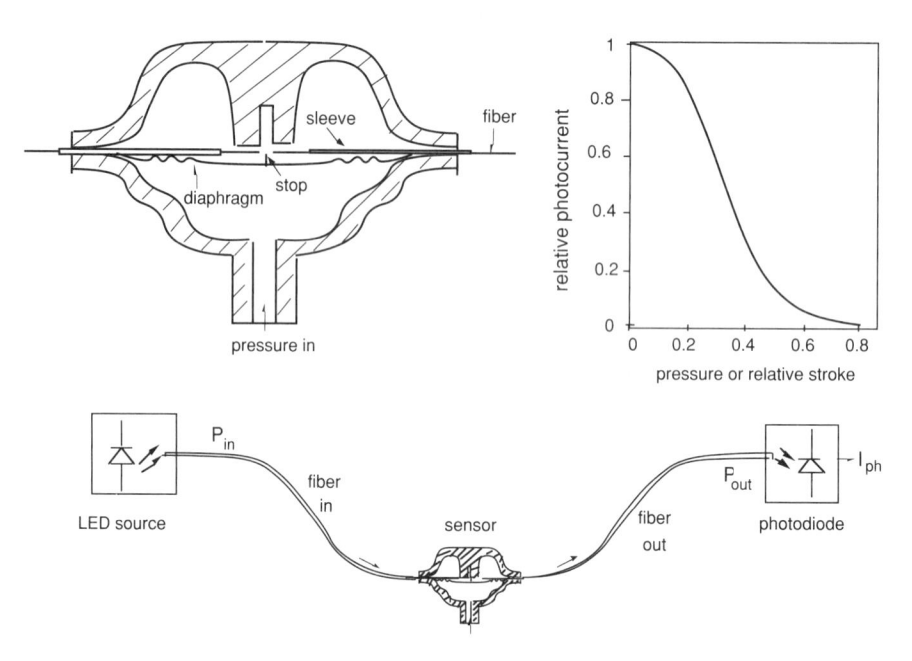

Fig.8-7 A pressure sensor (top left) uses a fiber joint with a stop actuated by a thin diaphragm. The response curve (top right) of detected current vs. stop displacement replicates the ratio of optical power (or loss) P_{out}/P_{in}. The sensor is read by the LED/photodiode combination (bottom).

8.3 Readout Configurations

Writing $P_{out} = AP_{in}$ and $I_{ph} = \sigma AP_{in}$, we can see that the photocurrent I_{ph} is a good replica of attenuation A, provided the input power and spectral sensitivity are constant. Actually, both P_{in} and σ are likely to vary with temperature and aging. In particular, the input power will be supplied by an LED, and we should check temperature coefficient and long-term drift of input power from manufacturer's data or by specific testing.

We may as well correct the measurement from drifts in P_{in} and σ, adding a circuit to compute the ratio of I_{ph} and P_{in}/σ. Of course, this would spoil the inherent simplicity and low component count of the intensity-readout approach.

An additional error affecting the scale factor is the transversal misalignment of the two fiber pigtails. If this error drifts with time, it will seriously impair the sensor response calibration further.

Another concern is the poor linearity of the pressure-to-current curve displayed in Fig.8-7 (right). We may improve linearity by limiting the range of the input pressure, but then we should require a calibration to set the working point at the middle of the dynamic range. In practice, the linearity error is not easily corrected unless we are ready to add sophistication and increase the price of the sensor substantially. Thus, we would probably end up with accepting poor linearity and limiting the sensor to the least demanding application, for example, an on/off alarm or the like.

In the design of Fig.8-7, we are using two fiber pigtails for sensor access and connection to the source (an LED) and the detector (a photodiode). Two fibers increase the cost of installation and are not that necessary. It would be better to access the sensor from just one fiber, used downstream to send optical power to the sensing region and upstream to collect the returning signal.

We may thus wish to modify the initial configuration of the intensity sensor to: (i) read in reflection; (ii) improve the linearity of response; (iii) have a self-aligned placement of the fiber. Two examples of the several possible solutions are shown in Fig.8-8.

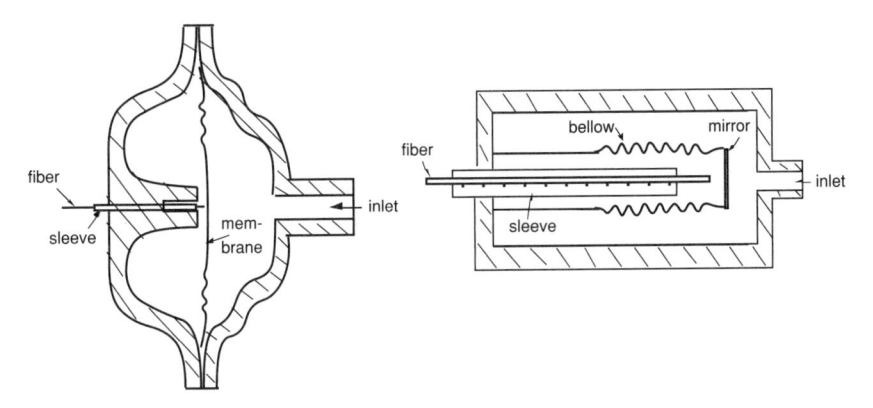

Fig.8-8 By using the membrane diaphragm as a reflector, the pressure sensor can be read in reflection by a single fiber (left). In place of the membrane, a bellow coaxial to the fiber can be used to reduce overall size (right).

In the examples, the end-face of the multimode fiber is placed at a suitable distance z_0 in front of the membrane subjected to external pressure.

The membrane is a thin metal foil with enough reflectivity to act as a mirror for the light exiting from the fiber. In another implementation, the membrane is the cap of a bellow co-axial to the fiber (Fig.8-8 right) and overall size is shrunk to a minimum.

The downstream and upstream optical signals can be separated with the aid of a 50% fiber coupler (Fig.8-9). The coupler is the all-fiber counterpart of the bulk-optics beamsplitter. It transmits a fraction T and reflects a fraction R=1-T of incoming power. To minimize the total loss T(1-T) experienced by the go-and-return path to the sensor, we must take T=0.5 or, have a 50% coupler.

The fiber and membrane combination is equivalent to a two-fiber joint with a gap twice the distance z_0, and the diagram of attenuation due to longitudinal displacement (Fig.8-6) is still applicable using $z = 2z_0$.

At the output of the measurement photodiode, the response of current I_{phm} is found close to a negative exponential (Fig.8-9), and thus not at all linear. However, we can convert this response into a linear one by computing the logarithm of I_{phm}.

In this way, we get the log-attenuation A (dB), a quantity which is about linear with longitudinal displacement or input pressure, as indicated by Figs.8-6 and 8-9.

We can perform the conversion at little extra cost by using the photodiode itself as a logarithm converter (Fig.8-9). As it is well-known [7], to perform this function we shall use an op-amp follower reading the photodiode in the open-circuit mode.

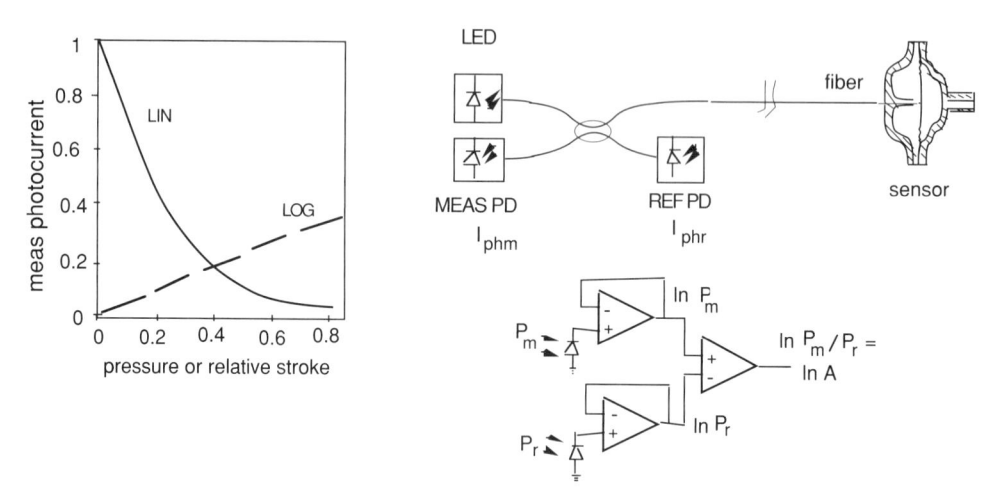

Fig.8-9 The current response of the reflective readout pressure sensor is a negative exponential (left), but we transform it into a linear one by a logarithm conversion performed by the photodiode itself. As the circuit performs the ratio of powers P_m and P_r (right), it also corrects the measurement from drifts in LED power and photodiode sensitivity.

8.3 Readout Configurations

Thus, the op-amp output is proportional to the logarithm of the photodetected current [7] $I_{ph}=\sigma P =\sigma AP$, and hence to the logarithm of attenuation A.

Working with a single measurement channel, however, the dependence of the log-signal from A would be readily masked by drifts in the power P emitted by the source as well as in the spectral sensitivity σ of the photodiode.

The logarithm circuit is important because it allows us to introduce a reference channel for the correction of drifts at a minor increase of circuit complexity and extra cost. As shown in Fig.8-9, we use a second photodiode at the lower port of the fiber coupler. This photodiode acts as a reference, and it yields a photocurrent $I_{phr}= \sigma_r P_r =(1-T)\sigma_r P_{LED}$ proportional to the power emitted by the LED and to the spectral sensitivity σ_r.

The voltages v_m and v_r at the outputs of the measurement and reference branches can be written as:

$$v_m = V_0 \log_{10} \sigma_m P_m / I_{m0} = V_0 \log_{10} \sigma_m A T P_{LED}/I_{m0}$$
$$v_r = V_0 \log_{10} \sigma_r P_r / I_{r0} = V_0 \log_{10} (1-T)\sigma_r P_{LED}/I_{r0} \qquad (8.1)$$

where V_0 is a scale factor (=59.6 mV, see [7]), $I_{m0,r0}$ are the dark currents, $\sigma_{m,r}$ are the sensitivity of the photodiodes, and $P_{m,r}$ are the received powers.

By subtracting the outputs of the measurement and reference channels, we get a signal v_{out} given by:

$$v_{out} = v_m - v_r = V_0 \log_{10} \sigma_m P_m / \sigma_r P_r = V_0 \log_{10} A + V_0 \log_{10} T\sigma_m I_{m0}/(1-T)\sigma_r I_{r0}$$
$$= V_0 \log_{10} A + \text{const.} \qquad (8.2)$$

The last term of the first line has been assumed to be a constant because we use matched photodiodes in the measurement and reference channels, and coupler transmission T is reasonably stable with aging.

In conclusion, with the log-conversion circuit we can obtain a response with a good linearity and are able to introduce a reference channel that cancels the drifts of LED and photodiode [7]. With minor modifications to the basic scheme of Fig.8-9, the concept of double-channel readout with measurement and reference for drift cancellation is utilized in a number of intensity-based sensors.

A temperature OFS based on the fluorescence of a crystal excited by an LED source is illustrated in Fig.8-10.

The crystal is a thin plate of GaAs cemented onto the fiber pigtail end-face, and we illuminate it with an LED emitting at a wavelength λ_{LED} below the photoelectric threshold λ_t of the GaAs material. As $\lambda_{LED}<\lambda_t$, the LED photons are absorbed by the crystal, generating electron-hole pairs. After a certain characteristic time, recombination of a pair takes place, with the emission of a fluorescence photon at wavelength λ_t.

Photons are radiated out of the crystal and a substantial fraction of them is collected by the fiber and brought back to the detector where we can measure the fluorescent power.

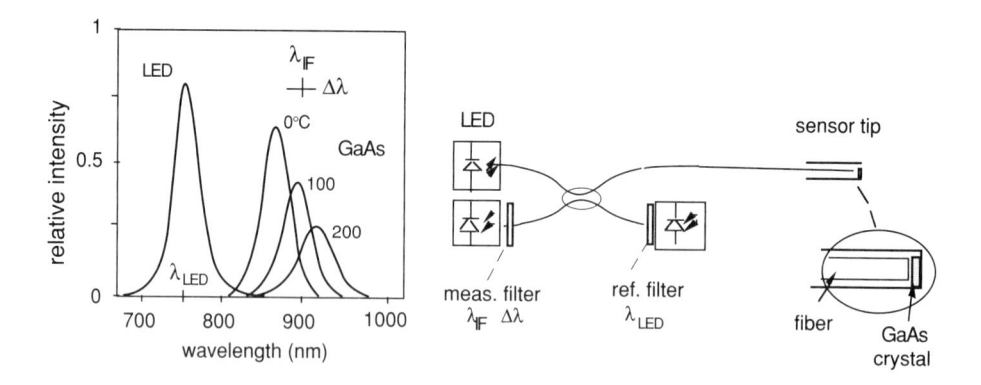

Fig.8-10 A temperature OFS based on fluorescence of a GaAs crystal mounted on the tip of an optical multimode fiber. The LED illuminates the thin GaAs crystal at a wavelength ($\lambda_{LED} \approx 750$ nm) of strong absorption. The crystal emits fluorescent light at a wavelength longer then λ_{LED}, with a precise temperature dependence of the peak wavelength (≈ 0.26nm/°C). The fiber guides the light emitted by the crystal back to the photodiode.

Now, both the wavelength of emission λ_t and line width of fluorescence change with temperature, as illustrated in Fig.8-10. By inserting an interference filter in front of the photodiode, with a suitable central wavelength λ_{IF} and bandwidth $\Delta\lambda$, we can maximize the response as a function of temperature and render it approximately linear.
A typical performance obtained by this approach [8] is a ≈ 0.2°C accuracy on a measurement range extended from -50 to $+200$°C.

In another approach aimed to cover a higher temperature range (0-800°C), we measure the temperature dependence of absorption $\alpha(\lambda)$ in a piece of special fiber (Fig.8-11).
The special fiber is a modified version of the normal silica fiber, prepared to withstand high temperature and to exhibit a temperature-dependent $\alpha(\lambda)$.
In an ordinary fiber, as we increase temperature, we find several failure mechanisms to fix. The first is burnout of the secondary coating, a plastic material that can't exceed 150 to 180°C. We simply don't use coating in the high-temperature fiber, and protect the brittle fiber tip with an appropriate end-cap. Second, at about 400 to 500°C, the silica fiber itself starts softening. We can improve mechanical resistance of the silica fiber up to ≈ 800°C by using quartz cladding in place of the ordinary silica cladding.
Finally, we get the desired temperature dependence of $\alpha(\lambda)$ by suitably doping the silica core. There is a variety of choices available, but a good one is Nd, the same rare-earth metal used to fabricate an optical-amplifier fiber for the 1060-nm wavelength. At a $\approx 0.1\%$ concentration in the core, the Nd-doped fiber has the absorption bands shown in Fig.8-11.
Of the two absorption bands, the one at λ_1 has attenuation strongly dependent on temperature and will be the measurement channel, whereas at λ_2 attenuation is nearly independent from temperature and will be used as the reference channel.

8.3 Readout Configurations 325

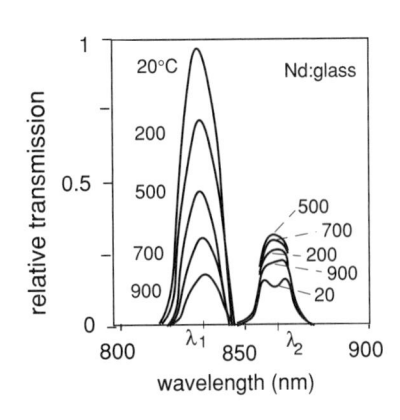

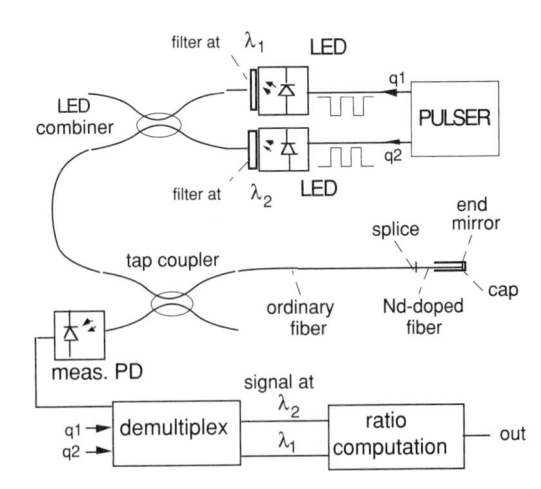

Fig.8-11 A high-temperature OFS uses a Nd-doped fiber that has two absorption bands at λ_1 and λ_2 (left): one with $\alpha(\lambda)$ strongly dependent on T, and the other nearly independent from T. Outputs from LEDs are filtered at λ_1 and λ_2, combined in the 50/50 coupler and sent to the doped fiber (right). The fiber end-face is metal-coated and reflects back the light signal to the photodiode. LEDs are pulsed in alternate phases so that both λ_1 and λ_2 channels are read by a single PD. From the PD output, a demultiplexer returns the $\alpha(\lambda_1)$ and $\alpha(\lambda_2)$ signals, from which temperature is computed.

To perform the attenuation readout, we use two LEDs emitting at the central wavelength λ=850 nm and with a line width $\Delta\lambda\approx$50 nm. By filtering the LED emission with narrow-band interference filters (for example with $\Delta\lambda_{IF}\approx$ 5-10 nm), we can obtain adequate power at λ_1 and λ_2 for the measurement of attenuation at the reference and measurement wavelengths. The sensing fiber can be used in transmission or in reflection. In transmission, we need to fold back the fiber, and this operation is objectionable because the fiber without secondary coating is brittle. We may read the fiber in transmission if we metal-coat the fiber end so that it works as a mirror (as shown in Fig.8-11).

In the receiver section, we can use two photodiodes with filters at λ_1 and λ_2, or a single photodiode with a multiplex of LED sources (Fig.8-11). To perform multiplexing, we drive the LEDs in alternate phases by signals q_1 and q_2. Thus, only one LED is switched on at a time to feed the sensing fiber. When the returned signal is detected by the photodiode PD, we simply demultiplex it to the λ_1 and λ_2 channels through the phases q_1 and q_2.

Last, we can compute the quotient of the signals, usually with the aid of a logarithm converter, and obtain a signal related to temperature.

A typical performance obtained by the doped fiber OFS [8] is $\approx$1°C accuracy in a measurement range up to 800°C.

Still higher temperatures can be measured using fibers. In Ref.[9], the principle of a channel-ratio or two-color radiant sensor is described and analyzed.
Based on this concept, a radiant OFS has been developed [8]. It works on the blackbody emission of a special fiber tip, made of a multimode sapphire core and coated with alumina black acting as the blackbody surface (Fig.8-12).

The sapphire fiber is ≈10 cm long to withstand temperature, and is spliced to a normal silica fiber for the down-lead connection to the measuring instrument.

Principle of operation

The radiance per unit wavelength emitted by the fiber tip is given by the blackbody expression [9] $r(\lambda,T)= (hc^2/\lambda^5)/(\exp hc/\lambda kT -1)$. Multiplying this quantity by the fiber acceptance $A\Omega$, A being the area and Ω the solid angle, we get the power per unit wavelength guided by the fiber down to the detection section. We use two photodetectors, looking at two bands of the emission at wavelengths λ_1 and λ_2, and compute the ratio of their currents as: $I_1/I_2=(\lambda_2/\lambda_1)^5 \exp[hc/kT(1/\lambda_1-1/\lambda_2)] (\Delta\lambda_1/\Delta\lambda_2)$. Taking the inverse logarithm of this ratio, we get a quantity $[\ln I_1/I_2]^{-1}=C_a+C_bT$, where C_a and C_bT are instrument constants. This quantity is linearly related to absolute temperature T without the need for any calibration, at least in principle.

The operating range of the radiant-emission OFS is 500 to 2000°C, the typical resolution is ≈0.05°C, and the accuracy is about 0.5°C.

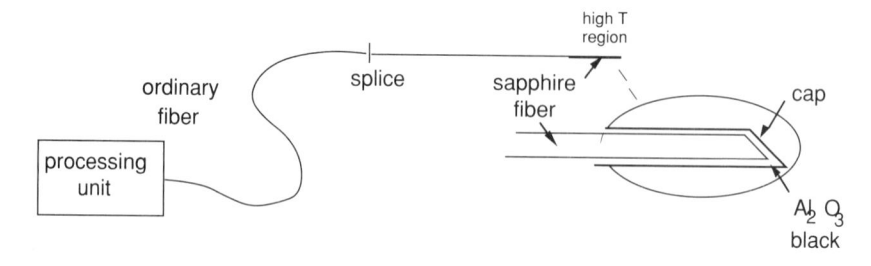

Fig.8-12 This OFS for temperature measurements in the range 500 to 2000°C employs a temperature-resistant sapphire fiber coated with alumina black. The fiber tip emits blackbody radiation, and power collected by the fiber is sent to the two-color processing unit.

In the field of chemical OFSs, a simple sorter of chemical species is readily implemented by a multimode fiber exposed to the surrounding environment [10,11]. To accomplish this, we remove the clad from a piece of fiber of suitable length (usually ≈10 cm is enough). The external medium determines the local numerical aperture $NA=\sqrt{(n_1^2-n_{ext}^2)}/n_0$, n_{ext} being the index of refraction, and hence the extra attenuation experienced by the fiber tip.

The attenuation depends on the angle θ_1 under which the source launches light in the fiber, the maximum value being that of fiber acceptance $\sin \theta_1 = \sqrt{(n_1^2-n_2^2)}/n_0$. Light launched at angles θ between θ_1 and $\theta_{ext} = \sin^{-1}NA$ is no more guided and the corresponding attenuation is a function of n_{ext}.

8.3 Readout Configurations

The typical attenuation read by a multimode OFS is shown in Fig.8-13 for several chemicals of interest in pollution monitoring. Attenuation can be read in reflection or transmission with the help of one of the several readout schemes discussed so far for intensity sensors.

Regarding this application, we actually get a large measurable signal from most species, as shown in Fig.8-13. However, we have no means to discriminate one substance from another, or, the measurement has no selectivity to species.

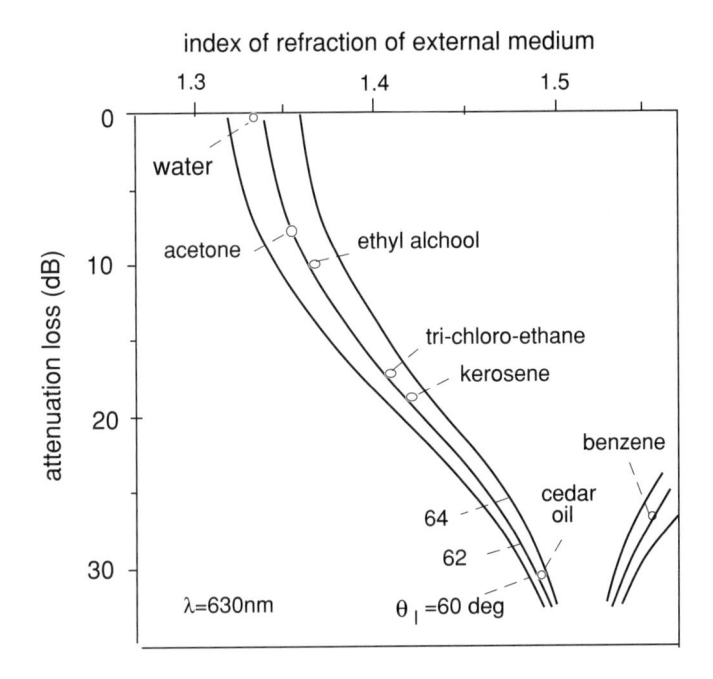

Fig.8-13 Transmission loss produced by an external liquid surrounding an uncoated silica fiber, as a function of the index of refraction and with the angle of launch θ_l as a parameter.

Another OFS used for measuring of chemical quantities is based on the optical change undergone by a special fiber tip, called *optrode* (a contraction of *optical electrode*). As exemplified in Fig.8-14, the optrode is made of a thin layer of a polymer incorporating a reactant substance. The reactant may change either its spectral attenuation or the fluorescence response when exposed to the chemical species under measurement.

For example, to measure the pH or acidity of a solution, the polymer layer is made of XAD-2 microspheres, treated for adsorption of phenol red.
Phenol red has a peak of absorption at λ=560 nm, which is strongly dependent on the pH, whereas at λ=475 nm we find a nearly constant attenuation (Fig.8-14).

Thus, we can measure the phenol red optrode with the aid of a scheme similar to that of Fig.8-11, and obtain a quantity related to the relative attenuation.

After a calibration, the pH is read with an accuracy of ≈0.02 in the range 6.8 to 7.9, and reproducibility is good. The measurement time is relatively long, about 60 s, because the external medium shall diffuse through the boundary membrane before being coming in contact with the reactant.

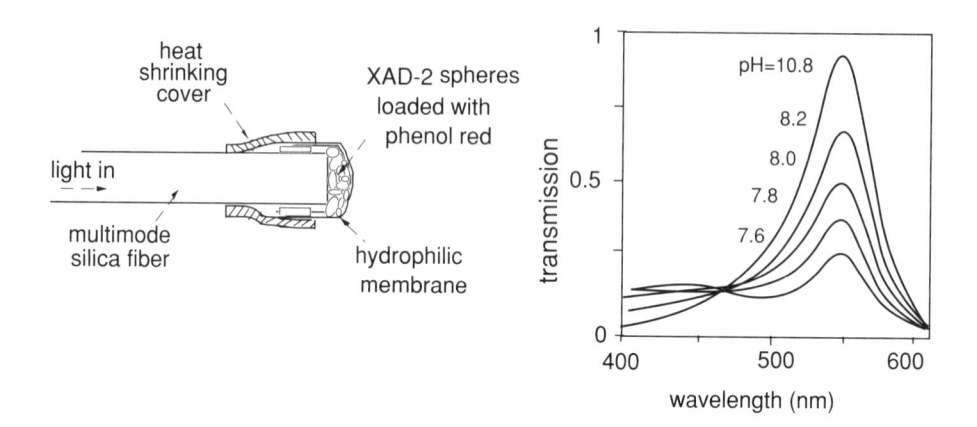

Fig.8-14 A pH optrode has its fiber tip covered with polymer microspheres (XAD-2) bearing the reactant phenol red (left). The absorption spectrum of phenol red shows peak sensitivity to pH at λ≈550 nm, and a reference at λ≈475 nm (nearly independent from pH).

Another reactant used for a somewhat wider range of measurement is bromothymol blue, a dye that has a maximum variation of absorption at λ=620 nm and a minimum at λ=500 nm. With it, a pH optrode has been developed for the measurement range 6.4 to 7.8. A number of other chemical quantities can be measured with optrodes [12]. They are pO_2, pCO_2, and a number of water pollutants like pesticides, hydrocarbons, heavy metals, etc.

For biomedical use, additional applications of OFS measurement include [13,14]: glucose, metabolites, haemoglobin, breathing and blood gases [15], enzymes and co-enzymes, inhibitors, lipids, drugs, immunoproteins, etc.

8.3.2 Polarimetric Readout

In a polarimetric OFS, the transducer mechanism affects the state of polarization and we perform a polarimetric readout of the birefringence induced by the measurand.

Birefringence effects may be classified as *linear* or *circular*.
We get linear birefringence when the measurand introduces an optical retardance Φ along a (slow) axis of the fiber with respect to the other, orthogonal (fast) axis. For example, in

8.3 Readout Configurations

strain and force sensors, the axis parallel to the force is the slow axis, and retardance is proportional to the force.

Linear birefringence is described as a difference in the propagation constants β_s and β_f along the slow and fast axes. Thus, down a propagation length L, we get a retardance $\Phi = (\beta_s - \beta_f)L$.

On the other hand, we get circular birefringence (also referred to as rotatory power) when the measurand affects the direction of a linear polarization state, making it rotate while it propagates through a piece L of fiber. The angle of rotation is then $\Psi = \Delta\beta L$.

Equivalently, circular birefringence can be described as retardance between right- and left-handed circular polarization states. We may then write $\Delta\beta = \beta_R - \beta_L$, where β_R and β_L are the propagation constants for right- and left-handed polarizations, respectively.

Different from linear birefringence, which can only be reciprocal, circular birefringence can also be non-reciprocal. Reciprocity has to do with the change of birefringence when the direction of propagation is reversed or, propagation vector $\underline{k}$ is changed to $-\underline{k}$.

Only two non-reciprocal effects of circular birefringence are known: the Faraday effect, which is due to a longitudinal magnetic field, and the Sagnac effect (see Sect.7.2), which is due to inertial rotation. All other effects (sugar rotatory power, crystal twist, etc.) are reciprocal, that is, rotation Ψ is unchanged when reversing the direction of observation (or the cell containing the material). Instead, reversing the direction changes rotation Ψ to $-\Psi$ in a non-reciprocal material or medium.

8.3.2.1 Circular Birefringence Readout

Among intrinsic polarimetric sensors, we find two significant examples that have been developed to take advantage of circular birefringence effects.

The first OFS is the torsion-bar balance, based on the reciprocal rotation induced by a twist of the fiber. The twist is applied by the small weight to be measured. Because we are able to resolve an angle of polarization rotation as small as $\approx 10^{-4}$ rad, the resulting OFS is a very sensitive balance, capable of appreciating a very small torque applied to the fiber (down to $\approx \mu g \cdot cm$).

The second OFS is a magnetic field or electrical current sensor, and is based on non-reciprocal rotation. In the following, we refer to this sensor to explain the readout of circular birefringence. However, the concept also holds for the readout of reciprocal circular birefringence as well as for extrinsic sensors.

Non-reciprocal rotation is also called the Faraday effect, and it is induced by the longitudinal component H_l of magnetic field applied to the fiber. The angle of rotation Ψ is proportional to the line integral of the field H_l along the propagation path. The elemental contribution is $H_l \, dl = \underline{H} \cdot \underline{dl}$, where the dot $\cdot$ indicates the scalar product of vectors $\underline{H}$ and $\underline{dl}$. The constant of proportionality V is called the *Verdet constant*.

Thus, we may write the rotation angle as [16]:

$$\Psi = V \int_L \underline{H} \cdot \underline{dl} \tag{8.3}$$

330 Optical Fiber Sensors Chapter 8

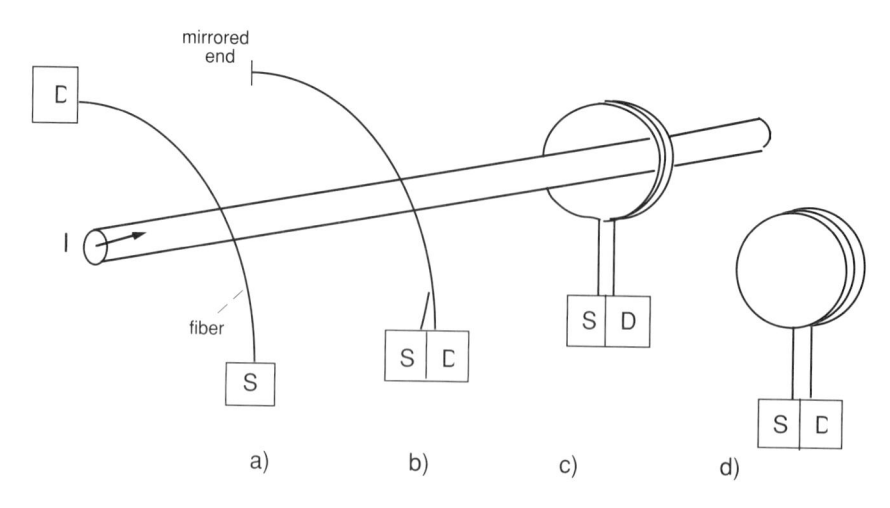

Fig.8-15 A piece of fiber or a coil can be used in connection with a polarized source S and a detector analyzer D to sense the Faraday rotation Ψ induced by a magnetic field. In a), with Ψ we read the field $\int \underline{H} \cdot \underline{dl}$, averaged along the fiber length. If the fiber has a mirrored end, the rotation doubles because it is non-reciprocal. With a fiber coil, rotation is proportional to the concatenated current I. Therefore in c), Ψ is proportional to the current I in the concatenated wire, whereas in d) Ψ is exactly zero.

Using a piece of fiber to sense the Faraday rotation, we ger a magnetic field OFS (Fig.8-15a). Indeed, in this case, rotation Ψ is proportional to the magnetic field $\int_L \underline{H} \cdot \underline{dl}$ averaged along the fiber length L.
Using a mirrored end (Fig.8-15b) on the fiber, we fold back the propagation and get twice the rotation because the Faraday circular birefringence is non-reciprocal. By contrast, applying the fiber a twist-induced, reciprocal circular birefringence, we get rotations with opposite signs along a go-and-return path, and total effect is zero.
With a fiber coil wound around the conductor (Fig.8-15c), the OFS is a true current sensor because the rotation becomes the circulation of magnetic field intensity. This quantity is known from elementary magnetism to be equal to the magneto-motive force, that is, $\int_L \underline{H} \cdot \underline{dl} = N_f I$. Thus, if we arrange the sensor in the form of a winding of N_f turns of fiber wound around an electrical conductor carrying a current I, Eq.8.3 becomes:

$$\Psi = V \int \underline{H} \cdot \underline{dl} = V N_f I \qquad (8.3')$$

It is interesting to note that the result holds independently from coil size or shape, the only requirement being that the coil is linked to the conductor carrying the current I.
Also, if the coil is not linked to the conductor (Fig.8-15d), the rotation Ψ is exactly zero.

8.3 Readout Configurations

This feature is very attractive for the application of the current OFS to three-phase power lines [16]. Here, the three wires are close and the sensor of each phase must be sensitive only to the current carried by its conductor and not to others.

For the readout of either reciprocal or non-reciprocal rotation, we can use the general scheme illustrated in Fig.8-16.

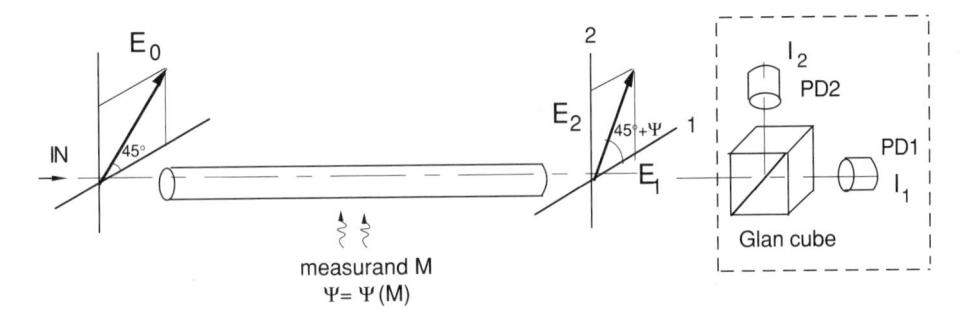

Fig.8-16 Schematic for the readout of circular birefringence. The input state of polarization E_0 is rotated by a small angle $\Psi=\Psi(M)$ by the measurand M. The Glan cube analyzer is oriented at 45° with respect to E_0, and splits the output field into components E_1 and E_2. The photodiodes provide output currents proportional to E_1^2 and E_2^2.

In this scheme, the fiber is read with a linearly polarized state, which can be shown [17] to be the optimal choice. As a reference system, let us take the axes (1 and 2 in Fig.8-16) of the output analyzer (the Glan cube), and assume that the input field E_0 is oriented at 45°. Along the fiber, the measurand M produces a rotation of the state of polarization, and the output field emerges from the fiber rotated by an extra angle $\Psi=\Psi(M)$ added to the initial 45°, ending up oriented at 45°+Ψ.

The Glan cube divides the output field in the components E_1 and E_2, and directs them on two separate photodiodes, PD1 and PD2. We may then write the fields E_1 and E_2 and the associated currents I_1 and I_2 as:

$$E_1 = E_0 \cos(45°+\Psi), \quad E_2 = E_0 \cos(45°-\Psi) = E_0 \sin(45°+\Psi)$$

$$I_1 = E_1^2 = E_0^2 \cos^2(45°+\Psi), \quad I_2 = E_0^2 \sin^2(45°+\Psi) \qquad (8.4)$$

As we can see from Eq.8.4, neither the fields nor the currents are linearly related to the measurand. To obtain a better response, we compute the ratio S of difference I_1-I_2 to sum I_1+I_2 of the detected currents. Doing so, we obtain the following with easy algebra:

$$S = (I_1-I_2)/(I_1+I_2) = E_0^2[\cos^2(45°+\Psi)-\sin^2(45°+\Psi)]/E_0^2[\cos^2(45°+\Psi)+\sin^2(45°+\Psi)] =$$

$$= \cos 2(45°+\Psi) = -\sin 2\Psi \qquad (8.5)$$

Eq.8.5 is a good result because, with the difference-to-sum ratio, we have obtained a signal linearly related to the measurand Ψ [$\sin 2\Psi \approx 2\Psi$ for small Ψ].

In addition, the measurement is no longer dependent on E_0^2, the power of the input beam. In other words, the common-mode fluctuations are eliminated thanks to the ratio-like structure of the signal S.

The range of linearity may also be improved further by computing the rotation angle as $2\Psi =$ arcsin $S \approx S + S^2/6$. However, this correction does not affect the measurement ambiguity that we encounter in the sine function when the rotation angle becomes large ($2\Psi > \pi/2$).

Similar to the treatment of interferometer signals, we need to supplement the measurement channel of Fig.8-16 with a second measurement channel to remove the ambiguity. This way, we make two orthogonal signals of the type $\sin 2\Psi$ and $\cos 2\Psi$ available.

We can obtain the second channel by deriving a fraction of the output and analyzing it through a Glan cube oriented at 0° and 90° with respect to the direction of the input state of polarization, as illustrated in Fig.8-17.

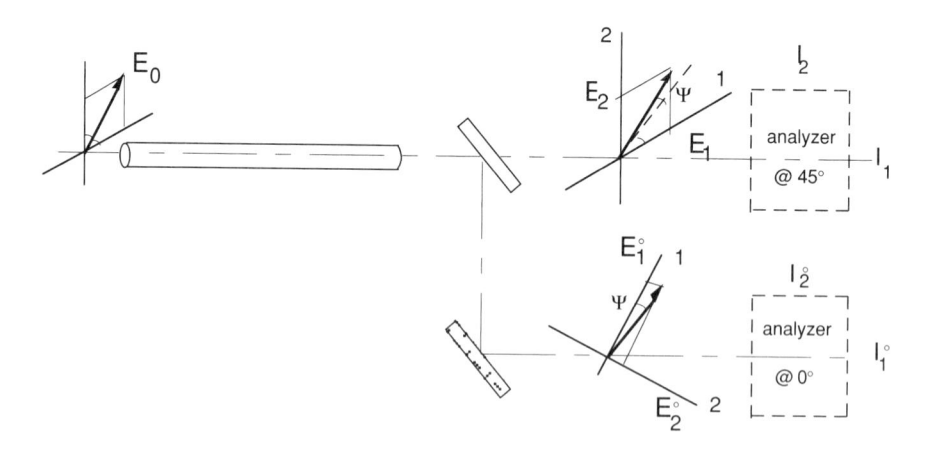

Fig.8-17 By using two Glan-cube analyzers of the output state of polarization, we can generate two signals, sin 2Ψ and cos 2Ψ, which allow us to compute the angle Ψ without ambiguity, also for $\Psi > \pi/2$.

Repeating the arguments leading to Eq.8.4 for signals $E_1°$ and $E_2°$ of the second channel, we can write:

$$E_1° = E_0 \cos \Psi, \quad E_2° = E_0 \cos(90° - \Psi) = E_0 \sin \Psi$$

$$I_1° = E_0^2 \cos^2 \Psi, \quad I_2° = E_0^2 \sin^2 \Psi$$

$$S° = (I_1° - I_2°)/(I_1° + I_2°) = \cos 2\Psi \tag{8.6}$$

8.3 Readout Configurations

The functions indicated in Fig.8-17 can be implemented with the all-fiber technology, as shown in the schematic of Fig.8-18. In it, we use a normal coupler C (polarization-insensitive) as the main divider of the output signal, and polarization-splitter couplers PS as the all-fiber versions of the Glan cube.

The fiber used in the coil and in the down-lead trunks is critical and requires brief discussion. In the coil, the bend of the fiber introduces linear birefringence that interferes with the small Faraday rotation. To quench the linear birefringence, a larger circular birefringence is added to the fiber of the sensing coil. This is done by twisting the fiber [16] with an appropriate period of twist (typically ≈1-3 cm). The obtained fiber is classified as high circular birefringence (hbc). It adds a reciprocal rotation Φ_{rec} to the useful Faraday rotation Ψ. This is not a problem with ac currents, however, because Φ_{rec} is a constant term and can be filtered out in the detected signal.

The same hbc fiber can also be used in the down-lead trunks connecting the coil to the instrument. Finally, in the upgoing lead, we can use a high linear birefringence fiber (hbl) in place of the hbc, as the hbl fiber is more insensitive to bends. [We can't use the hbl fiber in the downgoing lead, because the state of polarization out of light from the coil is unknown].

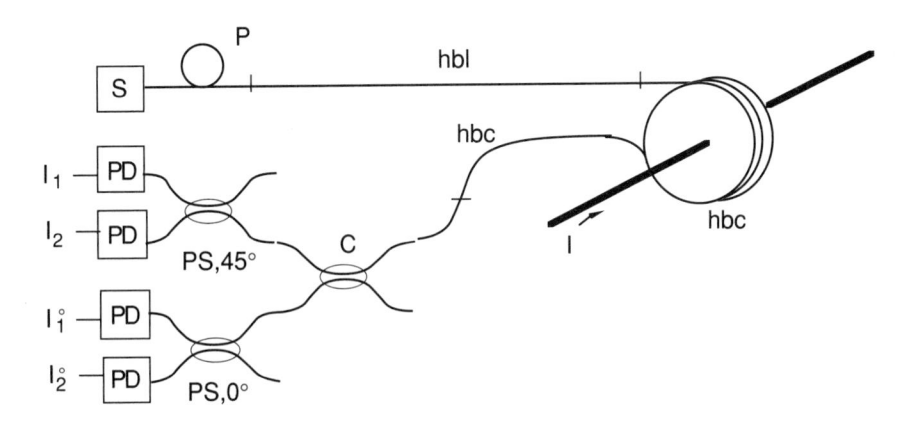

Fig.8-18 An all-fiber circuit for the measurement of circular birefringence. Light from source S is polarized by fiber polarizer P and sent up to the sensing coil through the high linear-birefringence (hbl) fiber to maintain polarization. The down-lead fiber is made of high circular-birefringence (hbc) fiber. Coupler C splits the output at 50% for analyzers PS. Analyzers are polarization-splitting couplers with slow axes oriented at 0° and 45°.

A final improvement to the all-fiber scheme of Fig.8-18 is to take advantage of the difference in reciprocity of the terms Φ_{rec} and Ψ. If we are able to cancel out Φ_{rec}, operation of the current sensor is extended down to the dc component. The reciprocal rotation Φ_{rec} of hbc fiber pieces is actually subtracted if we trace back propagation through the fiber, whereas the non-reciprocal rotation Ψ is doubled.

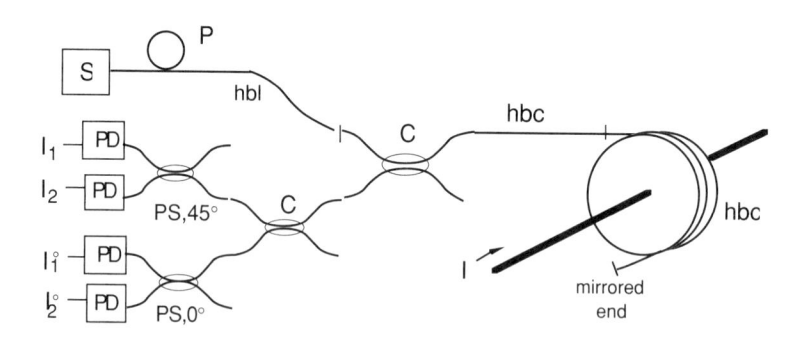

Fig.8-19 With a fiber coil ending on a mirrored end, we are able to cancel out the static rotation due to circular birefringence of the hbc fiber and extend the current measurement down to the dc component.

As shown in Fig.8-19, we obtain this condition by using a mirrored fiber coil end, reflecting back the radiation at the coil exit [16].

Developing the concept further, we might think of canceling the residual linear birefringence of the coil and access lead.

This can indeed be done using a Faraday mirror at the exit of the coil, in place of the normal plane mirror obtained by metal coating the fiber end face (Fig.8-19).

The Faraday mirror is a device combining a 45° Faraday rotator and a mirror. The rotator is simply made of a YIG crystal exposed to a magnetic field [16,18] that provides a non-reciprocal 45° rotation. Because of the total 90° rotation on reflection at the Faraday mirror, any input state of polarization is turned into its orthogonal on the Poincaré sphere [17]. Then, the go-and-return net change of state of polarization is zero, provided the effects generating the polarization changes are reciprocal. Therefore, all the linear and circular (reciprocal) birefringences are canceled out.

Unfortunately, cancellation does not affect the quenching effect discussed previously, and thus an eventual Faraday mirror can't free us from using the hbc fiber in the sensing coil to contrast bending birefringence [16].

Going back to consider the signals S and S°, we can reconstruct argument Ψ and current I also for large values of rotation ($>2\pi$). The processing of signals may be either analog or digital, as outlined in Fig.8-20.

In the analogue approach, we use the same method discussed in Sect.4.5.2.3 to process the $\sin\Psi$ and $\cos\Psi$ signals of an interferometer. We compute the derivatives of both signals, cross-multiply them with the other signal, and subtract the result. In this way (Eq.4.39), we obtain $d\Psi/dt$ and, after a final integration, the desired Ψ proportional to current I.

While relatively cheap to implement, the analog approach has some limitations at very small signals as the errors due to the analog multiplier become important. In addition, at low frequency (or in dc), the low-pass function of the derivative is not exactly canceled out by the final integration and a low frequency cutoff remains.

8.3 Readout Configurations

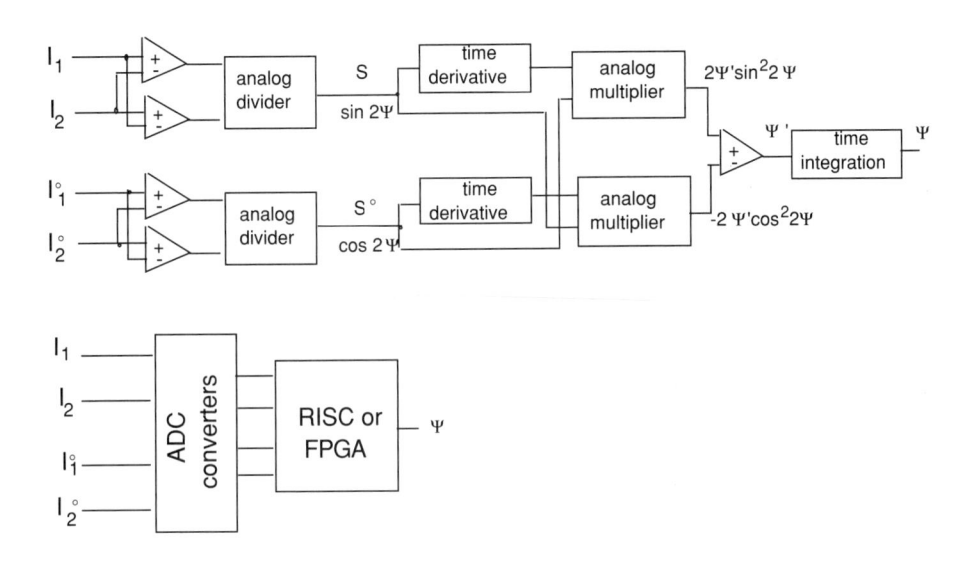

Fig.8-20 Analog (top) and digital (bottom) processing of photodetected currents to obtain rotation angle Ψ.

We may then prefer the digital processing (Fig.8-20), whose only limitations are associated with the offset and truncation errors introduced by the analog-to-digital converters (ADC). A reduced instruction set microcomputer (RISC), or even a fully programmable gate array (FPGA), may be adequate to carry out the modest computational load that is required.

8.3.2.2 Performance of the Current OFS

After the measurement of currents I_1 and I_2 and the calculation of the angle 2Ψ (Eq.8.5), we shall now consider the minimum detectable signal NEI= σ_I (Noise-Equivalent Current) of the sensor. At the quantum noise limit, this quantity is given [16] by the rms phase noise $(2eB/I_0)^{1/2}$ divided by the responsivity 2VN, or:

$$\text{NEI} = (1/2VN)(2eB/I_0)^{1/2} \tag{8.7}$$

In this expression, $V = \Psi/NI$ is the Verdet constant of the material [16], expressing the specific Faraday rotation (Ψ) per unit magneto-motive force (NI), and I_0 is the photodetected current in each photodiode. The 3-dB bandwidth B of the sensor is primarily limited by the transit time τ through the coil and is given by [16]:

$$B = 0.44/\tau = 0.44c/2\pi rN \tag{8.8}$$

where r is the coil radius and N the number of turns.

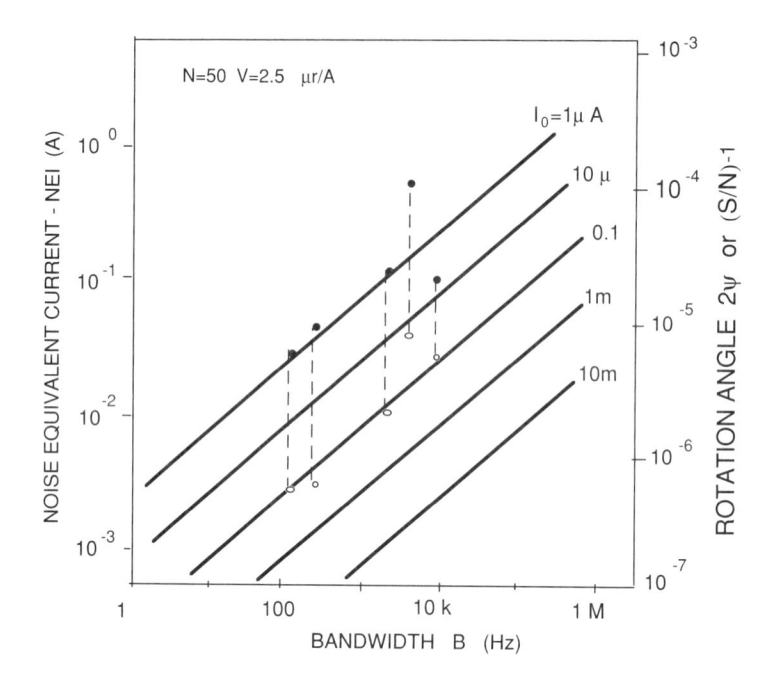

Fig.8-21 Minimum current (or NEI) that can be detected in a Faraday rotation OFS, as a function of bandwidth B and with photodetected current I_0 as a parameter. Experimental points are shown (full circle) along with the corresponding theoretical values (open circles).

In Fig.8-21, we plot the NEI of the current OFS versus the bandwidth of measurement B, with the photodetected current I_0 as a parameter [16], as given by Eq.8.7. In the diagram, we also report the results obtained by a few research groups [19-23]. These are the dotted lines connecting the experimental results to the expected theoretical values.

As can be seen, the theoretical NEI limit due to the quantum noise associated with the readout beam is approached by about a decade, in practice. This may look like a poor result, but actually it would be very satisfactory because currents of, let's say, 1-A can be detected with an adequate signal-to-noise ratio, up to bandwidth of ≈10 kHz.

In Fig.8-22, we report the actual waveforms obtained by the Faraday current OFS and the corresponding electrical waveforms. It can be seen that they are nearly the same.

When the waveform is quasi-sinusoidal, the electrical one is delayed by 90°, and this is correct because the secondary voltage $V_2 = j\omega M I_1$ lags the primary current I_1 of 90 degrees. Also shown in Fig.8-22 is the ability of the current OFS to reproduce fast transients, down to pulse widths in the range of 0.1 milliseconds or less.

8.3 Readout Configurations

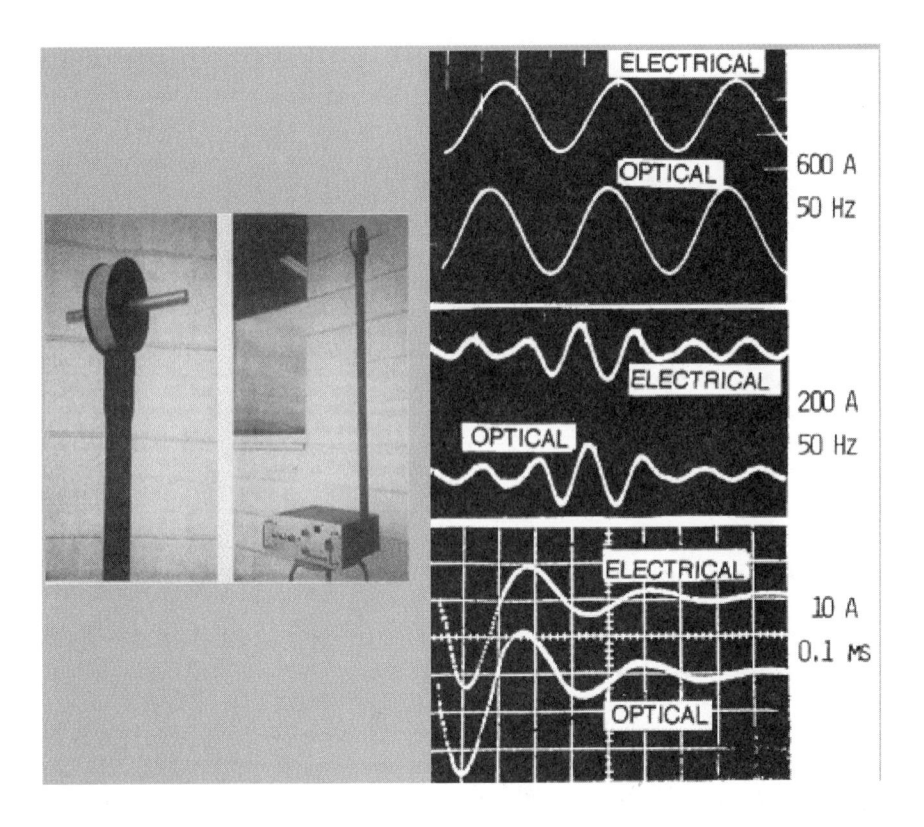

Fig.8-22 A laboratory prototype of the current sensor (left) and examples of detected waveforms (right). Top traces: currents measured by a normal electrical transformer, bottom traces: currents obtained with the OFS. Top: mains current 600 A, 50 Hz; middle: a transient with 200 A peak current at 50 Hz; bottom: a fast disturbance with 10-A peak on a 0.1-ms time scale.

Another interesting application of a Faraday OFS is magnetic field sensing. In principle, we should use a single, short piece of fiber as indicated in Fig.8-15a, to get a net non-zero rotation $\Psi = V \int_L H \cdot dl$.

The Verdet constant is rather small in ordinary silica fibers ($V \approx 2.5$ μr/A at 800 nm), however. We may wish to devise some special trick to use a coil and be able cumulate the Faraday rotation on a much longer length. However, a coil unlinked to the source of the magnetic field (Fig.8-15d) has a line integral $\int_{line} H \cdot dl = 0$.

We can take advantage of bending birefringence to have a non-zero line integral for our coil. Using an appropriate curvature radius for the winding [21,24], we may obtain a linear birefringence of 2π per turn. We need to select a $\approx$1-cm radius for that [25].
Then, the state of polarization changes its orientation by π each half-turn. The Faraday elemental rotation, $+d\psi$ and $-d\psi$ on opposite half-turns, becomes summed with the same sign

and we have a total rotation $\Psi=VH\pi r$, which is only half that corresponding to the total fiber length [25].

Moreover, the two in-plane components of the magnetic field, H_x and H_y, affect the state of polarization of the bending-birefringence and Faraday rotation coil in a different way.

Thus, with a proper processing of the optical signal out of the coil, we are able to measure both of them with a single sensor [25].

8.3.2.3 Linear Birefringence Readout

Several intrinsic and extrinsic OFSs are based on linear birefringence effects. Indeed, silica in the fiber and other optical materials are sensitive to external mechanical perturbations, such as compression and bending. This circumstance leads to the development of OFSs for mechanical measurands (Fig.8-3). In addition, birefringent crystals like lithium niobate ($LiNbO_3$) exhibit a significant and reproducible temperature coefficient, and based on that, a temperature OFS can be developed.

Through the elasto-optical effect, mechanical perturbations are translated into linear birefringence. As it is well known, linear birefringence consists in an optical delay Φ between the linear state of polarization propagating along fast and slow axes.

The optical delay or phase shift Φ is proportional to the measurand modulus, at least at small levels of perturbation, and the axes of birefringence are normally parallel and perpendicular to the axis along which the measurand is applied (Fig.8-23).

The basic scheme for the readout of linear birefringence is shown in Fig.8-23. Light from the source enters the fiber with a linear state of polarization, oriented at 45° with respect to the birefringence axes set by the measurand, that is, the x and y axes.

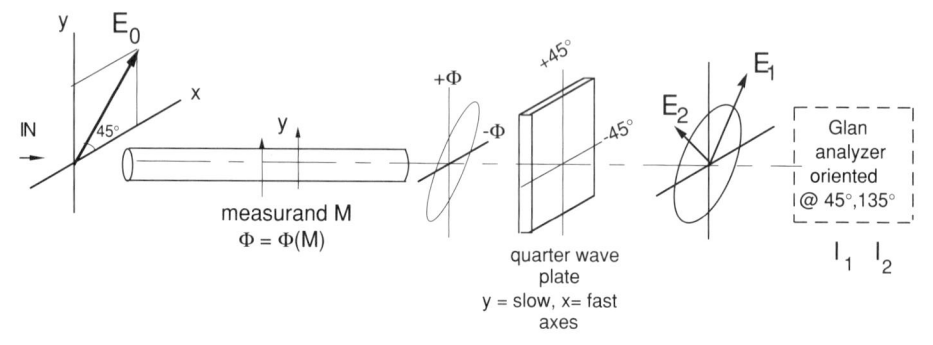

Fig.8-23 Schematic for the readout of linear birefringence. We enter the fiber with two equal components on the xy axes, using a linear polarization E_0 oriented at 45°. The measurand introduces a delay $+\Phi$ and $-\Phi$ between the components. At the fiber exit, we place a quarter wave plate for an additional +45° and −45° delay. We then analyze the field components E_1 and E_2 by means of a Glan cube, with the axes oriented at 45° and 135°.

8.3 Readout Configurations

At the fiber output, the two polarization components are phase shifted by $+\Phi$ and $-\Phi$. Thus, light emerges from the fiber with an elliptical state of polarization.

Before analyzing the state of polarization, we insert a quarter-wave ($\lambda/4$, or 45° birefringence) plate with the principal axes oriented along x and y. This way, the total phase shift between the two components becomes $+45°+\Phi$ and $-45-\Phi$.

Next, we analyze the output field with a Glan cube beam splitter, with the axes oriented at 45° and 135° in the x-y reference, that is, parallel and perpendicular to input field E_0. To write the two field components E_1 and E_2 supplied to the Glan cube, we note that the input field feeds with two equal amplitudes $E_0/\sqrt{2}$ the x and y components. These components are phase-shifted by $45°+\Phi$ and $-45°-\Phi$ as indicated in Fig.8-23, and then projected on the 45° and 135° axes. Adding the two contributions from the x and y axis, the fields of E_1 and E_2 are written as follows:

$$E_1 = (1/2)E_0 \exp i(-45°-\Phi) + (1/2)E_0 \exp i(45°+\Phi) = E_0 \cos(45°+\Phi),$$

$$E_2 = (1/2)E_0 \exp i(-45°-\Phi) - (1/2)E_0 \exp i(45°+\Phi) = -i E_0 \sin(45°+\Phi) \quad (8.9)$$

In the above equations, the factor of 1/2 in the amplitude comes from the double projection on x and y, and on 45° and 135° of the initial component E_0. In addition, to simplify the algebra, we have written delays in a symmetrical format, using -45° and +45° in place of 0 and $\pi/2$, and $-\Phi$ and $+\Phi$ in place of 0 and 2Φ. Thus, the total phase introduced by the measurand is assumed to be 2Φ.

Associated with the fields, we get two signals, I_1 and I_2, from the photodetectors, given by the square of the field modulus or:

$$I_1 = |E_1|^2 = E_0^2 \cos^2(45°+\Phi),$$

$$I_2 = |E_2|^2 = E_0^2 \sin^2(45°+\Phi) \quad (8.10)$$

It is now easy to proceed as in Sect.8.3.2.1 to derive a quantity linearly related to the measurand phase Φ. We compute the ratio S of difference I_1-I_2 to sum I_1+I_2 of the detected currents and obtain:

$$S = (I_1-I_2)/(I_1+I_2) = \cos 2(45°+\Phi) = -\sin 2\Phi \quad (8.11)$$

As in the previous section, we may wish to supplement the measurement by a second orthogonal signal, $\cos 2\Phi$, to remove the modulus-π ambiguity of the sine function. To accomplish this function, we take off the quarter-wave plate and analyze the output with the Glan cube, again oriented at 45° and 135° with respect to the x-y reference. We then obtain two field components, $E_1°$ and $E_2°$, given by:

$$E_1° = (1/2)E_0 \exp -i\Phi + (1/2)E_0 \exp i\Phi = E_0 \cos \Phi,$$

$$E_2° = (1/2)E_0 \exp -i\Phi - (1/2)E_0 \exp i\Phi = -i E_0 \sin \Phi \quad (8.12)$$

Computing the signal $S°=(I_1°-I_2°)/(I_1°+I_2°)$ gives as a result:

$$S° = \cos 2\Phi \qquad (8.13)$$

Also for a linear birefringence OFS we can extend the range of linearity beyond 2π.
The functions indicated in Fig.8-23 and the extensions required for S and S° can be implemented in all-fiber technology, as shown in the schematic of Fig.8-24.

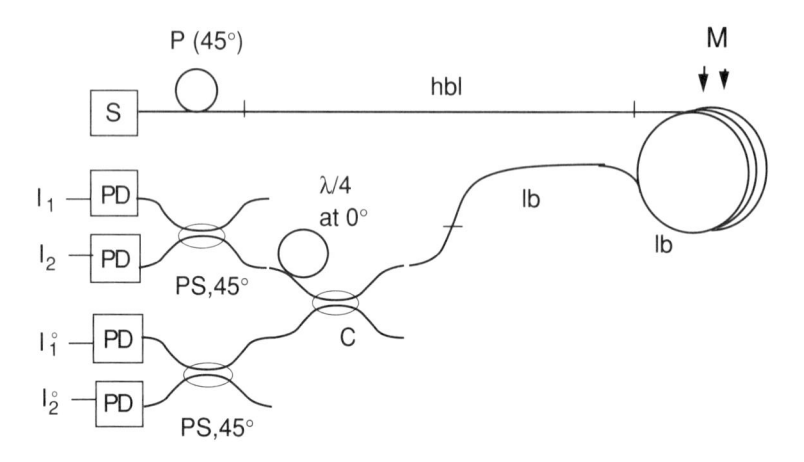

Fig.8-24 An all-fiber circuit for measuring linear birefringence. Light from source S is polarized by the fiber coil P at 45° with respect to the slow/fast axes of birefringence of the measurand M. Coupler C splits the output from the sensing region for the double-section analyzer. One section with the λ/4 plate and a Glan cube oriented at 45° gives the I_1 and I_2 signals for the sin2Φ term. The other section, without the λ/4 plate, gives the $I_1°$ and $I_2°$ signals for the cos2Φ term.

The schematic operates much like that for the readout of circular birefringence (Fig. 8-18). We need two sections for deriving the I_1, I_2 pair of signals associated with S, and the $I_1°$, $I_2°$ pair associated with S°. In one section, the analysis is made after a quarter-wave phase shift, whereas in the other section, the quarter-wave is missing.
In principle, the adduction fiber should have high linear birefringence (with axes oriented at 45° and 135°) to preserve polarization, whereas the down-lead fiber is a low-birefringence (or spun) fiber and be shielded by external mechanical disturbances.

From the photodetected outputs, the processing to compute S and S° follows as outlined in Fig.8-20, and the same circuits apply.
It is interesting to note that the double-section analyzer can be simplified to a single-section analyzer if we replace the λ/4 plate with a controlled phase delay device. This can be realized by a PZT (lead zirconate titanate ceramic) element with a short piece of fiber cemented on it.

8.3 Readout Configurations

When the piezo is fed by the appropriate voltage, it squeezes the fiber and introduces a linear birefringence via the elasto-optical effect. With ≈1-cm of fiber and a 1-mm thick PZT, we may need a voltage $V_S \approx 10$ V to impress the desired $\lambda/4$ birefringence. At $V_S=0$, the birefringence is missing and we can therefore use just a single analyzer leg, as indicated in Fig.8-25.

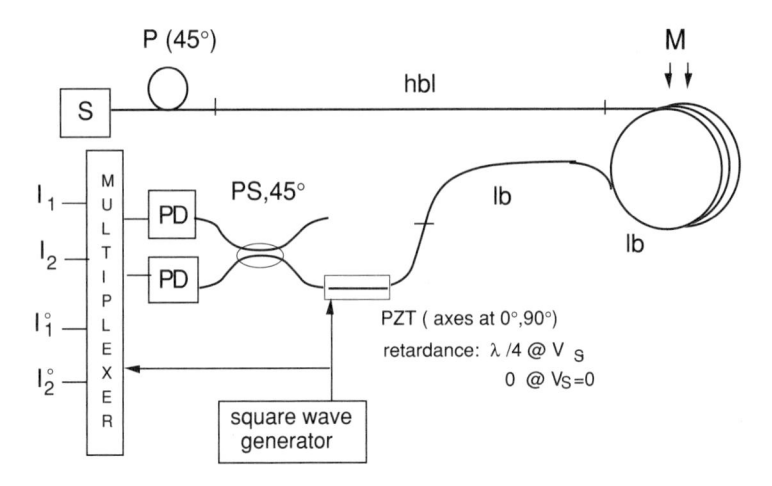

Fig.8-25 The linear birefringence readout can be made with a single analyzer if we use a controlled birefringence element PZT in place of the $\lambda/4$ in Fig.8-24. The PZT is driven by a square wave generator (amplitude V_S, frequency $f_S \approx$ kHz) and operates in combination with a multiplexer to recover the I_1, I_2 and $I_1°$, $I_2°$ signals.

The minimum amplitude of the measurand we can resolve with the linear birefringence readout depends on the responsivity $\Phi = \Phi(M)$ connecting M to the phase Φ. About the phase, the minimum detectable phase or NEΦ is ultimately determined by the shot noise associated with the detected currents I_1 and I_2. Thus, we have the same NEΦ as considered in the circular birefringence readout analyzed in Sect.8.3.2.2 and plotted in the diagram of Fig.8-21 (right-hand scale).

As we can see from the diagram, limiting resolutions or NEΦ as small as 10^{-5} rad are obtained experimentally and are actually representative of the state-of-the-art design. Values down to 10^{-7} to 10^{-6} rad are in the reach of the technique, and can be attained if bandwidth is not too demanding.

8.3.2.4 Combined Birefringence Readout

We may now wonder about the simultaneous measurement of linear and circular birefringence. In practice, it is uncommon to think of a double measurement on two measurands, one inducing circular birefringence and the other inducing linear birefringence in the same fiber. Yet, the case is one of interest as it allows us to evaluate the effect of disturbances in the sensor fiber as well as in the down-lead piece of fiber accessing the sensor.

To gain insight into this case, let us consider a fiber with two birefringences acting on the fiber (Fig.8-26), a circular one with rotation Ψ and a linear one with retardance Φ.

Let us suppose that we use the readouts illustrated in previous sections: those best suited for circular birefringence (Figs.8-16 and 8-17) and those best suited for linear birefringence (Figs.8-23 and 8-24).

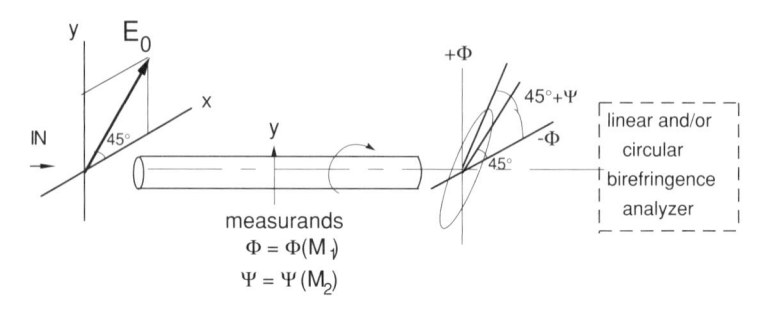

Fig.8-26   The case when circular and linear birefringences are simultaneously present in the fiber.

To distinguish the signals, let us call I_{1C} and I_{2C} the outputs of the circular birefringence scheme and I_{1L} and I_{2L} those of the linear birefringence scheme.

Now, when Ψ and Φ are present, the following relations hold in all cases:

$$I_{1C} + I_{2C} = I_{1L} + I_{2L} = I_{1C}^{\circ} + I_{2C}^{\circ} = I_{1L}^{\circ} + I_{2L}^{\circ} = E_0^2 \qquad (8.14)$$

These equations follow from the conservation of energy in the readout schemes. In addition, with easy calculations similar to those leading to Eqs.8.6 and 8.9, and omitted here for saving space, we get:

$$I_{1C} - I_{2C} = -E_0^2 \sin 2\Psi \qquad (8.15a)$$

$$I_{1L} - I_{2L} = E_0^2 \sin 2\Phi \, \cos 2\Psi \qquad (8.15b)$$

$$I_{1C}^{\circ} - I_{2C}^{\circ} = I_{1L}^{\circ} - I_{2L}^{\circ} = E_0^2 \cos 2\Phi \, \cos 2\Psi \qquad (8.15c)$$

These expressions tell us that, if we really wish to do so, we can perform a simultaneous measurement of Ψ and Φ, putting together the readout schemes already discussed. In fact, for small values of the measurands, $\Psi, \Phi \ll 1$, we get $\cos 2\Psi \approx 1$ in Eq.8.15b and the difference signals are $I_{1C} - I_{2C} \approx -2\Psi$, $I_{1L} - I_{2L} \approx 2\Phi$.

An example of application of the concept is provided by Ref.[25], in which two components, H_x and H_y of the magnetic field acting on a fiber coil, are simultaneously measured thanks to the interplay of Faraday (circular) and bending (linear) birefringence.

8.3 Readout Configurations

To combine the schematic of linear and circular birefringence readout, we can take advantage of the PZT controlled-phase elements discussed above, and multiplex a single channel of polarization analysis for the entire setup. A variety of useful configurations can be developed from this concept, and we will not detail them further, leaving the effort to the reader.

8.3.2.5 An Extrinsic Polarimetric Temperature OFS

Another example of an extrinsic polarimetric OFS is provided by the measurement of temperature in the range 0..500°C, as required in the application of geothermal mining [26]. Ordinary fibers do not work well at these high temperatures, and thus the OFS must be extrinsic. A suitable crystal, like lithium niobate (LiNbO$_3$), is chosen to withstand the required temperature and provide a linear birefringence with a sizeable and very reproducible temperature coefficient.

By measuring the phase shift $\Phi = \Delta\beta L$, where $\Delta\beta = \beta_x - \beta_y$ is the difference of propagation constants along the principal axes x and y, and L the crystal length, we obtain a $\Phi = \Phi(T)$ dependence for the measurand. Indeed, if $\alpha_{\Delta\beta} = d\Delta\beta/dT$ is the temperature coefficient of the birefringence, then we get $\Phi = \Delta\beta L = \alpha_{\Delta\beta} \Delta T L$.

As illustrated in Fig.8-27, the crystal is placed in a probe at the end of a long tube running parallel to the drill in search of the geothermal poll. The tube carries the adduction fiber and may be several km long. Thus, a requirement of the system is that the fiber shall be a rugged, multimode fiber with a large (50...100 μm) core diameter, so that mechanical tolerances of the mounting are minimized. We would also like to use a single fiber for both feeding the crystal with polarized light and bringing back the useful signal to be measured.

Because the multimode fiber does not maintain polarization, we must insert a polarizer (a calcite prism) in the gauge, in front of the sensing crystal. After the optical signal coming out the LiNbO$_3$ crystal is properly polarized, we only need that the adduction fiber has little attenuation and a negligible polarization-dependent loss. This constraint can be amply satisfied with ordinary multimode fibers.

We now need to modify the double channel strategy illustrated in last section and look for the sinΦ and cosΦ signals from a single amplitude measurement, performed external to the probe.

This can be done by taking advantage of the wavelength dependence of birefringence, $\Delta\beta = \Delta\beta(\lambda)$. Using a semiconductor laser diode, a wavelength sweep is easily achieved by applying a sweep to the drive current. Thus, in addition to the signal Φ, we have a time-dependent phase $\phi(t) = \Delta\beta(\lambda) L$. We manage $\phi(t)$ swings on a full 2π cycle.

The arrangement of Fig.8-27 has the crystal and the polarizer passed twice in the go-and-return path. With the arguments leading to Eq.8.9, it is easy to find the signal E_{1out} leaving the probe, and the result is:

$$E_{1out} = (1/2)E_0 \exp i[-\phi(t)-\Phi] + (1/2)E_0 \exp i[\phi(t)+\Phi] =$$

$$= E_0 \cos[\phi(t)+\Phi], \qquad (8.16)$$

344   Optical Fiber Sensors Chapter 8

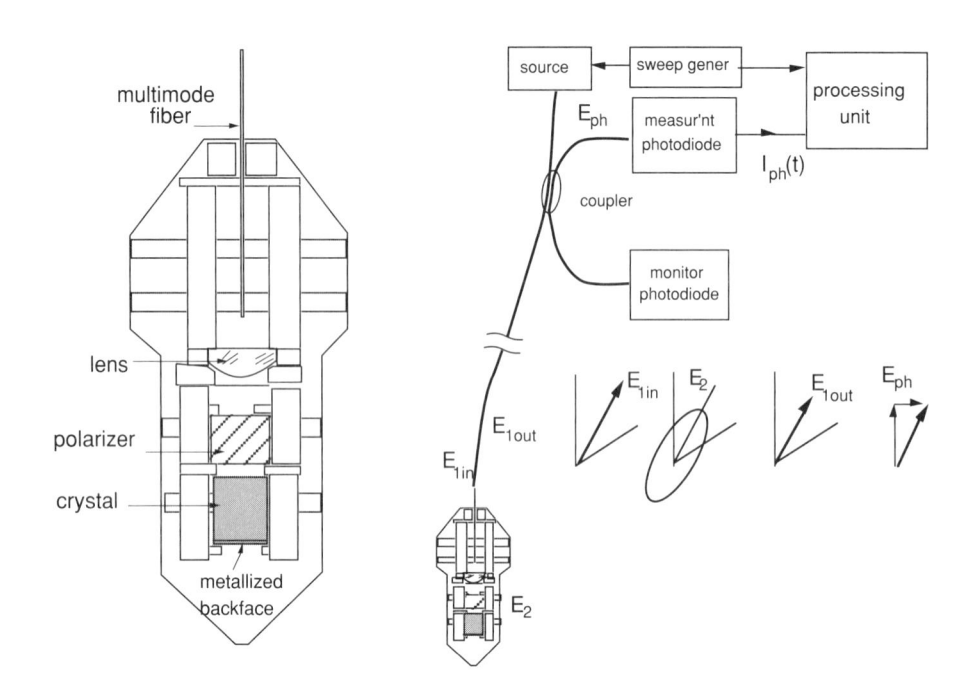

Fig.8-27 Left: layout of a temperature probe based on extrinsic linear birefringence. Light from the source is guided by the multimode fiber to the probe and passes through a polarizer oriented at 45° with respect to the LiNbO$_3$ principal axes. Light mirrored back from the crystal is analyzed by the polarizer itself and brought back to the detector location. Right: to recover the phase Φ of linear birefringence, the laser source is swept in frequency so that the optical path length $\Delta\beta L=\phi(t)$ undergoes a full 2π cycle. By calibrating the crystal response, we can trace back the phase $\Phi=\Phi(T)$ and hence the measurand T.

After propagation through the multimode fiber, the state of polarization of E_{1out} will suffer a major change, yet the total power contained will be the same E_{1out}^2 value leaving the probe.
Then, after detection, we have a current $I_{1ph} = E_0^2 \cos^2[\phi(t)+\Phi] = E_0^2\{1+\cos 2[\phi(t)+\Phi]\}/2$.
We may drop the constant term and consider the time-dependent signal $I_{1ph(td)}=\cos 2[\phi(t)+\Phi]$.
After calibrating the system (crystal, laser diode, and sweep amplitude), we know the times at which it is $\phi(t)=0$ and $\phi(t)=\pi/4$.
By sampling the output signal $I_{1ph(td)}$ at these times, we obtain the desired signals $\cos 2\Phi$ and $\cos 2[\pi/4+\Phi]=-\sin\Phi$, and we are brought back to the usual signal processing.

The measurement is incremental and we need to start with the FOS at a reference temperature (usually ambient temperature) and count the periods or fractions of them to get the

8.3 Readout Configurations

current temperature. The incremental constraint is not a problem in geothermal gauging, because the drill starts from ground.

In the practical instrument [26], a resolution of 0.02°C has been achieved in the range of 0 to 500°C, while the accuracy has been estimated of better than 0.1°C.

Modifications of the concept, with the use of a polarization-maintaining fiber in the adduction trunk, have been also reported [27].

8.3.3 Interferometric Readout

In an interferometric OFS, the measurand affects the optical path length of a piece of fiber exposed to it, and we obtain the path length or phase readout by means of a suitable configuration of interferometer.

In Chapter 7, we treated in detail the best-developed interferometric OFS, the FOG. Other interferometric OFSs share with the FOG excellent sensitivity, unparalleled by other electronic sensors. A feature common to all interferometric OFSs is the remarkable sophistication of the device structure.

Because of these features, interferometric OFSs have a very appealing potentiality of performance, but their rather high cost has been a serious obstacle to market penetration.

In general, any type of optical interferometer can be virtually employed to provide a measurement and a reference arm, thus we are left to make the fiber counterpart out of the conventional bulk-optics interferometers described In Appendix A2 (see Fig.A2-1).

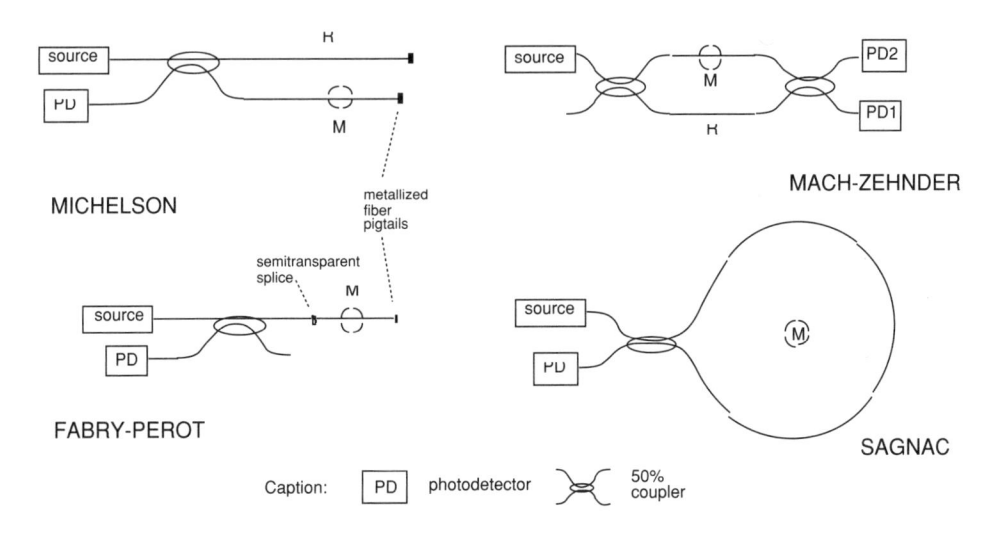

Fig.8-28 Commonly employed configurations of OFS interferometers are the all-fiber versions of conventional bulk-element interferometers. M and R indicate the measurement and reference arms.

We obtain the interferometric OFS illustrated in Fig.8-28 by replacing free propagation with fiber-guided propagation, beam splitters with fiber-couplers, and mirrors with metal-coated pigtails.

Of course, if we need a well-reproducible phase of the optical field after propagation in the fiber, this shall be monomode. Then we may add some special feature, for example polarization control (the hi-bi fiber of the FOG), dispersion control, nonlinear effects, active-ion doping, and so on, yet the basic requirements is the monomode character of the fiber. As a consequence, all the components in an interferometric OFS shall be monomode.

All the different configurations basically supply a signal of the type $\cos\Phi$. Interferometer performance differs from one configuration to another, however. As discussed in Appendix A2.1, the responsivity of the interferometer may vary from 1 (Mach-Zehnder and Sagnac) to 2 (Michelson) and to F (the finesse, $\gg 1$) in the Fabry-Perot interferometer (see Table A2-1).

In addition, a reference arm is available for the cancellation of extraneous disturbances in Michelson and Mach-Zehnder interferometers, whereas it is missing in the Fabry-Perot, and is not separately accessible in the Sagnac interferometer.

Other important features of the OFS are: the balance or unbalance of the arms, the co-location of source and detectors on the same side of the fiber, and last but not least, the retro-reflection into the source.

In view of these features, it turns out that the Mach-Zehnder configuration is best for a general-purpose OFS, when we want an easy-to-implement sensor with a readily available reference arm and no disturbance effects from retro-reflections. Additionally, the differential output (PD1 and PD2 in Fig.8-28) is useful for common mode cancellation. The price for this configuration is the increase of component count (two couplers) and the placement of the laser and detector on opposite sides of the fiber.

If we don't want to fold the fiber to get a probe-like sensor with input and output on the same side, we may resort to the Michelson configuration. This uses a single coupler and has twice the responsivity of the Mach-Zehnder, but the dual detector can't be fitted in, and retro-reflection may disturb the source. If the source is a laser with a coherence length larger than the go-and-return optical path length, then we need to protect the laser with an optical isolator placed at the laser output before entering the OFS fiber.

Because the isolator is a rather expensive component, we generally try to avoid it and limit the coherence length to the value strictly required, that is, the path length difference between the arms.

The Fabry-Perot is an attractive configuration when we want an OFS with the highest responsivity. In fact, a finesse up to F=50-100 can be attained in metal-coated fibers, and the improvement in response may be quite significant. Disadvantages of the configuration are (i) the criticality of metal coating required to build the cavity, and (ii) the sensitivity of the access lead to external disturbances.

The Sagnac configuration is definitely ideal when sensing a non-reciprocal effect, as already discussed in connection with the inertial rotation in the FOG and the Faraday effect in current and magnetic field sensing. For reciprocal effects, on the other hand, one cannot separately access the reference and measurement paths for disturbance cancellation.

8.3 Readout Configurations

The ultimate sensitivity of interferometric OFS can be expressed in terms of a NEM (Noise Equivalent Measurand).

If $\Phi=\Phi(M)$ is the relationship connecting the measurand and the optical phase impressed to the fiber, we can write the NEM as NEM= ϕ_n/R_ϕ. Here, ϕ_n is the phase noise of the interferometric measurement (given by Fig.4-18), and $R_\phi=\Phi'(M)$ is the phase responsivity given by the derivative of the phase respect to the measurand, calculated at the working point.

In practical OFS, we can attain phase noise ϕ_n in the range of 10^{-7} to 10^{-6} rad. Yet, these values require a careful design of both the optical layout and electronic processing. In addition, we should take into account the thermodynamic phase fluctuations of the fiber (Sect.4.4.5), as this limit is in the reach of OFSs.

8.3.3.1 Phase Responsivity to Measurands

The phase responsivity R_ϕ of a fiber is important to evaluate the signal that can be generated in an intrinsic OFS with interferometric readout.

Table 8-1 collects typical values that are illustrative of the responsivity of three types of fibers: (i) standard monomode fiber with standard polymer secondary coating; (ii) fiber with special coating, for enhanced sensitivity to measurand; (iii) fiber with special coating, to desensitize it to the measurand.

The reason for the three fibers is that ordinary fibers are, to our surprise, appreciably sensitive to lot of unexpected measurands.

We certainly will make our best effort to compensate for undesired measurands with the use of a reference arm, which greatly helps to desensitize it. However, the ideal sensor, sensitive only to a specific measurand and immune to any other, is surely best approached if we use special fibers.

TABLE 8-1   Typical Phase Responsivity of Fibers

Measurand	Ordinary fiber	Coated for max response	Desensitized fiber	(Unit)
Temperature	300	5000	0.5	rad/°C·m
Pressure (isotropic)	36	600	0.1	μrad/Pa·m
Force (compression)	0.55		0.01	rad/N
Bending on radius R	0.6		0.01	rad·cm
Current	3	220		μrad/A

In particular, a special fiber designed for sensitivity to the measurand is useful in the OFS sensing section (the measurement arm), whereas another special fiber desensitized to the measurand is best suited for the reference section (or arm).

Finally, the piece of fiber adducting the sensing region should be desensitized to all extraneous measurands. As we can see from the typical data reported in Table 8-1, the desensitized fiber is about 2...3 decades less sensitive than the most responsive fiber. This is a good result, but perhaps not enough to seriously challenge conventional electronic sensors, that may put forward a typical cross-immunity of 10^5 to 10^7 to undesired measurands.

8.3.3.2 Examples of Interferometric OFS

Several interferometric OFSs have been proposed in the literature, and the most exhaustive reference on this topic is perhaps the recently prepared Collection of Optical Fiber Sensors Proceedings [28].

If we exclude the FOG, which is by far the unrivaled interferometric OFS, the most interesting OFS is the hydrophone, from the application point of view. The hydrophone is a device intended for hydrostatic pressure sensing in underwater environments.

Using a properly jacketed fiber sensitive to isotropic pressure, and laying a section of it coiled so as to cumulate the measurand on an appropriate length, we get the schematic of Fig.8-29. Here, the pressure-sensitized fiber is coiled at the end of the measurement arm of the Michelson configuration.

The reference arm brings a phase modulator (with a PZT) to adjust the quiescent working point in quadrature, so that the photodetected signal is of the form $I_{ph} = I_0 [1+\cos\{\Phi(M)-\phi_r\}]$ = $I_0 [1+\sin\Phi(M)]$, where $\Phi(M)$ is the phase induced by hydrostatic pressure, and ϕ_r is the reference arm phase shift whose low-frequency (or average) component shall be dynamically locked to $\pi/2$ (compare to Fig.4-34 and Sect.4.6.1).

With several cm of coiled fiber, the sensitivity of the OFS hydrophone can reach 0 dB$_a$ or $p_a = 2 \cdot 10^{-5}$ Pa [29]. This is the reference level of acoustic pressure and corresponds to the threshold of audibility.

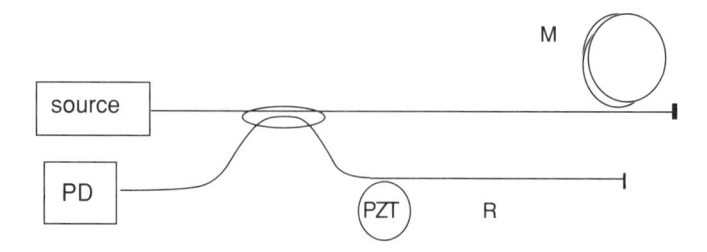

Fig.8-29 Schematic of an interferometric hydrophone using the Michelson readout configuration. The sensing fiber is a coil of pressure-sensitized fiber, whereas the adduction and reference arm fibers are desensitized.

8.3 Readout Configurations

Another schematic similar to that of Fig.8-29 applies to the measurement of temperature. In this case, we take advantage of the PZT modulator to perform two functions: (i) removing the ambiguity of the cosine function, and (ii) extending the resolution to a small fraction of wavelength (or 2π angle). To do so, we sample the phase of the measurand $\phi=\Phi(M)-\phi_r$ by three or four values of the reference ϕ_r. Three values are those already illustrated in Fig.4-11; some researchers prefer using four values.

For example, we may switch ϕ_r on a sequence of three values: $0, ..2\pi/3...-2\pi/3$. By sampling the $1+\cos[\Phi(M)-\phi_r]$ signal in correspondence, we triple the interferometric signal, as in the general scheme described in Sect.4.2.2.1. From the three signals, we can compute $\Phi(M)$ (see Eq.4.11) with a $\approx$nm resolution (equivalent to $\approx\pi/200$ rad) and get the appropriate up/down signals required for a range exceeding 2π.

Similar to an interferometer for displacement measurement, when the phase induced by the measurand exceeds 2π, we count up/down transitions in a counter, and the measurement is an incremental one. To work correctly, the sensor shall be reset at an initial temperature T_0 and allowed to count the periods induced by temperature between T_0 and the current temperature T. Temperature resolutions down to 10^{-4} degrees have been reported based on this configuration of interferometric OFS.

8.3.3.3 White-Light Interferometric OFS

The white-light interferometer described in Sect.4.8 is readily adapted to an OFS, as illustrated by the schematic in Fig.8-30.

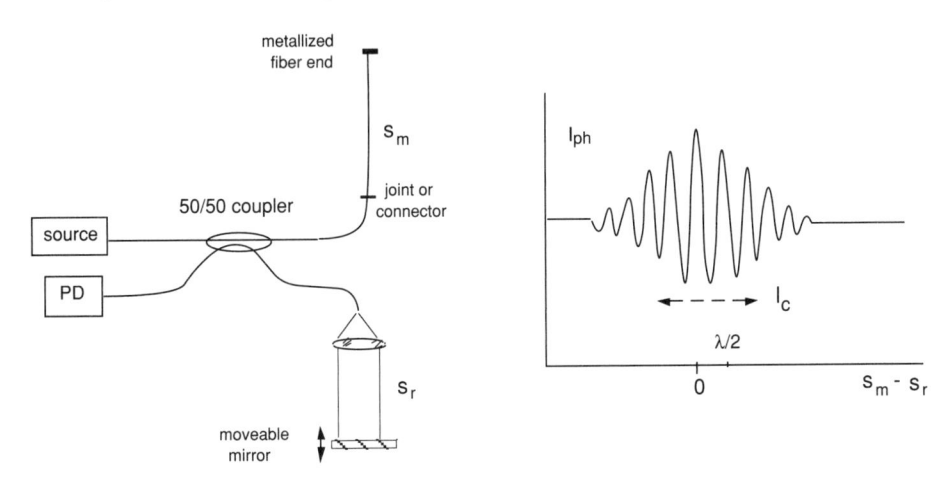

Fig.8-30 Schematic of a white-light interferometer OFS. Mirror M is mounted on a motorized stage, and when the arms are balanced ($s_m=s_r\pm l_c$), fringes appear.

In the Michelson version of the interferometer, the arm with in-fiber propagation is used to collect the measurand phase perturbation. In the other arm, we exit from the fiber with a collimating lens and project the beam onto the moveable mirror.

When the mirror position matches the condition $s_m = s_r \pm l_c$, fringes appear and we can measure the deviation from the condition $s_m = s_r$ looking at the fringe envelope. The resolution of this measurement is of order of the coherence length l_c (Fig.8-30).

Using a SLED source, we may typically have $\Delta\lambda \approx 25$ nm at a central wavelength $\lambda = 800$ nm, and the coherence length is $l_c = \lambda^2/\Delta\lambda = 0.8^2/0.025 = 26$ µm, a good starting value as resolution of a strain-measuring OFS. If we scan the response curve a few times and average the results, we may go down to resolve typically $\approx 0.2\, l_c = 5$ µm.

This value of resolution is obtained with a very simple optical setup and needs only very limited electronic processing. Because it is cheap, the OFS lends itself to interesting application in the field of construction, called either non-destructive testing (NDT) or smart sensors.

For example, we may start with a 1-m piece of fiber, with one end-face metal-coated and the other end-face ending on a connector. The fiber is then threaded inside a steel tube and glued to it, and this constitutes an arm of the OFS. The arm duplicates the strain $\Delta l/l$ applied by the structure external to the tube.

Now we can cast the fiber inside a concrete pillar to monitor its state of strain during the useful life of the pillar. The fiber tube is put in place while the concrete is poured to fabricate the pillar, leaving out just the connector for access.

After a period of time (e.g., months to years), we can go back to examine the sensor and check the strain applied to it and to the concrete structure. From this we get non-destructive diagnostics of the construction work. As a 10-µm resolution on a 1-m fiber length is a strain $\sigma = \Delta l/l = 10 \mu m/1 m = 10\,\mu strain$, we obtain a sensor amply capable of diagnosing incipient overload of the structure (usually found at 100…1000 µstrain).

About disturbances, temperature variations may actually affect the measurement of strain. Indeed, a typical sensitivity (Table 8-1) of 300 rad/°C·m amounts to about $300/2\pi \approx 50$ wavelength/°C. At $\lambda = 0.8 \mu m$, this quantity corresponds to a strain equivalent of $50 \cdot 0.8 = 40\,\mu m/°C \cdot 1m = 40\,\mu strain/°C$.

However, we can measure the temperature of the pillar and correct against thermal drift and associated errors if we characterize the response prior to installing the fiber.

The white-light approach is interesting and has led to successful products. Its only drawback is the mechanical scanning of the reference path with the moveable mirror.

8.3.3.4 Coherence-Assisted Readout

A more elegant configuration of a white-light interferometer can be devised starting from the idea that, casting in the experiment a full interferometer instead of just the measurement arm, we make available a reference to subtract the undesired disturbances.

To do so, we shall read the path length difference $s_{mS} - s_{rS}$, between the measurement s_{mS} and reference s_{rS} paths contained in the sensing (S) section of the item under test.

8.3 Readout Configurations

For the subtraction, we secure the measurement fiber to the enclosure tube and leave the reference fiber loose. Thus, the difference $s_{mS}-s_{rS}$ is sensitive to the measurand as before, whereas temperature and other common mode disturbances are canceled out.

Now, the problem is how to read the s_m-s_r difference having already used both arms. With a white-light interferometer, we have one more degree of freedom available, which is coherence.

Thus, we develop a coherence-assisted readout as shown in Fig.8-31.

In this scheme, we use two interferometers in series, in the sensing (S) and readout (R) section. We make the path lengths s_{mS} and s_{rS} differ by several coherence lengths l_c of the source so that the output fields from these arms will not beat at the photodetector. For example, we typically let $s_{mS}-s_{rS}=\Delta \approx 20 \cdot l_c \approx 0.5$ mm, for arms s_{mS}, $s_{rS} \approx 1$-m long.

In the readout section, we work with the same nominal lengths of the reference and measurement paths. For simplicity, let us take identical reference lengths $s_{rR}=s_{rS}$. Then, scanning the measurement path s_{mR} with the moveable mirror, we will find the readout interferometer in the same imbalance of the sensing interferometer when we reach $s_{mR}=s_{mS}$.

This condition corresponds to the balance of two special paths in the optical setup. One passes through the measurement arm of the sensing section and the reference arm of the readout section with a length $s_{mS}+s_{rR}$. The other passes through the reference arm of the sensing section and the measurement arm of the readout section with a length $s_{rS}+s_{mR}$. Along these two paths, the electric fields collect a phase delay $2k(s_{mS}+s_{rR})$ and $2k(s_{rS}+s_{mR})$, respectively.

When the two fields are collected at the receiver, they generate the following photodetected signal:

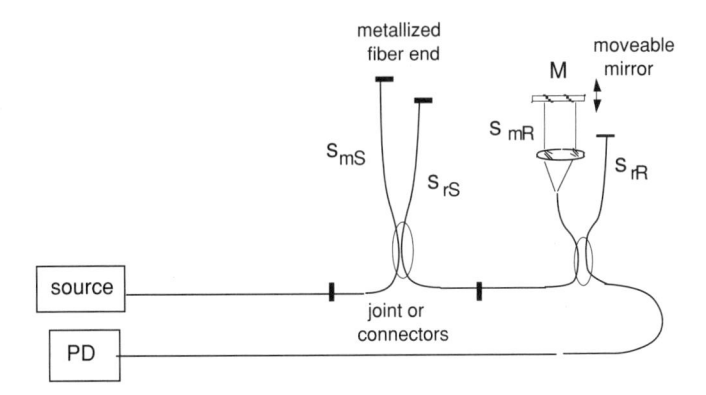

Fig.8-31 An optical fiber sensor using coherence-assisted readout employs two interferometer sections for sensing (S) and readout (R). The arm lengths of the interferometers differ by a quantity Δ much larger than the coherence length l_c. The moveable mirror is scanned, and when position $s_{mR} = s_{rR} +(s_{mS}-s_{rS})$ is reached, white-light fringes appear as in Fig.8-30.

$$E_0^2/4 \{1+ (1/2) \cos 2k[(s_{mS}+s_{rR})-(s_{rS}+s_{mR})]\} \qquad (8.17)$$

With a coherence length l_c, interferometric fringes similar to those of Fig.8-30 are generated in the interval $s_{mR}= s_{mS}+(s_{rR}-s_{rS})\pm l_c$ of moveable mirror position.
When the reference paths are identical ($s_{rR}=s_{rS}$), this gives $s_{mR}= s_{mS}\pm l_c$.

As there are four possible paths in the double interferometer schematic of Fig.8-31, we shall also examine the additional terms beating at the photodetector. These terms are easily found by writing the amplitude and phase of the field propagating down the fiber circuit in Fig.8-31. Developing the calculation, we find the additional terms and write the total current at the photodetector as:

$$I_{ph} = E_0^2/4 \{1+ \cos 2k(s_{mS}-s_{rS}) + \cos 2k(s_{mR}-s_{rR}) +(1/2) \cos 2k[(s_{mS}+s_{rR})-(s_{rS}+s_{mR})]$$
$$+ (1/2) \cos 2k[(s_{mS}+s_{mR})+(s_{rS}+s_{rR})]\} \qquad (8.18)$$

The first term in curl brackets has a zero mean value because the argument, $s_{mS}-s_{rS}\approx\Delta$, is much larger than coherence length. The second term can give a fringe pattern, but only when $s_{mR}-s_{rR}\approx\pm l_c$. We never get it, because the difference $s_{mR}-s_{rR}$ is about equal to $s_{mS}-s_{rS}\approx\Delta$. Thus the second term has a zero mean value, too. The third term is the desired one, beating for $s_{mR}-s_{mS}\approx\pm l_c$. The last term has a beat only for $s_{mS}-s_{mR}\approx-2\Delta$, or it has zero mean value.
In conclusion, the extra terms may contribute with their beatings as well, but far away from the main one if we keep $\Delta\gg l_c$, and we can easily reject them in the course of the scan.

Once we have been able to compensate the interferometer from thermal and other undesired drifts, we can push the resolution further. Indeed, when the white-light fringes are still under the envelope waveform, we can look for the one with the largest amplitude.
This is the central fringe, defined by position to the usual $\lambda/2$-displacement per period. Therefore, if we look at the central fringe we get a resolution of about 0.5µm, or 20 times better than the envelope (or coherence-based) position-sorting.

Important to note is that, near the balance condition, the white-light interferometer behaves exactly like a conventional interferometer, and its ultimate limits of performance are those already discussed in Sect.4.4. Thus, nm-resolution is also achievable.

8.4 MULTIPLEXED AND DISTRIBUTED OFSs

Despite the generally excellent technical performances of OFSs, their relatively high cost has so far prevented their widespread use in applications, especially in comparison with conventional electronic sensors.
One way to get around this problem is to offer a number of measurement points instead of a single point. We can multiply the number of measurement points substantially, either by allowing the measurement to be distributed in space or by multiplexing a number of individual sensors on the same line served by a single readout unit.

8.4 Multiplexed and Distributed OFSs

If the approach is to become the winning breakthrough, the cost of multiplexing or distributing the sensors should be low, about a fraction of the total OFS cost.
Several approaches have been studied through the years to provide a viable approach to the multiplexed or distributed architecture, and copious literature has been published on the subject [28,30,31].

8.4.1 Multiplexing

Multiplexing may be implemented by several methods; the most popular are (i) space-division; (ii) time-division; (iii) wavelength-division; and (iv) coherence-division.

Space-division uses a switch connecting each individual sensor to the readout unit. Each sensor is read in sequence, one right after the other. Because of cost requirements, the switch is usually a mechanical rotary device, and the operation time (switching from one sensor to the other) is about a few milliseconds.

For faster response, we may use a more sophisticated approach: for example, we may start with a Mach-Zehnder waveguide interferometer realized in $LiNbO_3$ or in silica-on-silicon (SoS). These are 2-way switches, connecting the input guide to one of the two output guides, and are actuated by an electrical signal applied to the control electrode. By arranging N stages in cascade, we realize a 1×2^N switch array. Price becomes prohibitively large at increasing N, however, and this solution is limited in practice to, say, $N \le 4$.

Time-division implies the use of a short-pulse laser source to interrogate sensors, which are located down a fiber at different lengths from the source. For example, Fig.8-32 shows an example of time-division multiplexing OFS.

Called ladder configuration, this multiplexing scheme employs a pulsed light source and a series of sensors connected with couplers to common rails of fiber for feeding the sensors and for collecting the returned signals. The time duration of the pulse τ is made shorter than the extra delay $2L/c$ experienced by a sensor with respect to the preceding one.

Under this condition, each sensor responds with a pulse falling in a well-defined time-slot (Fig.8-32) and thus we can demultiplex the responses at the receiver.

Each coupler takes a fraction of the incoming power to its sensor. To equalize responses, the coupling factor at the n-th sensor should be chosen as $k_n = 1/(N-n+1)$, where N is the total number of multiplexed sensors. By doing so, the first sensor has $k_1 = 1/N$ and leaves a through power of $1-1/N$. The second sensor has $k_2 = 1/(N-1)$, takes a fraction $[1-1/N]/(N-1) = 1/N$ of incoming power, and leaves a through power of $[1-2/N]$. The third sensor has $k_3 = 1/(N-2)$ and takes a fraction $[1-2/N]/(N-2) = 1/N$. This continues up to the N-th sensor that has $k_N = 1$ and takes a fraction $1/N$ of input power.

In the return path, the fraction of power deviated onto the common rail is the same value, $1/N$ of the power deviated from the feed rail. Thus, with respect to addressing individual sensors, the operation of multiplexing introduces an extra attenuation $1/N^2$.

With an appropriate design of source power and receiver sensitivity, attenuation limits the maximum number of sensors that can be multiplexed to $N \approx 15..20$.

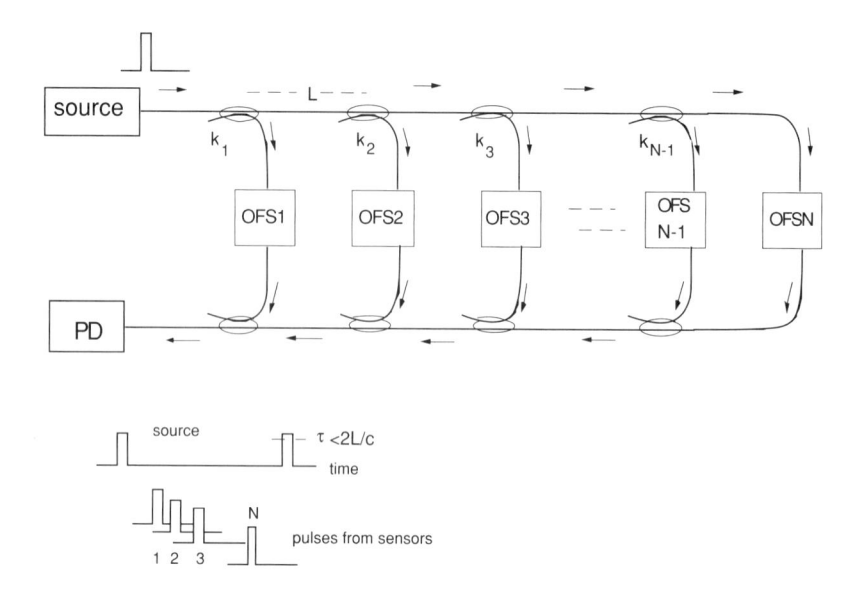

Fig.8-32 Time-division multiplexing of an OFS, using a pulsed source and individual sensors arranged in a ladder array. To distinguish the pulses returning from individual sensors, the pulse width τ should be shorter than the delay 2L/c added at each additional stage. Extra fiber length can be added between sensors to match the condition.

The time-division scheme has been employed for polarimetric as well as interferometric sensing [31] of temperature, strain, and acoustical emission, especially in connection with marine applications.

Wavelength-division multiplexing consists of assigning a wavelength slot to each individual sensor and reading the composite output with a narrow-band source tunable in frequency. To be suitable for cascade connection, the sensor should be frequency-selective, respond in transmission or reflection and let the out-of-band components pass unaltered in reflection or transmission.

These requirements are satisfied by Fiber Bragg-Gratings (FBGs), passive components well known in optical fiber communications. An FBG is a longitudinal grating, written along the axis of the fiber as a periodic variation Δn of the refraction index, produced by exposure to UV radiation [4].

If Λ is the spatial period of the grating, the Bragg condition $2n\Lambda = \lambda_B$ determines the wavelength of resonance λ_B. For wavelengths close to λ_B, the device reflects back the incoming radiation, whereas radiation off-resonance ($\lambda \neq \lambda_B$) passes unaltered through the FBG. We summarize this behavior by saying that the FBG is a band-reject filter in transmission or a band-pass filter in reflection.

8.4 Multiplexed and Distributed OFSs 355

The FBG can be used as a sensor because the resonant wavelength λ_B depends in a reproducible way on temperature and strain, the two quantities most commonly measured with FBGs. The relative change of resonance wavelength is written as:

$$\Delta\lambda_B/\lambda_B = \Delta n/n + \Delta\Lambda/\Lambda = (1-p_e)\varepsilon + (\alpha+\psi)\Delta T \qquad (8.19)$$

Here, p_e is the elasto-optic coefficient of the fiber (typically $p_e \approx 0.22$ for silica), $\kappa = \alpha+\psi$ is the total thermooptic coefficient (typ. $\kappa \approx 5 \cdot 10^{-6}$ in silica), and $\varepsilon = \Delta l/l$ is the strain applied to the FBG.

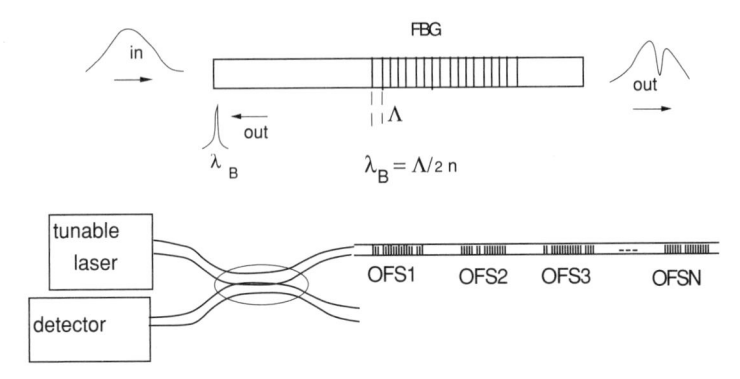

Fig.8-33 An FBG reflects the wavelengths close to the resonant wavelength λ_B and transmits the rest of the spectrum (top). A wavelength-multiplexed OFS is arranged by cascading a number of FBGs at different wavelengths λ_B and using a tunable laser source for reading the individual sensors.

Connecting a number of FBGs in series, as indicated in Fig.8-33, we get a wavelength-multiplexed sensor. The array can be read by a wavelength-tunable laser source because each FBG other than the selected one is transparent, whereas the sensor interrogated by the laser source will respond with a peak reflection in correspondence to its resonant frequency.

The shift in wavelength resonance $\Delta\lambda_B$ is dependent on temperature T and strain ε applied to the FBG. If we are to avoid cross-sensitivity to T when measuring ε, or vice versa, we should use a pair of FBGs, arranged in the experiment to have different sensitivity to T and ε. By combining the results of the measurements in a set of two equations, we can solve for both (T and ε) measurands.

A feature that is unsatisfactory in the λ-multiplexed sensor of Fig.8-33 is the use of a wavelength-tunable laser source. This is an expensive item and, if it is the only possibility, it is likely that an OFS incorporating it will hardly leave the laboratory.

Thus, we may want to modify the concept of λ-multiplexed readout. One of the several solutions reported in the literature is shown in Fig.8-34. Here, we use a broadband source to illuminate the sensors and duplicate each FBG in the sensing chain with another FBG in the measurement chain, with nominally the same resonant wavelength λ_B.

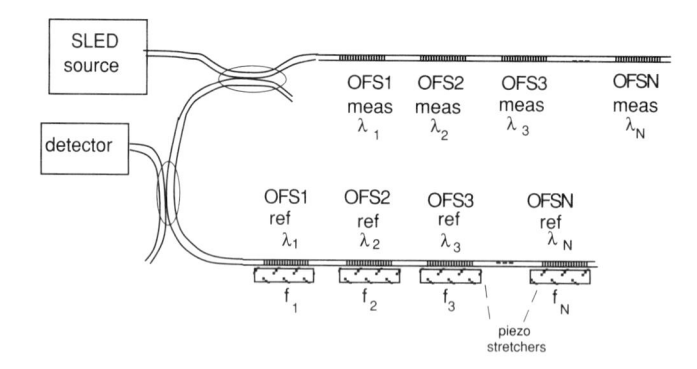

Fig.8-34 Schematic of a λ-multiplexed sensor that uses two sets of FBGs in the measurement and reference chain. Corresponding OFSs have nominally identical resonant wavelengths λ_B. The signal returning from the measurement OFS is brought to the reference OFS for response comparison. Modulating the reference λ_B with the piezo actuator allows us to demultiply the responses of individual sensors.

Upon application of the measurands, the k-th sensor changes its resonance wavelength by, say, a deviation $\Delta\lambda_{Bk}$. To measure the set of all $\Delta\lambda_{Bk}$ simultaneously, we apply to each k-th FBG of the reference chain a stress appropriate to produce the same $\Delta\lambda_{Bk}$ shift of the corresponding FBG in the measurement chain. This is done by the piezo actuators, the PZT slabs stretching the fiber of teach FBG (Fig.8-34).

To operate all sensors independently, each k-th piezo is actuated with a different electrical modulation frequency f_k. The photodetector output is filtered in k bands, centered at the f_k resonance, and in each band, we look for the maximum response of the k-th FBG sensor.

Finally, coherence-multiplexing is an interesting technique that has been recently proposed. We may think of it as a generalization of the concepts discussed in Sect.8.3.3.4.

As illustrated in Fig.8-35, we can multiplex several interferometers (for example, Mach-Zehnder interferometers) by using a different arm imbalance in each OFS, and by arranging all of them in a double chain, one for sensing and the other for readout.

In the interferometers, we let the imbalance be much larger than the coherence length l_c, that is, $L_k - l_k \gg l_c$. In addition, we let the imbalance of consecutive sensors be larger than the coherence length $L_k - L_{k-1} \gg l_c$. At each detector, say the k-th of the sensing section, we find the superposition of many field contributions propagated through different paths across the interferometers. All of them are unbalanced by more than the coherence length l_c, except the two that go through the corresponding k, OFSk and Rk, in a crossed fashion, that is, $l_{k(OFS)} + L_{k(R)}$ and $L_{k(OFS)} + l_{k(R)}$, and share the same path in the other interferometers.

Thus, we can simultaneously read all the multiplexed sensors because the beating is returned only at the appropriate interferometer.

Coherence is an elegant solution for multiplexing, but the drawback is a large attenuation.

8.4 Multiplexed and Distributed OFSs 357

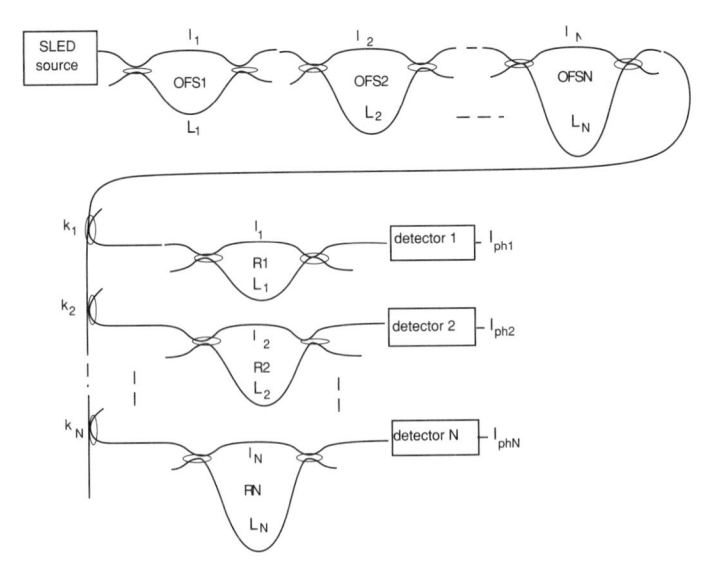

Fig.8-35 Schematic of a coherence-multiplexed OFS. Light from a low-coherence source is passed through a cascade of interferometers with different arm-length imbalance. In the readout section, light is split by means of couplers to a corresponding number of interferometers, duplicating the arm imbalance of the sensing section. The cross-path through the sensing and readout interferometers returns the useful signal.

First, we shall equalize the channel amplitudes by choosing $k_n=1/(N-n+1)$ as in the time-division scheme (Fig.8-32), and this causes an attenuation of $1/N$.
In addition, at each interferometer of the sensing section, we waste a factor of 2 because of the unused ports, and this is an extra 2^N attenuation. In practice, we can hardly go beyond $N \approx 4$ unless we introduce a costly optical amplifier.
In conclusion, when designing new systems we shall use the concept of multiplexing whenever possible to reduce the cost per point. But, the total cost of the array keeps high anyway. Because complexity increases with N, a practical limit of maximum number of points comes out. Thus, we are not really gaining new segment of the sensor market by multiplexing.

8.4.2 Distributed Sensors

If we are able to read an optical parameter of a fiber resolved in distance, then each interval we resolve defines an individual sensor and a measurement point for what becomes a distributed sensor. Most of the optical parameters we use for sensing are position-dependent. In other words, fibers are born essentially as distributed sensors, but we usually make an averaged or cumulative reading of the optical parameter affected by the measurand.

From the application point of view, distributed sensors offer a decisive advantage with respect to sensors based on conventional technologies. They can be regarded as an extreme form of multiplexed sensors in that a single readout apparatus serves to measure an arbitrarily large number of individual sensing points. The cost per point of a distributed sensor then really decreases as 1/N.

Even more important, we can develop a sensor with a very large number of points, many more than is reasonable with multiplexed sensors. This feature is unequalled by conventional technology and is a great advantage in applications.

In a distributed sensor, we should be able to remotely interrogate the fiber and have a parameter telling us the distance of the fiber piece being measured. Then we should read the measurand through an optical parameter and a readout scheme (Sect.8.3). Common measurands are strain and temperature, and common optical parameters are intensity and SOP of light scattered at the individual sensing point.

Several approaches have been reported in the literature [1-4] for the interrogation of distributed sensors, namely time-domain, frequency-domain, and coherence-domain.

The time-domain approach is readily adapted to sensing because it takes advantage of instrumentation and measurement techniques [32] developed in optical fiber communication, for the remote testing of losses along fiber lines. The instrument for such a distributed measurement of attenuation is known as Optical Time-Domain Reflectometer (OTDR).

The OTDR has been used to sense stress and temperature in a variety of fibers since the early times of OFSs [33]. The interaction mechanisms first tested were attenuation by microbends to measure local stress, and temperature coefficient of the Rayleigh scattering to measure temperature. Both effects are small in ordinary silica fibers, however, and therefore special fibers (with a liquid core or doped with rare-earth elements) were tested in an attempt to improve sensitivity and resolution [33,34].

Later, variants of the basic OTDR scheme rather than fiber were able to provide the improvement. One is the Polarization OTDR (POTDR), by which we get a distance-resolved measurement of the SoP and hence of the birefringence of the fiber [34-36]. As birefringence is sensibly related to the stress imparted to the fiber, the POTDR is a powerful tool to remotely test and localize mechanical stresses in the fiber or in a pipeline, for example, to which the fiber is cemented.

The basic scheme of OTDR readout for the distributed sensing of attenuation is shown in Fig.8-36. Similar in concept to the LIDAR discussed in Sect.3.5, the OTDR uses a short pulse of power at a suitable wavelength to interrogate the fiber. Wavelength is chosen in the windows of optical fibers (850, 1300, or 1500 nm), so that attenuation is at a minimum and the range covered is at a maximum.

The source is a pulsed laser, usually a semiconductor laser driven by a fast current pulse and emitting peak powers of a few tens of mW. The source is eventually followed by a booster optical amplifier and protected by reflections from the fiber with an optical isolator.

The time duration τ_p of the pulse determines, as discussed in Sect.3.5, the spatial resolution of the OTDR measurement. Assuming that the response time T of the detector is faster than the pulse duration $T \ll \tau_p$, the resolution is given by $\Delta z = c\tau_p/2n$, where $n \approx 1.5$ is the effective index of refraction of the fiber.

8.4 Multiplexed and Distributed OFSs

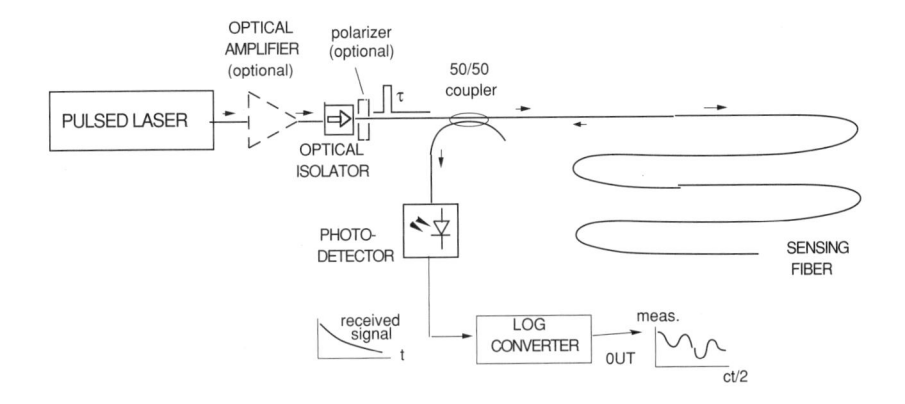

Fig.8-36 In a distributed fiber sensor, the local back-scattered power is read by an OTDR. The duration τ_p of the optical pulse used for interrogation determines the spatial resolution $c\tau_p/2n$ of the measurement, that is, the size of the virtual sensing element. The back-scattered signal looks the same as it does in Fig.3-26. After detection, the logarithm of the signal is computed, yielding the dependence from distributed attenuation and back-scattering.

Resolutions of $\Delta z \approx 1$m are readily achieved with pulses of ≈ 10-ns duration. The quantity Δz also represents the length on which the returning signal is averaged, and thus the equivalent length of the elemental sensor provided by the distributed OFS.

The dependence from fiber attenuation α_f and back-scattering σ_{bs} are described by the same equations as for a LIDAR (Eqs.3.33 through 3.36), if we change the acceptance term $(\pi D^2/4z^2)$ into the solid angle of the fiber πNA^2, and the neper-base attenuation in the base-10 attenuation, $\alpha = 2.3 \alpha_f$.

The power in transit down the fiber is attenuated exponentially with distance z (as $10^{-\alpha z}$ where α is the dB/km attenuation), and a fraction $\sigma_{bs}dz$ of it is back-scattered by the elemental length dz of fiber at z. The back-scatter coefficient $\sigma_{bs}(z)$ is dependent on the fiber material (and fiber imperfections) and on the interaction of the measurand with the fiber.

The back-scattered power returning to the source reaches the fiber coupler, and half of it is deviated to the photodiode (Fig.8-36). The detected output signal is log-converted and the result L(t) supplies the time dependence of attenuation α and the back-scattering σ_{bs} (Eq.3.36).

With typical powers and measurement time, an OTDR may resolve attenuation of tenths of dB, but we need a special fiber if such a resolution is to correspond to a meaningful temperature and strain range.

Alternatively, we need a different optical parameter such as polarization. The setup of Fig.8-36 becomes a POTDR by adding a polarizer in the launch section and a Stokes parameter analyzer in the detector section [37]. Using the POTDR with a normal monomode fiber at $\lambda=1500$ nm, a typical sensitivity to ≈ 100-µstrain with 100-m resolution can be obtained.

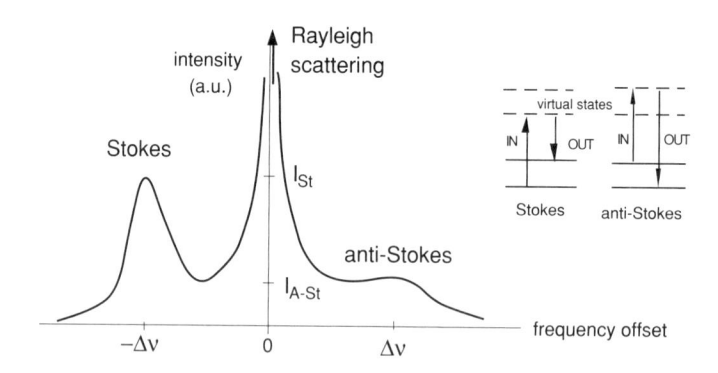

Fig.8-37 Spectrum of anelastic scattering with Stokes and anti-Stokes peaks located at the sides of the much larger Rayleigh peak of elastic scattering. The frequency shift Δν depends on the type of scattering.

Another approach to improve performance is using a scattering different from the normal Rayleigh-Glan elastic scattering related to attenuation. Raman and Brillouin anelastic scatterings provide the desired mechanism.

These have to deal with phonon-assisted scattering, which introduces a virtual energy level in the interaction (Fig.8-37). Because of the new level, scattered radiation gains or loses a small amount of energy, $E_{sc}=h\Delta\nu$, and thus the line corresponding to scattered energy splits into two lines, separated by E_{sc} from the central Rayleigh-Gans (or elastic scattering) line. The upper and lower frequency lines are called Stokes and anti-Stokes lines, respectively.

Raman and Brillouin scattering differ by the energy of the phonon they interact with. The Raman effect deals with the optical phonon, relatively energetic and such that, in silica, $\Delta\nu_R \approx 13.2$ THz (or $\Delta\lambda_R \approx 100$ nm). This large frequency separation makes it relatively easy to separate the weak Raman lines from the much larger Rayleigh scatter line, a clear advantage in developing the readout instrument.

The Brillouin effect deals with the acoustic phonon and has a much smaller frequency separation ($\Delta\nu_B \approx 11$ GHz, or $\Delta\lambda_R \approx 0.08$ nm). Thus, narrowline lasers and very selective filters are required to resolve the lines.

An interesting feature of anelastic scattering is the clean and repeatable dependence from temperature of the relative amplitudes of Stokes and anti-Stokes lines. This is given by:

$$I_{AS} / I_{St} = (\lambda_{St}/\lambda_{AS})^4 \exp -hc/\lambda kT \qquad (8.19)$$

A Raman OTDR can be developed starting from the wavelength-resolved measurement of back-scattered power done at the photodetector (Fig.8-36) and by inserting suitable optical filters to sort out the Stokes lines [38]. By measuring the amplitudes of the Raman lines, temperature T can then be computed. A typical distributed sensor based on the measurement of Raman lines ratio can resolve 1°C on individual 1-m elements out of a 10-km total length of a multimode fiber, using ≈500 mW of peak power and τ_p=10 ns.

8.4 Multiplexed and Distributed OFSs

In applications, the Raman-based approach is the undisputed choice when we need distributed temperature only, because the signal is stronger and more easily sorted out.

With Brillouin lines, however, in addition to the temperature dependence given by Eq.8.19, we can also measure the frequency dependence of the separation $\Delta\nu_B$ from strain $\Delta l/l$, a small but measurable quantity because the unperturbed $\Delta\nu_B$ is small, too.

The dependence is written as $\Delta\nu_B \approx \Delta\nu_{B0} + \zeta\Delta l/l$. In silica fibers the strain coefficient is about $\zeta \approx 50$ (kHz/µε), and $\Delta l/l$ is in microstrain (µε) units. An experimental setup based on the Brillouin scattering has demonstrated [39] the simultaneous distributed measurement of strain and temperature, resolving about 4°C and $\Delta l/l \approx 300$ µε on a 15-km long fiber.

REFERENCES

[1] B. Culshaw and J. Dakin, *"Optical Fiber Sensors: Principles and Components"* vol.1, Artech House: Norwood, MA, 1988; *"Optical Fiber Sensors: Systems and Applications"*, vol.2, Artech House: Norwood, MA, 1989; *"Optical Fiber Sensors"*, vol.3, Artech House: Norwood, MA 1996; *"Optical Fiber Sensors: Applications, Analysis and Future Trends"*, vol.4, Artech House: Norwood, MA, 1997.

[2] E. Udd (editor), *"Fiber Optic Sensors"*, J.Wiley & Sons, Chichester 1991; —, *"Fiber Optic Smart Structures"*, J.Wiley & Sons: Chichester, 1995.

[3] K.T.V. Grattan and B.T. Meggitt (editors), *"Fiber Optic Sensor Technology"*, Chapman & Hall: London, 1995.

[4] J.M. Lopez-Higuera (editor), *"Handbook of Fibre Sensing Technology"*, J.Wiley & Sons: Chichester, 2002.

[5] C.P. Sandbank, *"Optical Fibre Communication Systems"*, J.Wiley & Sons: Chichester, 1980, Chapter 3.

[6] S. Donati, *"Photodetectors"*, Prentice Hall: Upper Saddle River, 2000, Sect.5.2.

[7] see Ref.[6], Sect.5.3.1.3.

[8] T. Tambosso, *"Fiber Optics Temperature Sensors"*, Alta Freq. Riv. Elettr., vol.56 (1987), pp.143-155 (in Italian).

[9] see Ref.[6], Sect.6.3.1.

[10] M. Nagai, M. Shimitzu, and N. Ohgi, *"Sensitive Liquid Sensor for Long Distance Leak Detection"*, Proc. 4th Intl. Conf. on Fiber Sensors, OFS-4: Stuttgart, 1984, pp.207-210.

[11] S. Donati and M. Brenci, *"Fiber Optic Sensors for Ambient and Industrial Applications"*, L'Elettrotecnica, vol.79 (1992), pp.241-250.

[12] A. Harmer and A.M. Scheggi, *"Chemical, Biochemical, and Medical Sensors"*, in Ref.[1], vol.2, pp.599-651.

[13] G. Boisde' and A. Harmer (editors), *"Chemical and Biochemical Sensing with Optical Fibers and Waveguides"*, Artech House: Norwood, 1996.

[14] O.S. Wolfbeis (editor), *"Fiber Optics Chemical Sensors and Biosensors"*, vol.1, CRC Press: New York, 1991.

[15] A.G. Mignani and F. Baldini, *"In-vivo Medical Sensors"*, Ref.[1], vol.4, pp.289-326.

[16] S. Donati, V. Annovazzi Lodi, and T. Tambosso, *"Magnetooptical Fibre Sensors for the Electrical Industry: Analysis of Performances"*, IEE Proceedings part J, Optoelectronics, vol.135 (1988), pp.372-382.

[17] T. Tambosso and S. Donati, *"Influence of the State of Polarization on Interferometric and Polarimetric Measurement of Birefringence"*, Opt. Lett., vol.14 (1989), pp.476-478; see

also *Polarization*, edited by B.U. Billings, SPIE Milestone Series, vol.MS-23, pp.388-390.
[18] A.H. Rose and G. Day, *"Optical Fiber Current and Voltage Sensors for the Electrical Power Industry"*, in Ref.[4], pp.569-611.
[19] A. Papp and H. Harms, *"Magneto-Optical Current Transformers"*, Applied Optics, vol.19 (1980), pp.3729-3745.
[20] S.C. Rashleigh and R. Ulrich, *"Magneto-Optical Current Sensing with Birefringent Fiber"*, Appl. Phys. Lett., vol.34 (1979), pp.768-770.
[21] V. Annovazzi Lodi and S. Donati, *"Fiber Current Sensors for HV Lines"*, in Proc. SPIE Symp. Fiber Optics Sensors II, vol.798 (1987), pp.270-274; see also: Alta Frequenza, vol.53 (1984), pp.310-314.
[22] M. Berwick, J.D.C. Jones, and D.A. Jackson, *"Alternate Current Measurement and Noninvasive Data Ring Utilizing Faraday Effect in a Closed-Loop Magnetometer"*, Optics Lett., vol.12 (1987), pp.293-295.
[23] H. Ahlers and T. Bosselman, *"Complete Polarization Analysis of a Magnetooptic Current Transformer with a new Polarimeter"*, Proc. OFS-9, (1993), pp.81-84.
[24] V. Annovazzi Lodi, S. Donati, and S. Merlo, *"Vectorial Magnetic-Field Fiberoptic Sensor based on Accurate Birefringence Control"*, Proc. OFS-9 (1993), pp.293-302.
[25] V. Annovazzi Lodi, S. Donati, and S. Merlo, *"Coiled-Fiber Sensor for Vectorial Measurement of Magnetic Field"*, J. of Lightw. Techn., LT-10 (1992), pp.2006-2010.
[26] L. Fiorina, S. Mezzetti, and P.L. Pizzolati, *"Thermometry in Geothermal Wells: an Optical Approach"*, Appl. Optics, vol.24 (1985), pp.402-406.
[27] M. Corke, D. Kersey, K. Lin, and D.A. Jackson, *"Remote Temperature Sensing using a Polarization Preserving Fiber"*, Electronics Lett., vol.57 (1984), pp.77-80.
[28] The Complete Collection of *OFS Conference Proceedings*, from OFS-1 to OFS-13, a CD published by SPIE, vol.CDP-01 SPIE: Bellingham, 1997.
[29] C. Belove (editor), *"Handbook of Modern Electronics and Electrical Engineering"*, J.Wiley Interscience: Chichester, 1984, Ch.44.
[30] R. Kist, *"Point Sensor Multiplexing Principles"*, in Ref.[1], vol.2, pp.511-574.
[31] A. Dandridge and C. Kirkendall, *"Passive Fiber Optic Sensor Networks"*, in [4], pp.433-449.
[32] D. Derickson (editor), *"Fiber Optic Test and Measurement"*, Prentice Hall: Upper Saddle River, 1998.
[33] J.P. Dakin, *"Distributed Optical Fiber Sensor Systems"*, in Ref.[1], vol.2, pp.575-598.
[34] A.J. Rogers, *"Distributed Optical Fiber Sensing"*, in Ref.[4], pp.271-309.
[35] B.Y. Kim and S.S. Choi, *"Backscattering Measurement of Bending-Induced Birefringence in Single-Mode Fibers"*, Optics Letters, vol.17 (1981), pp.193-194.
[36] A. Galtarossa, L. Palmieri, M. Schiano, and T. Tambosso, *"Single-End Polarization Mode Dispersion Measurement using Backreflected Spectra and a Linear Polarizer"*, J. Light. Techn., vol.17 (1999), pp.1835-42; see also: J. Opt.Soc.Am. vol.16 (1999), pp.576-583.
[37] B. Hutter, B. Gisin, and N. Gisin, *"Distributed PMD Measurement with a P-OTDR in Optical Fibers"*, IEEE J. Light. Techn., vol.17 (1999), pp.1843-48.
[38] G.P. Lees at al., *"Advances in Optical Fiber Distributed Temperature Sensing using the Landau-Placzek Ratio"*, IEEE Phot. Techn. Lett., vol.10 (1998), pp.126-128.
[39] H.H. Kee, G.P. Lees, and T.P. Newson, *"Low-loss, Low-cost Brillouin-based System for Simultaneous Strain and Temperature Measurement"*, Proc. CLEO 2000, paper CthI4, p.432.

APPENDIX **A0**

Nomenclature

In this book, we frequently talk about the ultimate limits of sensitivity of a specific instrument or measuring method. Terms like *sensitivity, quantity of response, ultimate limit of performance*, etc. are not devoid of ambiguity. Indeed, in electro-optical instrumentation, we use results coming from different disciplines, like optics, electronics, and measurement science, where terms are not always intended in the same way. For the sake of clarity, in this appendix we will define a few terms used in the text.

A0.1 Responsivity and Sensitivity

In a classical paper [1], Jones considered the response of a generic sensor (actually, it was a photodetector, but the concepts had a more general applicability) and introduced the following quantities:
- The *responsivity* R, ratio of the output signal (let's say, a voltage signal V_u) to the input physical quantity M (or measurand) $R = V_u/M$. Clearly, this is not a sensitivity limit, because using an amplifier we can increase it at will. However, the responsivity is an interesting quantity because it supplies the scale factor of the conversion performed by the sensor.
- The *NEI*, or *noise-equivalent-input*, of the sensor. This is given by the output noise, usually taken as the rms value v_n of the fluctuation found at the sensor output, divided by the responsivity, or $NEI = v_n/R$.
- The *dynamic range* DR, which is defined as the ratio of the maximum signal V_{max} supplied at the output before the sensor saturates (or its response becomes appreciably distorted) to the noise seen at the output, that is $DR = V_{max}/v_n$.

363

- The *detectivity* D*, a figure of merit of sensor's ability to detect small signals [2]. With these definitions, the term *sensitivity* is no longer necessary. We think it helps to alleviate ambiguity because some researchers understand it as the NEI, whereas especially in electronic measurements [3,4], *sensitivity* is used with the meaning of responsivity.

A0.2 Uncertainty and Resolution

The uncertainty of the result of a measurement generally consists of several contributions, which have been traditionally classified as *accidental* (or random) and *systematic*. Recently, however, the NIST and other International Committees have adopted [5] the more correct definition of uncertainty as *Type A* (those evaluated by statistical methods) and *Type B* (those evaluated by other methods). The correspondence between the old and new classifications is not one-to-one, as is explained by the example below.

In the old classification, uncertainty is systematic when the deviation from the true value is repeatedly the same in successive measurements. Incorrect calibration and bias or offset of measurement are systematic. Systematic uncertainty cannot be reduced by averaging. To remove it, we must introduce data correction or act on the measurement method.

Uncertainty is accidental when the deviation from the true value is randomly varying from measurement to measurement. Because of the random nature, this uncertainty can be reduced by averaging the results of several successive measurements.
Usually, measurements affected by accidental effects obey Gaussian statistics and their rms deviation decreases as $1/\sqrt{N}$ with the number of measurements N being averaged.
In the old classification, *accuracy* was the accidental uncertainty of an instrument, whereas *precision* referred to the systematic uncertainty. Either of the two may prevail in a sensor.

One last quantity of interest is resolution. *Resolution* is defined as the minimum increment of response that can be perceived by a sensor. If the sensor has a digital readout, resolution is just one unit of the least significant digit (LSD) and represents the effect of the truncation (or round off error). Truncation is a Type B effect if the measurement randomness is much less than 1-LSD, whereas it is Type A if the randomness is much larger than 1-LSD (or, a type A effect is added). In this case, we can improve resolution by averaging.

Example. Let a 99.4-mm stick be measured with a sensor with 1-mm resolution. If the accidental uncertainty is $e_a < 0.1$ mm, every measurement will be rounded to the same result, 99 mm. But, if $e_a = 1$ mm, we will get different outcomes from the measurement, for example, 99, 100, 99, 99, 101, etc. By averaging, we will get a result approaching the true 99.4 at increasing N.

REFERENCES

[1] R.C. Jones, *"Performances of Detectors for Visible and Infrared Radiation"*, in Advances in Electronics and Electron Physics, vol.4, pp.2-88, Academic Press: New York, 1952.
[2] S. Donati, *"Photodetectors"*, Prentice Hall: Upper Saddle River, NY, 2000, Ch.3.
[3] H. K.P. Neubert, *"Instrument Transducers"*, 2nd ed., Oxford Univ. Press: London 1976.
[4] *"ISO Intl. Vocabulary of Basic and General Terms in Metrology"*, ISO: Geneva 1993.
[5] B.N. Taylor and C.E. Kuyatt, *"Guidelines for Evaluating and Expressing the Uncertainty of NIST Mesurement Results"*, NIST Technical Note 1297, 1994.

APPENDIX **A1**

Lasers for Instrumentation

*I*n all instrumentation applications, we want a low-cost, compact laser source capable of supplying the necessary power at the wavelength of operation with the quality of emission required by the application.

The quality of emission may be substantiated by parameters like modal composition, spatial and temporal coherence, wavelength accuracy and repeatability, immunity of emission to external disturbances, etc.

In this appendix, we will review the basic features of a few laser sources relevant to the instrumentation described in the text, that is, He-Ne lasers, semiconductor diode lasers, and diode-pumped, solid-state lasers. Of course, we have no pretense of completeness, but we will discuss parameters of the source affecting performance of the instrument, and special solutions specific to instrumentation applications. Readers interested in a more thorough discussion of lasers can find excellent treatment in a number of textbooks see, for example, Refs.[1-3].

In electro-optical instrumentation, lasers are central to the development and design of new, clever measurement concepts. The special emission characteristics they offer are unparalleled by conventional light sources, and lead to instruments outperforming the previous electronics counterparts. To attempt classification, we may list, as in Table A1-1, the most important characteristics exploited in each particular instrument.

Instruments for alignment and sizing applications require a moderate power preferably in the visible wavelength range. The beam is projected on the measuring target by a collimating telescope, so the beam should have a single-mode spatial distribution. Additionally, for use in alignment instruments, we require a good pointing stability of the beam.

These specifications are satisfied by the red He-Ne (helium-neon) lasers. Units commercially available from several vendors readily provide several mW of power in the red (λ=633 nm) with a single spatial mode (TEM_{00}) distribution, nearly free from spatial defects.

Table A1-1 - Classes of electro-optical instruments and the lasers they use

Application	Typical Laser	Main Characteristic
Alignment, Pointing and Tracking	He-Ne	collimation, beam quality
Diameter Sensors and Particle Sizing	He-Ne	collimation, beam quality
Laser Telemeters - geodimeter (d>1 km) - topograph (d<1 km)	solid-state (Nd, etc.) GaAs diode	high peak-power (Q-switched) high-frequency modulation
Interferometers for dimensional metrology	He-Ne	temporal coherence, 6-digit wavelength accuracy
Doppler VELOCIMETERS	He-Ne, Ar	spatial coherence, beam quality
ESPI vibration analyzers	He-Ne, Ar	spatial coherence, beam quality
FOG gyroscopes and Fiber Optics Sensors	GaAlAs diode and SLED	high radiance, controlled temporal coherence

Pointing stability is excellent in internal-mirrors, side-arm tubes (Sect. A1.1.2). These tubes may reach an angular stability of ≈ 0.1 μrad.

In interferometers, coherence and wavelength accuracy are the most demanding performance. Though the general discussion is valid also for other applications, in the next Sections we will describe frequency and wavelength stabilization.

Temporal coherence allows us operating the laser interferometer on a substantial target distance, that is, with a large arm-length difference. Wavelength accuracy and stabilization imply that the measurement we are performing is inherently calibrated and that several digits of the measurement are indeed meaningful.

Since the early times of lasers, the most popular choice to satisfy these requisites at low cost has been the He-Ne laser at $\lambda=633$ nm.

When frequency-stabilized, it easily supplies coherence lengths well in excess of km's and a wavelength precision easily going to the 6-7 decimal place.

Despite being a gas laser, He-Ne has a very satisfactory useful lifetime (in excess of 10 000 hours) demonstrated by use, is cheap and readily available in quantities by vendors all over the world. Decades of progress have rendered the He-Ne laser a well-reproducible device, with a plain and controlled behavior, ideal for a safe design of instruments with a long MTTR (mean time to repair).

Emission in the visible and moderate power (0.5 to 2mW), adequate in most applications) add the benefit of a generally eye-safe device (see Sect.A1.5) requiring no special effort to comply with laser safety standards.

Drawbacks of He-Ne lasers are:
- a relatively bulky size (typically the bore is 15 to 25 cm in length and 2 to 5 cm in diameter),
- a high-voltage supply (however, not so dangerous from the shock hazard point of view, as the level 5 to 10 mA is rarely surpassed),
- a relatively low optical power ($\approx$ 1mW) available with reasonable tube length (power is proportional to the length of the discharge).

On the other hand, semiconductor diode lasers are very compact and work with a low-voltage supply, but their wavelength accuracy and stability are a few orders of magnitude worse than those of He-Ne lasers. In recent years, however, diode lasers are catching up in these performance areas and may eventually become competitive also in interferometers, especially for low-accuracy versions.

Diode lasers based on the ternary compound GaAlAs (gallium aluminum arsenide) have been used as an optical source since the early years of laser telemetry, both in sine-wave modulated and pulsed versions for medium-distance and middle-accuracy applications. To push up the attainable range, power was increased using a stack of several diodes connected in series. The largest peak powers of solid-state lasers operating in the Q-switching regime have never been surpassed, however. Thus, very long-range telemeters have traditionally used solid-state lasers (glass and YAG-doped Nd, Er, Yb) optically pumped by flash lamps.

Recently, with the advent of high-efficiency, low-threshold diode lasers with quantum well active layers, arrays of laser diodes have reached power levels adequate to pump solid-state lasers. Because of their much greater efficiency and very good reliability, laser diode arrays have become the preferred choice for pumping, superseding high-power lamps.

A1.1 LASER BASICS

The He-Ne laser was one of the first types developed, dating back to 1961. Its production volume is second only to semiconductor lasers, reaching about 300,000 units per year. The He-Ne laser can oscillate on several lines, the most commonly used being the red at λ=633 nm, followed by the green at λ=543 nm, and by a few infrared lines at λ=1.15 µm, 1.52 µm, and 3.39 µm.

The medium is a low-pressure (a few torr) gas mixture of He and Ne, in $\approx$5:1 proportion. Neon atoms provide the active transitions through a number of energy levels, spaced in energy by $\Delta E = h\nu = hc/\lambda$, where λ is the wavelength of the line. The helium atoms are excited by collision with electrons of the discharge, and transfer excitation to neon atoms by collision between ions. The capillary tube is carefully sealed to attain a loss of <0.01 torr/year, and the wall thickness is kept $\approx$5 mm to limit He leakage through the glass.

The He-Ne active medium is a low-gain one, with a typical gain of $\gamma=0.5\text{-}1\%/\text{cm}$. Correspondingly, the optical gain per pass (that is, $\exp\gamma L$) is just $\approx 1.1\text{-}1.2$ in a typical tube of 20-cm length. The end-surfaces of the capillary tube are worked flat and parallel. The surfaces accommodate the mirrors or the Brewster-angle windows, in laser units with internal or external mirrors, respectively.

Mirrors of the optical cavity are made by deposition of dielectric multilayers working on interference. This allows to limit the optical loss to <0.1% or less, compared to several % of metal layers. To get the best cavity properties, usually the output mirror is made flat (radius of curvature $r_1=\infty$) and the other (rear mirror) is concave, with a radius of curvature larger than the tube length (typically $r_2=1$ m). Mirror reflectivity values in this example are $R_1=0.95$ (to optimize the output power) and $R_2=1$ nominally (it may be 0.998 in practice).

If the active medium is analogous to the amplifier in an electronic oscillator, the mirror cavity is the counterpart of the feedback loop, ensuring a positive narrowband feedback.

The mirror cavity is a frequency-selective element because it is a Fabry-Perot resonator (see also Appendix A2). Radiation in the cavity bounces back and forth between the mirrors. The resonance wavelengths λ are those at which the round-trip optical phase shift k2L is a multiple of 2π, so that the electrical field of the optical signal continues to add in phase in successive round-trips. Writing the condition $k2L = N2\pi$ in terms of $k=2\pi/\lambda$, we get the condition that the cavity length is an integer multiple of half-wavelength:

$$N (\lambda/2) = L \qquad (A1.1)$$

Here, the integer N is the order of the resonance, also indicated in the mode designation as TEM_{00N}. By writing the above condition for the orders N and N+1 and subtracting, we obtain the frequency spacing $\Delta\nu$ of the resonance as:

$$\Delta\nu = \nu_{N+1} - \nu_N = c[1/\lambda_{N+1} - 1/\lambda_N] = c[(N+1)/2L - N/2L] = c/2L \qquad (A1.2)$$

The c/2L spacing of resonance is typical of the so-called longitudinal modes, or TEM_{00} modes, characterized by an electric field distribution given by:

$$E(r) = E_0 \exp{-r^2/w_0^2} \qquad (A1.3)$$

In the Gaussian field distribution given by Eq.A1.3, the parameter w_0 has the meaning of a characteristic radius, called the *spot size* of the laser beam.

In particular, w_0 is the radius at which the field amplitude drops off to $1/e=0.37$ and the power density (proportional to E^2) drops off to $1/e^2=0.13$ of the maximum value. In addition, by integrating $E^2(r)$ on r, we find that the relative power contained within w_0 is $1-1/e^2 = 0.86$ of the total beam power.

Each longitudinal mode satisfying Eqs.A1.1-A1.3 is accompanied by a set of transversal modes. These modes are the distributions of the electric field that satisfy Eq.A1.1 by virtue of a nonvanishing transversal component of the wave vector. The field distribution of the transversal mode TEM_{pqN} of order p,q can be written in polar coordinates (r,ϕ) as [3]:

A1.1 Laser Basics

$$E_{pq}(r, \phi) = E_0 \, \Pi_p(r/w_0) \, \cos(q\,\phi) \, \exp{-r^2/w_0^2} \tag{A1.3'}$$

Π_p being a polynomial of order p [1-3] and q in an integer. From Eq.A1.3', we can see that the main spatial dependence is the same Gaussian of the longitudinal mode (Eq.A3.1). However, because of the multiplication by a polynomial of order p, the resulting distribution is considerably broader than the Gaussian, and wavier because of the zeroes of Π_p.

Transversal modes have a spot size, again defined as the radius containing 86% of the beam power, larger than the fundamental longitudinal mode w_0 and increasing with mode order.

In frequency, TEM$_{pqN}$ transversal modes are close to the lowest of them TEM$_{00N}$, the separation being much less than c/2L. Losses experienced by the transversal mode are larger than the fundamental mode (see Fig.A1-1), and this circumstance is useful for preventing modes from oscillation, so that the beam quality of a purely Gaussian mode is preserved.

Propagation [1-3] of Gaussian modes is such that the mode distribution (Eq.A1.3) keeps itself unaltered in free propagation of the beam and in imaging of it by lenses and mirrors. Only the spot size w(z) changes, and its relation to the beam waist or spot-size w_0 originated by the laser at z=0 is:

$$w^2(z) = w_0^2 + (\lambda z / \pi w_0)^2 \tag{A1.4}$$

In Eq.A1.4, when z=0 we get $w=w_0$, the beam waist of the laser. For z>0, w increases as the quadratic sum of two terms: the beam waist w_0 and a propagation term proportional to distance z. Writing this term as $\theta_0 z$, we can see that $\theta_0 = \lambda/\pi w_0$, and this is the angular divergence of the beam, equal to the diffraction angle of the aperture w_0.

When z is large enough (z>$\pi w_0^2/\lambda$), the propagation term prevails and $w(z) = \theta_0 z$. Then, at the angle $\theta = r/z$ off the beam axis, the electric field given by Eq.A1.3 becomes $E(\theta) = E_0 \exp{-\theta^2/\theta_0^2}$. This tells us that θ_0 duplicates in angle the spot-size meaning w_0 has in radius.

In a real laser with spot size w_0, the 86%-power angle θ_{eff} is larger than $\theta_0 = \lambda/\pi w_0$, the ideal single mode value. Then we define an M-squared factor as:

$$M^2 = \theta_{eff}^2 / \theta_0^2 \tag{A1.4'}$$

The M^2 factor is close to unity as emission approaches a single mode, and is therefore used to describe the angular (or single-mode) quality of a generic laser beam.

By analyzing propagation in the mirror cavity [1-3], we calculate the beam waist's dependence from the curvature radii of mirrors and distance L, as well as the position of the waist (or, z=0 origin) with respect to mirror position. The result reads:

$$w_0^2 = (\lambda/\pi) \, mL \tag{A1.5}$$

In this expression, m is a numerical factor that depends on the ratio of the mirrors' curvature radii to distance L. For example, a plano-concave mirror cavity ($r_1=\infty$, $r_2=KL$) has m=$\sqrt{(K-1)}$, a confocal cavity ($r_1=r_2=KL$) has m=$\sqrt{(K/2-1)}$, etc.

The beam waist is located on the plane mirror in a plano-concave cavity, whereas in the confocal cavity, it lays at the midpoint of the mirrors.

A1.1.1 Conditions of Oscillation

A picture of the conditions leading to laser oscillation in a He-Ne medium is depicted in Fig.A1-1. The drawing shows the atomic line providing optical gain, as well as the resonance of the mirror cavity. The width of the atomic line is Δv_{at} =1.5 GHz in the He-Ne mixture, and the spacing of the longitudinal modes is Δv =c/2L in frequency (or Δv =500 MHz for L=30 cm).

As described above, beside the main peaks corresponding to the TEM_{00N} longitudinal modes, we find smaller resonance due to the higher order spatial modes.

The medium provides a round-trip gain $exp2\gamma L$, where γ (≈ 0.01 cm^{-1}) is the gain per unit length and L the length of the active medium.

As in any oscillator, we shall apply the Barkhausen's conditions to find which frequency can oscillate and how amplitude (or gain) will adjust in the permanent regime of oscillation.

The first Barkhausen's condition is about the onset of oscillation and states that the round-trip gain shall have a modulus larger than 1. Once oscillations set in, the second Barkhausen's condition states that the round-trip gain shall become exactly equal to 1 and a phase shift equal to 0. The necessary gain reduction is achieved by saturation of the medium gain.

With reference to Fig.A1-1, when the gain $exp2\gamma L$ is large enough to overcome the losses 1/p [mainly due to the mirrors' reflectivity, p=$r_1 r_2$], the pattern of cavity modes determines the oscillating frequencies. Those modes falling within the $exp2\gamma L>1/p$ dotted line in Fig.A1-1 are allowed to oscillate, whereas the others are not. In the example, modes 2 and 3 will oscillate. With oscillation, the mode diminish the gain (dotted line), preventing the oscillation of further modes (including the undesired higher order modes).

Ideally, the oscillation frequency is centered on the cavity resonance peak. The medium itself introduces a phase shift, however. Then, the actual frequency moves a little under the cavity line toward the atomic line center. In this way, the atomic line phase shift is compensated by an equal and opposite amount of phase shift due to the cavity line (Fig.A1-1).

Because of this small shift, longitudinal modes are not spaced exactly by 500 MHz, but deviate a modest quantity from c/2L, equal to a fraction of the cavity line width, 50-500kHz, typically. This effect is called *frequency pulling*.

In a short-length (L=20 to 30 cm) He-Ne laser, we usually find 2 or 3 oscillating modes, the exact number depending on the actual position of the cavity pattern with respect to the atomic line center. However, the cavity pattern is far from being still in a non-stabilized laser. Any perturbation (mechanical, thermal, etc.) will cause ample drifts.

This can be appreciated by considering that a $\lambda/2 = 0.3$ µm variation of the mirror distance (on a L=30 cm) will shift the mode pattern of one full c/2L period, and eventually produce a change in oscillating modes.

Also, when the main mode is right under the atomic line center v_{lc}, the gain dip and its symmetrical dip become superposed, and there is a decrease in power with respect to frequencies slightly detuned off v_{lc}. This effect is the so-called Lamb's dip, and is useful for frequency stabilization.

A1.1 Laser Basics

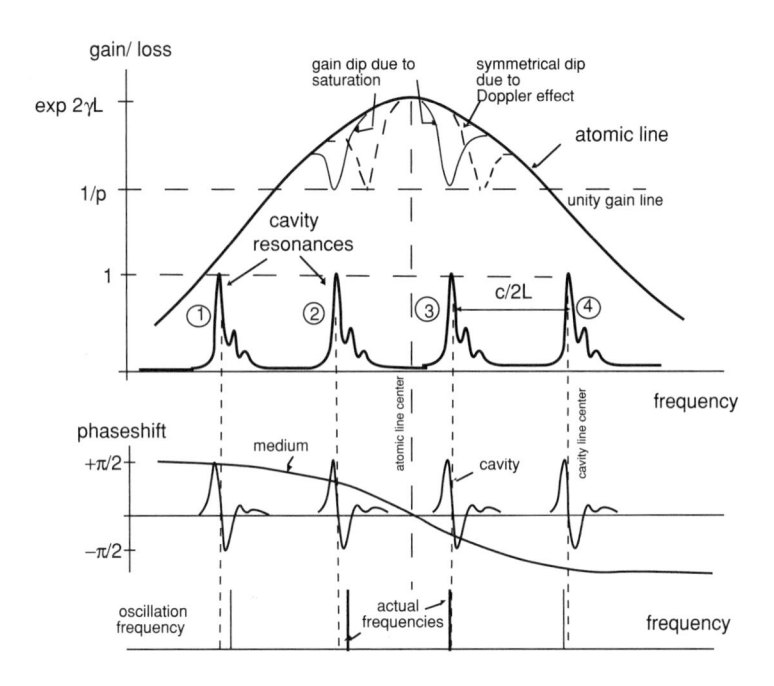

Fig.A1-1 Oscillations in a He-Ne laser. Under the atomic line, the modes labeled 2 and 3 can oscillate because their gain is larger than the loss. Because of oscillation, holes are burned in the atomic line by saturation and the Doppler effect. Mode-pulling makes the longitudinal mode spacing slightly different from c/2L.

A1.1.2 Coherence

Coherence is a very important feature of laser sources, one widely used in instrumentation (Table A1-1). In the following, we will recall some basic quantities about it.

In general, the term *coherence* is used to deal with the property of correlation between the optical fields in two points $P_1(x_1,y_1,z_1)$ and $P_2(x_2,y_2,z_2)$.

Correlation is expressed by the coherence factor μ_c defined in Eq.5.10. When μ_c is high (e.g., $\mu_c > 0.5$), we say fields P_1 and P_2 are well correlated, or that there is coherence between them. When μ_c is small or negligible, we say fields P_1 and P_2 are not coherent.

Usually, we distinguish two aspects of coherence, spatial and temporal.

Spatial coherence is when points P_1 and P_2 are taken on the wavefront, or transversal to the propagation direction ($z_1=z_2$). The region inside which μ_c is high is defined as the *coherence area*, and its size $r=\sqrt{[(x_1-x_2)^2+(y_1-y_2)^2]}$ is called the *coherence radius*. Along a fundamental spatial mode of a laser, we have $\mu_c=1$, and coherence dimension is as large as the spot size w_0. Instead, if some power is shared by higher order modes, we use the factor M^2 to describe the laser emission. Then, at increasing number of modes N, the coherence radius becomes smaller and smaller, and $r \approx w_0/\sqrt{N}$.

Temporal coherence is when points P_1 and P_2 are taken along the wavefront, say at a distance z. Then, the coherence factor is a function of z, $\mu_c=\mu_c(z)$, and the distance z at which μ_c is dropped to 0.5 is defined as the *coherence length* L_c. So, the coherence length is the length of the wave packet inside which the beating term $\langle E(P_1)E^*(P_2)\rangle$ is not less than half the maximum value, given by $[\langle |E(P_1)|^2\rangle \langle |E^*(P_2)|^2\rangle]^{1/2}$ (see Eq.5.10), or a good interference signal is developed.

Writing $L_c=cT_c$, where c is the speed of light and T_c is the time to cover L_c, unveils the *coherence time* associated with the source. The coherence time is interpreted as the time during which the phases of fields $E(P_1)$ and $E(P_2)$ maintain a definite difference or are coherent.

Time T_c is connected to the *line width* $\Delta \nu$ of the laser oscillator, being $T_c \approx 1/2\pi \Delta\nu$.

If the laser emission is multimode (in frequency), the atomic line is likely be filled with oscillating lines (Fig.A1-1), and the line width is about the whole atomic line value, $\Delta\nu \approx \Delta\nu_a$. For a He-Ne, $\Delta\nu_a \approx 1$GHz and then $T_c=1$ns and $L_c=0.3$ m.

By contrast, if the laser oscillates on a single longitudinal mode, the line width of oscillation is not much different from the cavity resonance width (Fig.A1-1), with typical values of $\Delta\nu \approx 1..10$ MHz in a He-Ne laser. This corresponds to a coherence time $T_c \approx 150..15$ ns and to a coherence length $L_c \approx 50..5$ m.

Lastly, if the laser is frequency-stabilized (Sect.A1.2) line width may go down to $\approx$kHz, coherence time up to $\approx$ms, and corresponding coherence length may be in excess of ≈ 50 km.

A1.1.3 Types of He-Ne Lasers

A number of typical, commercial He-Ne tubes are reported in Fig.A1-2. Depending on the application, we may choose an easy-to-mount, compact internal-mirror tube, or we may prefer to have access to the cavity, like in a Brewster's window tube with external mirrors.

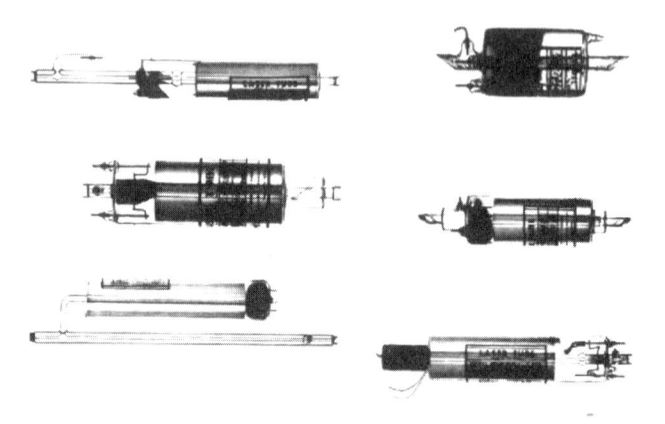

Fig.A1-2 Some typical He-Ne laser tubes for instrumentation. Left, with internal mirrors cemented to the capillary bore (unpolarized output). Right, with Brewster's windows, require external mirrors, and yield a linear polarization output. At the left bottom is a side-arm tube; other units are coaxial tubes.

A1.1 Laser Basics

The state of polarization of the emitted beam is different in the two cases. It is randomly polarized in internal mirror units (or, more precisely, it is a random mixture of linear states apparently behaving as unpolarized) and is linearly polarized in Brewster's window tubes, the polarization being parallel to the window's incidence plane. Windows are tilted at Brewster's angle $\alpha_B = \text{atan } n_{glass}$ to have zero reflection loss for the parallel polarization (while perpendicular polarization has a substantial loss and can't oscillate).

All the tubes in Fig.A1-2 are 15 to 30 cm long and their longitudinal mode spacing is accordingly c/2L=1000 to 500 MHz, a value suitable for near-single mode operation in a medium with an atomic line width of $\Delta \nu$ =1.5 GHz. Their typical gain per pass $\exp 2\gamma L$ is about 1.05 to 1.10, that is, barely in excess of unity but enough to tolerate a mirror loss of 0.95 to 0.98, respectively. These values call for good (multilayer interference) mirrors. Usually, one is chosen with maximum reflectivity (r_1=0.995, the rear mirror), the other with r_2=0.95 to 0.98 reflectivity (the output mirror) adjusted to maximize the output power.

The radius of curvature of the mirrors is as follows: one is flat (usually the output mirror), the other has R=2L to 5L. This choice matches the so-called stability condition of the Fabry-Perot cavity formed by the two mirrors, by which no extra loss in excess of the mirrors' reflectivity is found because of the back-and-forth propagation of the beam.

The power obtained by the specimen in Fig.A1-2 is typically 0.5 to 2 mW, increasing with tube length. Additional details of tube construction are supplied in Fig.A1-3. In coaxial tubes, the transversal size is substantially reduced with respect to side-arm types, at the expense, however, of a questionable self-alignment and beam pointing stability.

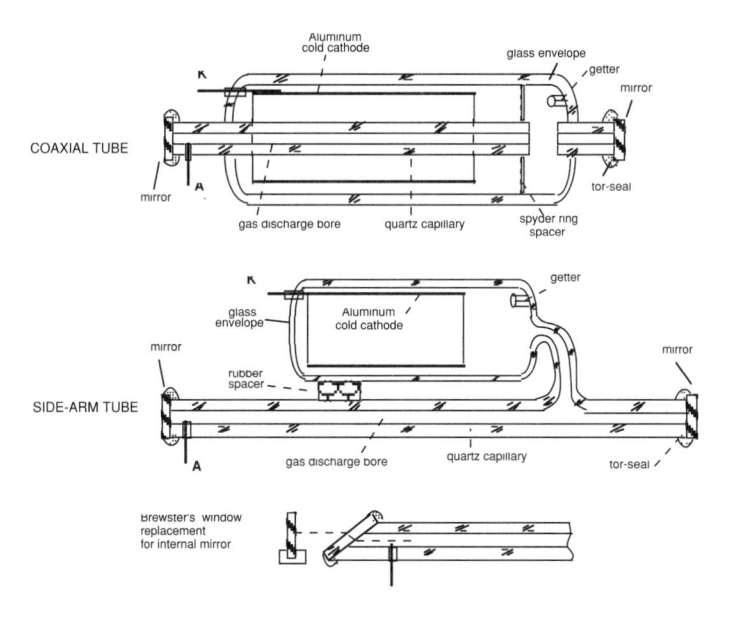

Fig.A1-3 Coaxial and side-arm construction of He-Ne laser tubes with internal-mirrors. Details of a Brewster's window are shown at the bottom.

Indeed, in a side-arm tube, mirrors are glued against the tube end-faces, which can be figured flat and parallel with very good accuracy (typically, a few arcsec) by normal optical tooling. As the capillary is kept rather thick (10 mm for a 1-mm bore) to ensure a low loss rate of He, a gas that can appreciably leak through glass on the long term, the structure is inherently stable and alignment-tight.

This is a better solution as compared to the coaxial tube, which requires a two-sections capillary with a gap in between, to allow the discharge going from cathode to anode. In coaxial tubes, pointing stability is left to the strength of a relatively thinner outer envelope.

To save filament-heating power, cold cathodes are used throughout He-Ne lasers, in the form of an Al tube. This means a ≈ 2 W savings with respect to a thermionic cathode. Because the current density obtained by a cold cathode is rather low (<0.1 mA/cm^2), we need a relatively bulky Al cylinder to get the 5 to 15 mA cathode current normally required by the discharge. Regarding the tube supply, the I-V characteristic of the He-Ne tube is that typical of a low-pressure gas discharge (Fig.A1-4). When we start applying a voltage to the tube, initially current is very low. We need a V_T=+15 to 20 kV to trigger the switch-on of the tube and get $I_a \approx 5$ to 10 mA to pass through it. Then, the voltage is lowered to the value (typically V_A=1500 V) required by the gas discharge. We can apply the correct sequence of supply voltage by means of the circuit shown in Fig.A1-4. Here, the rectifier DP is the main diode giving the V_A on-voltage. To get also the initial V_T trigger overdrive, other diodes are used in a diode Marx pump circuit, supplying a high voltage when current is low and being cleared by the current passing through them when the tube is on.

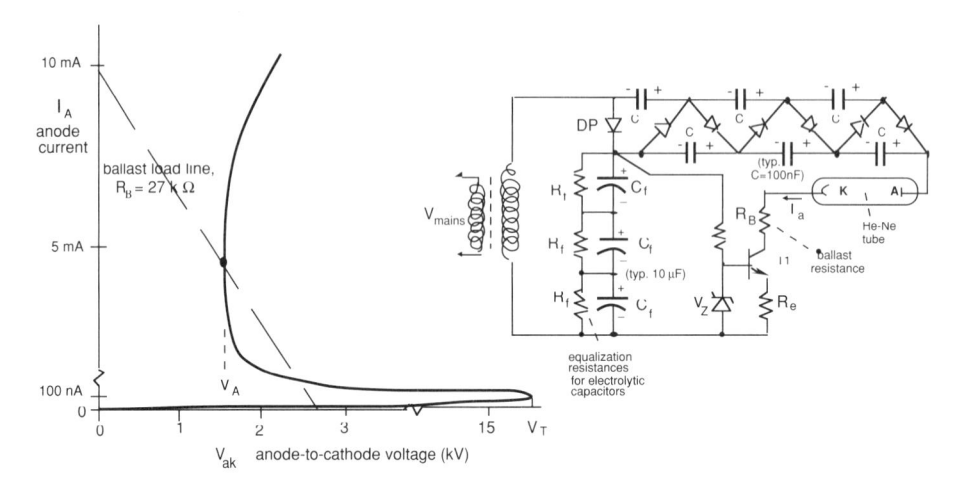

Fig.A1-4 Volt-Ampere characteristic of a typical He-Ne discharge tube (left). For discharge switch-on, we need a high voltage V_T (typically +15 to 25 kV, depending on tube length L), but with a very small current (less than 1µA). When the tube starts conducting, the anode voltage drops to about 1500 V. In the drive circuit (right), a diode pump supplies the trigger voltage peak, while transistor T1 serves to stabilize the tube current.

A1.2 Frequency Stabilization of the He-Ne laser 375

Further, to prevent rectifier ripple from reaching the discharge current, and affecting the emitted power with a mains ripple, a transistor is added in series to the tube so as to feed the discharge by a constant-current generator (equal to $I_a=V_Z/R_e$, see the schematic in Fig.A1-4).

The ballast resistance R_B put in series to the tube is because the tube differential resistance is negative (Fig.A1-4) at the quiescent point of operation (e.g., $\approx$ 5mA, 1600 V). To avoid spurious oscillations in the supply circuit, we allow for a R_B larger than the negative resistance of the discharge. Values of R_B in the range 10 to 50kΩ are adequate for the purpose, and we shall minimize the parasitic capacitance of R_B by proper wiring. Last, a dc/ac converter (not shown in Fig.A1-4) is sometimes included in the supply module for battery operation of the source in portable instruments.

A typical laser tube with its supply circuit module (rectifier, switch-on and current regulation) is shown in Fig.A1-5. We can see that the drive circuit does not add much to the total tube size.

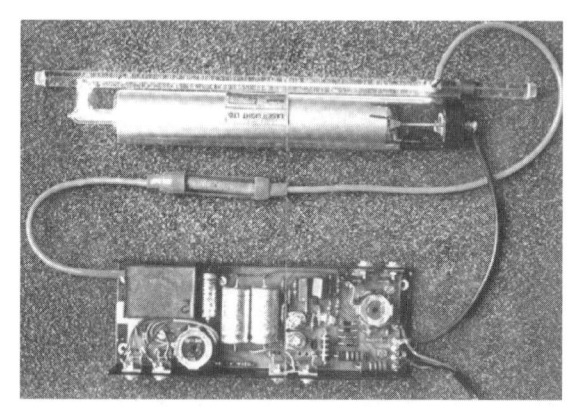

Fig.A1-5 A typical He-Ne side-arm laser complete with its supply and drive circuit (commercial unit fabricated by Lasertronics, Germany)

A1.2 FREQUENCY STABILIZATION OF THE HE-NE LASER

To stabilize the frequency of emission, three basic functions must be performed:
- a frequency reference (i.e., a frequency marking)
- a mechanism for generating a signal error
- an actuator to change the frequency (through the cavity length)

The reference determines how good the ultimate frequency stability of our laser will be. After choosing the reference, we get a signal indicating how far from the reference the actual frequency is, best if it is proportional to it. This signal will be used in a control loop to actuate the cavity length, thus changing the frequency (for a $\lambda/2$ change in length, we get a c/2L shift in frequency).

Several methods are available for implementing the frequency reference, including the following:
- Lamb's dip
- Two-mode, cross-polarized
- Zeeman splitting
- Iodine (external cell) signature

A1.2.1 Frequency Reference and Error Signal

The *Lamb's dip reference* consists of looking at the power amplitude P_m of the mode as its frequency f_m sweeps under the atomic line (Fig.A1-6). It works well with a single longitudinal mode regime of oscillation, as obtained with a not-too-long cavity length (say 15 to 20 cm), either with internal or external mirror units. When the detuning Δv is large, the active medium supplies less gain and power is small, while near to the line's center we get maximum power. However, because of the self-saturation (Sect. A1.1.1), right at the line's center we find a small dip, both in gain and in emitted power. The waveform of P_m versus detuning Δv is readily seen experimentally, during the laser warm-up following the tube switch-on (Fig.A1-6).

Now, if we control the cavity length by adjusting the power to be locked at the Lamb's dip minimum, frequency stabilization is achieved.

To generate the error signal, we cannot directly use the power, because near the peak $P_m(\Delta v)$ is about quadratic in Δv and does not tell us the sign of detuning. But, in accordance with a well-known technique used in control systems engineering, we can add a small ac modulation, ΔL, to the cavity actuator, and look to the corresponding power modulation ΔP_m.

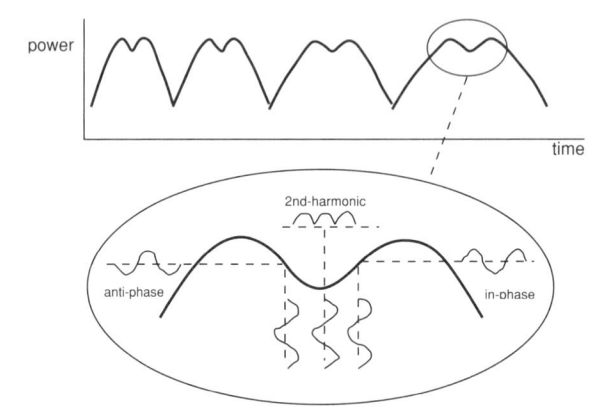

Fig.A1-6 At switch-on, the cavity warms up and power from the He-Ne laser undergoes cycles of variation, replicating the P_m vs. detuning waveform (top). Adding an ac modulation to the cavity length actuator allows us to get a signal proportional (with sign) to detuning from the Lamb's dip center.

A1.2 Frequency Stabilization of the He-Ne laser

Now, the ac signal ΔP_m is in-phase with $\Delta \nu$ (or with the ΔL drive signal) for $\Delta \nu < 0$, is in-phase opposition for $\Delta \nu > 0$, and for $\Delta \nu = 0$ carries only a second harmonic component. Thus, by phase detection of ΔP_m with respect to the ΔL drive signal used as a reference, we can obtain an error signal adequate for control.

Another popular technique, again taking advantage of the power dependence $P_m(\Delta \nu)$ on detuning, is that of the *two-mode, crossed-polarization regime* of oscillation.

The underlying principle is known as spatial hole-burning. When a mode to break into oscillation, a standing wave pattern is established in the medium, subtracting energy from the active atoms aligned with its particular polarization. If another mode is about breaking into oscillation, the gain for the state of polarization orthogonal to that already running is at its maximum, and is at its minimum for the same state of polarization.

The result is such that, when two adjacent longitudinal modes oscillate simultaneously, they put themselves in orthogonal states of polarization. In internal-mirror lasers, there is no theoretically preferred polarization to start with, but even very minute deviations from ideal symmetry in a practical structure will privilege a specific linear polarization.

Thus, in a two-mode laser tube ($L \approx 15$ to 30 cm for best operation), two adjacent longitudinal modes oscillate with linear orthogonal polarization states. If P_s and P_p are the powers they carry, we will find replicas of the normal $P_m(\Delta \nu)$ curves for each of them, and the diagram of powers as a function of detuning is that shown in Fig.A1-7.

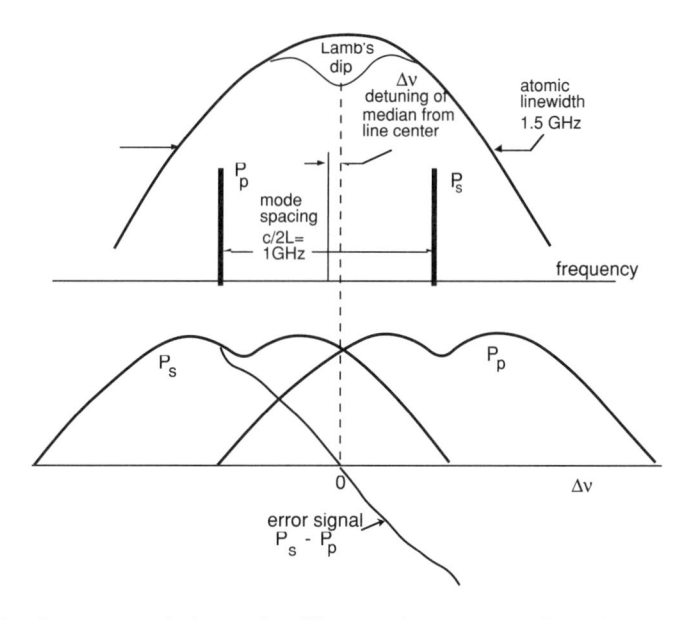

Fig.A1-7 In a two-mode laser, the difference in power amplitude is an easy marking of detuning with respect to the atomic line center. The two modes can be separated and detected because their polarizations are orthogonal.

Note that the power difference P_s-P_p is adequate as an error signal marking the symmetry in frequency of the two modes with respect to the atomic line center.
The signal P_s-P_p is easily obtained by placing two photodiodes at the output of a Glan cube polarizing beamsplitter receiving a fraction of the emitted power.
A convenient location for the Glan cube is at the laser's rear mirror (the one normally unused), where a small but sufficient power is available.

The advantage of the two-mode cross-polarized method is that two polarization modes are readily made available, separated by $c/2L = 500$ to 1000 MHz, typically.

The frequency stabilization performances obtained by the Lamb's dip and by the two-mode, cross-polarized methods are comparable.
The frequency rms deviation from the average, $\sigma_f(T)$, depends on the observation time T. Typical values relatively easy to obtain are in the range $\sigma_f \approx 1..5$ MHz for short times (T=1 ms), but the best results reported may be 10 to 50 times better.
At $\lambda=600$ nm, an rms stability $\sigma_f \approx 2$ MHz means a relative accuracy of frequency $\sigma_f/f = 2$ MHz/500THz = $4 \cdot 10^{-9}$, and a coherence length $L_c = c/\sigma_f = 150$ m. These are very good figures indeed, largely adequate for industrial applications of interferometry. To get still better results, and also a precision of the frequency reference, many efforts can be pursued. Perhaps the first point to care about is the composition of the active Ne gas. Natural Ne is mixture of ^{20}Ne and ^{22}Ne isotopes, which have slightly different atomic lines giving rise to a waveform distortion (not indicated in Fig.A1-1 to A1-7) that disturbs accuracy and repeatability. The remedy is to use the pure isotope ^{20}Ne.

The *Zeeman-splitting* method of stabilization is based on the application of a magnetic field to the atomic medium. If the magnetic field is applied *parallel* to the tube axis (or along the propagation direction of the oscillating mode), the atomic line becomes split in two frequency shifted lines, one supporting the right-handed circular state of polarization, the other supporting the left-handed circular state of polarization (Fig.A1-8).
The amount of frequency splitting is proportional to magnetic flux density B, and is given by $\Delta\nu_Z=C_Z B$, where $C_Z= e/4\pi m=1.4$ MHz/Gauss is the gyromagnetic ratio or Zeeman constant.
Applying a magnetic field *perpendicular* to the tube axis leaves the atomic line unperturbed for the linear polarization parallel to the magnetic field (say, the horizontal), while the other polarization (vertical) line becomes double-peaked, as indicated in Fig.A1-8.

The consequence of splitting (also called polarization degeneration removal) is that the active medium now can support two oscillating modes with orthogonal polarizations, each sustained by the atomic population matched to that polarization.
With an internal mirror cavity adding no extra polarization selectivity, the cavity line is the same for the two polarization modes.
But, because of frequency pulling (Sect.A1.1), the modes oscillate at a slight deviation from the cavity line's center (Fig.A1-8) to compensate for the medium phase shift, each being attracted toward the respective atomic line center.
Thus, we have a frequency difference $\Delta\nu_{pol}$ between the two polarization modes, which changes as detuning $\Delta\nu$ changes.

A1.2 Frequency Stabilization of the He-Ne laser

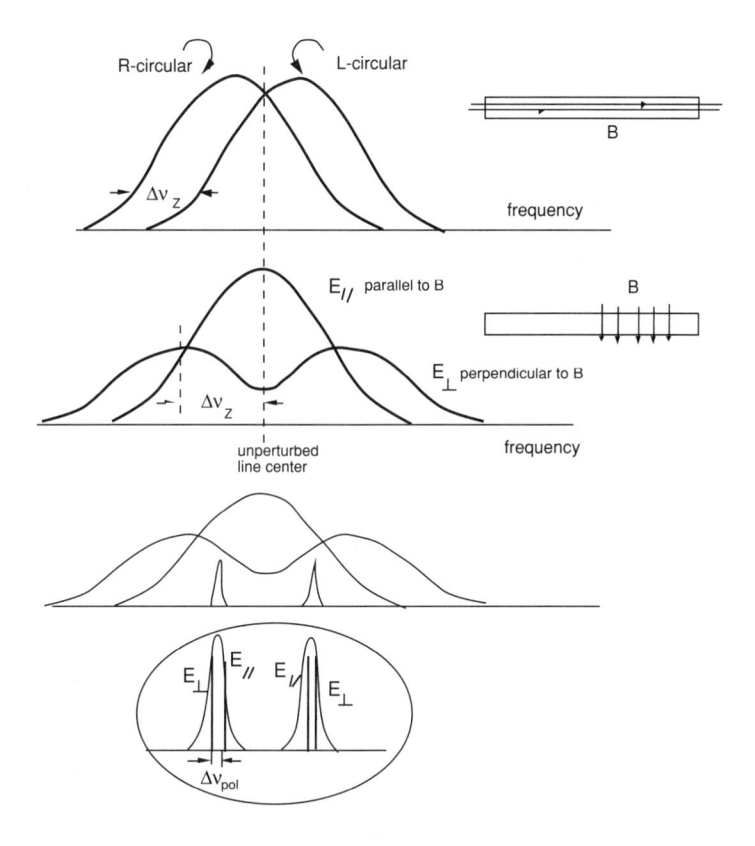

Fig.A1-8 Zeeman splitting of the atomic line, for a magnetic field applied parallel (top) and perpendicular (center) to the tube axis. Because of mode-pulling, the two polarization modes oscillate at slightly different frequencies (bottom), typically Δv_{pol}=50-200 kHz, dependent on the detuning of the cavity line from the original atomic line center. The difference Δv_{pol} provides the error signal for cavity control, and can be recovered by a photodiode and polarizer combination detecting the mode beating at the rear mirror.

This is a frequency marking of the atomic line and readily provides the error signal for cavity length control. To get the error signal, it is customary to use the rear mirror of the laser (opposite of the main output mirror), where a small but significant power is emitted. Here we place a polarizer and photodiode.

The polarizer is oriented at 45° with respect to linearly polarized modes so that they beat on the photodiode. The photodetected current is then a sinusoidal signal at frequency Δv_{pol} and, after a frequency-to-voltage conversion, the error signal is ready for the actuator.

Both longitudinal and transversal Zeeman effects may be used in practice.

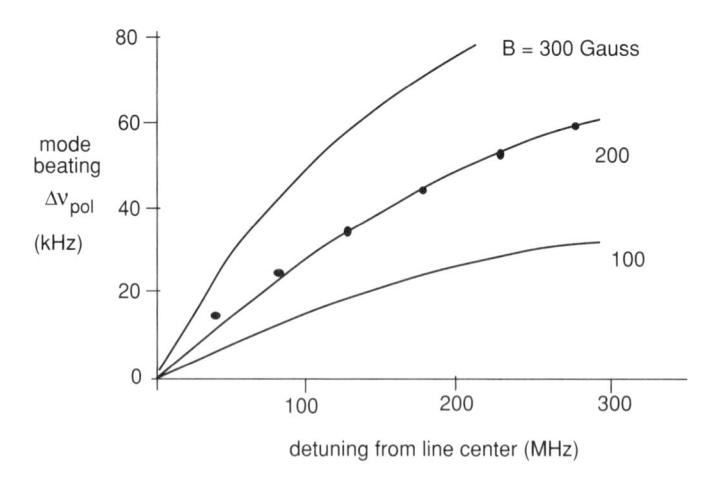

Fig.A1-9 The frequency difference between linearly polarized orthogonal modes in an He-Ne Zeeman laser with a transverse magnetic field, as a function of detuning and for several applied magnetic fields. Lines are the theoretical results, and points are the experimental values for B=200 Gauss.

An example of the dependence of $\Delta\nu_{pol}$ from the magnetic flux density and detuning is reported in Fig.A1-9, which shows that moderate fields (hundreds of Gauss) are sufficient.

This magnetic field is easily provided, either by a coil wound on the capillary tube (B parallel) or by Al-Ni-Co permanent magnets (B perpendicular). By actuation of the cavity length (see below), the typical frequency stability of the beating note obtained using a commercial laser He-Ne tube [4] is reported in Fig.A1-10. Translated to optical frequency, this amounts to a relative stability of $\sigma_f/f \approx 1 \cdot 10^{-11}$. However, this is not the actual precision of the wavelength because the beat note $\Delta\nu_{pol}$ depends on B and its eventual drifts.

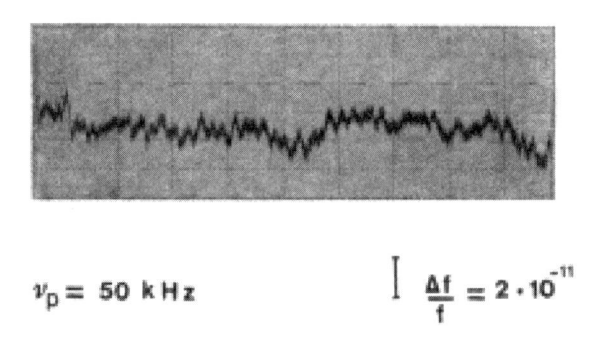

$\nu_p = 50 \text{ kHz}$ $\frac{\Delta f}{f} = 2 \cdot 10^{-11}$

Fig.A1-10 Typical stability of frequency difference in an He-Ne Zeeman laser with transverse magnetic field, on a time period of 1 hour.

With the *iodine (external cell) stabilization method*, we take advantage of the several lines of absorption that the $^{127}I_2$ vapor has under the Ne atomic line [1,4]. These lines are due to the hyperfine structure of iodine and are very narrow dips, typically ≈100 kHz wide.

We can probe the $^{127}I_2$ cell with the laser beam, sweeping its frequency by means of the actuator. If we use the same modulation technique explained for the Lamb's dip method, we can lock the frequency at one of the fine dips. Frequency stability down to ≈100 Hz has been reported with an external cell, the best achievable and of relevance for metrology [5]. Despite that, the absorption cell method is seldom used for interferometers, because so many lines falling under the atomic line require a sophisticated control strategy to lock on a particular one, adding complexity to the final source layout.

A1.2.2 Actuation of the Cavity Length

The two methods most frequently used for actuation of the cavity length are (i) thermal expansion of the capillary tube, best suited for internal mirrors lasers, and (ii) piezoelectric movement of the mirror, best suited for external mirrors lasers. Both can be used, of course, with any of the frequency reference schemes described previously.

Thermal expansion is accomplished by Joule dissipation in a resistive wire (e.g., Ni-Cr), directly wound on a substantial fraction of the capillary tube length so as to minimize the thermal power required to expand it and shorten response time.
Usually, a few Watts of dissipated power is enough to ensure a ≈100-λ expansion from 0 to full control.
The response time of thermal actuation is relatively slow, typically τ=20-50 ms.
This is fast enough to compensate for slow drifts and thermal transients at switch-on and from external temperature, but not enough to reduce vibration-induced fluctuations picked from the external environment (which may have components up to ≈100 Hz).

To improve the response time, the amplifier feeding the resistive winding will include a PID (Proportional-Integral-Derivative) controller. The amount of each PID action will be carefully trimmed to obtain the maximum open loop gain and the fastest closed loop response.

Piezo actuation is implemented by mounting one of the external mirrors on a PZT piezoceramic element, on its turn secured to the mounting basement carrying the other mirror and the tube.
A few-mm thick ceramic disk will usually provide a few μm of mirror movement when driven by a 500-1500 V supply (however, with small current demand), and will present capacitance load (typically 1000pF) to the drive amplifier.
The few μm amount of movement is enough for frequency stabilization when the basement has little thermal expansion, for example, when we are using an Invar bearing structure or enclose the structure in a thermostat. If not, we have to wait for thermal equilibrium at switch-on and at large ambient temperature changes, before the frequency regulation stops skipping from one mode to the next (for typically 30 min. or more in a normal He-Ne tube).
The advantage of the piezo compared to thermal actuation is its faster response time, down to 1..10μs, which makes it capable of canceling ambient-related vibration disturbances as well.

A1.2.3 Ultimate Frequency Stability Limits

We can analyze the performance of frequency-stabilized lasers with the aid of the block-scheme of Fig.A1-11, describing the functions implemented in the feedback control loop.

The frequency error signal is transferred into an electrical signal by a conversion factor G(ν), the power actuator amplifier has a voltage gain A, and the cavity control makes a ΔL variation for unit voltage with a factor K. Lastly, the ΔL variation turns into a Δν change with a ratio of Δν /ΔL=(c/2L)/(λ/2)=c/λL.

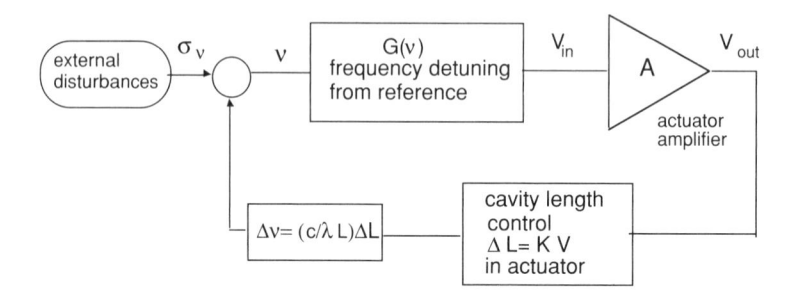

Fig.A1-11 Block scheme of frequency stabilization

The loop gain of the control circuit is accordingly G_{loop}= G(ν)AK(c/λL). Thus, following control theory basic results, any disturbance $σ_ν$ altering the frequency of oscillation is reduced by a factor 1+G_{loop}. Usually, in the above-quoted examples, we have a typical $G_{loop} \approx 10^3$-10^4. The resulting closed-loop rms frequency fluctuation is 1-10MHz, typically.

The ultimate limit to frequency stability, when $σ_ν$/(1+G_{loop}) is made negligible, is set by the quantum limit [1] as $σ_{ν(limit)}$ =[2hνB/P]$^{1/2}$/$τ_c$, where P is the power in the laser cavity and $τ_c$ =(1-$r_1 r_2$)c/2L is the decay time of radiation in the mirror cavity.
Values for $σ_{ν(limit)}$ are down to few Hertz for P≈1mW, but are seldom approached in practice.

A1.2.4 A Final Remark on He-Ne Lasers

He-Ne lasers continue to be preferred as the right way, no surprises, foolproof engineering choice in instrumentation, and especially in interferometers. Stabilization not only gives a narrow-line, long coherence length source, but also provides inherent frequency calibration, and not explicitly noted above but worth mentioning, amplitude stabilization.

While laser diodes are rapidly catching up, He-Ne lasers may maintain a significant role in applications and new designs because of the favorable performance-to-price ratio. Indeed, most commercial instruments using He-Ne lasers survive unsurpassed by laser diode counterparts.

A1.3 Semiconductor Narrow-Line and Frequency Stabilized Lasers

Semiconductor diode lasers are the best choice for compact size, low-drive voltage, and small power requirement. However, mode quality is poor as compared to gas lasers, both spatially and in frequency.

In general, we can say that if a plain, cheap diode laser can be found with sufficient performance for an application, it is undoubtedly the ideal choice for the instrumentation application.

On the other hand, performances of diode lasers are more difficult to improve than in gas lasers, especially when frequency stability is concerned, because complexity and cost increase fast. This circumstance has hindered the spread of diode lasers in applications so far, but the situation is slowly improving. In particular, diode lasers already offer narrow-line, long coherence length sources with reasonable wavelength accuracy as advantages in interferometer applications.

In this section, we discuss the performances of diode lasers relevant to applications in electro-optical instruments. The focus is on engineering issues rather than on device physics. A more basic introduction can be found in a number of excellent textbooks, see Refs.[1-3,6].

A1.3.1 Types and Parameters of Semiconductor Lasers

For ease of use and safety requirements (see Sect.A1.5), it is desirable a laser with emission at a visible or near infrared wavelength. For the range λ=620-800 nm, materials include the ternary semiconductors GaAlAs (the 780 nm being the best choice for high-yield CD lasers), InGaP (going down to 620 nm), and the quaternary InGaAsP.

The pn-junction structure is usually a heterojunction for optical and electrical confinements and, in the active layer, quantum wells are preferred for band confinement.

Band confinement greatly reduces the threshold current, increases the output power, and improves the single-mode operation by narrowing the line width and increasing the side mode suppression, all key points for applications in interferometers and electro-optical instruments.

The I/V characteristic is that of a normal diode (Fig.A1-12), and the P/I curve has a step increase in power in correspondence to the threshold current I_{th}.

A laser diode chip has a length L=100 to 300µm, typically. The corresponding mode spacing (for λ=800 nm and L=150 µm) is c/2nL=300 GHz (or 0.64 nm in λ). The chip ends with cleaved facets, used as mirrors of a plain parallel cavity without additional treatment. The Fresnel reflection at the facet gives a reflectance $\approx$30% (being n$\approx$3.3 in most semiconductors).

A point of concern in diode lasers is that emitted power (Fig.A1-12), as well as modal composition and wavelength (Fig.A1-13), depend markedly on current and temperature.

As shown in Fig.A1-13, when current is gradually increased from threshold, first the laser emits a multimode spectrum, then the mode with the largest gain increases faster than the adjacent and finally dominates over the others, the so-called side modes.

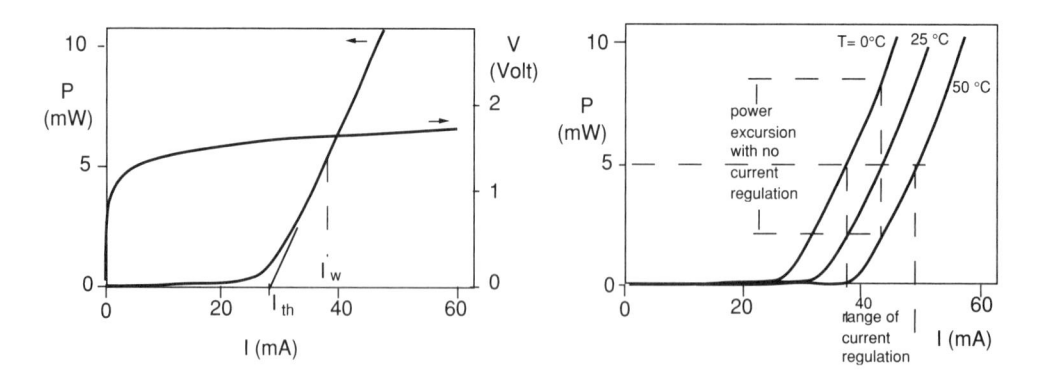

Fig. A1-12 Typical I/V and P/I characteristics of a small-power diode laser at 780 nm (left). As seen at the right, a current regulation is required to keep the emitted power at the desired value if temperature changes from 0° to 50°C.

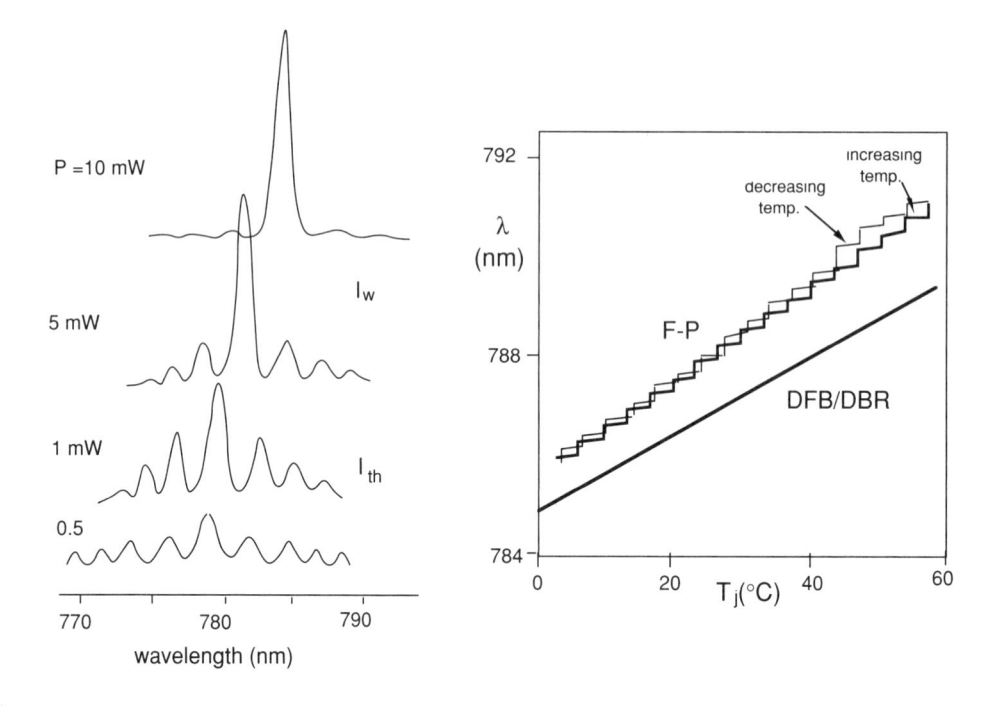

Fig. A1-13 Mode spectrum as a function of drive current (left, note the scale difference) and wavelength dependence on current (right) for a Fabry-Perot (exhibiting mode-hopping) and a DFB/DBR laser.

A1.3 Semiconductor Narrow-Line and Frequency Stabilized Lasers

This is because in laser diodes the ASE (Amplified Spontaneous Emission) is comparatively larger than in other structures. Filtered by the cavity mode resonances, the ASE looks like a multimode spectrum. Only when most of the power available is sunk by the main mode, the spectrum becomes monomode, hinting that the laser has to be driven with the largest permissible current.

The active region has a cross-section $w_{y0} \times w_{x0} = 0.5 \times 1.5$ µm (typically), and the spot size of the mode emerging from the output facet has about the same dimensions (Fig.A1-14). Spot size is the parameter of the Gaussian approximation to mode field distribution, that is, $E(x,y) = E_0 \exp -(x^2/w_{x0}^2 + y^2/w_{y0}^2)$. A fraction 0.86 of the total power is contained within an ellipse of semi-axes w_{x0} and w_{y0}.

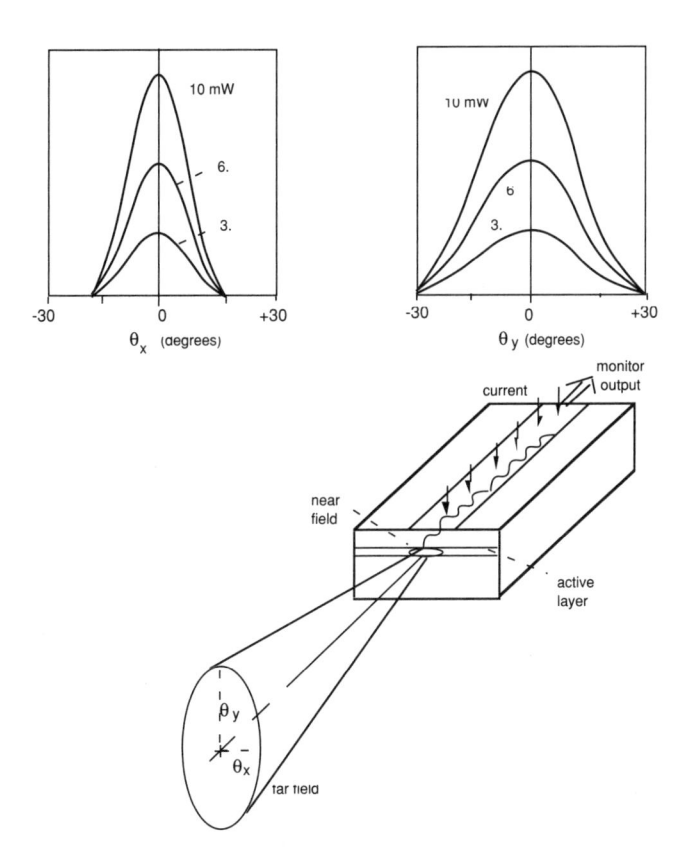

Fig.A1-14 Far-field distributions of the diode laser's emitted power (top). The major axis of the beam spot changes from horizontal to vertical in going from the near to the far field (bottom).

In the propagation out of the laser chip, the laser beam expands with divergence angles $\theta_x = \lambda/\pi w_{x0}$ and $\theta_y = \lambda/\pi w_{y0}$ (typically $\theta_x \approx 10°$ and $\theta_y \approx 30°$). At a distance z from the output facet, the spot-size has dimensions given by: $w_x^2 = w_{x0}^2 + (\theta_x z)^2$ and $w_y^2 = w_{y0}^2 + (\theta_y z)^2$.

The elliptical shape of the spot requires a correction. To transform it in a circular symmetry spot, anamorphic prisms or cylindrical (stigmatic) lenses are commonly used.

The laser chip is welded on a sub-mount (Fig.A1-15) of the package, which incorporates an access window or a collimating lens, and a monitor photodiode mounted on the rear to detect the emission from the back mirror. The monitor photodiode is used to control the dc drive current of the laser, which changes considerably with temperature at constant power (Fig.A1-12).

To add flexibility in laser operation and reduce thermal requirements, a Peltier thermo-electric cooler (or TEC) may be used, sandwiched between sub-mount and case. A thermistor is also added to monitor chip temperature. With a typical 1-W power spent to control the Peltier, we can adjust the chip temperature over a ±30°C swing from ambient temperature. A last option, more commonly found in laser diodes for optical fiber communications than in units intended for instrumentation, is the optical isolator used to prevent back-reflections from returning in the active cavity.

Other commonly available packages for laser diodes are illustrated in Fig.A1-16. Of course, chip and sub-mount versions of laser diodes allow the OEM (original equipment manufacturer) to optimize the key aspects of source design, as well as reduce costs (up to 50% is due to packaging), at the expense of a further effort being required on the component.

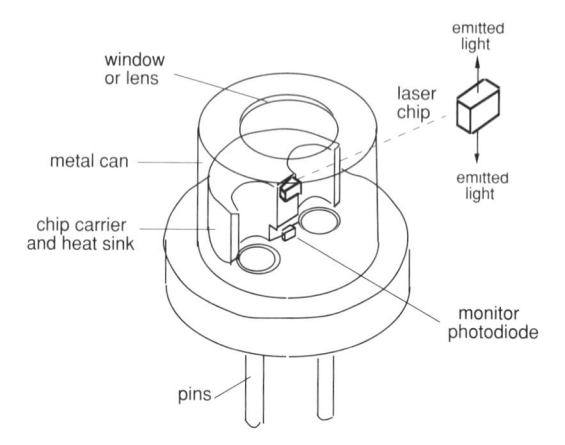

Fig.A1-15 In the typical package of a laser diode for instrumentation, the laser chip is mounted on a chip carrier that also accomodates the monitor photodiode at the bottom. A Peltier cell (not shown) is eventually mounted between the chip carrier and the package basement. The top contact to the chip is by thermocompression. A window (or lens) seals the metal package.

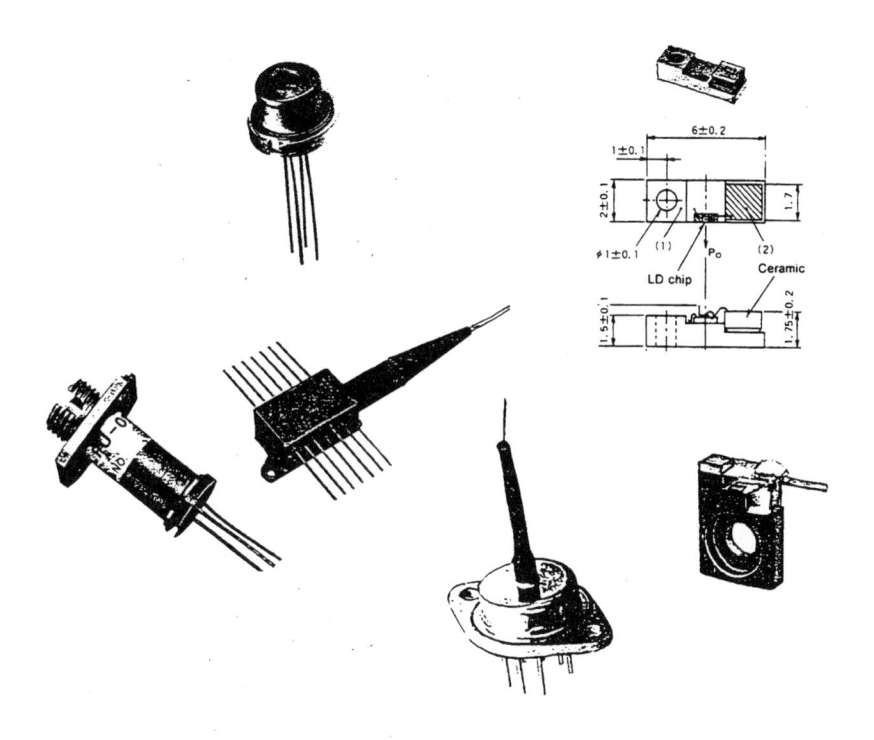

A1.3 Semiconductor Narrow-Line and Frequency Stabilized Lasers 387

Fig. A1-16 Several options for the packages of diode lasers are available: metal can with window or lens (top left), chip-carrier sub-mount (top right), connectorized (bottom left) and pigtailed (middle left) packages, power package (bottom center), and sub-mount version (bottom right).

About frequency response, laser diodes are very fast and easily exceed the GHz cutoff frequency even in plain devices unintended for modulation of the drive current at high frequency. This is a consequence of the heterojunction structure being very thin, and of the fast removal of carriers crowding the junction through stimulated emission.

The temperature and current coefficients of the wavelength (or frequency) of emission are considerably high in diode lasers, and this is due to the dependence of the index of refraction on temperature and carrier density.

In a single-mode Fabry-Perot structure with plane cavity mirrors, at $\lambda=800$ nm we may have $\alpha_\lambda = d\lambda/dT \approx 0.02$ nm/°C (or, ≈ 10 GHz/°C) and $\alpha_I = d\lambda/dI \approx 0.004$ nm/mA (or, ≈ 2 GHz/mA). These quantities amount to about one ppm (part-per-million) error in wavelength for a change of 0.04°C in temperature or of 0.2 mA in drive current (at $\lambda=800$ nm). Thus, if we want a 5-digit accuracy in wavelength, it is mandatory to stabilize the chip temperature (by a TEC, for instance) and use a well-regulated current supply to feed the diode.

However, in Fabry-Perot lasers, this is not enough because, in addition to the λ-drift, the laser also exhibits mode-hopping in the form of sudden jumps of wavelength (Fig. A1-13). Mode-hopping comes from the concurrent drift of the atomic line and the cavity line pattern, which have different rate of drift, however. When a cavity mode escapes from the atomic line and let the next adjacent mode to break into oscillation, a c/2L≈300 GHz (or ≈0.8 nm) jump in frequency (or wavelength) is generated. Some diode specimens may exhibit a large interval in temperature or current between jumps, but this does not help stabilize the emission because hopping has hysteresis. So, the real solution is to remove the physical cause of hopping, that is, the multi-resonance regime of the Fabry-Perot structure.

If it were not for the mode-hopping and the excessive $α_λ$ and $α_I$ dependence, Fabry-Perot lasers would be good for interferometers as they have long coherence length, typically >10m, when single-mode units are used well above threshold.

A better frequency behavior is obtained by other types of semiconductor laser structures, such as:
- DBR (distributed Bragg reflector)
- external FBG (fiber Bragg grating)
- external bulk-optics grating

A1.3.2 Frequency-Stabilized Semiconductor Lasers

Frequency-stabilized and narrow-line lasers are based on the use of a frequency-selective element possessing just one resonance in correspondence to the active medium wavelength.

The *DBR laser* incorporates a Bragg-grating reflector as one of its mirrors (see Fig. A1-17). The grating is obtained by etching a periodic corrugation in the semiconductor at one mirror location. Because of the corrugation, a small index of refraction step Δn is seen by the propagating mode at each grating period, giving a field contribution reflected back to the active region. When the number of periods N is of the order of 1/Δn, a large reflectivity is found, provided the individual contributions come back and add all in phase.
This is the Bragg condition, written as 2nΛ=kλ, where Λ is the spatial period and k is the order of the grating. For k=1, we need a grating period Λ=λ/2n, or Λ=120 nm for λ=800 nm, being n=3.3. Such a short period is a challenge to fabrication by photolitography, and it increases the cost of the device substantially, compared to a normal Fabry-Perot laser. For visible wavelength DBR lasers, the required period is even smaller, and we shall resort to the second order k=2 to keep Λ a reasonable value. With k=2, we get Λ=λ/n and it is Λ=200 nm for λ=680 nm and n=3.4.

Because the DBR structure has a single resonance in spectral range of the active medium gain, mode hopping is suppressed and wavelength is no more dependent on the current injected in the active region.
Yet, the temperature dependence remains as in the Fabry-Perot diode because in the grating, n is affected by temperature and Λ by thermal expansion. Using a temperature control, we can keep this drift as low as desired, at least in principle. As an example, using a Peltier cell, a Δλ/λ stability of a few ppm has been reported over a 1-year period [7], at λ=1500 nm.

A1.3 Semiconductor Narrow-Line and Frequency Stabilized Lasers 389

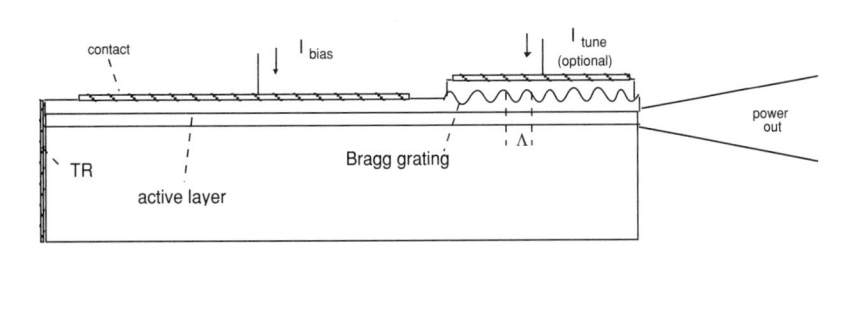

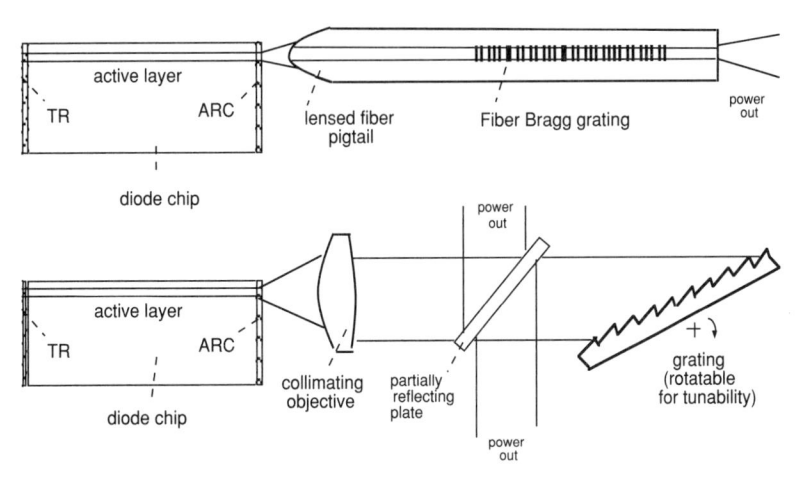

Fig.A1-17 DBR laser chip (top), external DBG laser (center), and external bulk optics (bottom). Drawing details are not to same scale.

Another interesting feature of the DBR is that wavelength can be finely tuned by a current injected in the Bragg grating region.

The general drawback of DBR lasers is that the device is difficult to fabricate at wavelengths in the visible range, and therefore it is expensive. For this reason, presently it is not included in the standard product list by manufacturers.

The *external-FBG laser* is the affordable version of the DBR. It uses a normal chip like that of Fabry-Perot diodes, but one facet is now ARC (anti-reflection coated) and the other is treated for total reflection (TR). Residual reflection may be <0.01% on the ARC facet, while the TR facet may have a reflectivity R=99.9%.

A fiber Bragg grating − that is, a Bragg grating written in the fiber core, of widespread use in fiberoptic communications − is butt-coupled to the ARC facet.

Thus, except for the longer cavity (a few cm of fiber are required in the practical implementation), this source is identical to the DBR, sharing the same narrow line width (100 kHz typically) and λ-temperature dependence (of the glass fiber).

Another variant uses a bulk-optics, *external grating* as the end-mirror of a cavity. The cavity is made by a Fabry-Perot chip with ARC/TR treatment on the facets and a collimating objective to fill the grating's useful area. This approach is the one realized in instruments known as tunable lasers, which are used in optical-fiber communications as a laboratory test-equipment. Most expensive of all choices, this approach provides the best performance of line width (down to 10 kHz) and frequency stability (repeatability of MHz).

A1.4 DIODE-PUMPED, SOLID-STATE LASERS

In pulsed telemeters, we need laser sources capable of emitting short pulses with high peak power so that the timing (and distance) accuracy is good and the maximum distance of operation is large. The typical pulse duration may range from a few nanoseconds to perhaps 10 ns and the peak power we require may go from kW to MW, while the pulse repetition frequency is generally low (a few Hz to a kHz).

These figures are typical of Q-switched solid-state lasers.

Solid-state lasers [2,3] employ a rare earth element such as Nd, Yb, Er, Cr as the active material. The active atom is embedded in an optically transparent host, a glass matrix or a garnet-like YAG (yttrium aluminum garnet) with a 0.01-0.5% concentration in weight, typically. The wavelengths of active-levels transition are located in the red and near infrared, depending on the active atom. For operation through the atmosphere, the wavelength is chosen to be inside regions of good transparence (see Fig.A3-3).

The useful absorption bands of the active atom, those with a fast decay to the active level, are located in the visible or near infrared, blue-shifted with respect to emission wavelength, and have a spectral width which determines pump efficiency.

Pumping is performed optically, and the only choice for pump sources until the 1990s was a high-pressure flash lamp, either fed in dc or, most frequently, pulsed by a capacitor discharge. Such a pumping scheme was bulky, had a modest efficiency, and the reliability was poor because of the strong electrical stress on the lamp.

Recently, with the advent of high-power, high-efficiency laser structures, the stack (or array) of semiconductor diodes has become the ideal candidate for pumping solid-state lasers. Stacks are made by pileup several bars, each filled with many individual diodes placed side by side (Fig.A1-18). One bar may contain typically n_d=500 diodes on a 1-cm wide chip. Each diode may emit p_l=5-20 mW, for a total power of 2.5-10 W. Up to n_b=5-20 bars may be piled one above the other, with a limit only due to thermal dissipation.

In the bar, a basic diode laser uses a comparatively long (400-800 µm) cavity and broad emitting area (4×6µm^2) to maximize the active volume [6]. It is designed for low threshold current and high saturation, for example using the QW-GRINSCH structure. The QW (quantum well) ensures confinement in the band structure, whereas the graded-index, separate confinement heterostructure (GRINSCH) optimizes optical confinement and mode size.

Wavelength of emission is determined by the composition of the active QW.

A1.4 Diode-Pumped, Solid-state lasers

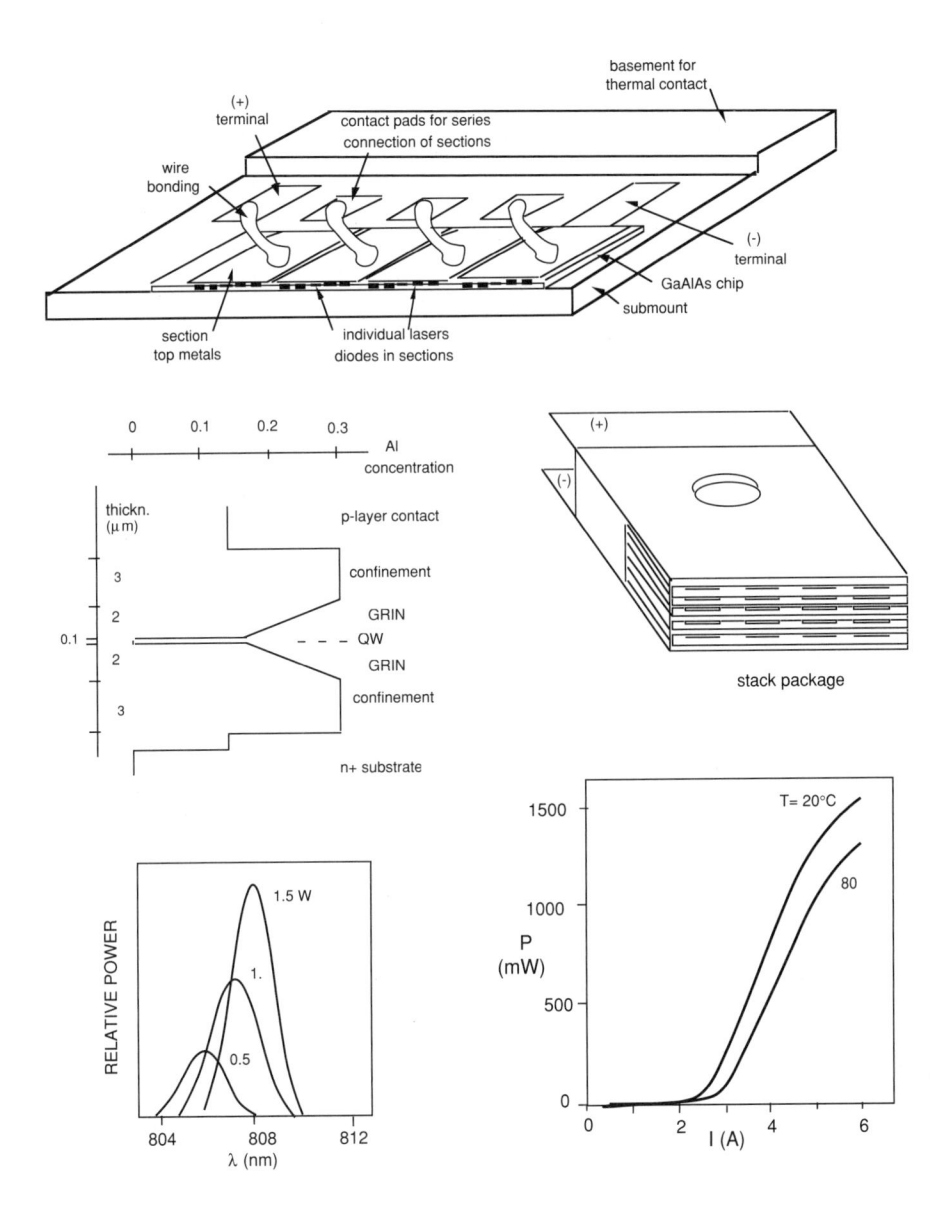

Fig. A1-18 Power diode stack for pumping solid-state lasers. Top: layout of the bar (drawing not to scale); middle: band diagram of the QW-GRINSCH structure (left) and typical package for a stack composed of 5 bars (right). Bottom: line width of the stack (left) and power/current characteristics of a 100-diode bar (right).

Using GaAs substrates and elements like Al and In in the active well and surrounding buffer layer, the wavelength of emission can go from 700 to 1000 nm, with λ=808 nm as the best value for pumping Nd doped or co-doped crystals.

A diode stack may typically deliver 1W to 300W of optical power in the CW (continuous working) regime. The actual power is proportional, of course, to the number of bars n_b in the stack and to the number of diodes n_d per bar. The limit to the total number of diodes $n_b n_d$ is set by thermal dissipation, whereas the limit to individual diode power is set by saturation and catastrophic optical damage (COD) of the output mirrors due to excessive power density. Both limits are not abrupt, but as we approach them, the useful life of the device is severely shortened. Thus, the nominal power specified by the manufacturer is usually a tradeoff between performance and reliability.

With minor modifications (increased modal area of the individual diode to reduce the COD), the stack may also operate in the so-called quasi-CW regime, which is actually a pulsed regime. When pulsed, a single laser diode may supply a peak power P_p=0.2..1 W at a duty cycle of η=2%.

A typical specified pulse width is τ_p=100µs and the corresponding pulse period is T=τ_p/η= 0.1m/0.02= 5ms (or f=200 Hz). From these figures, the average power emitted by the diode is P_pη=(0.2..1)·0.02= 4..20mW, nearly the same as the power from a CW laser diode, whereas the peak power is 1/η=50 times as much, at least for pulse duration shorter than the thermal time-constant τ_{th} of the device (milliseconds).

Thus, a diode stack may typically deliver P_p=50W to 15kW of peak optical power, with 2% duty-cycle and pulse repetition frequency of a few hundreds Hz.

Because of power dissipation, we cannot use pulse duration greater than about τ_{th}η (at constant duty cycle). At the opposite end, a pulse duration much shorter than ≈100µs (or B≈3.5 MHz) is difficult to achieve, because the active device has the combined requirement of high drive currents (see Fig.A1-18) and high-frequency switching.

Incidentally, the τ_p=100µs pulse duration is about optimal for pumping Nd crystals operating in the Q-switched regime. In Nd, the decay time τ_{21} of the active level is about 100µs. The excitation stored in the active level is proportional to $P_p \tau_{21}$. Thus, the optimal choice is a peak power maximized for the pulse duration $\tau_{21} \approx \tau_p$.

A general scheme for pumping crystal with diode-laser stacks is illustrated in Fig.A1-19.

A collimating lens is used to collect the diverging beam emitted from the stack and illuminate the crystal evenly. The crystal is placed in a cavity made by one plane mirror and one concave mirror (with long curvature radius). Both mirrors are multilayer, and offer different reflectivity at the pump λ_p and oscillation λ_{osc} wavelengths. The input mirror facing the diode stack has R=1 at λ_{osc} and R=0 at λ_p, while the output mirror has R=R_0 at λ_{osc} and R=1 at λ_p.

This way, the mirror cavity allows the pump power in for absorption in the active medium, while it is a resonator at the laser wavelength.

Properly designed, the diode-pumped Nd laser may have a low threshold (P_{th}≈10-50 mW typically) and a very good differential efficiency $\eta_{diff} = \Delta P_{out}/\Delta P_{pump}$ (η_{diff}≈60% typically). We may also operate the laser in the Q-switching regime by adding a suitable device in the

A1.5 Laser Safety Issues

mirror cavity, as shown in Fig.A1-19. The device switches the cavity from a low-Q (high-loss, non-oscillating state) to a high-Q (low-loss) state, in which the oscillation amplitude increases quickly and damps out all the available excitation stored during the pumping time.

The resulting pulse has a time duration given by $\tau_Q \approx (2L/c)/(1-R_0)$, that is, not much longer than the cavity go-and-return time, or in practice, $\tau_Q \approx$ 3-10ns.
In addition, the peak power is given by the stored level excitation $P_p \tau_{21}$ divided by the pulse duration τ_Q, or $P_Q = P_p \tau_{21}/\tau_Q$.

The Q-switched pulse can have a huge peak power. It exceeds the pump power by a factor of τ_{21}/τ_Q, a quantity that can be as high as 10^4, or P_Q=1MW for P_p=100W. This is the key point for preferring the Q-switched operation in pulsed telemeters, despite the increaed complexity.

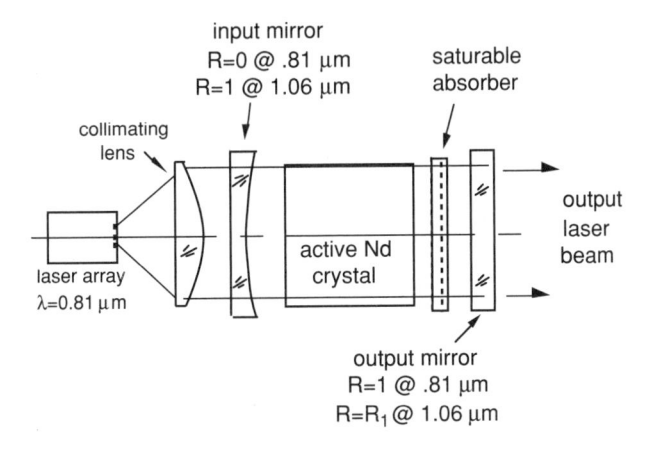

Fig.A1-19 Scheme of a diode-pumped, solid-state laser with passive Q-switching performed by a saturable absorber. A comparatively short ($\approx$1 mm long) crystal is used, and the mirror cavity is a good resonator at laser wavelengths while allowing input of pump power from the diode-laser array. For CW operation, the saturable absorber is omitted.

A1.5 LASER SAFETY ISSUES

All instruments incorporating a laser source must comply with laser safety standards. The actual legal regulations that enforce laser safety issues may vary considerably in each country, according to the applicable laws. However, rules always make reference to standards issued by international Organizations like IEC (International Electrotechnical Commission), ISO (International Organization for Standardization), or other national or learned societies like ANSI (American National Standardization Institute) or MIL-STD (Military Standards).

Both manufacturers and users shall take the appropriate actions in order to ensure that a laser-based product (or instrument) is prevented from being harmful. In particular, manufacturers must classify their products according to classes expressing the hazard about the laser used in the product. Users must use or install the product so that the physical regions where harm may occur are prevented from access.

Of course, the subject is much more involved than this and requires a careful study of the applicable standards [8] for a correct answer, because of the many parameters to be evaluated. Here, we simply report a brief overview to elucidate the general issues that one is likely to tackle when dealing with laser safety. Worth noting, the original 825 IEC standard is now being corrected and reissued, after about twenty years of application, and several changes are expected (formal rather than substantial, however).

Laser sources and laser-based equipment are categorized in Classes, from 1 to 4, according to their potential risk. Hazards include eye (sight) damage and skin (burning) damage, mainly.

One fundamental distinction for classification is between CW lasers and pulsed lasers. The main classification parameter is the optical accessible power (CW or peak). In addition, parameters influencing classification include: (i) wavelength, beam area, and beam divergence for all lasers, and (ii) pulse duration, pulse optical energy, and duty cycle for pulsed lasers.

As an illustration, we report in Fig.A1-20 the power versus wavelength diagram defining the safety classes for CW laser sources [8,9].

Class 1 is that of intrinsically safe sources. Safety is because the emitted power is below the threshold of harm, or because the laser is equipped with an automatic shutdown of emission, preventing the operator from being exposed to a dangerous level. Class 2 is that of lasers emitting in the visible range, where the blink reflex of the eye permits tolerating a larger CW power (1mW). No special safety measures are required for Class 1 and 2 laser equipment, just a label indicating the aperture from which radiation is emitted and a warning not to stare into the laser beam. Therefore, in all instruments where a power of $\leq$1mW is technically sufficient, we shall absolutely stay with such a low value to clear safety problems. Examples of Class 1 or 2 products are: bar-code laser scanners, alignment and pointing systems, laser interferometers, and sine-wave modulated telemeters.

Class 3A lasers and equipment are those safe to vision not assisted by optical instruments (for example, a microscope or magnifying glass). In the visible, the CW power limit of Class 3A lasers is 5mW, but in IR and UV the tolerable power is less because the blink reflex is absent. In Class 3B, the CW power in the visible goes from 5mW to 500mW, and this type of laser is always dangerous for direct vision. Class 3B lasers require a turn-on key for operation by authorized personnel only and warning sign in the area of operation.

Last, Class 4 lasers are the most dangerous. For a CW source, we get Class 4 when power is just $\geq$0.5 W. Even unintentional reflections (from rings, metallic objects, etc.) may scatter to the eyes a power exceeding the Maximum Permissible Exposure (MPE). Therefore, operation of a Class 4 laser requires a restricted access area, equipped with warning signs and with acoustic and red-light alarms indicating laser switch-on. This situation is typical of a CO_2 power laser for welding and other mechanical works, but is surely a big hindrance for a measurement instrument like a telemeter.

A1.5 Laser Safety Issues 395

Pulsed telemeters easily exceed Class 3B limits, and their operation falls under specifications for open-air and propagation through the atmosphere of laser-based equipment.
In this case, the standards prescribe that the Ocular Risk Zone (ORZ) be shielded by appropriate barriers from the reach of the public, which shall have a safety distance of 3 to 6 m from the reach of the beam.
Another quantity of interest is the DOR (Distance of Ocular Risk). This is defined as the distance at which the power or other related quantity falls below the level, and operation is safe.

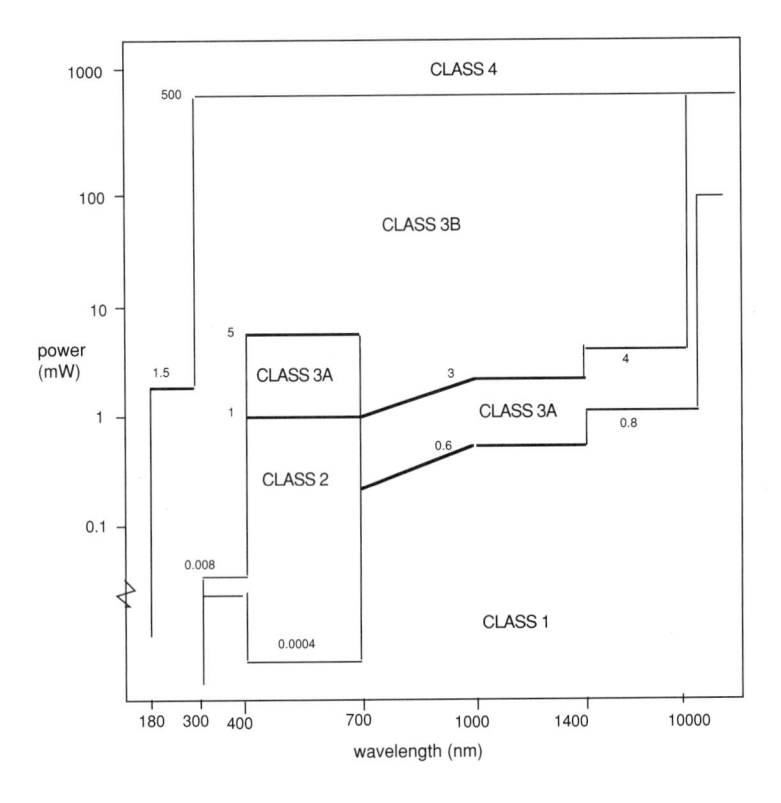

Fig.A1-20 Laser safety classification for CW sources, as a function of wavelength, according to IEC laser safety standard [8].

REFERENCES

[1] A.E. Siegman, "*Lasers*", University Science Books: Mill Valley, CA, 1986.
[2] J. Hecht, "*The Laser Guidebook*", McGraw-Hill: New York, 1986; "*Understanding Lasers, an Entry-Level Guide*", 2nd ed., IEEE Press and Wiley Interscience: New York, 1994;

"*Introduction to Laser Technology*", 3rd ed., IEEE Press and Wiley Interscience: New York 2001.

[3] O. Svelto, "*Principles of Lasers*", Kluwer Academic Press: Dodrecht, Holland, 1996.

[4] S. Donati, "*Laser Interferometry by Induced Modulation of the Cavity Field*", J. Appl. Phys., vol.49 (1978), pp.495-498; see also: Alta Frequenza, vol.47 (1978), pp.172-178.

[5] F. Bertinetto et al., "*Comparison of $^{127}I_2$ stabilized He-Ne Lasers at 633 and 612 nm*", IEEE Trans. Inst. Meas., IM-32, 1983, pp.72-76.

[6] P. Bhattacharya, "*Semiconductor Optoelectronic Devices*", Prentice Hall: Upper Saddle River, NY, 1996.

[7] R.S. Vodhanel et al., "*Long-term λ-drift of 0.01nm/yr for 15 running DFB lasers*", Opt. Fiber Conf. Techn. Digest, San Jose, Feb. 20-25, 1994, pp.103-104.

[8] IEC "*Laser Safety*", Standards 825, Geneva, 1984; Variants A1 and A2 of IEC 60825-1 (2001).

[9] D. Sliney and M. Wolbarsht, "*Safety of Lasers and other Optical Sources*", Plenum Press: New York, 1990.

APPENDIX **A 2**

Basic Optical Interferometers

In addition to the Michelson configuration used in displacement measurements, several other optical interferometers are useful in electro-optical instrumentation. All of them serve the purpose of dividing the beams in a reference and a measurement arm, let the measurand affect the optical path, and recombine the beam to yield the interferometric signal of the type 1+cosΦ at the photodetector.

In this appendix, we will review the properties of basic optical interferometers, namely Michelson, Fabry-Perot, Mach-Zehnder, and Sagnac (Fig.A2-1). All these configurations are basic and can be improved by the addition of extra components or by suitable variants, like exemplified by the Michelson interferometer that becomes the Twyman-Green interferometer when corner cubes are substituted with mirrors (Sect.4.2.1).

A2.1 CONFIGURATIONS AND PERFORMANCES

The Mach-Zehnder interferometer (Fig.A2-1) is used in fiber sensors, and compared to the Michelson, is a sort of unfolded optical interferometer with a second beamsplitter to recombine the propagated beams. It has two separate arms and makes two complementary outputs available. The Fabry-Perot interferometer has no reference path, and interference is between the back-and-forth contributions reflected by the two mirrors. The multipath propagation adds an increased sensitivity compared to the other configurations.
Last, the Sagnac has two counter-propagating beams (clockwise and counterclockwise) in a single propagation path, and is sensitive to non-reciprocal phase shifts only, as it is required in the gyroscope and in magnetic field sensors.

398 Basic Optical Interferometers Appendix A2

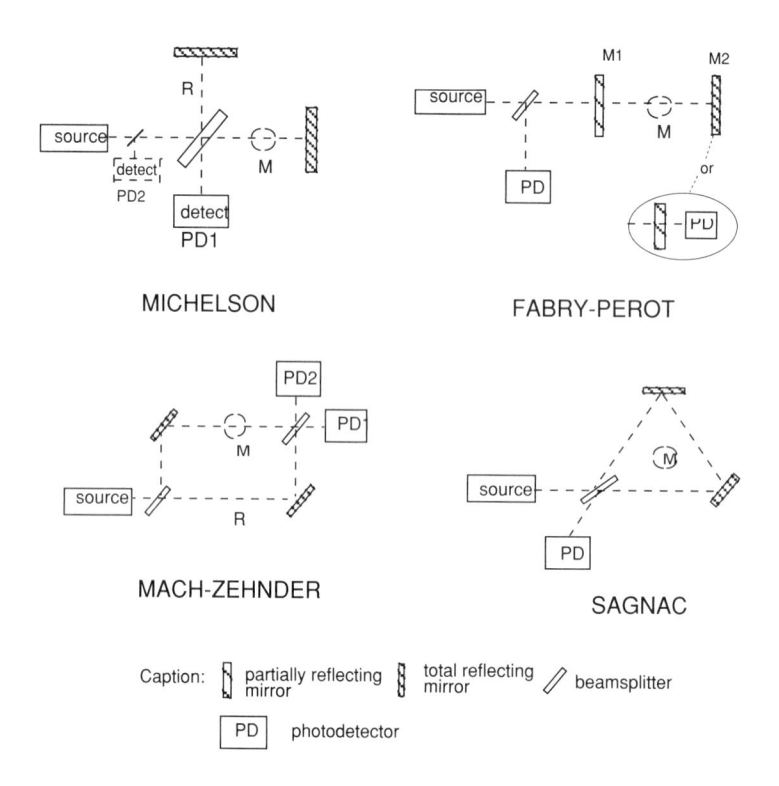

Fig.A2-1 Common optical interferometers and their bulk optics configuration. M and R are the measurand and reference arms, respectively.

Now, let's consider the signal obtained at the output of the interferometer. In general, the fields returned on the photodetector from the measurement and reference arms can be written in the form of rotating vectors, or $E_m = E_m \exp i(\omega t + \phi_m)$ and $E_r = E_r \exp i(\omega t + \phi_r)$.
The photodetected current is the average of the square modulus of the total field [1], or $I_{ph} = \sigma \langle |(E_m + E_r)|^2 \rangle$, where σ is a conversion factor and the brackets $\langle .. \rangle$ denote averaging.

Inserting the fields E_m and E_r in this expression and developing the modulus, we obtain $I_{ph} = \sigma \langle E_m^2 + E_r^2 + 2 E_m E_r \text{Re}\{\exp i(\phi_m - \phi_r)\}\rangle = I_m + I_r + 2(I_m I_r)^{1/2} \langle \cos(\phi_m - \phi_r)\rangle$, where I_m and I_r are the currents produced by the measurement and reference fields, respectively. For $I_m \neq I_r$, we can write the detected current as $I_{ph} = (I_m + I_r)\{1 + [2C^{1/2}/(1+C)]\langle \cos(\phi_m - \phi_r)\rangle\}$, where $C = I_m/I_r$ is the signal contrast.

Further, by developing the phase term $(\phi_m - \phi_r)$ in an average part $\langle \phi_m - \phi_r \rangle = \phi_m - \phi_r$ and a random deviation $\Delta\phi$, as $\phi_m - \phi_r = \phi_m - \phi_r + \Delta\phi$, the averaged cosine becomes $\langle \cos(\phi_m - \phi_r)\rangle = \mu \cos(\phi_m - \phi_r)$ [1]. The term $\mu = \langle \cos \Delta\phi \rangle$ has the meaning of the coherence factor of the measurement and reference fields, whereas $\phi_m - \phi_r$ can be written as $R \, k(s_m - s_r)$ in terms of the measurement and reference arm lengths s_m and s_r.

A2.1 Configurations and Performances

In summary, we may write I_{ph} in the form:

$$I_{ph} = I_0 \{1 + V \cos[R\, k(s_m - s_r)]\} \quad (A2.1)$$

where $V = 2\mu C^{1/2}/(1+C)$ is called the fringe visibility, or contrast factor [5], and R is the responsivity (or response factor) of the interferometer.

Ideally, we should have $V=1$, but if the coherence factor μ is not unity, polarization is not matched, the beam powers are unequal ($C<1$) or some stray light is collected, $V<1$. The factor V is also called the contrast factor because it is coincident with the difference divided by the sum of the maximum and minimum signal amplitudes:

$$V = (I_{phMAX} - I_{phMIN})/(I_{phMAX} + I_{phMIN})$$

The responsivity R is a measure of the interferometer's ability to supply a relative variation of the photocurrent I_{ph} to a variation of the optical path length in the measurement arm. In terms of the photocurrent signal, the responsivity is the maximum value of the relative derivative of I_{ph} with respect to ks when $V=1$:

$$R = \max_s \{(I_{ph}k)^{-1}\, dI_{ph}/ds\}$$

The reason for this definition is that, given the transfer ratio $\Phi = dks = HdM$ for the measurand M, the relative signal variation obtained in the photocurrent is readily found as:

$$dI/I = VR\, H\, dM \quad (A2.2)$$

In the Michelson interferometer, from Eq.4.3 of the text, the responsivity is seen to be:

$$R = 2$$

a value that is the simple consequence of the go-and-return path.

By inspection of the scheme in Fig.A2-1, it is easy to see that $R=1$ for a Mach-Zehnder interferometer. In the Sagnac, we usually find a non-reciprocal dependence on +HM and –HM for the two counter-propagating waves, and thus we take $R=2$ (see Table A2-1).

In the Michelson, we have two available outputs from PD1 and PD2 (Fig.A2-1), and they are complementary to unity. For $V=1$ they are written as:

$$I_{ph1} = I_{ph0}[1 + \cos 2k(s_m - s_r)], \quad I_{ph2} = I_{ph0}[1 - \cos 2k(s_m - s_r)] \quad (A2.3)$$

The two outputs are present also in the Michelson and Sagnac interferometers. Signal I_{ph1} is the output from PD in the scheme of Fig.A2-1, while I_{ph2} (not shown in Fig.A2-1) could be recovered from the beamsplitter output directed toward the source, if we insert an additional beamsplitter to pick it up.

TABLE A2-1 Comparison of basic optical interferometers

	Michelson	Fabry-Perot	Mach-Zehnder	Sagnac
Responsivity R	2	F	1	2
Balanced/Unbalanced configuration	B or U	U	B or U	B
No. of channels of basic setup	1	1	2	1
Reference available	yes	no	yes	no
Minimum number of beamsplitters (or partially reflecting mirrors)	1	1	2	1
Source and detector from the same side	yes	yes/no	no	yes
Retro-reflection to the source	yes	yes	no	yes
Metallization required (for fiber versions)	2	2	0	0
Cascadeability	no	yes	yes	no

Notes: Michelson and Mach-Zehnder may be operated as either unbalanced or balanced (the latter being of course the preferred choice). The Michelson and the Fabry-Perot can have source and detector on the same side if the output is taken by a beamsplitter added on the input path (Fig.A2-1). Responsivity is $R=2$ for the Sagnac, assuming equal and opposite effects on the two counter-propagating waves. Retro-reflection is for the basic configuration with mirrors; using corner cubes, it is avoided in the Michelson.

Lastly, in the Fabry-Perot, the photocurrent detected at either the front or rear mirror has a dependence of the type:

$$I_{ph}/I_{ph0} \propto 1/(1+F^2 \sin^2 ks) \qquad (A2.4)$$

where $F=2R^{1/2}/(1-R)$ is the finesse of the Fabry-Perot resonator and R is the mirror (power) reflectivity [2-4]. Looking for the maximum of the relative slope $d(I_{ph}/I_{ph0})/dks$ in Eq.A2.4, one can find that the responsivity of the Fabry-Perot is just equal to F.

A2.1 Configurations and Performances

Derivation of Eq.(A2.4).

Let R_1 and R_2 be the (power) reflectivity of M1 and M2, and $A=\exp-\alpha 2s$ the (power) attenuation suffered in the cavity of length 2s.

Just beyond the first mirror M1, the cavity field E_{cav} is the sum of two terms: one coming from the input field E_0 and transmitted with $\sqrt{(1-R_1)}$ through M1, and one coming after the round-trip of the cavity, with reflectance $(R_1R_2)^{1/2}$ attenuation A and delay 2ks. Thus, we can write $E_{cav} = E_0\sqrt{(1-R_1)} + E_{cav} A(R_1R_2)^{1/2} \exp i2ks$.

Solving for E_{cav}, we get:

$$E_{cav} \approx E_0\sqrt{(1-R_1)} / [1- A(R_1R_2)^{1/2}\exp i2ks] \quad (A2.5)$$

Both the outputs from M1 and M2 are proportional to E_{cav} and we can write them as:

$$E_{M1} = E_{cav} A [\exp i2ks] \sqrt{R_2} \sqrt{(1-R_1)}, \text{ and } E_{M2} = E_{cav} \sqrt{A} [\exp iks] \sqrt{(1-R_2)}.$$

Therefore, the outputs have the same selectivity, versus ks or λ, as the cavity field E_{cav}, and we can limit ourselves to study just E_{cav}.

By taking the square modulus of the field, we obtain the power. Letting $R^2 = A^2 R_1 R_2$, the power in the cavity is:

$$P_{cav} = P_0(1-R_1)/ |1- R\exp i2ks|^2 = P_0(1-R_1)/[1+R^2 -2R\cos 2ks]$$
$$= P_0(1-R_1)/[(1-R)^2 + 4R\sin^2 ks] \quad (A2.6)$$

where P_0 is the incident power, and the last passage is obtained by noting that $\cos 2ks = 1-2\sin^2 ks$. The output powers from the mirrors M1 and M2 are found as:

$$P_{M1} = 4 R_1 R_2 \sin^2 ks / [(1-R)^2 + 4R\sin^2 ks], \text{ and}$$
$$P_{M2} = (1-R_1)(1-R_2)A / [(1-R)^2 + 4R\sin^2 ks]$$

The diagram of P_{cav}/P_0 is shown in Fig.A2-2. It has a periodicity 2π in the argument 2ks, or it exhibits periodic resonance at $2(2\pi/\lambda)s = N2\pi$ or at $s = N\lambda/2$ (that is, at multiples of half-wavelength).

In frequency, the periodicity of resonance $\Delta\nu = \nu_{N+1} - \nu_N$ is obtained by letting $2(2\pi\nu/c)s = N2\pi$, and it reads: $\Delta\nu = c/2s$. For $2ks = N2\pi$, the resonance has a peak value $P_{cav}/P_0 = (1-R_1)/(1-R)^2$, even larger than one because of the buildup of power in the resonator cavity, while off-resonance the response has a minimum at $2ks = \pi$, where it is $P_{out}/P_0 = (1-R_1)/(1+R)^2$.

The half-width of the resonance curve is calculated from the condition: $P_{cav}/P_0 = 1/2[P_{cav}/P_0]_{peak}$, which from Eq.A2.6, becomes $(1-R)^2 = 4R\sin^2 ks$ and is solved as $\sin ks \approx ks = (1-R)/2\sqrt{R}$. In frequency, the half-height line width is: $\Delta\nu_{HW} = (c/2s)(1-R)P/\pi\sqrt{R}$.

Last, an important parameter of the Fabry-Perot is the finesse [2], defined as the ratio of half-height line width to period, or $F = (2/\pi)\Delta\nu/\Delta\nu_{HW}$.
From the above results, the finesse is given by $F = 2 R^{1/2}/(1-R)$.

Thus, we may substitute F for the factor at the denominator of Eq.A2.6 to obtain the more elegant expression Eq.A2-4.

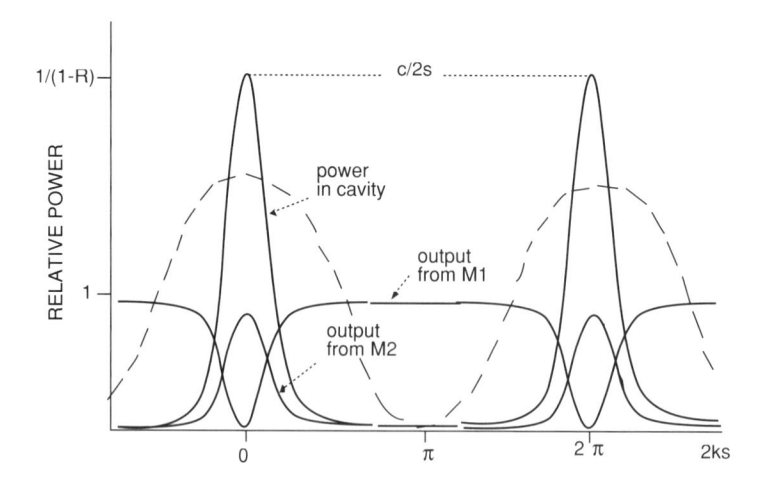

Fig.A2-2 The response of the Fabry-Perot is periodic by 2π in phase shift 2ks, or by c/2s in frequency. The selectivity is F (=finesse) times better than in other interferometers with a cosine-type response (dotted line)

When using interferometers, further details relevant to performance are the following [4,5] (see also Table A2-1):
- the availability of a reference path. The reference is very important as it allows compensation of the measurement against external disturbances other than the measurand (e.g., temperature, vibrations, EMI). For compensation, the reference path is laid close to the measurement path but shielded from the measurand so as to collect the same disturbance in both arms. Because of the subtraction in the interferometric process, disturbances are canceled out;
- the number of channels available. Basically, the Mach-Zehnder has two outputs readily available for detection, while the other interferometers have one. By an additional beamsplitter in the path going back to the source, a second channel is made available;
- the position of the detector with respect to the source. It is easier to have both on the same side to accommodate them in a single box;
- the retro-reflection from the optical interferometer into the source. Retro-reflections may severely impair the quality of the emitted field, adding amplitude fluctuations as well as widening the spectral line of the laser, especially in frequency stabilized sources;
- the balanced versus unbalanced paths. If reference and measurement paths are nearly the same, requirements on the coherence length of the source are alleviated;
- the number of beamsplitters and mirrors: this clearly affects the component count and cost of the final optical assembly;
- cascadeability. This feature has to deal with multiple-sensor operation (as in fiberoptic sensors) or when using the optical interferometer as a filter (frequency channel separation in fiberoptic communications).

A2.2 Choice of optical components

A last point to mention is alignment of the mirrors. All optical interferometers considered above are intended for operation with a laser source and basically work with a single-mode spatial distribution of the beams. Thus, their mirrors require an angle alignment accuracy within the diffraction limit $\theta=\lambda/\pi w_0$ associated with the Gaussian beam spot size w_0, otherwise an incomplete superposition (Sect.4.5.3) and a reduction of fringe visibility will occur. Of course, if we can use a corner cube in place of the mirror, like in the Michelson interferometer, the angular alignment problem is solved (provided that the corner cube dihedral error is $<\theta$).

Stray light suppression may also be of concern in the interferometer operation. When the laser line width is very narrow, an interference filter in front of the detector will usually be effective to suppress light from the environment.

Alternatively, we may use a spatial filter (see Fig.A2-3), consisting of a lens-and-pinhole combination placed in front of the detector. If the pinhole has a radius r and is placed in the focal plane of the lens with focal length F, only the rays contained in an angle $\theta=r/F$ can pass through the filter. Because the minimum θ is the diffraction value, this filtering is very effective.

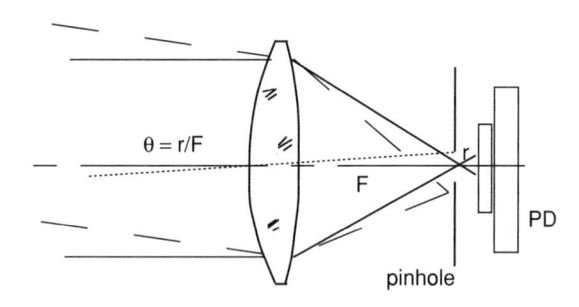

Fig.A2-3 A pinhole objective combination can be used for spatial filtering of the field reaching the photodetector. The acceptance angle is r/F.

A2.2 CHOICE OF OPTICAL COMPONENTS

When we move from conceptual schemes with ideal components as presented in the text, to practical implementation with real components, a number of non-idealities need to be taken into accounts, which if overlooked, may impair or severely degrade the performance we expect in principle.

To illustrate the problems that may be encountered, we will briefly discuss some issues concerning the choice of optical components with reference to a Michelson-based laser interferometer where the beamsplitter and the corner cube are the most critical components.

Beamsplitters

A plain glass-flat model of the popular BK7 glass can be used, provided that it has a reasonable flatness and few surface defects. But, unless we need just the R=4-8% reflection obtained at a low-incidence angle (<10°), we shall treat one surface by a single- or multiple-layer reflection coating to get the desired R. Also, as both front and rear surfaces contribute to reflection, we inadvertently have a sort of parasitic Fabry-Perot interferometer inserted in the propagation path. To avoid this effect, in a glass-flat beamsplitter we shall have one surface coated for the intended reflection, and the other be anti-reflection coated (ARC). Alternatively, to get rid of the reflection from the second surface, we may consider using wedge-flat and pellicle beamsplitter versions. As for any optical surface crossed by the wavefront, surface quality of the beamsplitter should be specified at least as $\lambda/4$ distortion and 60-40 scratch and dig, as a rule of thumb.

Polarizing beamsplitters

We may either choose from multireflection or birefringent versions. A multireflection cube is based on the reflection property at Brewster's angle incidence, by which $R_{//}=0$ for the polarization parallel to the incidence plane, while $R_\perp \approx 15\%$ (for n=1.5) for the polarization perpendicular to the incidence plane. Adding several layers of alternatively high and low n, we can obtain a reflection $R_\perp \approx 99\%$, while $R_{//}$ stays low (typ. $R_{//}<1\%$). The layers are deposited on the diagonal surface of a square prism, and the input/output surfaces will be treated ARC.
A better extinction ratio $R_\perp/R_{//}$ (up to 10^5-10^7) is obtained by birefringent cube polarizing beamsplitters that use calcite as the material of the prisms. The two halves of the cube are cut along properly chosen directions so as to obtain an index of reflection difference at the diagonal surface boundary, adequate for total reflection of one polarization and transmission for the other. Though one prism would suffice for polarization splitting, a pair of them is normally used to compensate for beam deviation, so that one beam is delivered parallel to the input and the other is about perpendicular. Several cube versions are available under the name of Glan, Glan-Thomson, Wollaston, Nicol, etc.

We can use a polarizer beamsplitter also as a polarizer, but if we only need to select a (linear) polarization state while blocking the other, we may prefer cheaper dichroic-sheet polarizers. Discovered by Kodak's researchers and widespread in applications [5], these sheets offer a reasonable extinction ratio (typically 10^2-10^3) in the visible and near infrared, which is often adequate for the use in front of the photodetector.

Corner cubes and retro-reflectors

These components can be understood by starting from the in-plane version of the 90° prism (Fig.A2-4). If the vertex angle is β, the angular deviation between the incident and reflected rays is easily found to be 2β, or 180° for a 90° prism. Incidentally, this general rule is exploited in the pentaprism, which has a vertex angle of 45° and deflects the beam of exactly 90° (Fig.A2-4). The angular deviation is irrespective of the incidence angle α, provided α is within an acceptance angle α_{max} determined by the total reflection angle internal to the cube. However, working with collimated beams, like in interferometry, even small α_{max} values (e.g., 5 to 15°) are adequate.

A2.2 Choice of optical components 405

To extend the concept in three dimensions, we need two dihedrals, or the corner cube structure depicted in Fig.A2-4, bottom, in which three reflections are used to deflect the incident beam in two orthogonal planes.

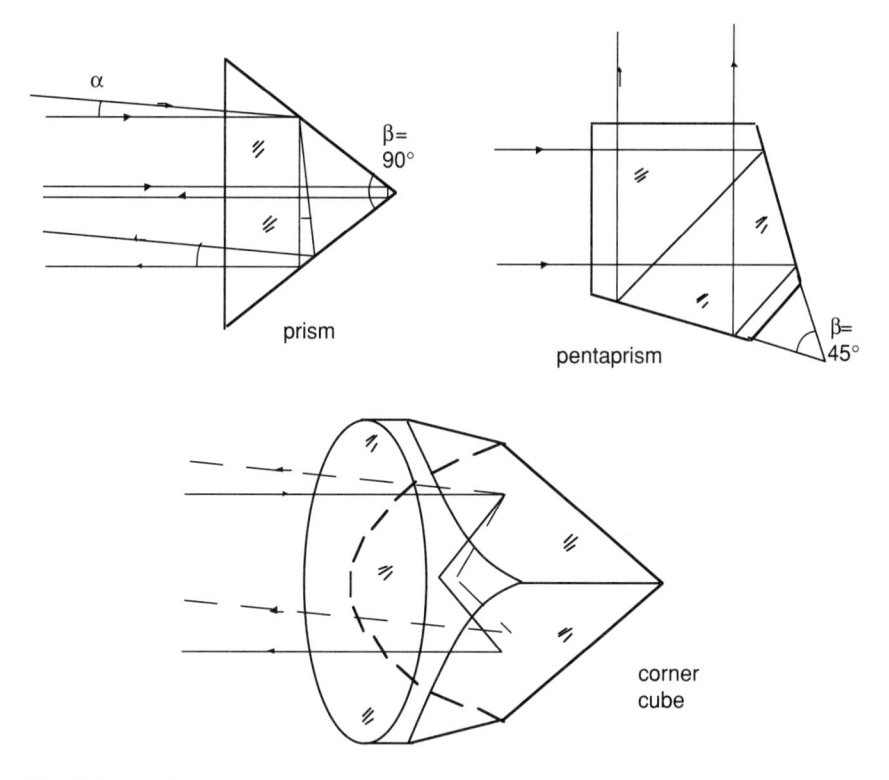

Fig.A2-4 Retro-reflection in a 90° prism and in a pentaprism with a 45° vertex angle (top), and in a corner cube retro-reflector (bottom)

Basically, we have solid-glass and hollow versions of corner cubes.
A solid-glass corner cube (as shown in Fig.A2-4) has an ARC front surface and may reach a reflection efficiency of 95% or better. The points of concern are (i) the fragility of the sharp edges and vertex, easily damaged even during handling of the device; and (ii) the polarization state change that radiation undergoes because of the phase shift at total internal reflection. Because of this effect, for an input linear state of polarization, the output is a mixture of elliptical states, dependent on the cube sector, as shown in Fig.A2-5. This can be disturbing if we rely on the state of polarization for the beam handling in our instrument. However, in a Michelson laser interferometer with two retro-reflectors, if the two corner cubes have the same edge orientation, the beams superpose on the photodetector with the same state of polarization in different sectors, and fringe visibility is unity.

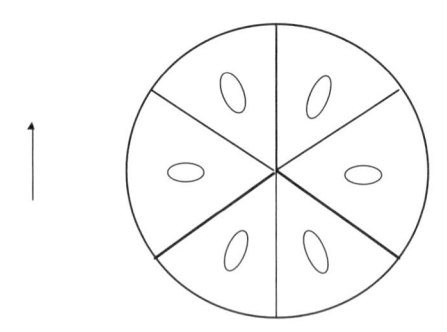

Fig. A2-5 Output polarization in a corner cube for an input linear polarization state

To make the corner cube less sensitive to polarization and protect the total reflection surfaces, these may be metallized (with a thin Au or Al film) and covered by a protection coating. Metallization slightly decreases reflection efficiency (to 90% typically), but increases the field-of-view or acceptance of the device [6].

Alternatively, we may use a hollow-type corner cube, made by three mirrors mounted in a trihedral arrangement at 90° angles. The hollow corner cube is cheaper and has the same performance as the metallized solid-glass cube, but the stability of dihedral angles (to be kept typically within 1 arcsecond for the useful device's life) may be questionable because of the demanding mechanical stability.

REFERENCES

[1] S. Donati, *"Photodetectors"*, Prentice Hall: Upper Saddle River, NY, 2000, Chapter 8.
[2] B. Rossi, *"Optics"*, Addison Wesley Publ. Co.: Reading, MT, 1957.
[3] W. Lauterborn, T. Kurz, and M. Wiesenfeld *"Coherent Optics"*, Springer Verlag: Berlin, 1995.
[4] B.E. Saleh and M.C. Teich, *"Fundamentals of Photonics"*, J. Wiley and Sons: Chichester, 1991.
[5] M. Francon, *"Optical Interferometry"*, Academic Press: New York, NY, 1966.
[6] B.H. Billings (editor), *"Selected Papers on Polarization"*, SPIE Milestone Series, vol. MS 23, SPIE: Bellingham, WA, 1991.

APPENDIX **A 3**

Propagation through the Atmosphere

When a laser beam is propagated through a substantial path length in the atmosphere, we experience a number of phenomena that affect the optical power carried by the beam and distort the wavefront. Usually, atmospheric effects are classified in two categories: (i) those associated with the non-ideal transparence of the atmosphere, notably absorption and scattering, called *turbidity effects*, and (ii) those coming from non-homogeneity of the index of refraction of the atmosphere, caused by thermal exchange and air mass motion, called *turbulence effects*.

A vast amount of literature is available on the subject (see, for example, Refs.[1-5]), and here we will report a few results to supply basic information useful for the evaluating performance of instruments treated in the text.

A3.1 Turbidity

In general, a beam propagated through a medium on a path length L (Fig.A3-1) suffers a power loss, with respect to the initial value P_0 described by the Lambert-Beer law of attenuation:

$$P = P_0 \exp{-(\alpha+s)L} \qquad (A3.1)$$

The quantities α and s are called the *absorption* and *scattering* coefficients of the medium, and their sum, a= α+s, is the (total) *attenuation* coefficient. Of course, a, α, and s are measured in m^{-1} or cm^{-1}. Their inverse values represent the propagation length after which the initial power is decreased to 1/e=37% and are called the absorption, scattering, and attenuation lengths, respectively.

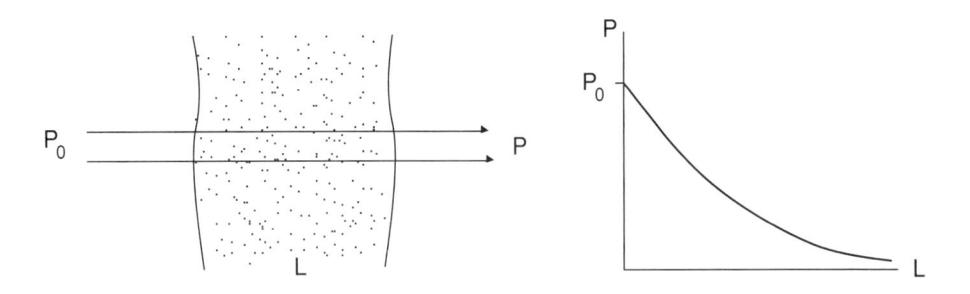

Figure A3-1 The attenuation of power propagated through a path length L follows the negative-exponential Lambert-Beer law

The quantity T= exp -(α+s)L is referred to as the *transmittance*.

In the atmosphere, absorption is due to constituent species (nitrogen, oxygen, carbon dioxide, etc.) and water vapor, while scattering is dominated by small-size (<1µm) particulate and eventually water droplets in the case of haze or fog.

The absorption contribution is fairly constant, whereas the scattering contribution strongly depends on weather conditions.

In Fig.A3-2 we report what is considered the most representative diagram [6] of atmospheric transmission T versus wavelength, measured at ground level on a normal, clear day on

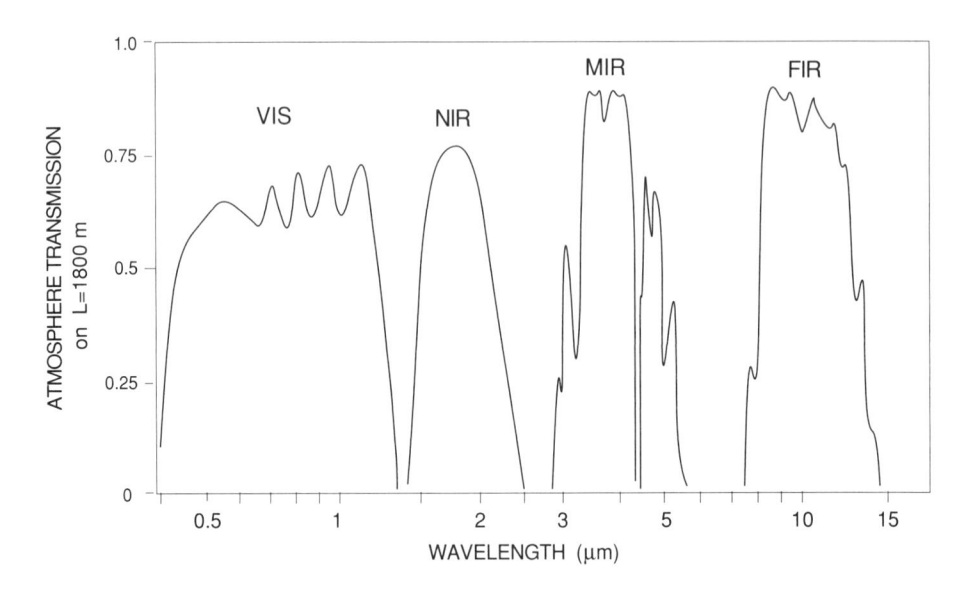

Figure A3-2 Standard atmospheric transmission (%) for a propagation on an L=1800 m length path at ground level

A3.1 Turbidity

a path length of L=1800 m.

As we can see from Fig.A3-2, the atmosphere has a few *transmission windows* of reasonable transparency, useful for operating laser instrumentation on medium distances (say, 100 to 1000 m).

Another diagram useful for evaluating atmospheric transmission on larger distances (tens of km) is shown in Fig.A3-3. The diagram is the plot of solar spectral irradiance received at ground level on a clear day at different elevation angles θ of the sun on the horizon. The quantity AM=(cosθ)$^{-1}$ is the air mass number, and represents the equivalent number of atmospheres crossed by the rays at elevation θ. The curve with AM0 in Fig.A3-3 is the irradiance outside the atmosphere. The ratios of the values read on the curves at different AM# and at AM0 are the transmissions of the atmosphere for different elevations or air masses.

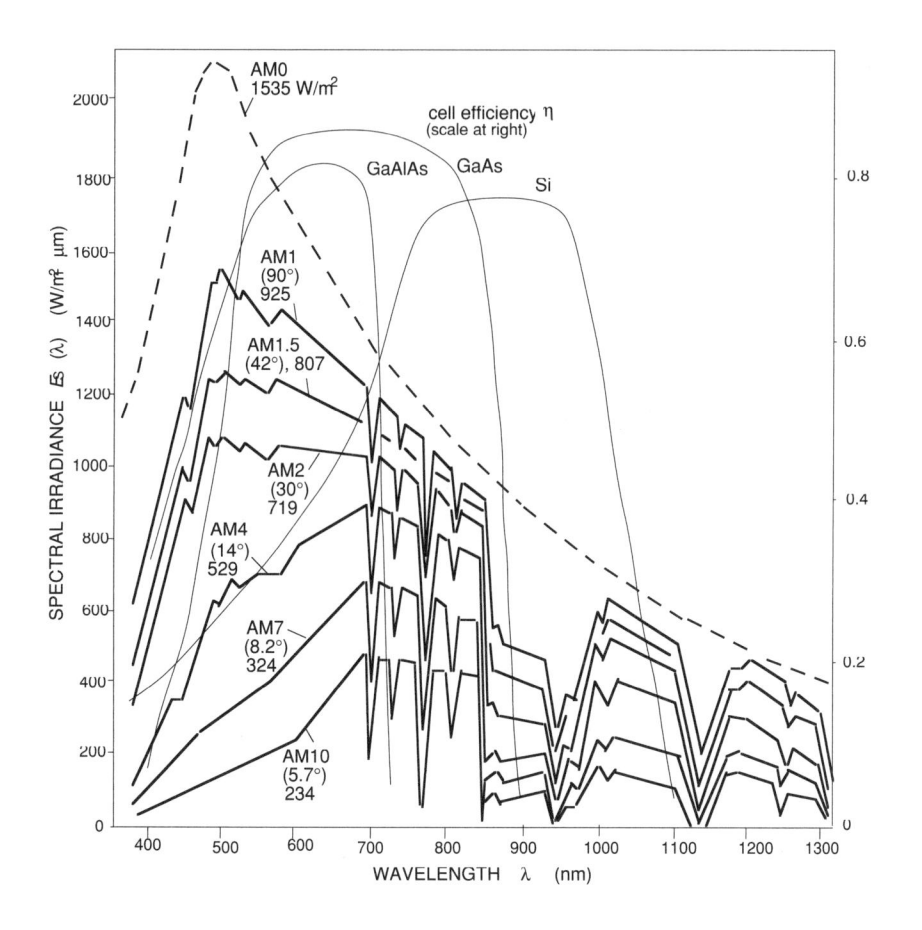

Figure A3-3 Spectral solar irradiance (W/m^{2}μm) received at ground level for several elevation angles or air masses AM

In addition to the VIS (visible) window, we find the three well-known IR (infrared) windows of the:
- near-infrared (or NIR, λ=1.5-2.3 µm),
- middle-infrared (or MIR, λ=3-5 µm), and
- far-infrared (or FIR, λ=8-14 µm).

Beyond the FIR window, no other window is found up to millimeter waves. For a very dry atmosphere (RH<5%), there is actually a modest window at λ=330±10µm with a peak transmission T=5% (L=300 m), but T rapidly decreases with increasing relative humidity RH, going down to T<0.1% at RH=70%.

In the UV, the atmospheric transmission gradually decreases at shorter wavelengths until we reach VUV (vacuum-ultra-violet), requiring vacuum for propagation because of an intolerable loss, even for path that is a few centimeters in length.

In the visible range, the attenuation coefficient of a standard atmosphere has been computed from constituents as the sum $a_{st} = \alpha_{oz} + s_R + s_a$ of the following: ozone contribution α_{oz} (important for λ<290 nm), Rayleigh scattering s_R (important for λ<400 nm), and aerosol scattering s_a, giving a value a=0.18 km^{-1} at λ=500 nm.

The term s_a is dominant for λ>400 nm and gives the trend $a/a_o = \sqrt{(\lambda/\lambda_o)}$ for total attenuation, a good approximation for λ= 0.5-5µm except for the narrow molecular absorption lines α_{mol} that must be added to the background value.

Data shown in Fig.A3-2 represent the transmission of the standard atmosphere and include the s_R and s_a terms. In addition to that, a much larger term, s_{par}, comes from scatterers extraneous to the clear atmosphere, such as particulate matter and small droplets.

Scatterers are small-size particles that can be schematized as spheres randomly distributed in space, with a certain radius r_p and an index of refraction n_p. The radius is found to range from sub-µm size to over hundreds of µm, and the index of refraction is always well in excess of n_{air}=1 (e.g., n_p =1.33 for water droplets), the index of the surrounding medium.

Small size and large index of refraction difference is the peculiarity of *turbidity*. Conversely, *turbulence* is the regime where scatterers are large in size (mm to m) and their index of refraction difference is very minute (10^{-6} to 10^{-3}) and may also fluctuate spatially and temporally.

Considering a small particle with a substantial Δn, the outgoing ray will largely deviate in angle on refraction inside and outside the particle (Fig.A3-4), whereas the deviation (or beam wandering) is very little if Δn is small. Also, as a large angular deviation brings the outgoing ray out of the beam, a scattered ray is equivalent to a lost ray, and thus scattering does contribute to attenuation. By contrast, the small deviation introduced by turbulence can be recovered by increasing the acceptance area of the detector and does not contribute to attenuation, at least in principle.

But, with turbulence, we will additionally have a random spatial mixing of contributions inside the beam, an effect called scintillation, which destroys spatial coherence.

The quantity used to describe the angular scattering process is the *scattering function*, $f(\theta)d\Omega = dP(\theta)/4\pi P$. It is defined as the relative power per unit solid angle Ω that we collect at the angle θ with respect to the θ=0 incidence (Fig.A3-4), 4π being a normalization factor.

A3.1 Turbidity

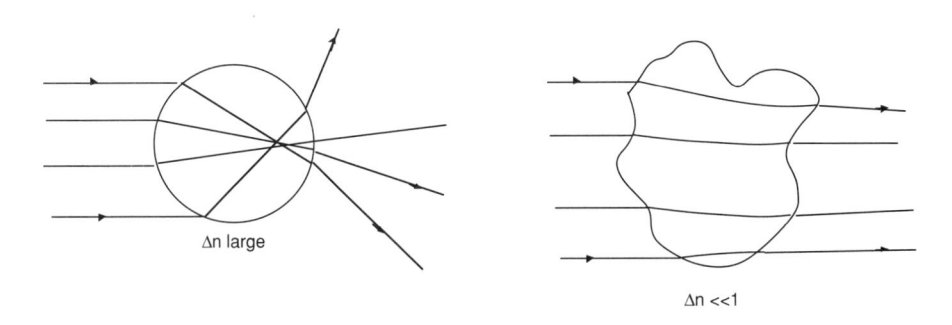

Figure A3-4 Ray tracing through scatterers with a large Δn (turbidity, left) and a small Δn (turbulence, right) reveals the difference in angle deviation

Regarding the shape of the scattering function f(θ), the critical parameter is the ratio r/λ, and we can distinguish between two limit cases:

(i) for small particles, i.e., $r \ll \lambda$, we are in the so-called Rayleigh scattering regime [3] and the function f(θ) is nearly isotropic. This means that we collect nearly as much power in the forward (θ≈0) as in the backward direction (θ≈π), the minimum being found at the right angle (θ≈π/2).

(ii) for very large particles, or $r \gg \lambda$, we are in the Mie regime [3] and the scattering function is strongly peaked in the forward direction, and an appreciable contribution in the backward direction is left (Fig.A3-5).

Another important parameter to describe the scattering process is the cross-section A_{sc} of the scatterer, intended as the effective area subtended to incoming rays.

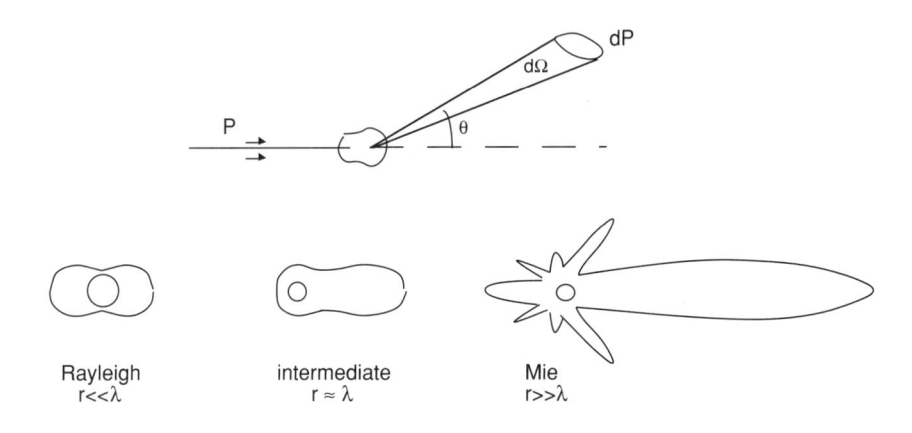

Figure A3-5 Top: definition of the scattering function f(θ). Bottom: the scattering function f(θ) in the Rayleigh and Mie regimes.

Like the scattering function f(θ), this quantity is calculated by an electromagnetic field analysis [3] of the scattering process. From the analysis, it is found that:

$$A_{sc} = Q_{ext} \pi r^2 \qquad (A3.2)$$

The factor Q_{ext} is called the extinction efficiency and is plotted in Fig.A3-6 versus the normalized radius r/λ. In the Rayleigh regime for small particles, the extinction efficiency Q_{ext} varies as $(r/\lambda)^{-4}$. The particles are seen from the electromagnetic radiation as if they were much smaller than their actual physical size. In the Mie regime for large particles, Q_{ext} =const ≈2 and particles are seen as twice as large as their actual radius. This apparently odd result is correct, because the diffraction contribution is found to be exactly equal to the physical obstruction produced by the particle.

Now we can sum up the effects of scattering. If we have a medium containing a concentration of c (cm^{-3}) particles per unit volume, each with an effective area A_{sc}, then the power remaining in the beam after propagation on a path length L will be:

$$P = P_0 \exp{-c\, A_{sc}\, L} \qquad (A3.3)$$

In addition, the power scattered in the far field at an angle θ..θ+Δθ from the incidence is $P_{sc} = P_0 [1 - \exp{-cA_{sc}L}]\, f(\theta)\, 2\pi\theta\Delta\theta$.

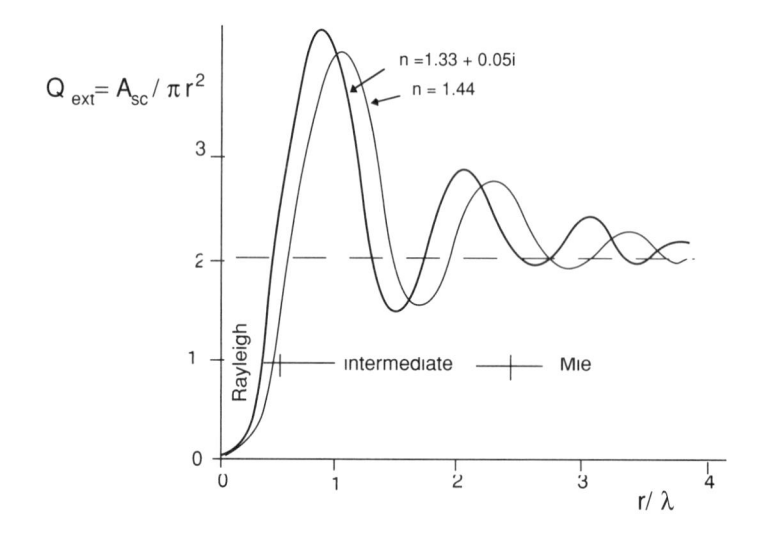

Figure A3-6 The extinction efficiency versus the normalized radius r/λ. In the Mie region (r>> λ), Q_{ext} tends asymptotically to a constant (≈2), whereas in the Rayleigh region (r<< λ), Q_{ext} and the attenuation coefficient vary as $(r/\lambda)^{-4}$. There is also a minor dependence on Q_{ext} by the index of refraction of the droplet.

A3.1 Turbidity

A quantity of importance, related to the attenuation coefficient, is the *visibility range*. Conventionally, the visibility range for vision through the atmosphere and other scattering media is taken as the value L_{vis} at which scene contrast is reduced to 2% of the initial value. This definition is related to the visual acuity of the eye, which can resolve a ≈2% contrast in good viewing condition [7]. As the contrast reduction is the same as the useful non-attenuated power, using Eq.A3.1 we find the visibility range as:

$$L_{vis} = (\ln P/P_0)/a = 3.92/a \qquad (A3.4)$$

Typical values of attenuation a, in km^{-1} are the following:
  0.18 for a clear atmosphere at sea level, (L_{vis} =22 km),
  0.5-1 for a light haze (L_{vis} =7.8-3.9 km),
  2-5 for thick haze (L_{vis} =1.9-0.78 km),
  50-200 for a medium fog (L_{vis} =78-20 m),
  300 for a thick fog (L_{vis} =13 m),
  0.02-0.1 for an exceptionally clear atmosphere, with Rayleigh diffusion limit, seldom reached, of s_R=0.012 (L_{vis} =326 km).

Looking at the dependence of Q_{ext} from r/λ in Fig.A3-6, we can see that visibility in fog and haze can be improved by moving the wavelength of operation to the infrared.

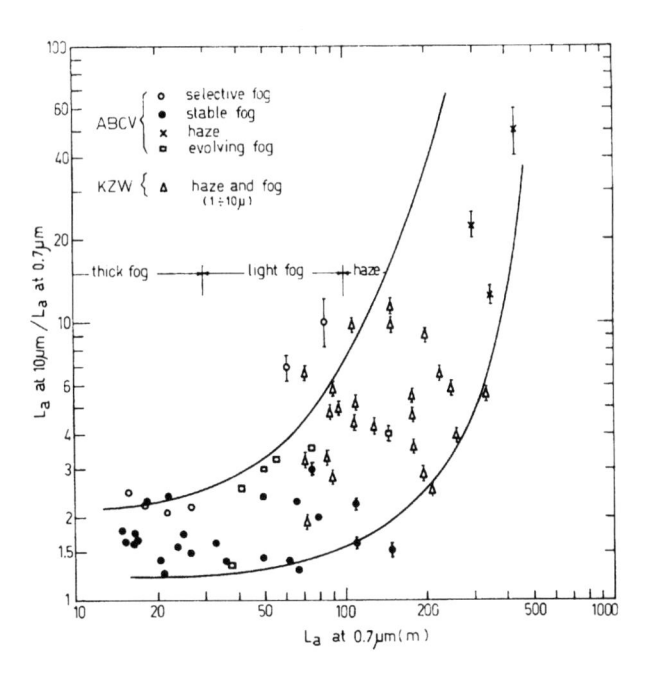

Figure A3-7 The ratio $L_{a10}/L_{a0.7}$ of attenuation lengths at 10 μm (FIR) and at 0.7 μm (visible) versus the attenuation length in the visible $L_{a0.7}$ (from Ref. [8])

With a smaller r/λ, we may go from the Mie regime with $Q_{ext} \approx 2$ to the Rayleigh regime with Q_{ext} and attenuation proportional to $(r/\lambda)^{-4}$. In Fig.A3-7, we report experimental results on the FIR-to-visible ratio of attenuation length $L_{a10}/L_{a0.7}$, as measured by several researchers [8].

Other things being equal, this ratio is the expected visibility improvement. The ratio depends on $L_{a0.7}$, because the average radius of a particle changes with the type of turbidity, being small (r≈0.5μm) in hazes and fairly large (r≈0.5-10μm) in fogs.

In common conditions, the ratio $L_{a10}/L_{a0.7}$ may range from ≈1.5 for $L_{a0.7}$=20m (L_{vis} = 78m) to ≈5 for $L_{a0.7}$=150m (L_{vis} = 600m), up to perhaps ≈10 for $L_{a0.7}$=100m (L_{vis} = 390m). These values are very attractive in several applications, including navigation aids in automotive, avionics, and military systems.

Unfortunately, the improvement is the smallest at the shortest visibility range, where it would be the most desirable: in very thick fog, it may be down to about unity. In addition, the actual visibility improvement is appreciably less than is expected from the $L_{a10}/L_{a0.7}$ ratio. The reason is that the scene appearance changes appreciably from the visible to the FIR regions [16].

In the visible, the scene contrast is dominated by the surface diffusion coefficient, whereas in the FIR, it is determined by temperature differences [7]. Thus, an infrared scene has details correlated to the visible image, but with significant deviations introducing a task of identification to the observer, whereas detection and recognition [7,16] are easier.

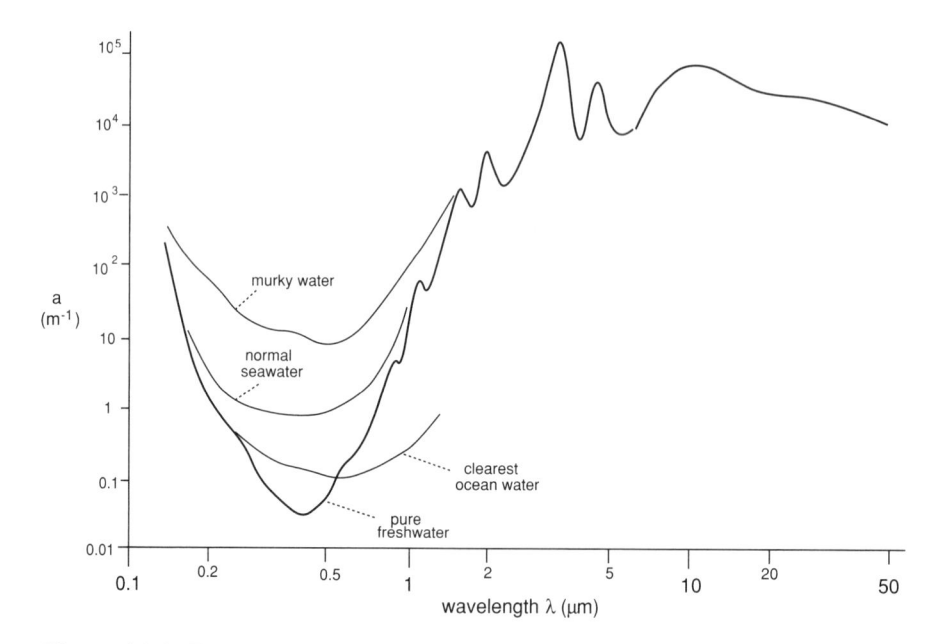

Figure A3-8 Typical dependence of attenuation vs. wavelength in water

A3.2 Turbulence

In detection, improvements of visibility up to a factor 50 have been actually observed [9] on 10-km ranges with FIR imaging. For identification tasks, however, we should take account a difficulty factor [7]. This can be roughly estimated to decrease by a factor 2-3 the actual improvement with respect to the attenuation ratio $L_{a10}/L_{a0.7}$ [9].

Another natural medium of importance for electro-optical instrumentation is water. As shown in Fig.A3-8, there is just one window in water, centered at 450 nm (blue-green), where the attenuation coefficient reaches the ultimate value of a=0.025 m^{-1} for distilled water (L_{vis}=156 m). In very clear deep ocean waters, attenuation may reach, at λ=440 to 510 nm, a=0.05-0.1 m^{-1} (L_{vis} = 78-39 m, near Madeira island) [10], but usually it is a≈1 m^{-1} and up to a≈10 m^{-1} depending on particulate. In the maximum transparency window, the relative weight of diffusion s and absorption in attenuation a=s+α are 0.6 and 0.4, respectively, whereas outside (Fig.A3-8), absorption is dominant (α>>s).

A3.2 TURBULENCE

Propagation over a substantial distance (say, >10 m) in the atmosphere is affected by turbulence. Thermal exchanges with the ground and the sun's heating are responsible for a daily temperature excursion producing convection and vortex in the air.

Because of the associated density fluctuations, the local index of refraction Δn has a small randomness superposed on the average value, and the propagated beam suffers a small angular deviation (Fig.A3-4).

If Δn were constant across the beam, there would be just a deflection, or a random *wandering* of the beam around the unperturbed direction. But, most frequently, the Δn scale of correlation is smaller than the size of the propagated beam after a substantial distance, and different portions of the beam are deflected differently.

As the result, the wavefront is distorted and the spatial distribution deviates from the Gaussian emitted from the laser, with dark and bright spots known as *scintillation* (Fig.A3-9). For such a distribution, the coherence factor μ_{sp} (Sect.4.5.3) of superposition at the detector is decreased. Also, when the wandering deviation is about the same as the spot size (Fig. A3-10), in addition to intensity fading, we face a serious speckle phase error (Sect.4.4.7).

The study of atmospheric turbulence is segmented into three steps: (i) assessing the statistical properties of Δn fluctuations, (ii) translating them into the statistics of the propagated optical field, and (iii) describing the consequent effect on the detected signal.

The details of the theoretical development are very complex, and here we will report just a few basic results, derived under a number of assumptions. An excellent review of turbulence is provided in the work of Hufnagel [11] and by other authors [12-14].

From the engineering point of view, nearly all the quantities of interest are related to $C_n^2 \propto \langle \Delta n^2 \rangle$, called the *structure constant*, and associated with the correlation function defined as $D_n(\rho) = \langle [n_1(0) - n_2(\rho)]^2 \rangle$, where ρ is the distance between two points P_1 and P_2. Tatarski [15] has found that $D_n(\rho) = C_n^2 (\rho)^{2/3}$, so the quantity C_n^2 is measured in m$^{-2/3}$.

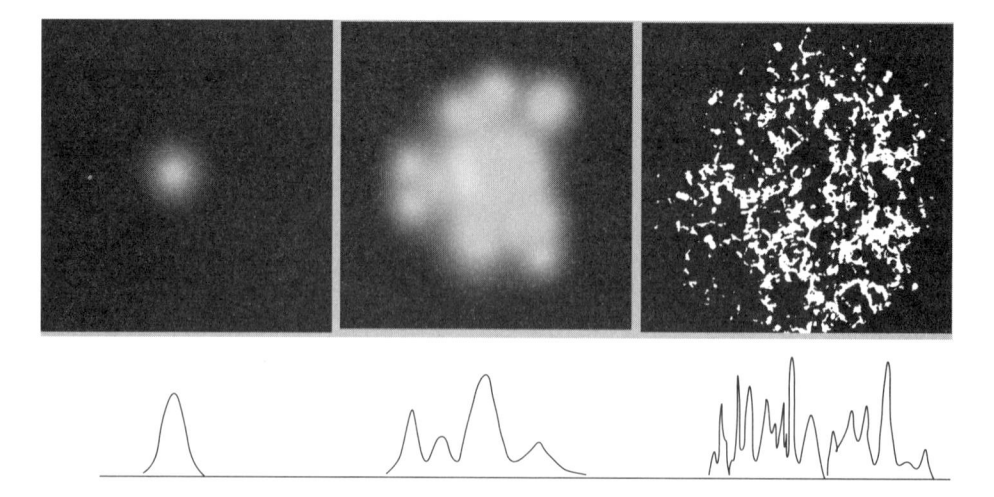

Fig.A3-9 Scintillation of a 20 mW, w_0=1 mm He-Ne beam propagated through air at H=5m height from ground. Top: the beam spatial distributions (not to same scale) as they appear at L=0 (left), 40 m (center), and 200 m (right); exposure is 0.1-s. Bright spots change their position randomly in the beam cross-sections with a typical 0.1 to 1-s time scale. Bottom: profiles of intensity along a diameter exhibit an increasing number of spikes with L.

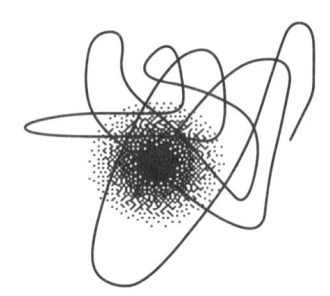

Fig.A3-10 Wandering of the propagated beam for an He-Ne laser (as in Fig.A3-8). The centroid of the beam cross-section (dotted circle) moves randomly in time around the average line-of-sight position.

The constant C_n^2 is easily connected to the temperature variance $C_T^2 \propto \langle \Delta T^2 \rangle$ by the error propagation rule $C_n^2 = (dn/dT)^2 C_T^2$. The temperature coefficient is $|dn/dT| \approx 1$ ppm at sea level [see Eq.4.27] and decreases exponentially with height as $|dn/dT| \approx \exp{-(H_{(km)}/12.6)}$ [11].

According to the model of Kolmogorov [15], the kinetic energy of vortex is dissipated by friction on an inertial sub-range scale, spanning from a few mm to a few m. In this range, temperature fluctuations have a correlation $D_T(\rho) = C_T^2(\rho)^{2/3}$, and Δn fluctuations have $D_n(\rho) = C_n^2(\rho)^{2/3}$, provided that $l_0 < \rho < L_0$, where l_0 ($\approx$mm) and L_0 ($\approx$m) are called the inner and outer scales of turbulence.

A3.2 Turbulence

Experimentally, the constant C_n^2 can range from $\approx 10^{-15}$ to $\approx 10^{-13}$ m$^{-2/3}$ for weak to strong turbulence, respectively, at sea level. The lowest values are found in the early morning and at dusk, in the absence of wind. With height, C_n^2 varies as H$^{-2/3}$ (low H or high wind) or H$^{-4/3}$ (high H or still air), with a break point at H$\approx$1-10m.

To assess the effect of C_n^2 on the propagating optical beam, following Rytov [11,15], we can represent the electric field E= $\langle E \rangle$ exp[i(kr-ωt)+χ(r,t)+iφ(r,t)] as composed of an average part $\langle E \rangle$ with the usual propagation term exp i(kr-ωt), and of a random exponential attenuation χ(r,t) and phase disturbance iφ(r,t), both fluctuating and due to turbulence.

From the Rytov analysis, it is found that these random variables have zero mean values, $\langle \chi \rangle$=0, $\langle \phi \rangle$=0, and a normal distribution. Thus, the distribution of the attenuation exp χ is log-normal.

In the simple case of a horizontal path propagation ($C_n^2 \approx$const) of a plane wave on a distance L, the variance σ_χ^2 of the log-normal attenuation χ has been calculated [11] as $\sigma_\chi^2 = 0.131 \cdot C_n^2 k^{7/6} L^{11/6}$, with k=2π/λ being the wave number. For a spherical wave, the numerical factor is 0.546. Experimentally, the predicted values nicely match the measurements for $\sigma_\chi^2 < 0.3$, while for larger values a saturation is found at about $\sigma_\chi^2 \approx 0.5$, which is typical for L$\approx$ 300m or larger.

Correspondingly, the variance σ_E^2 of amplitude fluctuations is $\sigma_E^2 = \langle E \rangle^2[(\exp 4\sigma_\chi^2)-1]$ and the variance of the intensity is I=$\langle E \rangle^2$ is $\sigma_I^2 = 4I^2(\exp\sigma_\chi^2 - 1)$.

About the phase variance σ_ϕ^2, Ishimaru [12] has reported that it is equal to σ_χ^2 or twice as much in the two cases $l_c \ll \sqrt{\lambda L}$ and $l_c \gg \sqrt{\lambda L}$, i.e., in the case of a turbulence scale of correlation l_c much smaller or much larger than the Fresnel-zone size $\sqrt{\lambda L}$, respectively. Conservatively, we may take $\sigma_\phi^2 \approx C_n^2 k^{7/6} L^{11/6}$ and find the maximum distance L at which the turbulence effect is negligible by equating σ_ϕ^2 to the value of accuracy required in our experiment. For example, letting $\sigma_\phi^2 = 0.01$(rad^2), at λ=1μm we get L=47 m for a weak turbulence ($C_n^2 = 10^{-15}$), but just L=3.8 m for a strong turbulence ($C_n^2 = 10^{-13}$).

Beam wandering can be characterized by a radius ρ_w, defined as the rms value of the beam centroid deviation off the expected position. An approximate expression for a collimated beam is [12]: $\rho_w = w/(1 - 2.45 C_n^2 k^{1/3} L^{11/6} w^{-5/3})$, where w is the unperturbed beam size at distance L, or explicitly $w^2 = w_0^2 + (\lambda/\pi w_0)^2 L^2$ for a Gaussian beam.

At a relatively short distance, the beam shape is substantially unchanged and its centroid wanders randomly by a rms deviation ρ_w. The wandering has a characteristic time τ_w (given by $\approx w/v_w$, v_w being the wind velocity), and σ_χ^2, σ_ϕ^2 are negligible. On a time-scale slower than wandering time, the apparent beam size is $w_{turb}^2 = w^2 + \rho_w^2$. At a large distance, where $\rho_w \gg w$, the beam no longer wanders appreciably, appears widened to a size w_{turb} with scintillation effects now showing up, and σ_χ^2, σ_ϕ^2 are no longer small.

A quantity of interest to interferometer operation through a turbulent atmosphere is the transverse coherence length S_t of the propagated beam, which corresponds to the speckle transversal size discussed in Sect.4.4.7 and is roughly the size of the individual grains found in both the scintillation and beam wandering regimes.

For propagation on a length L, the transverse coherence length is given by an approximate relation [16] as: $S_t = (0.5 k^2 C_n^2 L)^{-3/5}$. For example, at λ=1μm and L=200 m, we get S_t=40 cm for a weak turbulence ($C_n^2 = 10^{-15}$) and S_t=2.6 cm for a strong turbulence ($C_n^2 = 10^{-13}$).

All these statistical results describe the point-like detection of amplitude and phase of the propagated field. Now, we can consider the effect of detector size. When we increase the detector diameter from $d_{det} \ll S_t$ up to the coherence size S_t, the average signal increases and the relative variance improves, such as $\sigma_I^2/I^2 = 4(\exp\sigma_\chi^2 - 1)$ for the intensity. For $d_{det} > S_t$, we collect uncorrelated samples of the χ and ϕ random variables, and the relative variance of intensity weakly improves, only as the logarithm of the number of coherence areas, or $\ln(1+d_{det}/S_t)$, while the phase variance is substantially unchanged.

Finally, regarding the frequency distribution of fluctuations, the power spectrum $S(f)$ is related to the variance σ^2 by the Wiener-Khintchin theorem as $\int_{0-\infty} S(f)df = \sigma^2$. If the spectrum falls off at a corner frequency f_{tur}, we can assume $S(f) = \sigma^2/f_{tur}$ as a reasonable first-order approximation. The corner frequency f_{tur} for both amplitude and phase fluctuations depends on the wind transversal velocity v_t as $f_{tur} = l_c/v_t$, where l_c is the turbulence scale of correlation. Typical values of the frequency corner are in the range of $f_{tur} \approx 10$ to 200 Hz.

REFERENCES

[1] W.L. Wolfe and G.J. Zissis (editors), *"The Infrared Handbook"*, Environmental Research Institute of Michigan: Ann Arbor, MI, 1988.
[2] V.I. Tatarski, *"Propagation of Waves in Turbulent Medium"*, McGraw-Hill: N.Y., 1961.
[3] H.C. Van de Hulst, *"Light Scattering by Small Particles"*, J. Wiley: New York, 1957.
[4] Burle Industries Inc. *"Electro-Optics Handbook"*, Burle vol.TP-135, Lancaster, PA, 1992.
[5] OSA Staff, *"Optics Handbook"*, Opt. Soc. of Am.: Washington DC, 1992, vol 1.
[6] H.A. Gebbie et al., *"Atmospheric Transmission in the 1-14 micrometer Region"*, Proceedings of the Royal Society, vol.A206 (1951), pp. 87-96.
[7] S. Donati, *"Photodetectors"*, Prentice Hall: Upper Saddle River, NY, 2000, App.A3.
[8] S. Donati, *"Optoelectronic Techniques for Navigation Aids in Poor Weather"*, Alta Frequenza, vol.43 (1974), pp.725-732.
[9] S. Donati, *"Thermal Imaging Through Hazes and Fog: Experimental Results"*, Alta Frequenza, vol.42 (1973), pp.101-105.
[10] S. Duntley, *"Light in the Sea"*, J. Opt. Soc. of Am., vol.53 (1963), pp. 214-221.
[11] R.E. Hafnagel, *"Propagation through Atmospheric Turbulence"*, in The Infrared Handbook, edited by W.L.Wolfe and G.J. Zissis, IRIA: Ann Arbor, MI, 1978.
[12] A. Ishimaru, *"Wave Propagation in Scattering Media"*, vol.1 and 2, Academic Press: New York, 1978.
[13] A. Ishimaru, *"Wave Propagation in Turbulent Media"*, Proc. IEEE, vol.79 (1991), pp.1359-1366.
[14] R.L. Fante, *"Electromagnetic Beam Propagation in Turbulent Media"*, Proc. IEEE, vol.68 (1980), pp.1424-1443.
[15] V.I. Tatarski, *"Propagation of Waves in a Turbulent Medium"*, McGraw Hill: New York, 1961.
[16] G.R. Osche and D.S. Young, *"Imaging Laser Radar in NIR and FIR"*, Proc. IEEE, vol.84 (1996), pp.103-125.

APPENDIX **A 4**

Optimum Filter for Timing

The optimum filter derived in this appendix is just an example of the time-domain optimization of signal processing, a well-developed method found useful in a number of measurement applications [1].
To derive the optimum impulse response, we will represent the statistical properties of the signal at the second order, using mean value S(t) and covariance K(t,t') [1,2].

For a process dominated by shot-noise fluctuation, we may write the covariance as K(t,t')= S(t) δ(t'-t), where δ is the Dirac-delta function.
When the signal is passed through a filter with impulse response h(t), the output mean and covariance are given by [2]:

$$S_{out}(t) = S(t)*h(t) \qquad K_{out}(t,t')= = S(t)**h(t)h(t') \qquad (A4.1)$$

The time variance σ^2_t of a threshold crossing is accordingly, using a linear regression:

$$\sigma^2_t = |S(t)**h(t)h(t')| \, / \, |S'(t)*h(t)|^2 \qquad (A4.2)$$

To find the minimum of σ^2_t with respect to the function h(t), we use a method due to Hilbert [1,2]. We let h(t)=h_0(t)+αη(t), where h_0(t) is the desired optimum function to be found, α is a free numerical parameter, and η(t) is an arbitrary function. If h_0(t) is the function that minimizes σ^2_t, then the second term in Eq.A4.2 will be a minimum for α=0, irrespective of the chosen η(t). Thus, we may substitute h(t)=h_0(t)+αη(t) in Eq.A4.2 and look for the minimum with respect to α. In this way we obtain:

$$(d/d\alpha)\sigma^2_t = (d/d\alpha)\{\int_{0-\infty} S(t-\tau)[h_0(\tau)+\alpha\eta(\tau)]^2 d\tau / |\int_{0-\infty} S'(t-\tau)[h_0(\tau)+\alpha\eta(\tau)]d\tau|^2\}$$

By developing the derivatives and equating to zero, we get:

$$\int_{0-\infty} S(t-\tau)2[h_0(\tau)+\alpha\eta(\tau)]\eta(\tau)d\tau \} \times |\int_{0-\infty} S'(t-\tau)[h_0(\tau)+\alpha\eta(\tau)]d\tau|^2 -$$

$$-2 \int_{0-\infty} S'(t-\tau)[h_0(\tau)+\alpha\eta(\tau)]d\tau \times \int_{0-\infty} S'(t-\tau)\eta(\tau)d\tau \times \int_{0-\infty} S(t-\tau)[h_0(\tau)+\alpha\eta(\tau)]^2 d\tau = 0$$

Now, letting $\alpha=0$ in the above expression, we obtain:

$$\int_{0-\infty} S(t-\tau)2h_0(\tau)\eta(\tau)d\tau \} \times |\int_{0-\infty} S'(t-\tau)h_0(\tau)d\tau|^2 -$$
$$-2 \int_{0-\infty} S'(t-\tau)h_0(\tau)d\tau \times \int_{0-\infty} S'(t-\tau)\eta(\tau)d\tau \times \int_{0-\infty} S(t-\tau)h_0^2(\tau)d\tau = 0$$

It can be verified that this condition is satisfied for any arbitrary function $\eta(t)$ by taking:

$$h_0(\tau) = S'(t-\tau) / S(t-\tau) \tag{A4.3}$$

Thus, the optimum filter function is the time-reverse of the relative slope S'/S. The time t of reversal is not specified in the calculation, but to make sense, it should be larger than the time duration of the pulse $S(t)$. Taking $t=T_m$ according to this criterion, we have:

$$h(\tau) = S'(T_m-\tau)/S(T_m-\tau), \ S_{out}(T_m) = \int_{0-\infty} S(T_m-\tau)h(\tau)d\tau = \int_{0-\infty} S'(T_m-\tau)d\tau = 0 \tag{A4.4}$$

or, the mean signal passes through zero at time T_m, i.e., we have to perform a zero-crossing measurement. The timing variance supplied by the optimum filter at $t=T_m$ is:

$$\sigma^2_t = 1/ \int_{0-\infty} S'^2(\tau)/S(\tau) \, d\tau, \tag{A4.5}$$

a value is obviously smaller than $\sigma^2_t = 1/[S'^2(\tau)/S(\tau)]$, the value of a fixed threshold crossing.

As a final point, let us account for background and electronic circuit noise. We can summarize the signal-independent noise by adding a term ρ to the amplitude variance, where ρ can be written in terms of $P_{bg}+P_{ph0}$ (see Sect.3.2.3). With such a position, it is easy to repeat the calculations leading to Eq.A4.3 and we find the optimum filter response to be:

$$h(\tau)=S'(T_m-\tau)/[S(T_m-\tau)+\rho] \tag{A4.4'}$$

Again, the output signal passes through zero at measurement time T_m, or $S_{out}(T_m)=0$, and the optimum time variance is:

$$\sigma^2_t = 1/ \int_{0-\infty} S'^2(\tau)/[S(\tau)+\rho] \, d\tau \tag{A4.5'}$$

By inserting a normalized waveform function in Eq.A4.5' and after some algebraic manipulation, we can arrive at a time accuracy expression coincident to Eqs.3.16' and 3.16'' of this text. This result confirms the $1/\sqrt{N_r}$ and $1/N_r$ dependence of accuracy from the received number of photons, for the shot noise and background or electronics noise respectively.

REFERENCES

[1] S. Donati, E. Gatti, and V. Svelto, "*Advances in Electronics and Electron Physics*", vol.26, edited by L. Marton, Academic Press: New York, 1969, pp.251-307.
[2] E. Gatti and S. Donati, "*Optimum Signal Processing for Distance Measurements with Lasers*", Applies Optics, vol.10 (1971), pp.2446-2451.

APPENDIX **A 5**

Propagation and Diffraction

*M*aking no pretence of completeness, but to provide a useful reference for the reader, we will recall here some results concerning propagation and diffraction equations we have used in the text when dealing with size measurements (Ch.2) and speckle pattern instrumentation (Ch.5). Several textbooks are available for a more complete treatment (see Ref. [1-3]).

A5.1 PROPAGATION

Let's begin with a fundamental equation relating the amplitudes of the optical field in the propagation from a source plane (ξ,η) to a receiving plane (x,y) (Fig.A5-1).
The source can be divided into elemental areas $d\xi d\eta$, each with an amplitude $A_1(\xi,\eta) \, d\xi d\eta$. Each contribution is a source radiating as a spherical wave in view of the Huygens-Fresnel principle, and thus comes to the point (x,y) in the image plane with a phase delay kr_{12}, where $k=2\pi/\lambda$ is the wave number and r_{12} is the distance between point $P1(\xi,\eta)$ in the source plane and point $P2(x,y)$ in the receiving plane. The component perpendicular to the (x,y) plane is $A \cos\theta$, where $\cos\theta = z/r_{12}$ is called the obliquity factor. Summing all the elemental contributions, we have:

$$A_2(x,y) = (1/\lambda z) \iint_{-\infty,+\infty} A_1(\xi,\eta) \, (z/r_{12}) \exp(ikr_{12}) \, d\xi d\eta \qquad (A5.1)$$

This is the Rayleigh-Sommerfeld diffraction formula, valid with little loss of generality in the case $z \gg \lambda$ [1,2].
We can write the explicit expression for r_{12} as:

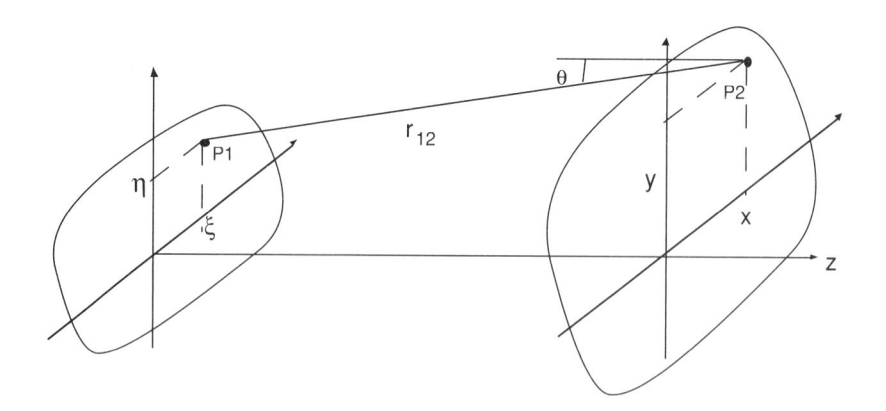

Fig.A5-1 Geometry for analyzing the propagation from a source plane (ξ,η) to a receiving plane (x,y)

$$r_{12} = [z^2+(x-\xi)^2+(y-\eta)^2]^{1/2} = z\{1+[(x-\xi)/z]^2+[(y-\eta)/z]^2\}^{1/2} \quad (A5.2)$$

By inserting in Eq.A5.1, we obtain:

$$A_2(x,y) = (1/\lambda z) \iint_{-\infty,+\infty} A_1(\xi,\eta) \frac{\exp ikz\{1+[(x-\xi)/z]^2+[(y-\eta)/z]^2\}^{1/2}}{\{1+[(x-\xi)/z)]^2+[(y-\eta)/z]^2\}^{1/2}} d\xi d\eta \quad (A5.3)$$

Eq.A5.3 is a convolution integral of the form $A_2(x,y) = \iint A_1(\xi,\eta)H(x-\xi,y-\eta)d\xi d\eta$, and can then be written in a compact form as:

$$A_2(x,y) = A_1(x,y) ** H(x,y) \quad (A5.4)$$

Thus, the function $H(x,y)=(1/\lambda z) \exp ikz\{1+(x/z)^2+(y/z)^2\}^{1/2}/\{1+(x/z)^2+(y/z)^2\}^{1/2}$ assumes the meaning of the impulse response of the propagation between input (source) and output (receiving) planes.

The Fourier transform $H_F(k_x,k_y)$ of the impulse response $H(x,y)$ in the spatial frequency k_x and k_y domain, conjugate to coordinates x and y is written as:

$$H_F(k_x,k_y) = \iint_{-\infty,+\infty} H(x,y) \exp i(xk_x+yk_y) \, dx \, dy \quad (A5.5)$$

The function $H_F(k_x,k_y)$ is called the transfer function between input and output planes, and it is found to be given by the following expression [1]:

$$H_F(k_x,k_y) = \exp\{ikz[1-\lambda^2(k_x^2+k_y^2)]^{1/2}\} \quad (A5.6)$$

A5.2 The Fresnel Approximation

Now, if we wish to compute the propagation problem in the frequency domain, in terms of the Fourier transforms $A_1(k_x,k_y)$ and $A_2(k_x,k_y)$ of the spatial domain distributions $A_1(x,y)$ and $A_2(x,y)$, we can take advantage of the convolution in Eq.A5.4 becoming the product of the transforms in the frequency domain, or:

$$A_2(k_x,k_y) = A_1(k_x,k_y) \, H_F(k_x,k_y) \quad (A5.7)$$

A5.2 THE FRESNEL APPROXIMATION

Eqs.A5.1 to A5.3 are insightful from a physical point of view, but are difficult to treat in most cases. A very helpful approximation is provided by the well-known Fresnel condition $L_{\xi,\eta}$, $L_{x,y} \ll z$, in which the transverse dimensions $L_{\xi,\eta}$ and $L_{x,y}$ of the source and receiving apertures are small as compared to distance z.
With this condition, we can approximate the obliquity factor to unity, $\cos\theta \approx 1$, and write the distance as:

$$r_{12} = z\{1+[(x-\xi)/z]^2+[(y-\eta)/z]^2\}^{1/2} \approx z\{1+{}^1\!/_2[(x-\xi)/z]^2+{}^1\!/_2[(y-\eta)/z]^2\}$$
$$\approx z + (x-\xi)^2/2z + (y-\eta)^2/2z \quad (A5.8)$$

Further, we may develop the square factor and obtain:

$$r_{12} = z + (x^2+y^2)/2z + (\xi^2+\eta^2)/2z - (x\xi+y\eta)/z \quad (A5.9)$$

Going back to the propagation equation (A5.3), we can see that, in the Fresnel approximation, the impulse response H becomes:

$$H(x,y) = (\lambda z)^{-1} \exp ikz \times$$
$$\times \exp i[k(x^2+y^2)/2z] \exp i[k(\xi^2+\eta^2)/2z] \exp -i[k(x\xi+y\eta)/z] \quad (A5.10)$$

The first exponential term in this equation gives the propagation delay on the distance z, a constant term independent from x,y. The second is a field curvature term and comes out of the integral (A5.3). The third term is a field curvature of the source field, and can eventually be incorporated in the source distribution $A(\xi,\eta)$. The last term is the kernel of a Fourier double-transform integral that conjugates the spatial variables ξ and η to the transform-domain variables kx/z and ky/z.
By rewriting Eq.A5.3 with the previous terms, we have in the Fresnel approximation:

$$A_2(x,y) = (\lambda z)^{-1} \exp i[k(x^2+y^2)/2z] \, FT \, \{A_1(\xi,\eta) \exp i[k(\xi^2+\eta^2)/2z]\} \quad (A5.11)$$

Note on terminology. In the Fourier transform (abbreviated above with *FT*), the variables conjugated to ξ and η are the spatial frequencies (Eqs.A5.5-A5.7) $k_x = kx/z$ and $k_y = ky/z$, and so we should write the left-hand term in Eq.A5.11 as $A_2(k_x,k_y)$. However, as $x=zk_x/k$ and $y=zk_y/k$ are connected to k_x and k_y, the notation $A_2(x,y)$ is also correct and can be used when we regard the propagation process as one conjugating the spatial domains (ξ,η) and (x,y). Additionally, we may interpret $x/z=\psi_x$ and $y/z=\psi_y$ as angular coordinates of diffraction, and write the left-hand term in Eq.A5.11 as $A_2(\psi_x, \psi_y)$ to indicate the Fourier conjugation from the spatial to the angular domain.

When distance is very large, so that $z>>kL_{\xi,\eta}^2$, $kL_{x,y}^2$, the curvature terms become negligible and we are in the Fraunhofer region (or far field). Eq.A5.11 can be written as the Fourier transform integral relating the source and image fields:

$$A_2(x,y) = \int\int_{-\infty,+\infty} A_1(\xi,\eta) \exp -i[k(x\xi+y\eta)/z] \, d\xi \, d\eta$$

$$A_2(x,y) = FT\,[A_1(x,y)] \qquad (A5.12)$$

Last, it is useful to recall that the irradiance, or power per unit surface, $I(x,y)$, associated with the electrical field is in general given by $I(x,y)= E^2(x,y)/Z_0$, where Z_0 is the free-space impedance. In the following, we will consider the irradiance as proportional to the square of the electric field amplitude, or simply by $I \propto E^2$.

A5.3 EXAMPLES

Rectangular aperture of width D across x, and infinite along y

In this case, we have $A_1=A_0 \, \text{rect}(-D/2,+D/2)$ and Eq.A5.12 gives:

$$A_2(x,y) = FT\,[A_1(x,y)] = A_0 \int_{-D/2\,..\,+D/2} A_1 \exp -ik(x\xi)/z \, d\xi$$

$$= A_0 (\sin kxD/2z)/(kxD/2z) = A_0 \, \text{sinc} \, xD/\lambda z \qquad (A5.13)$$

In the last equation, we used $k=2\pi/\lambda$ and have introduced the sinc function:

$$\text{sinc } x = (\sin \pi x) / \pi x \qquad (A5.14)$$

When defined with the normalization of Eq.A5.14, both the sinc function and its square, $\text{sinc}^2 x$, have unity area [1] when integrated on x from $-\infty$ to $+\infty$.
Letting $x/r_{12} \approx x/z = \sin\theta$ for the direction of diffraction, we have for the diffraction distribution in the far field:

$$E(\theta) = \text{sinc} \,(\sin\theta D/\lambda).$$

For small angle θ, it is $\sin\theta \approx \theta$, and then we have:

$$E(\theta) \approx \text{sinc} \,(\theta D/\lambda) \qquad (A5.15)$$

This is the well-known formula for the diffraction from a slit.
In addition, if P_0 is the radiant power intercepted by the slit and rediffused, and in view of the

A5.3 Examples

normalization to 1 of the sinc²x area, the density I(θ) of radiant power diffracted at the angle θ is written as:

$$I(\theta) = P_0 \, \text{sinc}^2(\theta D/\lambda) \qquad (A5.15')$$

The functions sinc x and sinc²x are plotted in Fig.A5-2. The HWHM (half-width at half-maximum) of the field (sinc x) is at x=0.6, whereas that of irradiance (sinc²x) is at x=0.44. As it is x=θD/λ, we can solve for the angle of diffraction at half-power as θ=0.44 λ/D. Other features of the sinc² function are: a first zero located at x=±1 (whence $\theta_{zero}=\pm\lambda/D$) and secondary maxima located at x=±1.43, ±2.47 and ±3.47, with amplitudes of 4.7, 1.6 and 0.8% of the maximum, respectively.

Comment on notation. We use the notation of Gaskill [1] for the functions sinc and somb (see below). Other authors prefer incorporating the factor π inside the argument.

Circular aperture of diameter D

In Eq.A5.12, Cartesian variables dξdη are changed to the radial variables ρdρdψ, and the integral is evaluated on the limits ρ=0-D/2 and ψ=0-2π. Letting r²=x²+y², the result reads:

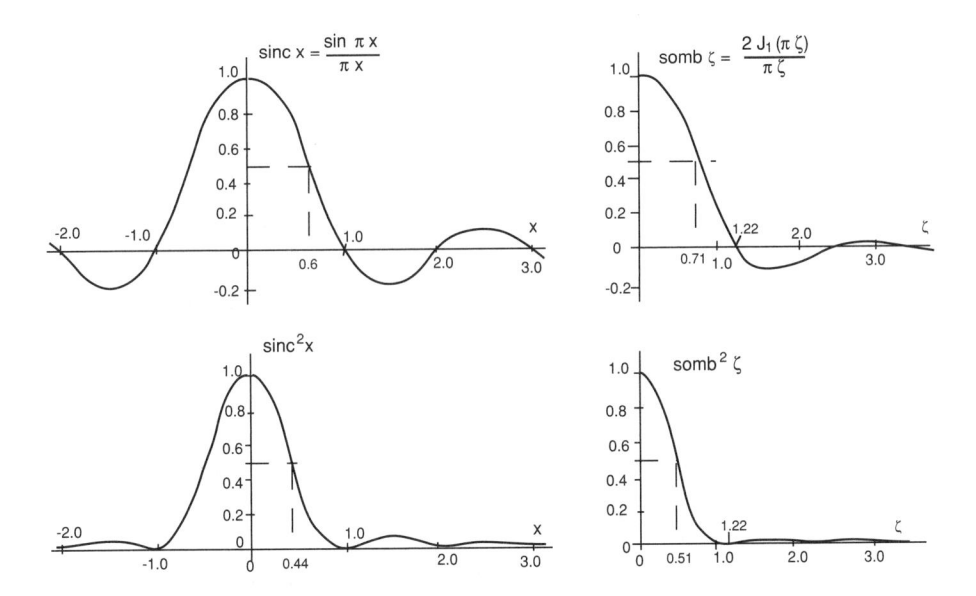

Fig.A5-2 Left: the sinc x function describes the relative amplitude of the field diffracted at x=(D/λ) sinθ from a slit aperture of width D. Its square, sinc² x, gives the associated radiant power density. Right: the somb and somb² functions describe the relative amplitude of the field and the power density diffracted at ζ=(D/λ) sinθ from a sphere of diameter D.

$$A_2(r) = A_0 \text{ somb } rD/\lambda z \qquad (A5.16)$$

In this expression, we have used the notation somb (from sombrero, the bidimensional generalization of the sinc function to radial coordinates) [1], defined as:

$$\text{somb}(\zeta) = 2J_1(\pi\zeta)/(\pi\zeta) \qquad (A5.17)$$

J_1 is the Bessel function of the first kind. Again letting $r/r_{12} \approx r/z = \sin\theta$ for the direction of diffraction and squaring A_2, we obtain the well-known Airy's diffraction formula of the radiant power density $I_2(\theta)$ diffracted at the angle θ:

$$I_2(\theta) = I_0 \text{ somb}^2 \sin\theta\, D/\lambda \qquad (A5.16')$$

The functions somb and somb^2 are plotted in Fig.A5-2. They are standardized so that the maximum value is unity and the volume over the x-y plane, from $-\infty$ to $+\infty$, is $4/\pi$ [1].

The half-width at half maximum (HWHM) of somb^2 is x=0.51, which corresponds to a diffraction angle θ =0.51 λ/D.

The first zero of the relative radiant power is at x =1.22, leading to the well-known diffraction formula θ_{zero}=1.22 λ/D.

Secondary maxima are found at x=±1.63, ±2.67, and ±3.69, with amplitudes of 1.7, 0.4 and 0.15% of the maximum, respectively.

Gaussian-distributed spot

This is the case of a real mode representing the field distribution E(r) of the laser beam along the radial coordinate $r=(x^2+y^2)^{1/2}$:

$$E(r) = E_0\, [(2\pi)^{1/2} w_0]^{-1} \exp -r^2/w_0^2 \qquad (A5.18)$$

The spot size parameter of the Gaussian beam is w_0, the rms deviation. This quantity is the radius at which the field has dropped at 1/e=0.37 of the maximum value, and the intensity $I \propto E^2$ at $1/e^2$=0.13 of the maximum.

The HWHM of the Gaussian is $1.18 w_0$ for the field and $0.84 w_0$ for the intensity. Additionally, w_0 is the radius carrying 86% of the total power.

Using Eq.A5-12, the Fourier transform E(r) is found as [1]:

$$E_2(r) = E_0 \exp -w_0^2 k^2 r^2/4z^2 \qquad (A5.18')$$

As a function of the angular coordinate of observation $\theta=r/z$, this equation can be written as:

$$E_2(\theta) = E_0 \exp -\theta^2/\theta_{w0}^2 \qquad (A5.18'')$$

The parameter $\theta_{w0} = \lambda/\pi w_0$ introduced in this equation represents the diffraction angle associated with the spot size w_0. As we can see from Eq.A5-18", the diffracted field in the Fraunhofer region has a Gaussian distribution with angular (spot) size θ_{w0}.

A5.3 Examples

This is the angular radius containing 86% of the total power.

For comparison to the Airy disk $\theta = 0.51\lambda/D$, we can use the spot diameter $D_{w0}=2w_0$ in θ_{w0} and obtain $\theta_{w0} = 2\lambda/\pi D_{w0} = 2\lambda/e = 0.63\lambda/D_{w0}$.

Sinusoidal amplitude grating of spatial period Λ.

In this case we get $A_1 = \sin 2\pi\xi/\Lambda$. By substitution in Eq.A5.12, we get:

$$A_2(x,y) = FT[A_1(x,y)] = (1/2i)\, FT\,[\exp i(2\pi\xi/\Lambda) - \exp -i(2\pi\xi/\Lambda)]$$

$$= [\delta(kx/z+2\pi/\Lambda) - \delta(kx/z-2\pi/\Lambda)]/2i \qquad (A5.20)$$

The result shows that we have two diffracted beams at $kx/z = \pm 2\pi/\Lambda$, or at angles $\sin\theta = x/z = \pm 2\pi/\Lambda k = \pm\lambda/\Lambda$, the first order of diffraction, a well-known result in optics.

Sinusoidal phase grating of spatial period Λ

It is $A_1 = \exp i[\phi_0 \cos 2\pi\xi/\Lambda]$, and by developing the exponential in series, we get the field in the output plane as:

$$A_2(x,y) = [1+J_0(\phi_0)]/2 + \Sigma_{n=1,+\infty}\,(i^n/2)J_n(\phi_0)[\delta(kx/z+n2\pi/\Lambda) - \delta(kx/z-n2\pi/\Lambda)] \quad (A5.21)$$

From this expression, we can see that for $\phi_0 \ll 1$, there is only the first order of diffraction as in Eq.A5.20, whereas when ϕ_0 is not so small, all the orders of diffraction are generated.

Grating of period Λ and a finite width D

We have $A_{1f} = A_1(x,y)\,\text{rect}(-D/2,D/2)$, where D represents the width of the reticle. Using Eq.A5.12 gives:

$$A_{2f}(x,y) = A_2(x,y) * FT\,\text{rect}(-D/2,D/2) \qquad (A5.22)$$

If $A_2(x,y)$ is the output of an amplitude grating (Eq.A5.16), we get:

$$A_{2f}(x,y) = \text{sinc}\,(kx/z+2\pi/\Lambda)D/2 - \text{sinc}\,(kx/z-2\pi/\Lambda)D/2. \qquad (A5.21')$$

Arbitrary profile illuminated with a divergence Θ

This is the case of a target positioned not exactly in the beam waist of the illuminating laser. Let the illuminating beam arrive at the ξ,η plane with an angle Θ or, given a diameter D of the target, with a curvature of the wavefront $R=D/\Theta$.

At point ξ,η, the phase due to the spherical wavefront is:

$$\varphi = k(R^2+\xi^2+\eta^2)^{1/2} - kR = kR\,[1+(\xi^2+\eta^2)/2R^2] - kR \approx k(\xi^2+\eta^2)/2R$$

Thus, in addition to the function $A_1(\xi,\eta)$ describing the target, in Eq.A5.12 we have a multiplying term $A_1^{\#}=\exp ik(\xi^2+\eta^2)/2R$.

We know that the Fourier transform of a product is the convolution of the transforms. Thus, our result is $A_2(x,y)**FT[\exp ik(\xi^2+\eta^2)/2R]$.

The transform of $A_1^{\#}$ is [1]:

$$FT[A_1^{\#}] = \exp\text{-}i[kR(x^2+y^2)/2z^2] = \exp\text{-}i[\theta^2/(2/kR)]$$

where $\theta = (x^2+y^2)^{1/2}/z$.

The argument of the transform is small enough to be safely neglected as long as the angular variance $2/kR$ is small with respect to the diffraction angle θ_2 associated with the distribution $A_2(x,y)$.

This condition is written as $2/kR << \theta_2^2 = (\lambda/D)^2$ in the case of a particle of diameter D. Solving for R, we get $R >> (2/k)(D/\lambda)^2 = D^2/\pi\lambda$, the well-known Fresnel distance [1].

REFERENCES

[1] J.D. Gaskill, "*Linear Systems, Fourier Transforms and Optics*", J.Wiley and Sons: New York, 1978, see Chapter 10.

[2] J.W. Goodman, "*Statistical Properties of Laser Speckle Patterns*", in Laser Speckle and Related Phenomena 2nd ed., edited by J.C. Dainty, Springer Verlag: Berlin, 1981, pp.9-74.

[3] J.W. Goodman, "*Introduction to Fourier Optics*", Mc Graw-Hill: New York, 1968.

APPENDIX **A 6**

Source of Information on Electro-Optical Instrumentation

*R*eaders who want to keep abreast of the latest news and scientific and technical developments in the field of electro-optical instrumentation may consult the journals and magazines listed in this appendix.

Scientific Journals

IEEE publications, 445 Hoes Lane, Piscataway, NJ (see web site at *ieee.org*):

IEEE Transaction on Instrumentation and Measurements (monthly)
IEEE Sensors Journal (bimonthly)
IEEE Journal of Quantum Electronics (monthly)
IEEE Photonics Technology Letters (monthly)
IEEE Selected Topics in Quantum Electronics (monthly)
Proceedings of the IEEE (monthly)
IEEE Spectrum (monthly)

The IEEE also provides access and download privileges to a large library ($\approx 10^6$ titles) of PDF documents, including journal papers, conference proceedings, and standards published by IEEE and IEE since 1980 (see *ieeexplore.com*)

OSA Publications (see web site at: *osa.org*):

Optics Letters (semi-monthly)
Journal of the Optical Society of America (monthly, issued in two parts)
Applied Optics (monthly, issued in three parts)

Optics and Photonics News (monthly)
Optics Express (monthly, web only)

IEE, the Institution of Electrical Engineers (London, UK) publications:
Electronics Letters (weekly)
IEE Proceeding part J: Optoelectronics (bimonthly)
IEE Proceeding part E: Instrumentation (bimonthly)

IOP, the Institute of Physics (Bristol, UK) publications (search engine *iop.org/EJ/welcome*):
Journal of Optics, part A and B (monthly)
Journal of Measurement Science (monthly)

SPIE (Bellingham, WA) publications (search engine *spie.org*):
Optical Engineering (monthly)

AIP, the American Institute of Physics publications (search engine *aip.org/japo*):
Journal of Applied Physics (monthly)
Review of Scientific Instruments (monthly)
Applied Physics Letters (semi-monthly)

Journal of Optical and Quantum Electronics (Kluwer, Dodrecht, Holland, bimonthly)
Sensors and Actuators (North Holland, Dodrecht, Holland, monthly)

Magazines

Photonics Spectra (Laurin Publications, monthly) search engine *photonics.com*
Laser Focus World (Pennwell, Nashua, NH, monthly) search engine *lfw.pennNET.com*
Europhotonics (Laurin Publications, NL, monthly) search engine *photonicsonline.com*

Societies

Several societies deal with laser and electro-optical instrumentation as topics of interest. They are:
IEEE – IM Society (Instrumentation and Measurement Society), see *im.ieee.org*
IEEE - LEOS Society (Lasers and Electro-Optical Society), see *leos.ieee.org*
OSA (Optical Society of America), see *osa.com*
SPIE (Society of Photonic Instrumentation Engineers), see *spie.org*

Meetings

Annual international meetings and conferences on electro-optical instrumentation are:
CLEO and CLEO Europe (Conference on Laser and Electro-Optics),
 promoted by IEEE and OSA
IMTC (International Measurement Techniques Conference), promoted by IEEE
IEEE LEOS Annual Meeting, promoted by IEEE
IMEKO (International Measurement Conference), promoted by IMEKO

Index

A

Absolute distance measurement, 163
Acceptance, 191, 200, 231, 261, 326, 359, 403-406, 410
Accuracy, 364, 378, 383, 403, 417
AGC, 202, 209
AHRS, 277
Air mass, 407, 409
Alignment, 11-16, 24-25
 error, 233
Allan's variance, 117
A-lNi-Co, 380
Amplitude fading, 201
Angle measurement, 16, 19
Apollo 11, 2
ASE, 386
Atmospheric transmission, 45, 81, 408,
 attenuation coefficient, 407, 414
 visibility, 411

B

Babinet's principle, 28
Background noise, 48-50
Backscattering, 285
Ballast resistance, 373
Barkhausen criteria, 137, 368
Beam waist, 12-14, 367
Beamsplitter prism 234
Bessel function expansion, 280
Birefringence readout, 331, 338
Bragg cell, 237
Brewster window, 372

C

Cascadeability of interferometers, 399
CCD, 29, 43, 81, 216
CD readout, 170-171
Cervit, 266-268
Chahine method, 34
Classes of laser safety, 395
Coherence 188-190, 371-72
 factor, 188, 234, 371, 415
 length, 366, 372, 388, 402
 multiplexing, 357
 spatial 366, 410
Cooperative gain, 46
Cooperative target, 44
Coriolis' force, 301, 308
Corner cube, 2, 44, 78, 94-95, 106-112, 404-406
Current sensor, 330

D

Detectivity, 363
DFB, 170, 384
DIAL, 86
Diameter sensor, 27
Diffraction, 421
 -based measurements, 27-30
Diffuser, 44, 101, 203
Dispersion, 119
Distributed Sensors, 357
Dither, 205, 221
Doppler effect, 89, 227, 239, 252
DOR, 393
Dove's prism, 234

431

DRLG, 267
Dynamic range, 251, 363

E

EDFA source, 287, 297
EMI, 251, 312
ESPI, 214
Extinction coefficient, 412

F

Fabry-Perot, 345, 368, 387, 398, 402
Faraday rotation, 330
    mirror, 334
FBG sensor, 354
FBG, external cavity laser, 389
Feedback regimes, 128, 136, 141
Fiber, hi-bi, 284, 263, 333
    coupler, 288, 298, 300
    Nd-doped, 325
    polarizer, 285
    sapphire, 326
Fiber sensors, 311
    classification, 311
    distributed, 354
    types of, 314
    intensity, 318
    interferometric, 345
    multiplexed, 352
    polarimetric, 328
    readout of, 318
    responsivity, 347
    SMR, 299
    white-light, 349
Filter, spatial, 403
Finesse factor, 400
Fizeau effect, 255
FOG, 7, 248, 259, 263
    open-loop, 277
    closed loop, 291-294
    resonant, 295
    3x3, 297
Fourier
    conjugation, 423
    reverse transform illumin., 35

Fraunhofer approximation, 32
Fredholm's integral, 33-37
Frequency pulling, 272, 370, 379
Fresnel approximation, 423
Fringe visibility, 234, 314

G

Gaussian
    beam, 12, 226
    envelope, 239
    spot, 369, 426
Glan cube, 101, 125, 273, 317, 331, 332,
    338, 378, 404
GRINSCH, 391
Gyroscope, 6, 248
    basic configuration, 261
    FOG development, 277
    locking range, 262
    MEMS, 301-306
    piezoelectric, 307
    recombiner prism, 273
    RLG development, 265
    RLG performance, 274
    scale factor, 280

H

Hilbert transform, 211
Hydrophone, 348

I

IFOG, 291
Induced modulation, 126
Interferometers, 345, 397
    comparison of, 400
Interferometry, 8, 89
    absolute distance, 163
    balanced vs unbalanced, 399
    configurations of, 112
    dual beam, 94
    diffusing target, 101
    external, 125
    internal, 124
    injection, 126

Index

injection with He-Ne, 131
injection with laser diode, 141
three-mirror model, 137
Michelson, 94
multi-axis, 108
nm-extension, 105
readout of OFS, 345
speckle regime, 201, 203
white-light, 182
INU, 7, 248
IO tecnology, 290
IOG, 249, 290
Integrated optics, 290
Iodine cell, 383

J

Jones, 363
Johnson noise, 164, 397, 399

K

Kramer-Kronig relation, 211

L

LAELS, 29-38
Lamb's equations, 128, 261
    dip, 371, 376
Lang-Kobayashi equations, 140
Laser
    cavity, 368-370
    CO_2, 2, 73, 82, 157
    diode-pumped, 390
    Doppler velocimeter, 223-245
    frequency stabilized, 376
    gain, 370
    He-Ne, 8, 11-15, 27, 30, 93, 125, 157, 165, 224, 265, 367
    level, 24
    longitudinal modes, 370, 378
    Ne-isotopes, 265, 378
    Q-switched, 3, 81
    ruby, 2

    safety, 15, 393
    semiconductor, 75, 82, 141, 170, 251, 367, 383
    wavelengths, 367
    YAG, 2, 4, 82, 390
Latex spheres, 238
LDA, 223
LDV, 223-245
    direction discrimination, 237
    particle seeding, 238
    signal processing, 239
LED, 183, 299
LIDAR, 4, 5, 81, 358
LIGA, LIGO, 8, 9, 93, 372
$LiNbO_3$, 80, 289
Linear birefringence readout, 338
Lock-in, 262, 270
Locking range, 125, 262
LURE, 2, 4

M

Mach-Zehnder, 286-287, 345, 398, 402, 412
Magnesium fluoride, 29
Marx multiplier, 345-346
Marx pump rectifier, 374
MDS, 258, 274
Mechanical gyro, 247-251, 307
MEMS, 155, 249, 251
Michelson, 345, 348, 398
Micro-optics, 289
Mie scattering regime, 36, 235, 238, 411
MIL-STD, 393
MOEMS, 7, 304
Mode
    Gaussian, 368
    hopping, 384
MOLA, 4, 5, 68
MPE, 395
M^2-factor, 369

N

NEI, 335

NED, 114, 121, 149, 158, 202, 257
NEΩ, 299, 305
Newton formula, 12
Noise in interferometers
    quantum limit, 114
    thermodynamic, 119
    Brownian, 120
    speckle-related, 120

O

ODIMAP, 189
OFS, 312-360
OOK, 256
OPO, 82, 86
Optical isolator, 158
Optrode, 315, 327
OTDR, 358

P

Particle seeding, 238
Particle sizing, 30
Pentaprism, 26, 110
Peltier, 388
PFV, 223
Phase noise, 115, 202, 255
Photodiode
    quadrant, 12, 26
    position-sensing, 19
Piezo, *see PZT*
PMD, 288
Polarimetric sensors, 313
Polarization-mantaining fiber, 285
Position sensing 16
    reticles, 22
Profilometry, 175
Propagation
    atmosphere, 407
    and diffraction, 421
PZT, 30, 150, 169, 204, 266, 268, 280, 282, 293, 296, 307, 340, 348, 381
PSD, 19, 24

Q

Q-switching, 393
Quadrant photodiode, 16
Quantum noise, 113, 139
Quartz plate, 272

R

Rayleigh, 32, 36, 235, 285, 411-12
    scattering regime, 235, 407
Readout, combined birefringence, 341
Receiver noise, 47, 49
Responsivity, 363
    of interferometers, 399
Reticles, 22-23
Retroreflection in interferometer, 400
R-FOG, 295
RLG, 6, 248, 256, 265, 274
Ruby laser, 2

S

Sagnac effect, 6, 252, 278-280
    phaseshift, 248, 255, 259, 278
    and relativity, 253
    interferometer, 345, 398
SAW, 148
Scattering regimes, 235, 407, 411
    Rayleigh, 360, 411
    Raman, 360
    Brillouin, 360
SEAS, 37
Sensitivity, 364
Scattering function, 410
    regimes, 411
Schmitt trigger, 74
Scintillations, 415
Scotchlite, 202
Scroll sensor, 12
SEC, 305, 306
Self-mixing, 126
SHG, 82
SLED, 182, 249, 286, 280
SMR fiber, 299

Index

Solar irradiance, 409
Somb function, 31
SoS, 290
Spatial coherence, 118, 147
Spatial filter, 403
Speckle effects, 112, 120, 186, 197
Speckle pattern
    properties, 182
      joint statistics, 194
      phase errors, 197
    size, 184
      statistics, 183, 186, 206
    subjective, 195
    target errors, 198
    tracking, 204
Spot size, 12, 231, 368

T

TEC, 384
Technologies for OFS, 289
Telemetry, 2-4, 39-88
    accuracy, 51, 59
      ambiguity problem, 61
    calibration, 63
    chirp, 73
    frequency sweep, 74
    imaging, 80
    power balance, 46
    pulsed, 66,
    time-of-flight, 43, 64
    vernier, 70
TEM_{00}, 12, 363, 366
Temperature sensor, 324-326, 359, 436
Temporal coherence effects, 46
Thermodynamic noise, 119, 288
Threshold level for timing, 52
Timing, 51-55
    CFT, 58
      optimum filter, 55, 419
      start/stop, 57-59
TPN, 119, 288
Tracking of speckles, 206
Transmission
    clear atmosphere, 413
    fog, 413,
    freshwater, 414
    haze, 413
    seawater, 414
Three-D telemetry, 80
Three-mirror model, 137
Triangulation, 40
Turbidity, 407
Turbulence, 415
    inner/outer scale of, 416
    structure constant, 415
Twyman-Green interferometer, 94

V

VCO, 243, 295
Velocimetry
    accuracy, 230
    laser Doppler, 223
    scale factor, 229
    sensing region, 232
Verdet constant, 330
Virgo antenna, 257
Visibility, 413

W

Wandering, 415-417
Wegel, 119
White light interferometer, 172
White light OFS, 349
Wiener-Khintchine law, 243, 418
Windows, He-Ne, 372

X

XPM, 288-89

Y

YIG crystal, 287

Z

Zeeman splitting, 131, 267, 271-3 378
Zero bias of gyro, 266
ZRLG, 266, 270, 272-276

About the Author

Silvano DONATI is a Professor of Optoelectronics at the University of Pavia (Italy), Faculty of Engineering, Department of Electronics. Upon earning a degree in Physics with honours from University of Milano, he joined CISE, a research center now merged with the Electrical Generating Board of Italy, and started working on photodetectors and electro-optical instrumentation.
In 1975 he joined the University of Pavia as a full-time lecturer of courses in Electronics Circuits Design, Electronic Materials and Technologies, and Electro-Optical Systems. In 1980, he became a full professor in Optoelectronics, and since then he has been in charge of courses on Photodetectors, Electro-optical Instrumentation, and Optical Communications.
At the University of Pavia, has been proactive in starting a curriculum in Optoelectronics for electronic engineers since 1975, and he has created a research group in Optoelectronics now totaling 12 researchers. Under his guidance, 120 Masters students and 25 PhDs have graduated.
Through the years, he has contributed to electronics (noise in CCDs, and coupled oscillators) and electro-optical instrumentation (laser interferometers, fiber gyros, and fiberoptic current sensors). His recent researches have dealt with all-fiber passive components for communications, noise in photodetection, optical chaos and cryptography, and photomixing. He has cooperated in several R&D programs with national companies active in the areas of communications, instrumentation, and avionics.
He has organized several national and international meetings and schools, as a member of the Steering and Programme Committees, and as the Chairman of "Fotonica" (Roma, 1997) and "Elettroottica" (Pavia, 1994).
From 1986 to 1992, he was the Director of the Italian scientific review in electronics "Alta Frequenza - Rivista di Elettronica" of the Italian Electronics Association (AEI).
He and his Group were awarded seven prizes (one from Philip Morris and six from AEI, including the Guglielmo Marconi gold medal for scientific careers).
He was the Chairman of the Optoelectronics Society of AEI from 1992 to 1996. In 1997, he founded the IEEE-LEOS Italian Chapter, which he chaired until 2002.
He promoted and chaired the WFOPC (Fiber Optics Passive Components) International Conference in 1998 and 2000. He also chaired ODIMAP II and III, the International Conference on Optical Distance Measurements and Applications, held in Pavia, May 20-22, 1999 and Sept.20-22, 2001.
He has authored or co-authored over 250 papers, and holds 12 patents. He is the author of the book "Photodetectors", published by Prentice Hall in 2000.
He is a Fellow Member of IEEE, Fellow Member of OSA, and Meritorious member of AEI. He is also a member of several other Societies, including SPIE, IMAPS, and IoP. You can visit his web site http://ele.unipv.it/~donati or e-mail your questions to silvano.donati@unipv.it.

informIT

www.informit.com

YOUR GUIDE TO IT REFERENCE

Articles

Keep your edge with thousands of free articles, in-depth features, interviews, and IT reference recommendations – all written by experts you know and trust.

Online Books

Answers in an instant from **InformIT Online Book's** 600+ fully searchable on line books. Sign up now and get your first 14 days **free**.

POWERED BY
Safari

Catalog

Review online sample chapters, author biographies and customer rankings and choose exactly the right book from a selection of over 5,000 titles.

Wouldn't it be great

if the world's leading technical publishers joined forces to deliver their best tech books in a common digital reference platform?

They have. Introducing **InformIT Online Books powered by Safari.**

■ Specific answers to specific questions.
InformIT Online Books' powerful search engine gives you relevance-ranked results in a matter of seconds.

■ Immediate results.
With InformIt Online Books, you can select the book you want and view the chapter or section you need immediately.

■ Cut, paste and annotate.
Paste code to save time and eliminate typographical errors. Make notes on the material you find useful and choose whether or not to share them with your work group.

■ Customized for your enterprise.
Customize a library for you, your department or your entire organization. You only pay for what you need.

Get your first 14 days FREE!
InformIT Online Books is offering its members a 10 book subscription risk-free for 14 days. Visit **http://www.informit.com/onlinebooks** for details.

POWERED BY safari

informit.com/onlinebooks

Keep Up to Date with
PH PTR Online

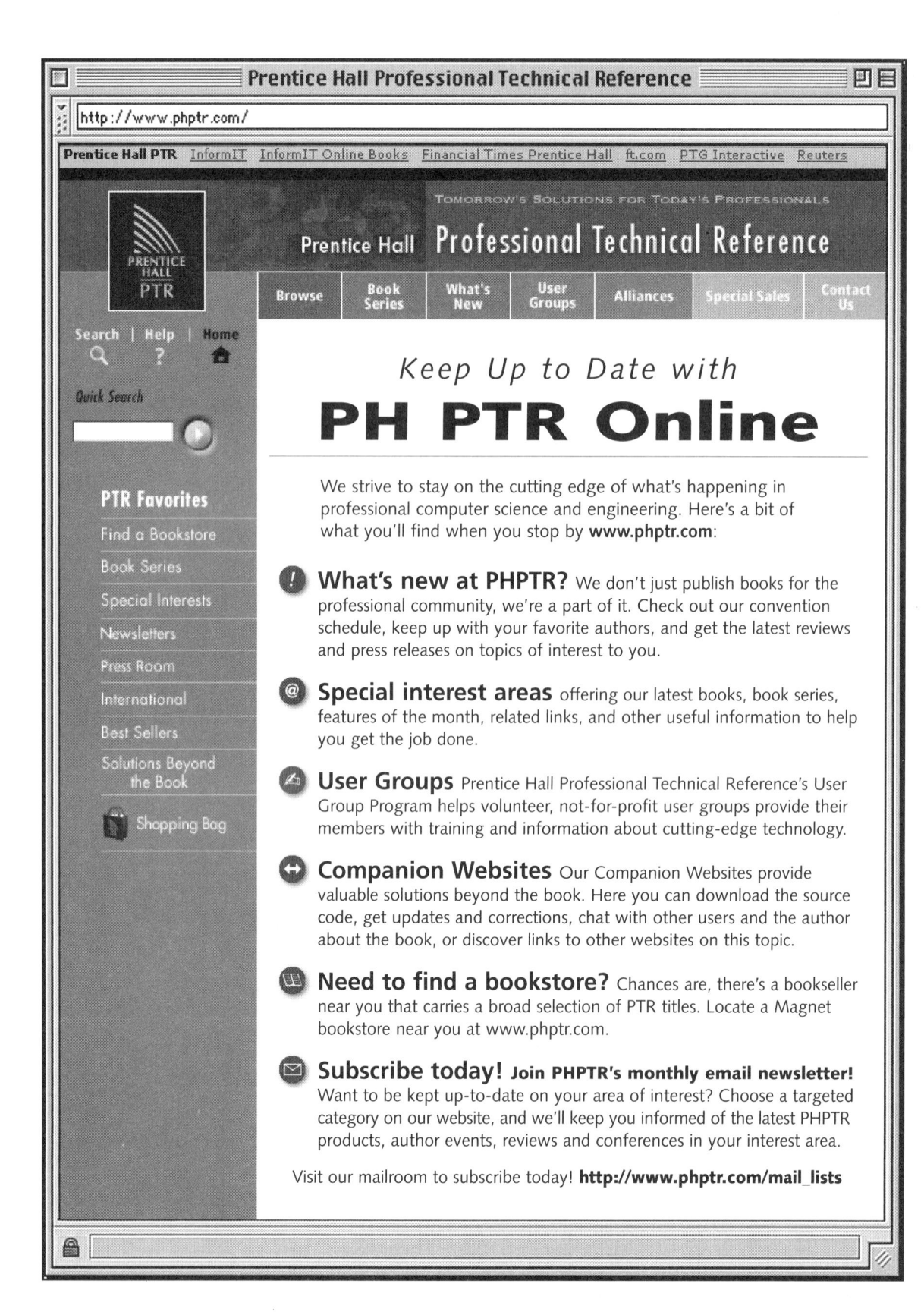

We strive to stay on the cutting edge of what's happening in professional computer science and engineering. Here's a bit of what you'll find when you stop by **www.phptr.com**:

What's new at PHPTR? We don't just publish books for the professional community, we're a part of it. Check out our convention schedule, keep up with your favorite authors, and get the latest reviews and press releases on topics of interest to you.

Special interest areas offering our latest books, book series, features of the month, related links, and other useful information to help you get the job done.

User Groups Prentice Hall Professional Technical Reference's User Group Program helps volunteer, not-for-profit user groups provide their members with training and information about cutting-edge technology.

Companion Websites Our Companion Websites provide valuable solutions beyond the book. Here you can download the source code, get updates and corrections, chat with other users and the author about the book, or discover links to other websites on this topic.

Need to find a bookstore? Chances are, there's a bookseller near you that carries a broad selection of PTR titles. Locate a Magnet bookstore near you at www.phptr.com.

Subscribe today! Join PHPTR's monthly email newsletter! Want to be kept up-to-date on your area of interest? Choose a targeted category on our website, and we'll keep you informed of the latest PHPTR products, author events, reviews and conferences in your interest area.

Visit our mailroom to subscribe today! **http://www.phptr.com/mail_lists**